Molecular Themes in DNA Replication

To my parents, Albert and Shirley Cox, for their unstinting love and support.

Molecular Themes in DNA Replication

Edited by

Lynne S. Cox
Department of Biochemistry, University of Oxford, Oxford, UK

RSC Publishing

The front cover shows the structure of PCNA (PDB 1W60), an essential DNA replication factor that encircles DNA and acts as a processivity factor for DNA polymerases and as a loading platform for other replication and repair factors, as described in Chapter 3. The surface fill of one subunit of the trimer shows interaction sites with DNA polymerase p66 subunit (cyan, amino acids 452–466, PDB 1U76), Fen1 nuclease (magenta, amino acids 331–350, PDB 1U7b), p21 (yellow, amino acids 139–160, PDB 1AXC) and Gadd45 (red, amino acids 196–215, P. A. Hall, J. M. Kearsey, P. J. Coates, D. G. Norman, E. Warbrick and L. S. Cox, Characterisation of the interaction between PCNA and Gadd45, *Oncogene*, 1995, **10**, 2427–2433).

ISBN: 978-0-85404-164-0

A catalogue record for this book is available from the British Library

Published by The Royal Society of Chemistry,
Thomas Graham House, Science Park, Milton Road,
Cambridge CB4 0WF, UK

Registered Charity Number 207890

For further information see our web site at www.rsc.org

Preface

DNA replication, the process of copying one double-stranded DNA molecule to form two identical copies, is highly conserved at the mechanistic level across evolution. In eukaryotes, the process includes also copying of epigenetic information in terms of chemical modifications to DNA and chromatin structure, and preparation of the sister chromatids ready for separation at mitosis. It comprises a highly complex set of biochemical reactions carried out by intricate enzyme assemblies and coordinated within the cell division cycle and in response to external and internal cellular signals. This book attempts to cover DNA replication from a biochemical perspective, relating the structures of key proteins involved in replication to their function: it does not attempt to deal with all the complexities of the biology of replication which have been so excellently discussed elsewhere.[†] The focus is very much on eukaryotic replication though archaeal paradigms and, where relevant, lessons from prokaryotes are also discussed. The book is not intended to cover every aspect and detail of replication, but rather aims to examine conserved structural and mechanistic themes in DNA replication, providing a suitable background for biochemical, chemical and pharmaceutical studies of this huge and exciting field.

Insights into the process at the molecular level provide opportunities to modulate and intervene in replication: rapidly diving cells need to replicate their DNA prior to dividing, and targeting components of the replication process is potentially a very powerful strategy in the treatment of cancer. Ageing may conversely be associated with loss of replication activity: restoring replication capacity to cells (particularly attending to the 'end replication problem' at telomeres) may be a route for moderating some of the diseases and frailty associate with ageing.

[†] M. L. DePamphilis (ed.), *DNA Replication and Human Disease*, Cold Spring Harbor Laboratory Press, New York, 2006.

Molecular Themes in DNA Replication
Edited by Lynne S. Cox

Published by the Royal Society of Chemistry, www.rsc.org

The importance of copying DNA in subcellular organelles is often overlooked. In this book, replication of the mitochondrial genome is examined and the clinical consequences of errors explored. Targeting DNA replication of pathogenic bacteria and viruses is already a clinical reality; the possibility of exploiting novel domains and motifs in replication factors of the malaria parasite is highlighted here as an important route for possible future therapies against malaria. DNA replication is therefore fundamental to a huge range of molecular biological and biochemical applications, and provides many potential targets for rational drug design in the treatment of hyper- and hypo-plastic diseases, and those caused by pathogens. Such targets, and current and future therapeutic approaches, are covered wherever relevant.

Interesting in its own right as a major feat of biochemical regulation and coordination, DNA replication is at the heart of modern advances in molecular biology. Without understanding the chemistry of DNA replication, Sanger would not have been able to develop his ground-breaking technique of chain termination dideoxy sequencing of DNA that has led to the explosion of molecular biology and revolutionized our understanding of, and ability to manipulate, the genome. As well as DNA sequencing (both dideoxy and next generation pyrosequencing), cloning of genes requires their replication; the polymerase chain reaction (PCR) used in almost every sphere of modern biochemistry is simply region-specific DNA replication on a grand scale. An understanding of DNA replication at both the biological and chemical level is essential to developing such technologies and using them effectively.

By providing an overview of the core mechanistic aspects of DNA replication and relating structure to function, this book is intended to serve as an introduction to higher level undergraduates and graduate students, as a source of information for those approaching DNA replication from other disciplines, and as a useful resource for researchers already in the field.

Contents

Molecular Themes in DNA Replication
Edited by Lynne S. Cox

Published by the Royal Society of Chemistry, www.rsc.org

Acknowledgements

I would like to thank Laurent Chaminade, formerly of the Royal Society of Chemistry, for his confidence in entrusting to me the task of editing this book, and to Janet Freshwater and Sue Humphreys of the RSC for their invaluable help and advice. I am indebted to the authors of individual chapters for their commitment and professionalism, and for their patience in the face of editorial interference. I would like to thank the members of my research group for their forbearance and understanding during the production of this book. Special thanks go to Professor Ron Laskey for introducing me to the field of DNA replication in the first place, and to Professor Sir David Lane for his support and encouragement through my postdoctoral years. Thanks are also due to Oriel College, who allowed me sabbatical leave from teaching to get the project moving, and to my colleagues in the Biochemistry Department at Oxford for providing a stimulating intellectual environment in which to work—and a wonderful new building to boot.

Finally, I would like to thank my husband, Charles Riddell, and my daughters, Sophie and Lucy, for putting up with me throughout the long drawn-out process of bringing a book to fruition. This book would not have been possible without their support and encouragement.

List of Abbreviations

2D-NAGE	two dimensional neutral agarose gel electrophoresis
ABC	ATP binding cassette
ABO	histone acetyl transferase binding to ORC1
ADP	adenosine diphosphate
ADOA	autosomal dominant ectopc atrophy
adPEO	autosomal dominant progressive external ophthalmoplegia
ALT	alternative lengthening of telomeres
AMP	adenosine monophosphate
AP	apurinic/apyrimidinic (ie abasic)
APB	ALT-associated PML body
APC	anaphase-promoting complex
APC/C	anaphase promoting complex/cyclosome
ATM	Ataxia Telangiectasia mutated (DNA damage response kinase)
ATP	adenosine triphosphate
ATR	atm and Rad3-related
BER	base excision repair
BIR	break-induced replication
BLM	Bloom syndrome helicase (RecQ family)
BrdU	bromodeoxyuridine
BS	Bloom syndrome (cancer prone progeroid syndrome)
CAF1	chromatin assembly factor 1
CDK	cyclin-dependent kinase
CenpA	centromeric histone
CFTR	cystic fibrosis transmembrane conductance regulator
CPD	cyclobutane pyrimidine dimer
DDK	Dbf4-dependent kinase
dGK	deoxyguanosine kinase
DNA	deoxyribonucleic acid
DSB	DNA double strand break
DUE	DNA unwinding element
EA	elenic acid
ECTR	extrachromosomal telomeric repeat
EM	electron microscopy

FACT	facilitates chromatin transcription
FFA1	Focus forming activity 1-Xenopus homologue of WRN
FHA	forkhead associated domain
FRET	fluorescence resonance energy transfer
GCM	GINS-Cdc45-Mcm2-7 complex
GCR	gross chromosomal rearrangement
GG-NER	global genome nucleotide excision repair
GINS	Go Ichi Nii San (Japanese for 5, 1, 2, 3)
HEAT	Huntington elongation A subunit TOR
HIRA	histone H3.3 chaperone
HIV	human immunodeficiency virus
hnRNP	heterogenous nuclear ribonucleoprotein
HO	yeast mating type endonuclease
HR	homologous recombination
HRDC	helicase and RNaseD C-terminal domain of RecQ helicases
HU	Hydroxyurea
ICL	interstrand cross link
ISM	initiator specific motif
ISWI	imitation switch
KAT	lysine acetyl transferase
LOH	loss of heterozygosity
MCM	minichrosmosome maintenance
MCM3AP	MCM acetylating protein
MDS	mitochondrial depletion syndrome
MFA	marker frequency analysis
MIRAS	mitochondrial recessive ataxia syndrome
MMS	methyl methanesulphonate
MRN	Mre11/rad50/Nbs1 double strand break/repair complex
MRX	Mre11/Rad50/XRs2 in yeast
mtDNA	mitochondrial DNA
NBD	nucleotide binding domain
NCR	non coding region
NDPK	nucleoside diphosphate kinase
NER	nucleotide excision repair
NHEJ	non-homologous end joining
NLS	nuclear localisation sequence
NMR	nuclear magnetic resonance
NRTI	nucleoside reverse transcriptase inhibitors
NTPase	nucleotide triphosphate hydrolysing enzyme
OBD	origin DNA-binding domain (SV40 large T antigen)
OBR	origin of bidirectional replication
ODN	oligonucleotide
OFP	Okazaki fragment processing
ORC	origin recognition complex
PAD	polymerase associated domain
PCNA	proliferating cell nuclear antigen

PCR	polymerase chain reaction
PEO	progressive external ophthalmoplegia
PNA	peptide nucleic acid
PDB	protein data bank (http://www.rcsb.org/pdb/home/home.do)
PIP	PCNA-interacting peptide
PLK	polo-loke kinase
PML body	promyelocytic leukaemia body
PNA	peptide nucleic acid
pol	polymerase
POLRMT	mitochondrial RNA polymerase
PPA2A	protein phosphatases 2A
PRR	post replication repair
PTM	post-translational modification
RBD	RNA binding domain (TERT)
RFB	replication fork barrier
RFC	replication factor C (sliding clamp loader)
RI	replication independent
RITOLS	ribonucleotide incorporation throughout the lagging strand
RNA	ribonucleic acid
RNP	ribonucleoprotein
RPA	replication protein A (single strand DNA binding protein)
RQC	RecQ C terminal domain (binds G4 DNA)
RT	reverse transcriptase
SCE	sister chromatid exchange
SMC	structural maintenance of chromosomes
SIM	SUMO-interacting motif
shRNA	short hairpin RNA
siRNA	short interfering RNA
SLBP	stem loop binding protein
snoRNA	small nucleolar RNA
SQDG	sulfoquinovosyldiacylgylcerol
SQMG	sulfoquinovosylmonoacylgylcerol
SSB	single strand DNA binding protein
S/TQ kinase	serine/threonine glutamine kinase
TBE	template boundary element (in TER)
TCA	tricarboxylic acid
TCR	transcription-coupled repair
TEN	telomerase essential N terminal domain
TER	telomerase RNA
TERT	telomerase reverse transcriptase
TEV	tobacco etch virus
TFAM	transcription factor A of mitochondria
TIF	telomere dysfunction-induced foci
TK	thymidine kinase
TLS	translesion synthesis
TRAP	telomerase repeat amplification protocol

TOR	target of rapamycin
T-SCE	telomere sister chromatid exchange
TSG	tumour suppressor gene
UBD	ubiquitin binding domain
UV	ultraviolet light
wH	winged helix
WhiP	winged helix interacting protein
WRN	Werner syndrome helicase/exonuclease (RecQ family)
WS	Werner syndrome (premature ageing syndrome)
WSTF	William's syndrome transcription factor
XP-V	Xeroderma Pigmentosum variant

CHAPTER 1

Conserved Steps in Eukaryotic DNA Replication

XIN QUAN GE AND J. JULIAN BLOW

Wellcome Trust Centre for Gene Regulation and Expression, University of Dundee, DD1 5EH, UK

1.1 Overview: the Biochemistry of DNA Synthesis

The genome of mammals comprises $\sim 6 \times 10^9$ nucleotides arranged in extremely long linear polymers—the chromosomes. Accurate copying of this amount of genetic information in a biologically relevant time frame (often a few hours, though sometimes as little as a few minutes) requires highly accurate enzyme machines together with complex molecular coordination and feedback.

The DNA template is a long polymer of four types of deoxyribonucleotide arranged as a double-stranded anti-parallel helix (Figure 1.1A). The backbone of each strand consists of phosphodiester linkages between the 3′ and 5′ carbons of deoxyribose. The 1′ carbon of the deoxyribose is linked to one of four different bases: the purines (adenine and guanine) or the pyrimidines (thymidine and cytosine). The two strands are held together by hydrogen bonds between complementary bases. The two-ringed heterocyclic purines always base pair with single ring pyrimidines, maintaining the linear axis of the helix and avoiding backbone distortion; specifically, guanine forms three hydrogen (H) bonds with cytosine and adenine makes two hydrogen bonds with thymidine (Figure 1.1B). Whilst each H bond is relatively weak, the huge number of H bonds in an average mammalian chromosome ($>10^9$) ensures that the duplex is extremely stable. As noted by Watson and Crick in 1953,[1] each of the two

Molecular Themes in DNA Replication
Edited by Lynne S. Cox

Published by the Royal Society of Chemistry, www.rsc.org

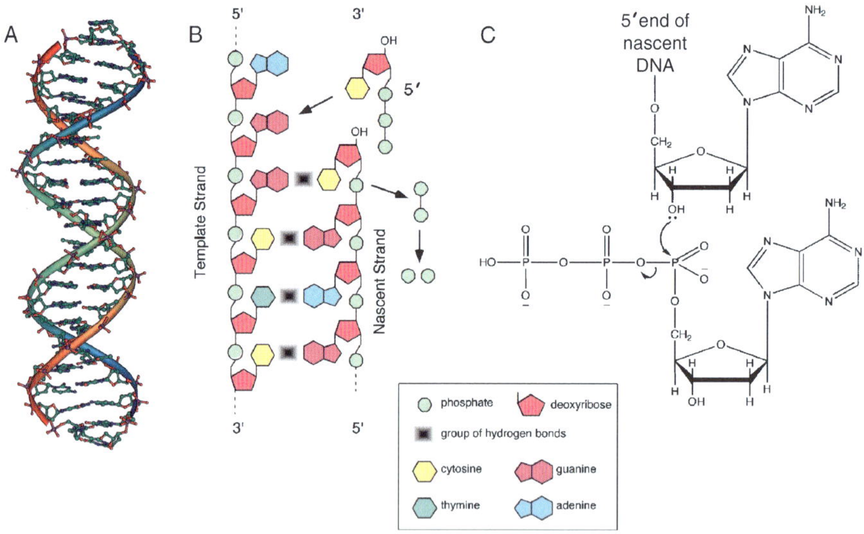

Figure 1.1 **The chemistry of DNA synthesis**. (**A**) The duplex DNA template showing the phosphodiester backbone, with the hydrophobic bases facing inwards, forming complementary base pairs. Reproduced from 'An overview of the structure of DNA', created by Michael Ströck 2006, released under the GNU Free Documentation Licence (GFDL).[a] (**B**) The four nucleotides of DNA form hydrogen bonds with their complement (note opposite polarities of the two strands). Adenine pairs with thymidine *via* two H bonds, while guanine pairs with cytosine through three H bonds. (**C**) Phosphodiester bond formation through nucleophilic attack from the 3′ OH of a newly incorporated nucleotide onto the α phosphate of an incoming nucleotide. The pyrophosphate released is rapidly hydrolysed by pyrophosphatase to inorganic phosphate, with a highly negative ΔG (~40 kJ mol^{-1}). [a]

single strands of DNA contains all the information necessary to produce new second strands through complementary base pairing.

During DNA replication, the two strands are opened up and a nascent strand, complementary to the template strand, is synthesised by a complex of many proteins called the replication fork or replisome. Unwinding of the two strands of DNA to expose bases during template-directed DNA synthesis requires the input of chemical energy to break the hydrogen bonds. This energy is derived from hydrolysis of ATP by helicases, which act as DNA-dependent ATPases that run ahead of the replication fork (Figure 1.2). In eukaryotes, the major replicative helicase is almost certainly the hexameric Mcm2-7 complex[2–5] (Figure 1.3, see also Chapter 3). At the same time, torsional stress (positive supercoiling) is caused by unwinding the DNA and this is relieved by topoisomerases (nicking-closing enzymes) (Figure 1.2). Then DNA is synthesised by enzyme-mediated polymerisation of deoxyribonucleotide triphosphates

[a] http://en.wikipedia.org/wiki/File:DNA_Overview.png

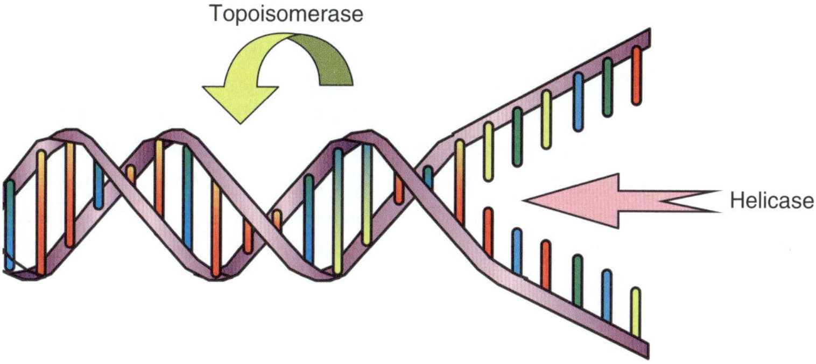

Figure 1.2 **DNA duplex being unwound during DNA replication.** Double-stranded DNA is being separated into two single strands by a helicase moving from right to left. The topological interlinks between the two strands are removed by a topoisomerase.

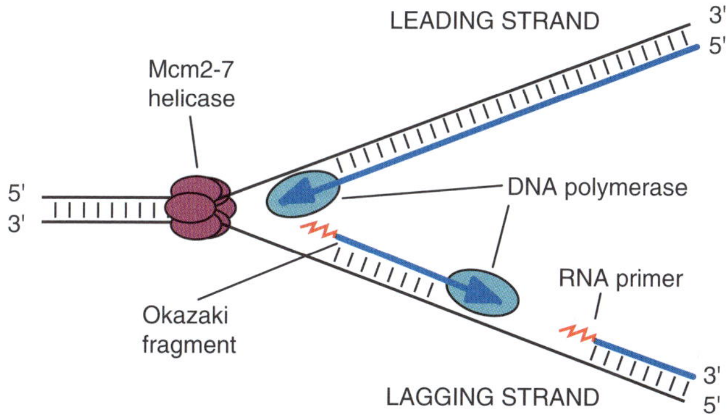

Figure 1.3 **Leading and lagging strand replication.** A replication fork is shown moving from right to left. The double-stranded template DNA is unwound by the Mcm2-7 helicase. The leading strand is synthesised continuously in the 5′–3′ direction. The lagging strand is also synthesised in the 5′–3′ direction with respect to the nascent strand, but since this is opposite to the overall direction of fork movement, it is synthesised discontinuously in Okazaki fragments. Each Okazaki fragment is started by a small RNA primer which is subsequently removed before the fragments are ligated together.

(dNTPs) complementary to the sequence of the exposed bases on the template strand (see Chapter 4). New daughter duplexes thus consist of one parental template strand base paired to a complementary daughter strand. This is semi-conservative replication and was first demonstrated experimentally by Meselson and Stahl,[6] who showed that newly synthesised DNA is composed of one template strand plus one nascent strand.

During polymerisation, nucleophilic attack by the lone pair of electrons on the 3′ hydroxyl (OH) of deoxyribose onto the 5′ phosphate of an incoming dNTP results in the formation of a phosphodiester bond with the elimination of pyrophosphate (Figure 1.1C). The subsequent, and very rapid, hydrolysis of pyrophosphate to two inorganic phosphates releases energy, and it is this that drives the polymerisation reaction forwards. An important consequence of this reaction is that DNA must always be synthesised in a 5′ to 3′ direction; all known DNA polymerases act 5′–3′ with respect to the newly synthesised (nascent) DNA molecule. However, while the double-stranded DNA template exists as an anti-parallel double helix, replication of both template strands is coordinated at the replication fork, which moves in a net direction away from the start point (replication origin). To overcome this conflict of directionality, on only one strand (the 'leading strand') can DNA be polymerised in the same direction as the fork is moving. On the other strand (the 'lagging' strand), nascent DNA is synthesised in short sections called Okazaki fragments, typically ~150 nucleotides long in eukaryotes (Figure 1.3). Okazaki fragments are started by short RNA primers which are subsequently removed before the fragments are joined together. Thus the fork can move away from the start site while co-coordinating synthesis of both nascent strands and without contravening the energy requirements of 5′–3′ synthesis.

In eukaryotes, the chromosomal DNA is located within the cell nucleus where it is associated with proteins to form a DNA-protein complex called chromatin (see Chapter 10). The basic building block of chromatin is the nucleosome core particle, which contains 147 base pairs of double-stranded DNA wrapped in 1.65 left-handed superhelical turns around the surface of histone octamer comprising two central H3–H4 dimers flanked on either side by two H2A–H2B dimers (Figure 1.4). A variety of other proteins also bind to DNA and regulate its activity. For replication to occur, pre-existing nucleosomes and other DNA-bound proteins that are located ahead of replication forks need to be transiently disrupted. After fork passage, those proteins are deposited back on parental as well as nascent DNA so that the chromatin status is reproduced in daughter strands[7] (see also Chapter 10).

1.2 Where and When Does DNA Replication Take Place?

1.2.1 Cell Cycle Control

In eukaryotes, DNA replication takes place during a distinct phase of the cell cycle called S phase, during which time the entire genome is precisely duplicated (Figure 1.5). The replicated DNA molecules are segregated to the two daughter cells during a subsequent cell cycle phase called mitosis (see Chapter 9). S phase and mitosis are separated by two 'gap' phases, G1 and G2. Progression through each stage of the cell cycle is very tightly regulated by a complex interplay of kinases (enzymes that phosphorylate proteins), phosphatases (enzymes that remove phosphate groups from proteins) and proteases (which degrade proteins into shorter polypeptides or constituent amino acids).

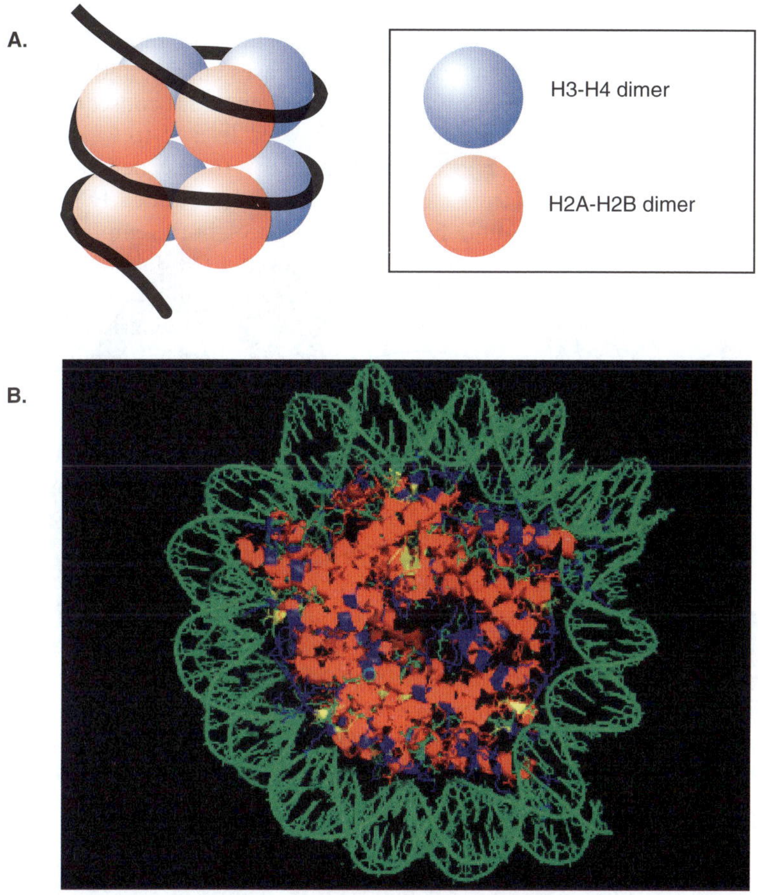

Figure 1.4 **Structure of the nucleosome**. (**A**) Cartoon of the nucleosome, showing 1.65 turns of DNA wrapped around an octamer consisting of two H2A-H2B dimers and two H3-H4 dimers. (**B**) Crystal structure of the nucleosome with DNA: DNA in turquoise and brown, core histones H3 (blue), H4 (green), H2A (yellow) and H2B (red). Reprinted by permission from Macmillan Publishers Ltd: K. Luger, A. W. Mader, R. K. Richmond, D. F. Sargent and T. J. Richmond, Crystal structure of the nucleosome core particle at 2.8Å resolution, *Nature*, 1997, **389**, 251–260, copyright (1997).[84]

During S phase, pairs of replication forks are initiated bidirectionally from chromosomal loci called replication origins. The large size of eukaryotic chromosomes (each of which can be tens or hundreds of megabases long) means that in order for them to be replicated in a reasonable period of time, a large number of replication origins are needed. Although the initiation of a pair replication forks at a replication origin is a tightly controlled process, each fork will typically then move along the DNA ('elongate') until it encounters a fork moving

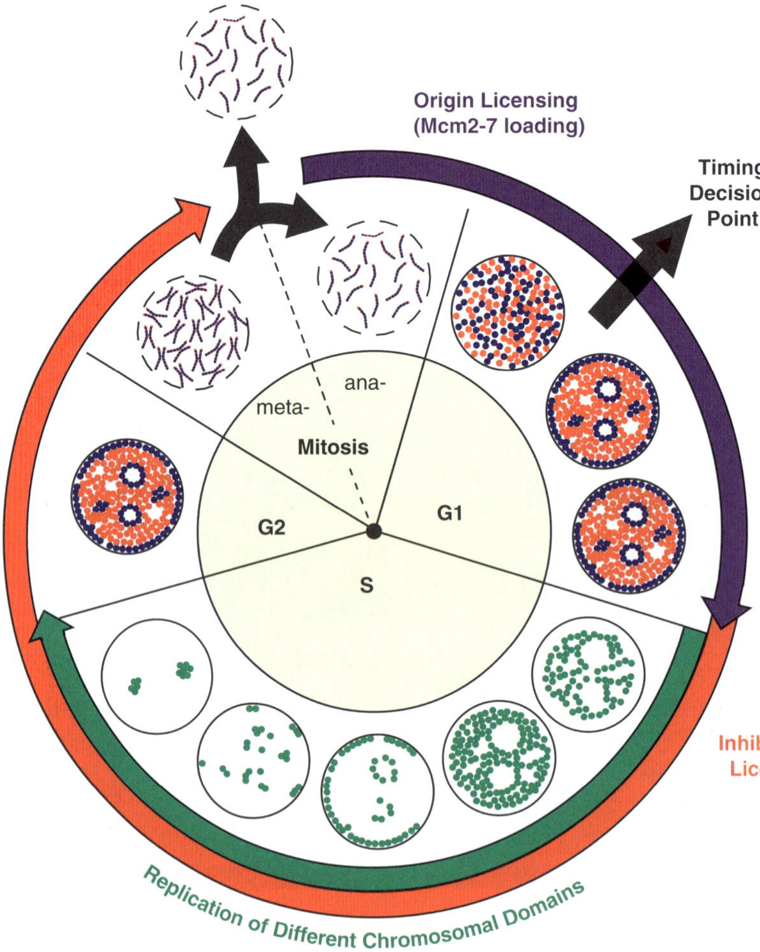

Figure 1.5 **DNA replication and the cell cycle (schematic view of events)**. In the metaphase of mitosis (M, meta-) condensed chromosomes (consisting of paired chromatids) are aligned on the metaphase plate by the mitotic spindle. During anaphase (M, ana-) the two sister chromatids are pulled into two daughter cells. During late mitosis and G1, replication origins are licensed by loading Mcm2-7 complexes. Origin licensing is inhibited at other cell cycle stages. During early G1, specific regions of chromosomal DNA take up specific positions in the nucleus, with open chromatin (red) tending to be positioned internally, and more condensed chromatin (blue) tending to be positioned at the nuclear periphery and in larger internal structures. At this time (the timing decision point), these different chromosomal regions become programmed to replicate at different stages of S phase (green).

in the opposite direction, at which stage both forks will disassemble ('terminate'). When DNA is visualised during the S phase, replicated DNA can be observed as a series of 'bubbles' with replication origins near their centres (arrowheads in Figure 1.6). The stretch of DNA replicated by forks emanating from a single origin is

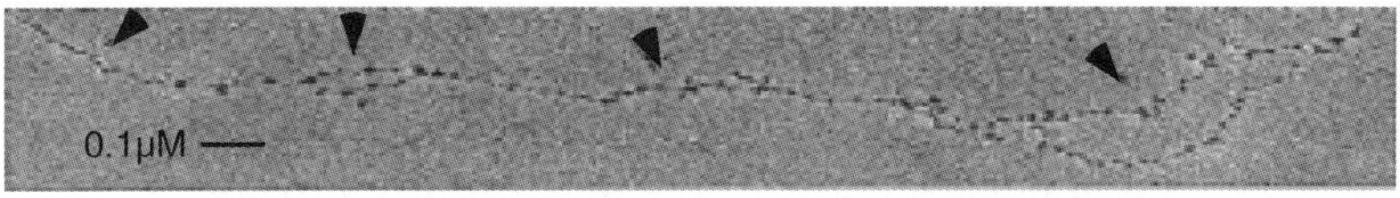

Figure 1.6 **Replication bubbles.** Electron microscopic image of replication origins in developing *Drosophila* embryos. Replication bubbles are indicated by the arrowheads. Scale bar 0.1 µM. Reprinted from: G. Micheli, C. T. Baldari, M. T. Carri, G. Di Cello and M. Buongiorno-Nardelli, An electron microscope study of chromosomal DNA replication in different eukaryotic systems, *Experimental Cell Research*, **137**, 127–140, copyright (1982), with permission from Elsevier.[85]

referred to as a replicon. Replicon sizes can vary significantly, both among different organisms and among different cell types in the same organism. Rapidly dividing cells typically have small replicon sizes (for example, cells in the early *Xenopus* embryo has an average replicon size of ~10 kb, whilst mammalian somatic cells typically have replicon sizes of 50–150 kb[8–10]).

1.2.2 Origin Clusters and Replication Foci

In metazoans, adjacent origins (typically 2–5) are organised into clusters which initiate synchronously while different origins clusters are activated at different stages of S phase.[11] One or more clusters of origins are organised into a discrete replication focal site, which has been estimated to comprise about 1 Mb of DNA and 6–12 replicons. Each focus is thought of as a factory for DNA replication and contains a range of replication fork proteins (forming so-called replisomes).[12] It is possible that replisomes are anchored to a fibrous network within the nucleus (the 'nuclear matrix' or 'nuclear scaffold') through which multiple replication forks are spooled; alternatively, the physical organisation of the chromosomal DNA into higher order chromatin structures could provide the framework on which replication foci are built.[13–15] DNA replication is typically completed in each focus within 30–120 minutes,[16] and during this time, live cell imaging of the replication fork protein PCNA[i] (see Chapters 3 and 7) has shown that replication foci do not merge, divide or have directional movement,[17,18] thus arguing that replication foci are achieved by the coordinated assembly and disassembly of replisomal proteins at sites that are more or less fixed.

1.2.3 The Replication Timing Programme

Eukaryotes replicate their genomic DNA according to a specific temporal programme, with different clusters of origins firing at different time during an S phase that lasts from minutes in yeast to hours in metazoans. Several pieces of evidence have suggested that chromatin context is a critical determinant of origin initiation time. The replication timing programme is re-established in

[i] Proliferating cell nuclear antigen

each cell cycle shortly after mitosis.[19] Transcriptionally active regions tend to have open chromatin structure and replicate early, whereas gene-poor regions and the more condensed heterochromatin replicate late.[20,21] Transcriptional silencing can reprogramme an origin from initiating early to late, as well as by promoting a more compact chromatin structure around the region.[22]

The timing decision point in early G1 (Figure 1.5) is the time when specific regions of the chromosome become programmed to replicate at specific stages of S phase. This takes places coincidently with the repositioning of chromosomes in the nucleus and the formation of immobile structures in the nucleus that restrict chromosome mobility.[19,23] It has been proposed that chromatin regulators might be concentrated into subnuclear compartments by a clustering of related chromosomal domains, which may influence the timing of origin firing within a chromatin domain. For example in yeast, late replicating origins reside close to the nuclear periphery in G1, whereas early replicating origins are apparently randomly localised within the nucleus throughout the cell cycle.[24]

Many other factors could also contribute to determining the timing of origin firing. For example, in *Saccharomyces cerevisiae*, the timing of replication in certain origins is shown to be affected by the origin sequence.[25] Precise levels of cyclin-dependent kinase (CDK) activity present at various stages of S phase are important for executing the temporal programme. In budding yeast, two S phase cyclins have differential roles in activation of early and late origins: Clb5 activates both early and late origins, while Clb6 activates only early origins.[26]

The replication timing programme determines the differential firing time of large sequence blocks containing replication origin clusters, but why has the cell evolved such a sophisticated programme for DNA replication? The grouping of replication forks into factories that are activated at different times might provide an environment whereby newly replicated DNA could be assembled into specific chromatin states, thus maintaining the epigenetic information that is important for regulation of other nuclear activities (such as transcription).[10,27] It may also allow for tight regulation feedback, for example blocking firing of late origins when replication from early origins is halted.

1.3 Origins of DNA Replication

The number of origins ranges from a few hundred in a yeast cell to tens of thousands in a human cell. The extent to which conserved DNA sequence elements determine origins differs significantly among eukaryotic species. Replication origins in the budding yeast *Saccharomyces cerevisiae* contain highly conserved sequence elements called A, B1, B2 and B3 boxes of the autonomously replicating sequence (ARS).[28] These conserved DNA sequences are required for binding of the initiator protein ORC (origin recognition complex, see Chapter 2). However, not all DNA segments containing the conserved sequence elements are recognised by ORC *in vivo*. Other sequences distributed over 100 bp also contribute to replication origin function, possibly by providing binding sites for proteins that can enhance the recruitment of ORC to DNA or by providing DNA sequences that can be easily unwound.[29]

The origins in most other eukaryotes are much less stringent in terms of sequence requirement. In the fission yeast, *Schizosaccharomyces pombe*, the required origin sequences are distributed over large DNA segments (500–1000 bp) and are AT rich.[30] It appears that it is the number of AT tracts in a given segment of DNA that determines its probability of binding ORC and functioning as an origin of DNA replication.

The nature of origins in metazoans is even less well defined than in yeasts and the origins appear not to contain any consensus sequence. Replication origins occur at frequent and nearly random intervals along metazoan chromosomal DNA, and only a fraction of them are utilised in each cell cycle with a wide variation of efficiency. A typical pattern of origin initiation in metazoans is broad zones containing many relatively inefficient origins, one or a few of which are selected stochastically and the rest are suppressed.[31,32] However, at some origins, such as lamin B2 and β-globin origins, replication starts from tightly-defined sites.[33,34]

Several interacting components may influence the location and efficiency of initiation in any given cell cycle, such as:

(1) **DNA sequences.** Sequences rich in AT could facilitate ORC binding or DNA unwinding.
(2) **Local chromatin structure.** It has been shown that the positions of nucleosomes near origins are important for origin function.[35] Whilst histone acetylation has been shown to affect origin specification in *Xenopus* and *Drosophila*,[36,37] in mammalian cells, an ATP-dependent chromatin remodelling complex is required for efficient replication of heterochromatin.[38]
(3) **Transcription.** Transcription has been shown to interfere with origin activity and indeed, replication origins are almost never found within actively transcribed DNA.[10,20,39,40]
(4) **Protein–protein interactions.** The presence of other proteins could help recruit ORC and enhance origin efficiency. For example, Abf1 and the Myb protein complex bind to origins and can affect the efficiency of origin utilisation in yeast and *Drosophila*.[41,42]
(5) **Origin interference**. It has been observed that in an initiation zone, firing of one replication origin appears to inhibit initiation at nearby origins, but is coordinated with neighbouring origins at more distant sites.[43] This may suggest some sort of long range interaction between origins.

1.4 Licensing of DNA for Replication

It is essential for a cell to replicate its genome only once per cell cycle and this is regulated by the ability of cells to load the Mcm2-7 protein complex onto the origins (see Chapters 2 and 3). Mcm2-7 form a clamp around DNA and provide helicase activity to separate the double helix ahead of replication forks[2–5] (see Chapter 3). During late M and G1 phases of the cell cycle, Mcm2-7 are loaded

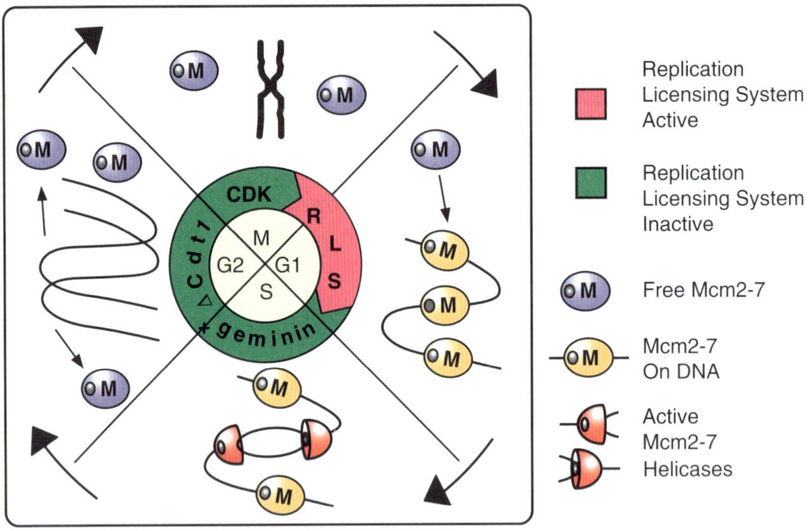

Figure 1.7 **The replication licensing cycle**. The replication licensing system (RLS) is activated in late mitosis, promoting the loading of Mcm2-7 double hexamers onto DNA. During S phase, DNA-bound Mcm2-7 becomes activated as helicases. When replication forks terminate, Mcm2-7 are released from DNA. In metazoans, the licensing system is shut down in S phase and G2 by degradation of Cdt1 and activation of the Cdt1 inhibitor, geminin; in early mitosis, high CDK levels also inhibit licensing. In yeasts, licensing inhibition in S, G2 and early mitosis is directly mediated by high CDK levels.

onto the DNA, which probably involves the clamping of the proteins around origin DNA without activation of their helicase activity (Figure 1.7). This 'licenses' the origin for use in the subsequent S phase. Mcm2-7 loading requires the recognition of the origin DNA by the origin recognition complex (ORC) (Figure 1.8). ORC in turn recruits proteins Cdc6 and Cdt1, which load Mcm2-7 onto DNA by hydrolysing ATP[44] (see Chapter 2). The complex of ORC, Cdc6, Cdt1 and Mcm2-7 at replication origins is termed the pre-replicative complex or pre-RC. It is not clear whether ORC, Cdc6 and Cdt1 open the Mcm2-7 ring and load it around DNA, or whether they facilitate the assembly of the Mcm2-7 hexamer on DNA from different Mcm subcomplexes present in the nucleoplasm.

As a licensed origin initiates during S phase, the Mcm2-7 complex becomes activated as helicase, possibly by binding other replication fork proteins including the GINS complex.[45,46] Since Mcm2-7 proteins travel with the replication fork,[47,48] this means that an origin becomes unlicensed after it initiates. To prevent DNA being replicated a second time in a single cell cycle, it is therefore important to prevent re-licensing of replicated origins during S and G2 phases of the cell cycle. The mechanisms for achieving this vary in different eukaryotes.

In yeasts, CDKs which are active from late G1 to mid-mitosis, prevent licensing outside late M and G1 phase by ORC inactivation, Cdc6/Cdt1

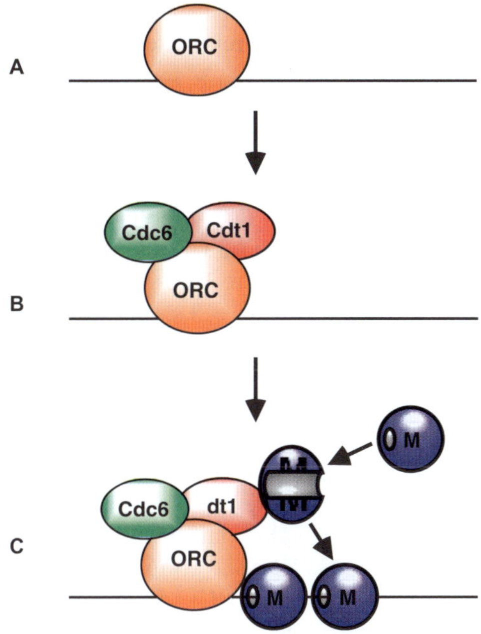

Figure 1.8 **Origin licensing.** Cartoon showing steps in the licensing of a replication origin. (**A**) ORC association with DNA at the replication origin. (**B**) ORC recruits Cdc6 and Cdt1. (**C**) ORC-Cdc6-Cdt1 allows the loading of multiple Mcm2-7 complexes onto the DNA.

degradation and Mcm2-7 export.[49] In metazoans, the main route by which licensing is prevented during S and G2 is the downregulation of Cdt1 activity. This is brought about both by degradation of Cdt1 protein and activation of a Cdt1 inhibitory protein, geminin. Cdt1 is degraded at the end of G1 and early S phase in a process dependent on SCF-class ubiquitin ligase and cul-4 ubiquitin ligase.[50–53] When geminin builds up during S, G2 and M phase, Cdt1 is stabilised by binding to geminin. As a result, licensing is inhibited and Cdt1 is protected from degradation, so that when geminin is degraded in late mitosis and G1, Cdt1 is ready for licensing.[49,54]

ORC can load multiple copies of Mcm2-7 complexes onto DNA, and Mcm2-7 are in ~20-fold excess over replication origins used in S phase.[55–60] Cells synthesise DNA at normal rates when the level of Mcm2-7 is reduced[61,62] and, in *Xenopus* egg extracts, normal replication rates are maintained when Mcm2-7 complex is reduced to ~two per origin.[59,60,63,64] This suggests that each of the loaded Mcm2-7 complexes could act at an origin to initiate DNA replication and up to 90% of them remain dormant in a single S phase. It has recently been shown that a biological role of the excess Mcm2-7 complexes loaded during licensing is to maintain genomic stability.[65] When forks stall during DNA replication, the dormant Mcm2-7 complexes can initiate and rescue DNA replication between two stalled forks, allowing the intervening DNA to be replicated. If a replication fork encounters an unfired (dormant) origin, the Mcm2-7 must be removed from it to prevent re-replication from occurring.

1.5 Initiation of DNA Replication

Mcm2-7 is loaded in an inactive form at replication origins during G1 (Figure 1.9A), and then activated to initiate DNA replication during S phase. Activation of Mcm2-7 requires both S phase CDK (cyclin E/A-Cdk2) and Dbf4-dependent kinase (DDK) activity (Figure 1.9B). DDK and CDK are expressed at relatively constant levels during the cell cycle, but the expression of regulatory subunits (Dbf4 and cyclin respectively) is increased in S phase. One of the known substrates of Cdc7-Dbf4 is the Mcm2-7 complex, the phosphorylation of which is thought to change its interaction with other replication fork proteins.[66] Recently, it has been shown in budding yeast that S phase CDKs phosphorylate two replication proteins, Sld2 and Sld3[46,67] (Figure 1.9B). By contrast, the critical S phase substrates of CDK-cyclin activity in metazoans have not yet been identified. Phosphorylation of yeast Sld2 induces its interaction with Dbp11, and facilitates its association with the GINS complex (Sld5, Psf1, Psf2, Psf3) and Dbp11 (Figure 1.9C). At the same time, Sld3 is phosphorylated by CDK and recruited to DNA by binding to Cdc45, where phospho-Sld3 recruits Dbp11 and Sld2. Current evidence suggests that the binding of Cdc45 and GINS to Mcm2-7 activates the helicase activity of Mcm2-7.[45,46] Consequently GINS and Cdc45 remain associated with Mcm2-7 and travel with the active replication forks (see Chapter 3).

1.6 Elongation of Replication Forks

Mcm2-7 and associated complex unwind the DNA in a bidirectional manner away from origins, and the single-stranded DNA (ssDNA) becomes coated with a binding protein called Replication Protein A (RPA). *Via* interactions with the helicase and RPA, DNA polymerase α (polα) is loaded onto the template (Figure 1.9D). A subunit of the polα holoenzyme provides primase activity and synthesises short RNA primers (8–12 nucleotides long), which are then extended by the DNA polymerase activity of polα to synthesise a short initiator DNA (iDNA) of about 30 bases. Because polα does not have a proofreading exonuclease activity, the iDNA synthesised only serves as a DNA primer for more extensive DNA synthesis by DNA polymerases with proofreading activity after polymerase switching (Chapter 4). The primer-template DNA structure is recognised and bound by a clamp-loading heteropentameric protein complex, replication factor C (RFC). This promotes structural changes in RFC, which uses the energy from ATP hydrolysis to open the ring of the trimeric sliding clamp PCNA (Chapter 2), and clamp it around the DNA, while at the same time polα is displaced. PCNA acts as a processivity factor for the elongation DNA polymerases polδ and polε by forming a ring that tethers them to the template DNA (Chapters 3, 6 and 7). Current data suggest that polε is on the leading

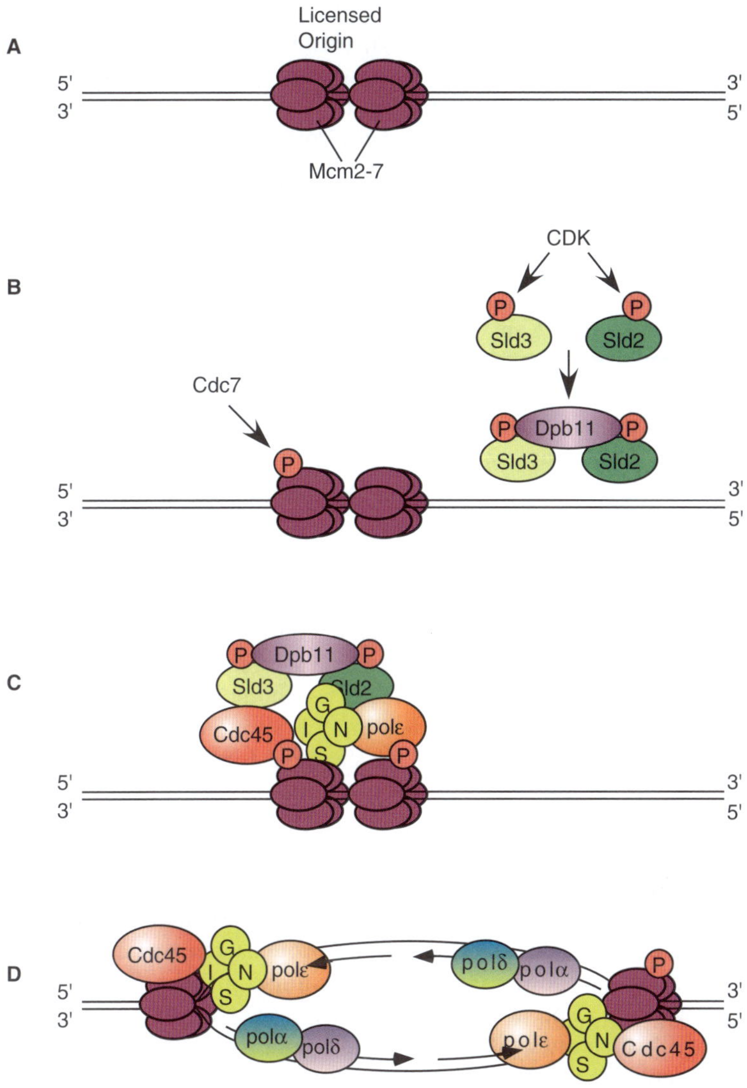

Figure 1.9 **Initiation of replication forks.** A model for events occurring during S phase as an origin initiates DNA replication. (**A**) A licensed origin, loaded with two Mcm2-7 hexamers. (**B**), Cdc7 phosphorylates members of the Mcm2-7 complex, whilst CDKs phosphorylate Sld2 and Sld3. Phosphorylated Sld2 and Sld3 associate with Dpb11. (**C**) The Sld2-Sld3-Dpb11 complex associates with phosphorylated Mcm2-7, possibly *via* interactions between Sld3 and Cdc45 and between Sld2 and GINS/polε. (**D**) Mcm2-7 helicase activity unwinds the origin DNA, allowing polα to initiate synthesis of the two nascent strands. These nascent strands are elongated by polε to form the leading strand of the fork. Behind this, polα and polδ act together to synthesise Okazaki fragments on the lagging strand.

strand and polδ is on the lagging strand.[68,69] When the lagging strand polymerase encounters the 5′ end of the adjacent Okazaki fragment, the 5′ end is displaced to form a 5′ flap, which is degraded by endonuclease FEN1 (Chapter 5). Then the two Okazaki fragments are ligated by DNA ligase and the DNA polymerase is recycled to a newly loaded clamp on the lagging strand.[70]

Recent work in budding yeast indicates that Mcm2-7 associate with a number of proteins at forks to form the 'replisome progression complex' (RPC) during elongation,[71] perhaps to ensure that fork progression is coordinated with DNA synthesis and other processes. In addition to Cdc45 and GINS, the RPC also contains Ctf4 which is important for establishing cohesion between sister chromatids, Tof1-Csm3, which mediate pausing of forks at DNA replication fork barriers, the checkpoint mediator Mrc1, the histone chaperone FACT, topoisomerase 1 and Mcm10.[72] In addition, PCNA acts as a landing pad for other proteins during replication such as the CDK inhibitor p21, cytosine methyltransferase, the chromatin assembly factor CAF-1, DNA ligase, FEN1 and other proteins involved in DNA repair[73–75] (see Chapter 3).

1.7 Termination of DNA Replication

Replication forks terminate when they encounter another replication fork coming from the opposite direction. In most cases, this occurs without the need for any special DNA sequences. In some cases, though, replication fork barriers at specific DNA sequences slow replication forks so that these sites are likely to become sites of termination. One such example is in the heavily transcribed ribosomal DNA genes, where the replication fork barrier is positioned to inhibit replication forks from moving through the gene in the opposite direction from transcription.[76]

The exact mechanism of how replication machinery is displaced from the DNA during termination is poorly understood. At termination, the replication forks must be disassembled. Most of the proteins released from terminated replication forks can be recycled to newly initiating forks. The Mcm2-7 proteins, however, are a special case. They are released from DNA at termination, but are not reloaded onto DNA until the next mitosis in order to prevent DNA from being replicated more than once in a single cell cycle (Figures 1.5 and 1.7). Similarly, if an active fork encounters inactive Mcm2-7 bound to a dormant replication origin, the inactive Mcm2-7 will be displaced from DNA.

1.8 Replication of Chromatin

The histones around which the DNA is wrapped (Figure 1.4) have to be displaced from the chromatin as the DNA replication forks pass, and the newly synthesised DNA has to be reassembled into chromatin (see Chapter 10). Nucleosome disruption is likely to be facilitated by ATP-dependent chromatin remodelling enzymes, such as WSTF which is targeted to replicating DNA through direct interaction with PCNA and in turn recruits ISWI-type nucleosome-remodelling factor SNF2.[77] At the same time, histone chaperones facilitate the disruption of parental nucleosomes by acting as histone acceptors and hence aid the transfer of the histones onto the nascent strand.[78] FACT is complexed with Mcm proteins during fork movement, and facilitates nucleosome disruption and re-deposition of H2A-H2B.[79,80] CAF-1 associated with PCNA is aided by Asf1 to deposit H3-H4 onto replicating DNA.[81,82]

Chromatin also contains epigenetic information in addition to that from the DNA sequence which affects, amongst other things, the level of gene expression. This information is encoded by covalent modifications to histones (such as acetylation and methylation) as well as methylation of cytosine bases. When DNA is replicated, the epigenetic information must be copied too. This is achieved by association of a large number of chromatin-modulating enzymes with PCNA during replication such as DNA methyltransferase I, CAF-1 and histone deacetylase (Chapters 3 and 10). These enzymes either themselves have catalytic activity or can recruit other enzymes implicated in chromatin modification.

1.9 Chromatid Cohesion and Segregation

After replication, it is essential that the sister chromatids are identified and each of them is sent to a different daughter cell. To achieve this, replicated chromosomes remain physically attached to each other by cohesion until anaphase, when they are separated by microtubule pulling force. Sister chromatid cohesion is established by cohesin, a complex consisting of at least four proteins (Smc1, Smc3, Scc1, Scc3) that form a ring structure loaded at discrete sites along the entire length of the chromosome in G1 phase. During S phase, the cohesin complex establishes a physical link (cohesion) between replicated sister chromatids by several factors, including Eco1, Ctf4 and Ctf18. Several models have been proposed to explain how cohesin contributes structurally to sister chromatid cohesion, one of which is that the cohesin ring establishes cohesion by embracing both sister chromatids (see Chapter 9).

At the metaphase-to-anaphase transition, the separation of sister chromatids is triggered by the removal of cohesin from chromosomes. This is achieved by activation of a protease called separase, which cleaves the cohesin ring. Separase is inhibited by protein securin and, at the metaphase-to-anaphase transition, securin is degraded after APC/C dependent ubiquitination.[83] Thus a complex interplay of cell cycle regulatory factors establishes cohesion concomitant with synthesis of sister chromatids and ensures separation only at the metaphase-anaphase transition (Chapter 9).

Acknowledgements

The authors are supported by Cancer Research UK grants C303/A4416 (XQG) and C303/A7399 (JJB).

References

1. J. D. Watson and F. H. C. Crick, Molecular structure of nucleic acids, *Nature*, 1953, **171**, 737–738.
2. K. Labib and J. F. Diffley, Is the MCM2-7 complex the eukaryotic DNA replication fork helicase?, *Curr. Opin. Genet. Dev.*, 2001, **11**, 64–70.
3. S. L. Forsburg, Eukaryotic MCM proteins: beyond replication initiation, *Microbiol. Mol. Biol. Rev.*, 2004, **68**, 109–131.
4. T. S. Takahashi, D. B. Wigley and J. C. Walter, Pumps, paradoxes and ploughshares: mechanism of the MCM2-7 DNA helicase, *Trends Biochem. Sci.*, 2005, **30**, 437–444.
5. M. L. Bochman and A. Schwacha, The Mcm2-7 complex has *in vitro* helicase activity, *Mol. Cell*, 2008, **31**, 287–293.
6. M. Meselson and F. W. Stahl, The replication of DNA in *Escherichia coli*, *Proc. Natl. Acad. Sci. U.S.A.*, 1958, **44**, 671–682.
7. S. E. Polo and G. Almouzni, Chromatin assembly: a basic recipe with various flavours, *Curr. Opin. Genet. Dev.*, 2006, **16**, 104–111.
8. J. J. Blow, P. J. Gillespie, D. Francis and D. A. Jackson, Replication origins in *Xenopus* egg extract are 5-15 kilobases apart and are activated in clusters that fire at different times, *J. Cell Biol.*, 2001, **152**, 15–25.
9. M. L. DePamphilis, Replication origins in metazoan chromosomes: fact or fiction?, *Bioessays*, 1999, **21**, 5–16.
10. D. M. Gilbert, Making sense of eukaryotic DNA replication origins, *Science*, 2001, **294**, 96–100.
11. R. Berezney, D. D. Dubey and J. A. Huberman, Heterogeneity of eukaryotic replicons, replicon clusters, and replication foci, *Chromosoma*, 2000, **108**, 471–484.
12. A. J. McNairn and D. M. Gilbert, Epigenomic replication: linking epigenetics to DNA replication, *Bioessays*, 2003, **25**, 647–656.
13. X. Wei, J. Samarabandu, R. S. Devdhar, A. J. Siegel, R. Acharya and R. Berezney, Segregation of transcription and replication sites into higher order domains, *Science*, 1998, **281**, 1502–1506.
14. A. A. Philimonenko, D. A. Jackson, Z. Hodny, J. Janacek, P. R. Cook and P. Hozak, Dynamics of DNA replication: an ultrastructural study, *J. Struct. Biol.*, 2004, **148**, 279–289.
15. E. Kitamura, J. J. Blow and T. U. Tanaka, Live-cell imaging reveals replication of individual replicons in eukaryotic replication factories, *Cell*, 2006, **125**, 1297–1308.
16. R. Berezney, Regulating the mammalian genome: the role of nuclear architecture, *Adv. Enzyme Regul.*, 2002, **42**, 39–52.
17. H. Leonhardt, H. P. Rahn, P. Weinzierl, A. Sporbert, T. Cremer, D. Zink and M. C. Cardoso, Dynamics of DNA replication factories in living cells, *J. Cell Biol.*, 2000, **149**, 271–280.
18. A. Sporbert, A. Gahl, R. Ankerhold, H. Leonhardt and M. C. Cardoso, DNA polymerase clamp shows little turnover at established replication

sites but sequential de novo assembly at adjacent origin clusters, *Mol. Cell*, 2002, **10**, 1355–1365.
19. D. S. Dimitrova and D. M. Gilbert, The spatial position and replication timing of chromosomal domains are both established in early G1 phase, *Mol. Cell*, 1999, **4**, 983–993.
20. D. M. MacAlpine, H. K. Rodriguez and S. P. Bell, Coordination of replication and transcription along a *Drosophila* chromosome, *Genes Dev.*, 2004, **18**, 3094–3105.
21. Y. Jeon, S. Bekiranov, N. Karnani, P. Kapranov, S. Ghosh, D. MacAlpine, C. Lee, D. S. Hwang, T. R. Gingeras and A. Dutta, Temporal profile of replication of human chromosomes, *Proc. Natl. Acad. Sci. U.S.A.*, 2005, **102**, 6419–6424.
22. D. C. Zappulla, R. Sternglanz and J. Leatherwood, Control of replication timing by a transcriptional silencer, *Curr. Biol.*, 2002, **12**, 869–875.
23. F. Li, J. Chen, M. Izumi, M. C. Butler, S. M. Keezer and D. M. Gilbert, The replication timing program of the Chinese hamster beta-globin locus is established coincident with its repositioning near peripheral heterochromatin in early G1 phase, *J. Cell Biol.*, 2001, **154**, 283–292.
24. P. Heun, T. Laroche, M. K. Raghuraman and S. M. Gasser, The positioning and dynamics of origins of replication in the budding yeast nucleus, *J. Cell Biol.*, 2001, **152**, 385–400.
25. K. Sharma, M. Weinberger and J. A. Huberman, Roles for internal and flanking sequences in regulating the activity of mating-type-silencer-associated replication origins in *Saccharomyces cerevisiae*, *Genetics*, 2001, **159**, 35–45.
26. A. D. Donaldson, M. K. Raghuraman, K. L. Friedman, F. R. Cross, B. J. Brewer and W. L. Fangman, CLB5-dependent activation of late replication origins in *S. cerevisiae*, *Mol. Cell*, 1998, **2**, 173–182.
27. A. Taddei, F. Hediger, F. R. Neumann and S. M. Gasser, The function of nuclear architecture: a genetic approach, *Annu. Rev. Genet.*, 2004, **38**, 305–345.
28. C. Cvetic and J. C. Walter, Eukaryotic origins of DNA replication: could you please be more specific?, *Semin. Cell Dev. Biol.*, 2005, **16**, 343–353.
29. A. K. Bielinsky and S. A. Gerbi, Where it all starts: eukaryotic origins of DNA replication, *J. Cell Sci.*, 2001, **114**, 643–651.
30. R. K. Clyne and T. J. Kelly, Genetic analysis of an ARS element from the fission yeast *Schizosaccharomyces pombe*, *EMBO J.*, 1995, **14**, 6348–6357.
31. M. L. DePamphilis, Origins of DNA replication that function in eukaryotic cells, *Curr. Opin. Cell Biol.*, 1993, **5**, 434–441.
32. Y. J. Machida, J. L. Hamlin and A. Dutta, Right place, right time, and only once: replication initiation in metazoans, *Cell*, 2005, **123**, 13–24.
33. I. Lucas, A. Palakodeti, Y. Jiang, D. J. Young, N. Jiang, A. A. Fernald and M. M. Le Beau, High-throughput mapping of origins of replication in human cells, *EMBO Rep.*, 2007, **8**, 770–777.
34. M. I. Aladjem, The mammalian beta globin origin of DNA replication, *Front. Biosci.*, 2004, **9**, 2540–2547.

35. J. R. Lipford and S. P. Bell, Nucleosomes positioned by ORC facilitate the initiation of DNA replication, *Mol. Cell*, 2001, **7**, 21–30.
36. B. D. Aggarwal and B. R. Calvi, Chromatin regulates origin activity in *Drosophila* follicle cells, *Nature*, 2004, **430**, 372–376.
37. E. Danis, K. Brodolin, S. Menut, D. Maiorano, C. Girard-Reydet and M. Mechali, Specification of a DNA replication origin by a transcription complex, *Nat. Cell Biol.*, 2004, **6**, 721–730.
38. N. Collins, R. A. Poot, I. Kukimoto, C. Garcia-Jimenez, G. Dellaire and P. D. Varga-Weisz, An ACF1-ISWI chromatin-remodeling complex is required for DNA replication through heterochromatin, *Nat. Genet.*, 2002, **32**, 627–632.
39. S. Saha, Y. Shan, L. D. Mesner and J. L. Hamlin, The promoter of the Chinese hamster ovary dihydrofolate reductase gene regulates the activity of the local origin and helps define its boundaries, *Genes Dev.*, 2004, **18**, 397–410.
40. L. D. Mesner and J. L. Hamlin, Specific signals at the 3′ end of the DHFR gene define one boundary of the downstream origin of replication, *Genes Dev.*, 2005, **19**, 1053–1066.
41. R. Li, D. S. Yu, M. Tanaka, L. Zheng, S. L. Berger and B. Stillman, Activation of chromosomal DNA replication in *Saccharomyces cerevisiae* by acidic transcriptional activation domains, *Mol. Cell. Biol.*, 1998, **18**, 1296–1302.
42. E. L. Beall, J. R. Manak, S. Zhou, M. Bell, J. S. Lipsick and M. R. Botchan, Role for a *Drosophila* Myb-containing protein complex in site-specific DNA replication, *Nature*, 2002, **420**, 833–837.
43. R. Lebofsky, R. Heilig, M. Sonnleitner, J. Weissenbach and A. Bensimon, DNA replication origin interference increases the spacing between initiation events in human cells, *Mol. Biol. Cell*, 2006, **17**, 5337–5345.
44. P. J. Gillespie, A. Li and J. J. Blow, Reconstitution of licensed replication origins on *Xenopus* sperm nuclei using purified proteins, *BMC Biochem.*, 2001, **2**, 15.
45. S. E. Moyer, P. W. Lewis and M. R. Botchan, Isolation of the Cdc45/Mcm2-7/GINS (CMG) complex, a candidate for the eukaryotic DNA replication fork helicase, *Proc. Natl. Acad. Sci. U.S.A.*, 2006, **103**, 10236–10241.
46. S. Tanaka, T. Umemori, K. Hirai, S. Muramatsu, Y. Kamimura and H. Araki, CDK-dependent phosphorylation of Sld2 and Sld3 initiates DNA replication in budding yeast, *Nature*, 2007, **445**, 328–332.
47. O. M. Aparicio, D. M. Weinstein and S. P. Bell, Components and dynamics of DNA replication complexes in *S. cerevisiae*: redistribution of MCM proteins and Cdc45p during S phase, *Cell*, 1997, **91**, 59–69.
48. J. M. Claycomb, D. M. MacAlpine, J. G. Evans, S. P. Bell and T. L. Orr-Weaver, Visualization of replication initiation and elongation in *Drosophila*, *J. Cell Biol.*, 2002, **159**, 225–236.
49. J. J. Blow and A. Dutta, Preventing re-replication of chromosomal DNA, *Nat. Rev. Mol. Cell. Biol.*, 2005, **6**, 476–486.

50. T. Senga, U. Sivaprasad, W. Zhu, J. H. Park, E. E. Arias, J. C. Walter and A. Dutta, PCNA is a cofactor for Cdt1 degradation by CUL4/DDB1-mediated N-terminal ubiquitination, *J. Biol. Chem.*, 2006, **281**, 6246–6252.
51. J. Hu and Y. Xiong, An evolutionarily conserved function of proliferating cell nuclear antigen for Cdt1 degradation by the Cul4-Ddb1 ubiquitin ligase in response to DNA damage, *J. Biol. Chem.*, 2006, **281**, 3753–3756.
52. D. Y. Takeda, J. D. Parvin and A. Dutta, Degradation of Cdt1 during S phase is Skp2-independent and is required for efficient progression of mammalian cells through S phase, *J. Biol. Chem.*, 2005, **280**, 23416–23423.
53. T. Kondo, M. Kobayashi, J. Tanaka, A. Yokoyama, S. Suzuki, N. Kato, M. Onozawa, K. Chiba, S. Hashino, M. Imamura, Y. Minami, N. Minamino and M. Asaka, Rapid degradation of Cdt1 upon UV-induced DNA damage is mediated by SCFSkp2 complex, *J. Biol. Chem.*, 2004, **279**, 27315–27319.
54. M. L. DePamphilis, J. J. Blow, S. Ghosh, T. Saha, K. Noguchi and A. Vassilev, Regulating the licensing of DNA replication origins in metazoa, *Curr. Opin. Cell Biol.*, 2006, **18**, 231–239.
55. R. Burkhart, D. Schulte, D. Hu, C. Musahl, F. Gohring and R. Knippers, Interactions of human nuclear proteins P1Mcm3 and P1Cdc46, *Eur. J. Biochem.*, 1995, **228**, 431–438.
56. M. Lei, Y. Kawasaki and B. K. Tye, Physical interactions among Mcm proteins and effects of Mcm dosage on DNA replication in *Saccharomyces cerevisiae*, *Mol. Cell. Biol.*, 1996, **16**, 5081–5090.
57. A. Rowles, J. P. Chong, L. Brown, M. Howell, G. I. Evan and J. J. Blow, Interaction between the origin recognition complex and the replication licensing system in *Xenopus*, *Cell*, 1996, **87**, 287–296.
58. S. Donovan, J. Harwood, L. S. Drury and J. F. Diffley, Cdc6p-dependent loading of Mcm proteins onto pre-replicative chromatin in budding yeast, *Proc. Natl. Acad. Sci. U.S.A.*, 1997, **94**, 5611–5616.
59. H. M. Mahbubani, J. P. Chong, S. Chevalier, P. Thommes and J. J. Blow, Cell cycle regulation of the replication licensing system: involvement of a Cdk-dependent inhibitor, *J. Cell Biol.*, 1997, **136**, 125–135.
60. M. C. Edwards, A. V. Tutter, C. Cvetic, C. H. Gilbert, T. A. Prokhorova and J. C. Walter, MCM2-7 complexes bind chromatin in a distributed pattern surrounding the origin recognition complex in *Xenopus* egg extracts, *J. Biol. Chem.*, 2002, **277**, 33049–33057.
61. D. Cortez, G. Glick and S. J. Elledge, Minichromosome maintenance proteins are direct targets of the ATM and ATR checkpoint kinases, *Proc. Natl. Acad. Sci. U.S.A.*, 2004, **101**, 10078–10083.
62. C. C. Tsao, C. Geisen and R. T. Abraham, Interaction between human MCM7 and Rad17 proteins is required for replication checkpoint signaling, *EMBO J.*, 2004, **23**, 4660–4669.
63. M. Oehlmann, A. J. Score and J. J. Blow, The role of Cdc6 in ensuring complete genome licensing and S phase checkpoint activation, *J. Cell Biol.*, 2004, **165**, 181–190.

64. A. M. Woodward, T. Gohler, M. G. Luciani, M. Oehlmann, X. Ge, A. Gartner, D. A. Jackson and J. J. Blow, Excess Mcm2-7 license dormant origins of replication that can be used under conditions of replicative stress, *J. Cell Biol.*, 2006, **173**, 673–683.
65. X. Q. Ge, D. A. Jackson and J. J. Blow, Dormant origins licensed by excess Mcm2 7 are required for human cells to survive replicative stress, *Genes Dev.*, 2007, **21**, 3331–3341.
66. C. F. Hardy, O. Dryga, S. Seematter, P. M. Pahl and R. A. Sclafani, mcm5/cdc46-bob1 bypasses the requirement for the S phase activator Cdc7p, *Proc. Natl. Acad. Sci. U.S.A.*, 1997, **94**, 3151–3155.
67. P. Zegerman and J. F. Diffley, Phosphorylation of Sld2 and Sld3 by cyclin-dependent kinases promotes DNA replication in budding yeast, *Nature*, 2007, **445**, 281–285.
68. Z. F. Pursell, I. Isoz, E. B. Lundstrom, E. Johansson and T. A. Kunkel, Yeast DNA polymerase epsilon participates in leading-strand DNA replication, *Science*, 2007, **317**, 127–130.
69. S. A. Nick McElhinny, D. A. Gordenin, C. M. Stith, P. M. Burgers and T. A. Kunkel, Division of labor at the eukaryotic replication fork, *Mol. Cell*, 2008, **30**, 137–144.
70. R. T. Pomerantz and M. O'Donnell, Replisome mechanics: insights into a twin DNA polymerase machine, *Trends Microbiol.*, 2007, **15**, 156–164.
71. K. Labib and A. Gambus, A key role for the GINS complex at DNA replication forks, *Trends Cell Biol.*, 2007, **17**, 271–278.
72. A. Gambus, R. C. Jones, A. Sanchez-Diaz, M. Kanemaki, F. van Deursen, R. D. Edmondson and K. Labib, GINS maintains association of Cdc45 with MCM in replisome progression complexes at eukaryotic DNA replication forks, *Nat. Cell Biol.*, 2006, **8**, 358–366.
73. L. S. Cox, Who binds wins: competition for PCNA rings out cell-cycle changes, *Trends Cell Biol.*, 1997, **7**, 493–498.
74. E. Warbrick, The puzzle of PCNA's many partners, *Bioessays*, 2000, **22**, 997–1006.
75. G. Maga and U. Hubscher, Proliferating cell nuclear antigen (PCNA): a dancer with many partners, *J. Cell Sci.*, 2003, **116**, 3051–3060.
76. G. Krings and D. Bastia, Molecular architecture of a eukaryotic DNA replication terminus–terminator protein complex, *Mol. Cell. Biol.*, 2006, **26**, 8061–8074.
77. R. A. Poot, L. Bozhenok, D. L. van den Berg, S. Steffensen, F. Ferreira, M. Grimaldi, N. Gilbert, J. Ferreira and P. D. Varga-Weisz, The Williams syndrome transcription factor interacts with PCNA to target chromatin remodelling by ISWI to replication foci, *Nat. Cell Biol.*, 2004, **6**, 1236–1244.
78. A. Groth, W. Rocha, A. Verreault and G. Almouzni, Chromatin challenges during DNA replication and repair, *Cell*, 2007, **128**, 721–733.
79. T. Formosa, Changing the DNA landscape: putting a SPN on chromatin, *Curr. Top. Microbiol. Immunol.*, 2003, **274**, 171–201.

80. B. C. Tan, C. T. Chien, S. Hirose and S. C. Lee, Functional cooperation between FACT and MCM helicase facilitates initiation of chromatin DNA replication, *EMBO J.*, 2006, **25**, 3975–3985.
81. A. Gerard, S. Koundrioukoff, V. Ramillon, J. C. Sergere, N. Mailand, J. P. Quivy and G. Almouzni, The replication kinase Cdc7-Dbf4 promotes the interaction of the p150 subunit of chromatin assembly factor 1 with proliferating cell nuclear antigen, *EMBO Rep.*, 2006, **7**, 817–823.
82. A. Groth, D. Ray-Gallet, J. P. Quivy, J. Lukas, J. Bartek and G. Almouzni, Human Asf1 regulates the flow of S phase histones during replicational stress, *Mol. Cell*, 2005, **17**, 301–311.
83. J. J. Blow and T. U. Tanaka, The chromosome cycle: coordinating replication and segregation. Second in the cycles review series, *EMBO Rep.*, 2005, **6**, 1028–1034.
84. K. Luger, A. W. Mader, R. K. Richmond, D. F. Sargent and T. J. Richmond, Crystal structure of the nucleosome core particle at 2.8Å resolution, *Nature*, 1997, **389**, 251–260.
85. G. Micheli, C. T. Baldari, M. T. Carri, G. Di Cello and M. Buongiorno-Nardelli, An electron microscope study of chromosomal DNA replication in different eukaryotic systems, *Exp. Cell Res.*, 1982, **137**, 127–140.

CHAPTER 2

The Action of AAA + ATPases in Loading Replication Factors

CHRISTIAN SPECK[a] AND JERZY MAJKA[b]

[a] DNA Replication Group, MRC Clinical Sciences Centre, Imperial College Faculty of Medicine, Hammersmith Hospital Campus, Du Cane Road, London, W12 0NN, UK; [b] Department of Biochemistry and Molecular Biophysics, Washington University School of Medicine, St Louis, MO 63110, USA

2.1 Introduction

AAA + proteins bind and hydrolyse ATP for regulatory purposes, acting either as molecular switches, or to remodel client proteins, or to use the energy from ATP hydrolysis to fuel molecular motors. These three key principles are seen in the initiation and elongation phases of DNA replication, where several AAA + proteins use their ATPase function to load ring-shaped replication factors onto DNA (see also Chapter 3). In this chapter, we focus on two reactions:

(i) AAA + DNA replication initiator-assisted loading of a ring-shaped helicase onto origin DNA;
(ii) AAA + clamp loader-assisted loading of a ring-shaped polymerase processivity clamp onto a primer junction.

These processes involve ATP binding by the loader, formation of the loader–ring complex in the presence (i) or absence of DNA (ii), ATP-assisted opening of the ring complex, transfer of DNA into the ring and finally release of the ring from the loader that coincides with activation of ATPase. The precise order of

Molecular Themes in DNA Replication
Edited by Lynne S. Cox

Published by the Royal Society of Chemistry, www.rsc.org

events depends on the system and will be discussed later. Nevertheless, the cycle of ATP binding, ring and DNA remodelling, and finally ATP hydrolysis to disassemble the complexes creates an energetic directionality that drives the loading reaction and hence the formation of a functional replication complex.

Licensing the DNA in preparation for DNA replication occurs during a period of low cyclin-dependent kinase (CDK) activity, in late M and early G1 phase (see Chapter 1). During this time, a helicase loading complex can assemble and load the helicase on origin DNA. The helicase loader consists of the origin recognition complex (ORC), a six subunit protein complex which recognises origin DNA, together with Cdc6, which binds specifically to an ORC-origin DNA complex, and Cdt1, which delivers the helicase to the ORC-Cdc6-DNA complex. The large ORC-Cdc6-DNA complex can then recruit and load the MCM2-7 helicase with the help of Cdt1 onto DNA (see Chapter 1). The complex of helicase loader and helicase is referred to as the pre-replication complex (pre-RC) and was first observed by *in vivo* DNase I footprinting.[1] The key step in the formation of processive replication forks is the activation of MCM helicase by an increase in CDK activity during S phase (Chapter 1). At the same time, CDK inactivates the ORC-Cdc6 loading complex by phosphorylation of several subunits. This recurring pattern of low and high CDK activity guarantees that DNA replication can only be initiated once every cell cycle (Chapter 1). Consequently, chromosomes can be maintained and duplicated in an ordered process that hinders genetic degradation and generation of cancerous cells.

2.2 ORC Assembly at the Replication Origin

2.2.1 The DNA Replication Origin—a Binding Site for the Initiator ORC

The Replicon Model presented by Jacob *et al.* in 1963[2] described a replicator, now called an origin of replication, and an initiator protein which interacts with the replicator to initiate DNA replication. Since then we have learned a great deal about initiators and sequence recognition, with bacteria[3] serving as useful models for our understanding of archaeal[4] and eukaryotic[5] replication. Bacterial DNA replication follows the simplest model, where the initiator, DnaA, a AAA+ protein, forms a multiprotein complex at a single origin to initiate DNA replication.[6] In *E. coli*, the best-studied bacterium, DnaA binds to a 9 bp non-palindromic repeat and promotes DNA unwinding in a neighbouring AT-rich region.[7] Here ATP hydrolysis functions as a switch, since ATP-DnaA is active in unwinding the origin while ADP-DnaA is not.[8] In contrast to the small bacterial chromosome, the large size of eukaryotic chromosomes suggested that their replication would require multiple start sites. Indeed only a few years after the Replicon Model was proposed, the first eukaryotic origins of DNA replication were identified.[9] In *Saccharomyces cerevisiae* (Sc), these autonomously replicating sequences (ARS) are short 100–150 bp sequences,[10] containing several defined genetic elements that are essential (A element), or important (B1, B2 elements) for replication activity,[11,12] together with a B3 element, which is not conserved between different ARSs. Based on this knowledge, the protein complex ORC was discovered and was shown to bind specifically to the A and B1 elements.[13–15] Homologues of ORC were subsequently identified in several eukaryotic species.[16–18] However in eukaryotes, A and B elements do not exist beyond budding yeast; instead origin recognition seems to be relatively independent of DNA sequence.[19]

2.2.2 ORC is an AAA+ Protein Complex

The origin recognition complex contains six subunits (Orc1–6);[13] Orc1–5 belong to the AAA+ family of ATP binding proteins,[20,21] while Orc6 has an unrelated structure.

AAA+ proteins are characterised by a specific structure, including three domains (D1, D2 and D3 in Figure 2.1) that can be fused to accessory domains. Domain 1 contains the Walker A ATP-binding, Walker B ATP-hydrolysis, and ATP sensing sensor-1 motifs (denoted W-A, W-B and Sen-1 in Figure 2.1, respectively). In addition, domain 1 contains a conserved arginine, known as arginine finger, (ARG-F in Figure 2.1) which acts in *trans* on a neighbouring

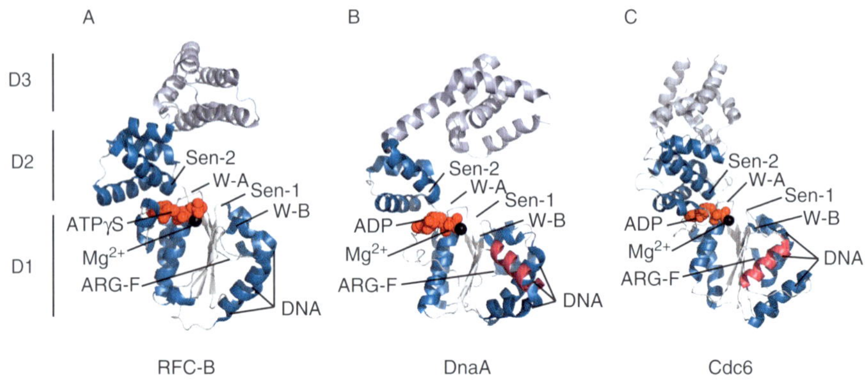

Figure 2.1 **Structure of a clamp loader and two initiator proteins**. The basic AAA+ structure is well conserved between the clamp loader and origin finding AAA+ proteins. Helices of domain 1 (D1) and domain 2 (D2) are shown in marine blue and beta sheets in grey. The C-terminal domain 3 (D3) is shown in light blue. (**A**) The structure of *Saccharomyces cerevisiae* RFC-B (PDB accession number 1SXJ). (**B**) The structure of *Aquifix aeolicus* DnaA (PDB accession number 1L8Q). (**C**) The structure of *Pyrobaculum aerophilum* Cdc6 (PDB accession number 1FNN). Conserved motifs are labelled as follows: Walker-A (W-A), Walker-B (W-B), sensor-1 (Sen-1), sensor-2 (Sen-2), arginine-finger (ARG-F). Insertions present in DnaA and Cdc6, not present in RFC, are shown in magenta. Recently identified DNA interaction regions within domain-1 of DnaA and Cdc6 are marked (DNA).

subunit to stimulate ATPase activity. Domain 2 constitutes a lid domain, which sits on top of domain 1. ATP is bound within the interface of domain 1 and domain 2. Depending on their nucleotide binding state (ATP or ADP bound), domains 1 and 2 undergo conformational changes. Furthermore, these conformational changes can be transmitted to connected domains or to interaction partners[22,23] such as proliferating cell nuclear antigen (PCNA) or the replicative helicase MCM2-7. In return, it is also possible for interaction partners to alter the interface between domain 1 and 2, thereby activating ATP binding or ATPase activity.[24] AAA+ proteins in DNA replication often contain a third domain, which is implicated in protein–DNA and protein–protein interactions.

AAA+ proteins are frequently organised as multimeric assemblies and can adopt different shapes, *e.g.* spiral pentamers or ring shaped hexamers. In these complexes, ATP binding sites are oriented in such a way that the nucleotide binds to a cleft formed between neighbouring subunits. This structural organisation ensures that the ATP-binding status of one subunit can be relayed to neighbouring subunits *via* ATP-induced conformational changes.[25] Among the eukaryotic Orc1-5 proteins, only three subunits, Orc1, 4 and 5, contain functional Walker A and B motifs, with the exception of ScOrc4 (Orc4 in *Saccharomyces cerevisiae*), which has a non-functional Walker A motif.[21] Secondary structure predictions for Orc2 and Orc3 suggest a fold similar to AAA+ proteins, although Walker A and B motifs are mutated. It is therefore fair to assume that ORC can adopt a typical AAA+ superstructure, where subunits interact closely to form a ring or spiral shape.[20,21]

2.2.3 The Structure of ORC and the Clamp Loader RFC

AAA+ proteins in DNA replication fall into three different classes: clamp loader, initiator, and helicase.[26,27]

The clamp loader family of proteins—including the small subunits of RFC, RFC B–E (Table 2.1 and Section 2.4.1 for more details)—possesses the most compact domain organisation, containing domains 1 and 2 required for ATP binding and an additional C-terminal wing helix domain (WHD) (domain 3), responsible for subunit oligomerisation and interaction with DNA (see RFC-B in

Table 2.1 RFC subunits.

RFC subunit	*S. cerevisiae*[a]	*Human*	*E. coli pol III subunits*[b]
Rfc1	RFC-A	p140	δ wrench
Rfc2	RFC-D	p37	γ motor
Rfc3	RFC-C	p36	γ motor
Rfc4	RFC-B	p40	γ motor
Rfc5	RFC-E	p38	δ' stator

[a]The nomenclature given for *S. cerevisiae* is based on the position of the subunit in the crystallised RFC complex[56] and is used throughout this chapter.

[b]Possible homologues to subunits of *E. coli* polymerase III holoenzyme are given, as proposed in the wrench–stator–motor model,[83,85] but are less clear in the light of more recent experimental evidence (see Section 2.4.3.3).

Figure 2.1A). Unlike RFC B–E, RFC-A (the large subunit of RFC) has additional N- and C-terminal extensions.

The structure of DnaA (Figure 2.1B), the bacterial initiator protein,[28] is similar to that of the clamp loader, but contains an insertion in domain 1.[26] DnaA is also characterised by an additional N-terminal protein interaction domain and a C-terminal domain 3, which contains a helix-turn-helix DNA binding motif. Archaeal Cdc6[29] serves as an archetype of the eukaryotic initiator proteins Orc1–5 and Cdc6. Like DnaA, Cdc6 has an insertion in domain 1, resulting in an extra extension (Figure 2.1C), but its C-terminal domain 3 contains a winged helix DNA binding domain, which can also mediate protein-protein interactions[30,31] (see Chapter 12). ORC and RFC are structurally, as well as functionally, similar; RFC forms a pentameric AAA+ spiral while ORC also contains 5 AAA+ subunits, although the structural organisation of the Orc1-6 subunits is less clear.

By contrast, the MCM2-7 helicase proteins have a significantly different AAA+ domain organisation compared with that of RFC, Orc1-5 or Cdc6. Although the structure of full length MCM helicase is not known, it is clear that it contains several extensions in domain 1 and an unusual configuration of domain 2.[26] The divergence of the AAA+ fold of RFC/Cdc6 and MCM most probably reflects a functional divergence. MCM proteins are motor proteins that use the energy of ATP hydrolysis to unwind DNA at the replication fork, while RFC and ORC both function like chaperones, altering the structure of their respective client proteins, PCNA and MCM2-7.

2.3 Pre-RC Assembly

2.3.1 Regulation of ORC Assembly

ORC subunits forms stable complexes in both budding yeast and *Drosophila melanogaster*.[13,17] However, *Homo sapiens* (Hs) ORC seems to be more flexible.[32–35] Human ORC consists of an Orc2-3-5 complex that binds Orc4 in an ATP-dependent manner to form the ORC2-5 core complex.[33,34] This complex associates in a cell cycle regulated manner with Orc1.[36] Binding of HsOrc1 to the Orc2-5 complex is probably dependent on the arginine finger in Orc4, as seen in *Saccharomyces cerevisiae* ORC (ScORC).[37] During S phase, HsOrc1 is released from the ORC core complex and is destroyed *via* ubiquitin-mediated proteolysis.[36,38] This regulated destruction of HsOrc1 probably hinders re-initiation of DNA replication in higher eukaryotes.[39] Orc6, although necessary for DNA replication, binds the core complex very weakly *in vitro* and *in vivo*.[32–35,40] The interaction of Orc6 with the ORC core complex could depend on a specific modification or the activity of other proteins. Indeed, it was recently reported that ScOrc6 serves as a docking station for the Cdt1–MCM2-7 complex,[41] so potentially HsOrc6 depends on Cdt1–MCM2-7 for binding to ORC. In addition, Orc6 has essential functions outside of DNA replication, *i.e.* it is involved in chromosome segregation and cytokinesis.[40,42] This dual functionality could link the progress of DNA replication to mitotic events (see also Chapter 9).

2.3.2 ORC–DNA Interaction

It was found early on that ATP binding to Orc1 is required for sequence-specific DNA binding of ORC in *S. cerevisiae.*[13] ATP binding to ORC also stimulates nonspecific DNA binding of both *Drosophila* (Dm)[43] and human ORC.[32] In contrast, *Schizosaccharomyces pombe* (Sp) ORC binds AT-rich DNA in an ATP-independent manner.[44]

DNA binding of ORC can probably be separated in two types: a sequence-nonspecific and a sequence-specific component. The sequence-nonspecific component generally involves DNA backbone contacts and may include wrapping of the DNA around the ORC complex,[45] while the sequence-specific component is realised in different ways in each eukaryotic system. ScORC, similarly to classical initiator proteins like DnaA[3] or papillomavirus E1,[46] binds with sequence specificity to DNA, potentially through direct interactions with the winged helix domains of Orc1, 2 and 4.[13,14,30,31] In other eukaryotes, ORC DNA binding mechanisms are divergent. In *S. pombe*, nine copies of an AT-hook motif on Orc4 are responsible for the interaction with origin DNA.[44,47] DmORC can bind to AT-rich DNA using the basal transcription factor IIB TFIIB- like domain in Orc6[48] and HsORC can bind to chromatin *via* interaction with adaptor proteins such as HMGA1a or TRF2[49,50] (see also Chapter 10). In addition, the structure of chromatin seems to be of major importance in restricting access for ORC.[51–53] Finally, the tertiary structure of the chromatin, loops and the organisation of the DNA replication machinery into clusters generates superstructures, which are considered to provide a framework for replication origins[54,55] (see Chapter 10).

Another emerging DNA interaction, probably involving single-stranded DNA (ssDNA), is mediated *via* the AAA+ domain, specifically helix 4–6 of domain 1 (Figure 2.1). In the case of DnaA and Cdc6, this interaction is further enhanced by the insertion of an additional helix (Figure 2.1, magenta), while in RFC, the AAA+ domain 1–DNA interaction involves probable contacts with partially ssDNA.[56] For DnaA, this interaction was reported to be necessary for DNA unwinding and ssDNA interaction.[57] In archaeal Cdc6, a AAA+ domain 1–DNA interaction results in significant deformation of the DNA.[30,31] We do not know if the AAA+ domain 1–DNA interaction of ORC promotes DNA unwinding, but ORC does bind to ssDNA.[58] The functional relevance for this ssDNA interaction is unknown.

Given that replication origins in higher eukaryotes are specified by a combination of sequence, chromatin state or superstructure, rather than tightly defined by DNA sequence (*e.g.* ARS in *Saccharomyces cerevisiae*), the loss of sequence specificity of ORC binding to DNA in higher eukaryotes is probably an advantage, and allows more flexibility in targeting ORC—with the help of adaptor proteins—to replication start sites within the DNA.

2.3.3 Cdc6 Binds in a DNA Sequence-dependent Manner to ORC and Remodels the Complex

During late M to early G1 phase of the cell cycle, Cdc6 and ORC assemble on DNA into a hexameric AAA+ complex.[29,59–61] The ScORC-Cdc6 complex forms a ring-like structure,[21] (Figure 2.2) a common shape among AAA+ proteins such as helicases (Chapter 3), chaperones and proteases.[5–27] However, the ring of ORC-Cdc6 is not as perfectly formed as in other AAA+ hexamers. In particular, Cdc6 within the ORC-Cdc6 complex is arranged at an angle to the plane of the ring (Figure 2.2D). This makes the ORC-Cdc6 complex more similar to the non-symmetric pentameric AAA+ RFC.[25] RFC, in contrast to perfectly round hexameric AAA+ proteins, is a spiral-shaped pentamer with an analogous function to the ORC-Cdc6 complex in that it loads a ring-shaped clamp (PCNA) onto DNA (see Section 2.4). Similarly to RFC, the structure of the ORC-Cdc6 complex has been obtained in the absence of DNA. Structural studies of a DNA-bound ORC-Cdc6 complex will be paramount to determine if conformational changes take place upon DNA binding that change the shape of the complex to make it perhaps more similar to RFC. (The spiral shape could be used by RFC as an interaction surface for the PCNA clamp and this interaction could in turn result in clamp opening—see Section 2.4.) The diameter of the ORC-Cdc6 ring is similar in dimension to that of the MCM ring[21] (see Chapter 3) and therefore represents a good interface for binding of MCM2-7. Ultimately, it remains to be seen whether the ORC-Cdc6 complex uses a similar mechanism to open the MCM ring as that employed by RFC to open the PCNA trimer.

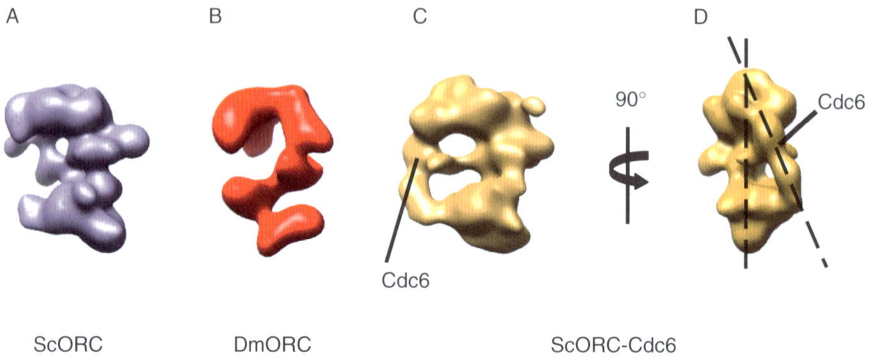

Figure 2.2 **Electron microscopy structures of ORC.** (**A**) ScORC structure in complex with ATPγS (EM data bank accession[i] EMDB-1156). (**B**) DmORC structure in complex with ATP (accession number EMDB-1253). (**C**) ScORC-Cdc6 structure in complex with ATPγS (accession number EMDB-1157). The proposed location of Cdc6 is indicated (see also supplementary Figure 8 in ref. 21). (**D**) The ring-shaped structure as shown in (**C**) has been rotated 90° anticlockwise. Cdc6 has an unusual conformation; it is not in the plane ring structure of ORC-Cdc6, but tilted diagonally.

[i] http://emnavi.protein.osaka-u.ac.jp/

ORC-Cdc6 complex formation is dependent on the presence of ATP and its stability is regulated by the ATPase activity of Cdc6.[21,62] In the absence of DNA, ORC and Cdc6 can interact, but this interaction stimulates the ATPase activity of Cdc6 and, as a consequence, the complex falls apart.[62] On origin DNA, the Cdc6 ATPase activity is downregulated, which stabilises the complex. However, mutations within the primary ORC binding site, the A element, or to lesser extent in neighbouring B elements, result in Cdc6-mediated ATP hydrolysis and complex breakup, suggesting that Cdc6 is probing the structure of the ORC-DNA complex.[62] Mutant Cdc6 lacking ATPase activity forms complexes with ORC on mutant origin DNA with an intermediate stability,[62] since these Cdc6 ATPase mutants lose their ability to destabilise the complex. Similarly, in the *Xenopus laevis in vitro* DNA replication system, blocking Cdc6 ATPase by addition of ATPγS results in a very stable, non-functional ORC-Cdc6 complex.[63] It is therefore not surprising that ScCdc6 ATPase mutants have a temperature-sensitive phenotype or are dominant lethal when over-expressed *in vivo*.[64–66] This could be due to complex formation on non-origin DNA,[62] or because Cdc6 ATPase activity is required at an early step during MCM loading.[67] Orc1 ATPase mutants also have a reduced ability to form the extended ORC-Cdc6 footprint on origin DNA, indicating that ATP hydrolysis by Orc1 is important for faithful DNA binding of the ORC-Cdc6 complex.[21]

The *in vitro* DNase I footprint of ScORC spans about 48 bp,[21] and is similar in size to the footprint observed *in vivo* during the G2 phase of the cell cycle.[1] This footprint expands to about 82 bp during G1 and represents pre-RC formation[1] (see Chapter 1). The pre-RC consists of ORC, Cdc6, Cdt1 and MCM2-7[68,69]. It was thought that the presence of MCM2-7 could be the reason for the expansion of the footprint.[1] However, *in vitro* reconstitution of ORC and Cdc6 on origin DNA, in the absence of MCM2-7, has produced a similar extended footprint,[21,62] indicating that MCM2-7 binding does not cause the expanded footprint. It is now assumed that ORC and Cdc6 reorganise the DNA into a specific structure, possibly to prepare for MCM2-7 loading onto DNA. Since the footprint of ORC-Cdc6 is significantly longer than its protein 3D structure,[21] DNA probably wraps around the round-shaped ORC-Cdc6 complex (Figure 2.3). This DNA interaction could be mediated *via* the winged-helix-domain or the AAA+ domains of the complex.[30,31] The wrapping of DNA around ORC-Cdc6 could be required to facilitate MCM helicase loading onto each strand to establish later bidirectional replication forks.[70] Therefore, DNA structure and DNA flexibility might be the ultimate determinants of origin specificity in higher eukaryotes.

2.3.4 ORC and Cdc6 Cooperate to Load Cdt1-MCM2-7

Pre-RC formation was first reconstituted in *Xenopus* egg extracts, where it was shown that ORC, Cdc6, Cdt1, MCM2-7 and nucleoplasmin are required to license chromatin for subsequent DNA synthesis.[68] In addition to nucleoplasmin, a further histone chaperone, nucleolin, remodels chromatin[71] to allow

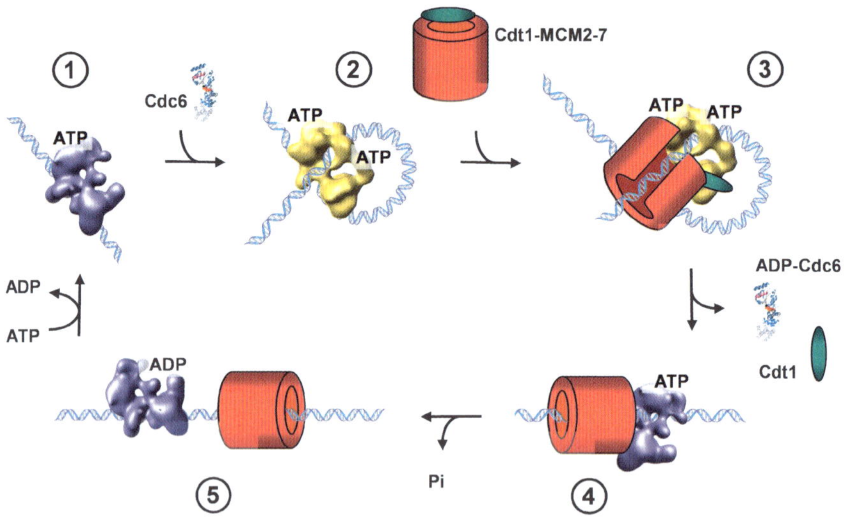

Figure 2.3 **Speculative model of pre-RC formation.** (1) ORC (blue) in the ATP-bound state can bind specifically to origin DNA. (2) ATP-Cdc6 can interact with ORC on DNA to assume an extended footprint (yellow) that may involve wrapping of the DNA around the ORC-Cdc6 complex. ATPase activity of Cdc6 is downregulated when bound to origin DNA. (3) The Cdt1-MCM2-7 complex is recruited by Orc6 to the ORC-Cdc6-DNA complex. We speculate that the large interaction surface between ORC-Cdc6 and Cdt1-MCM2-7 can destabilise MCM2-7 and open the helicase ring. The unusual wedge shape of Cdc6 (see Figure 2.2D) suggests that Cdc6 could contribute to opening the MCM2-7 ring. (4) ATP hydrolysis by Cdc6 may lead to Cdc6 and Cdt1 release from the ORC-MCM2-7 complex, resulting in closure of the MCM ring. The departure of Cdc6 could be associated with a change in DNA structure. (5) ATP-hydrolysis in Orc1 leads to release of MCM2-7 and formation of ADP-Orc1. ATP exchange in Cdc6 and Orc1 is required before another cycle of MCM loading can start.

access for ORC. ORC binding is also dependent on adenine nucleotide. Cdc6 and Cdt1 can bind to ORC independently of each other and require ATP. Loading of MCM2-7 is dependent on ATP hydrolysis, since the non-hydrolysable ATP analogue ATPγS was not competent to load MCM2-7. Loading of MCM2-7 can be impeded by geminin, a natural inhibitor of DNA replication in higher eukaryotes,[72] which stabilises Cdc6 and Cdt1 on DNA.

Recently pre-RC formation has also been reconstituted in yeast,[69] but an extract-based system[37,41,67,73] has provided the most informative approach. In this system, multiple origin DNA fragments were joined in an array that serves as a basis for pre-RC assembly. This approach revealed an ATP-hydrolysis requirement for pre-RC assembly,[73] similar to that seen in the *Xenopus* model. The mechanism of MCM loading was studied in more detail by using ATP hydrolysis mutants of ORC and Cdc6. A Cdc6 ATP hydrolysis mutant[64] binds MCM2-7, but fails to load the MCM ring onto DNA and instead traps Cdt1 on DNA. This indicates formation of an intermediate complex poised to trigger ATP hydrolysis and Cdt1 release. Cdt1 forms a complex with MCM2-7[69] and was shown to transport MCM2-7 from the cytoplasm to the nucleus in late M phase.[74] Nuclear Cdt1 cooperates with Orc6,[41,75] and to a lesser extent with Orc2,[75] to load MCM2-7 onto DNA. A fusion of Cdt1 and Orc6 cannot repeatedly load MCM2-7,[41] indicating an essential role for complex disassembly in repetitive loading. Orc1 ATP hydrolysis is probably involved in the last step of MCM-loading. A mutation in the arginine finger of Orc4 has been shown to inhibit Orc1 ATP hydrolysis, but this does not hinder successful MCM2-7 loading. Instead the release of MCM2-7 from the ORC-Cdc6 complex was hindered, suggesting that only a single MCM complex can be loaded onto DNA.[37]

In summary, Orc1 and Cdc6 ATPase activities force disassembly of very stable intermediate complexes and consequently drive the reaction forward (Figure 2.3) to ensure formation of a functional pre-RC.

2.3.5 A Model for Pre-RC Formation

Based on current knowledge supported by our findings and those of others,[76] we can now propose a speculative model of pre-RC formation (Figure 2.3). The ORC complex is chromatin-bound throughout the cell cycle. This interaction is ATP-dependent and might involve DNA contacts by the WHD and AAA+ domains of Orc1, Orc2 and Orc4. This structure represents the entry site for Cdc6, which binds to chromatin during late M early G1 phase of the cell cycle. Binding of Cdc6 to origin-bound ORC will inhibit Cdc6 ATPase activity. At the same time, the complex assumes an extended footprint, which probably involves wrapping of the DNA around the ORC-Cdc6 complex. This complex is now competent to load MCM2-7 onto DNA. Cdt1 interacts with Orc6, and potentially Orc2, to transfer the MCM complex onto the ORC-Cdc6 ring. The ORC-Cdc6 ring has an irregular shape but might assume a more spiral shape upon DNA binding. The large protein interaction surface between ORC-Cdc6 and MCM2-7 could twist the MCM ring into a structure that allows entry of DNA into the MCM ring. At this time, Cdt1 will be released from the complex, potentially assisted by Cdc6-ATP hydrolysis. This signals successful loading of the MCM ring, which is held in place by ORC. ATP hydrolysis by Orc1 can then finish the cycle and trigger the release of the MCM ring. Loading of another MCM complex will require Orc1 ADP to ATP exchange, so it can start the next cycle of MCM loading. During pre-RC formation about 20 MCM helicases will be loaded onto each origin.

2.4 RFC and Loading of PCNA

2.4.1 Structure of the RFC Complex

RFC is an archetype of the eukaryotic clamp loader family. This hetero-pentameric complex loads PCNA at the template–primer DNA junction of the DNA replication fork. The nomenclature of the RFC subunits differs between yeast and human; here, we use nomenclature based on the position of the subunit in the crystallised RFC complex[56] (see Table 2.1).

The four small subunits from all organisms fall in a narrow size range of 36–41 kDa and form the core complex. This core complex is also present in alternative eukaryotic clamp loader complexes, such as that involved in DNA damage and meiotic checkpoints,[77] where RFC-A is replaced by Rad24, Ctf18 or Elg1 (see also Chapter 3). The large subunit RFC-A shows a considerable variation in size between organisms: 77 kDa in *Candida albicans*, 95 kDa in *S. cerevisiae*, and 128 kDa in humans. Most of this size variability is confined to the N-terminal domain, which is not required for its function as a clamp loader.[78–80] The structure of this domain remains unknown since, in the crystal structure, an N-terminal deletion construct of RFC-A was utilised which was nonetheless fully functional in the clamp loading process.[56] However, it has been shown in *S. cerevisiae* RFC-A that this N-terminal domain contains a second DNA binding site in addition to the DNA binding site localised in domain 1.

All five RFC subunits contain a stretch of approximately 220 amino acids comprising seven regions of homology (RFC boxes II–VIII). Box III and V are nucleotide-binding and hydrolysis motifs Walker A and Walker B, respectively[81] (Figure 2.1A).

Electron microscopy images and pairwise subunit–subunit interactions studies—as well as homology modelling to the crystallised clamp loader γ complex from *Escherichia coli*—suggested that RFC adopts a circular arrangement of interacting subunits.[78,80–85] The crystal structure of the RFC-PCNA complex revealed that, in the presence of ATP-γS, the five AAA+ modules (composed of domain 1 and 2; Figure 2.1) of the clamp loader subunits form a right-handed spiral (Figure 2.4). The AAA+ module of each subunit is connected by a flexible linker to a C-terminal, helical domain [domain 3 (D3) in Figure 2.1]. Within the RFC subunits, these C-terminal domains are structurally more diverse than the AAA+ modules. The third domains of all five RFC subunits are tightly packed against each other to form a cylindrical collar which caps the AAA+ spiral (Figure 2.4A). The large subunit, RFC-A, contains an additional domain (domain 4), which in the crystal structure of the RFC-PCNA complex is localised between the ATPase domains of RFC-A and RFC-E. Thus domain 4 of RFC-A constitutes a joining element connecting both ends of the spiral.[56]

2.4.2 RFC Binding to DNA

Biochemical analysis suggests that the DNA recognition properties of RFC are distributed over at least three of the five subunits. Isolated RFC-D subunit has been shown preferentially to bind primed single stranded DNA.[86] In addition, yeast RFC-C has an ATPase activity that is stimulated by single-stranded but not double-stranded DNA (dsDNA), suggesting DNA binding by this subunit. The human orthologue of RFC-C also preferentially binds primed single-stranded DNA (ssDNA).[86,87] Two DNA-binding domains have been identified in the human RFC-A subunit.[80] The N-terminal region of RFC-A shows homology to prokaryotic DNA ligase and poly(ADP)-ribose polymerase.[88] The isolated N-terminal domain of human RFC-A has been shown to bind partially double-stranded DNA substrate in which at least one of the 5′ ends is phosphorylated and the phosphate is either recessed or at the blunt end, suggesting a possible functional significance of this domain at DNA ends. However, this domain is not essential for PCNA loading, and truncation of the N-terminal domain of RFC-A actually increases clamp loading activity of RFC.[78–80,89]

Since the crystal structure of the RFC-PCNA complex was obtained in the absence of DNA, blunt-ended DNA was modelled into the structure (Figure 2.4B, C) to obtain information about the interaction between RFC and DNA.[56] This modelling was based on the finding that the five AAA+ modules of RFC subunits form a right-handed spiral with a pitch very similar to that of B-form DNA. For the modelling, blunt-ended DNA was passed through the central channel of PCNA and along the screw axis of RFC. In the energetically most stable configuration of the RFC-DNA complex, each of the RFC subunits tracks the minor groove of the double helix. Within the AAA+ domain 1, the N-terminal regions of three α-helices (α4, α5, α6), as well as a loop preceeding helix α4 (Figure 2.1), interact with the backbone of DNA (Figure 2.4B). This model explains also the specificity of the PCNA loading onto 3′ recessed ends. By tracking the minor groove of DNA, the RFC complex would simply push a PCNA clamp onto DNA, threading like a screw cap onto the last turn of the double helix. Further extension of this movement would be blocked by a physical barrier imposed by the C-terminal collar on the 3′ end of the primer. In this scenario, the template strand would end up near the junction between subunits RFC-E and RFC-A, where the gap is wide enough for the single-stranded DNA to leave the RFC complex.

2.4.3 Loading of PCNA by RFC

2.4.3.1 PCNA–RFC Complex

In complete contrast to the DNA-dependent manner in which the ORC-Cdc6 loading complex binds the MCM2-7 ring, the order of events during the PCNA loading process is reversed. The RFC loader and PCNA ring first form a complex, and only subsequently do they bind to DNA. Once PCNA encircles the double helix, ATP hydrolysis is triggered, which results in release of PCNA from RFC.[90] The crystal structure of the binary RFC-PCNA complex revealed the details of interaction between these two protein factors. All five AAA+ modules of RFC are arranged in a right-handed spiral. As a result of this spatial organisation, only three RFC subunits located at the 'bottom end' of the spiral (RFC-A, RFC-B, RFC-C) make contact with the clamp[56] (Figure 2.4). This observation is in good agreement with biochemical analysis, which showed that only these three subunits interact with PCNA.[87,91–93] Each of the three PCNA subunits has a hydrophobic cleft in the interdomain connector loop (see Chapter 3), and the subunits RFC-A and RFC-C interact with two of these hydrophobic grooves. The RFC-B subunit, positioned between RFC-A and

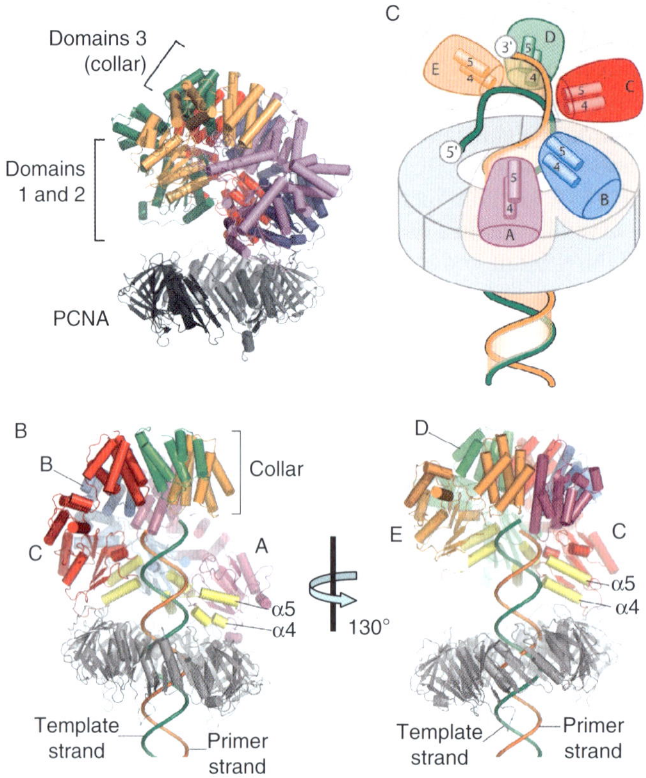

Figure 2.4 **The crystal structure of the RFC-PCNA complex and a model of its interaction with the primed DNA**. (**A**) Cartoon representation of the RFC-PCNA complex. The interaction between the RFC spiral and PCNA is mediated by the N-termini of three RFC subunits (domains 1 and 2), which contact two subunits of PCNA. The rotation axis of the clamp is tilted with respect to the screw axis of the RFC spiral resulting in a deep, wedge-shaped cleft separating PCNA and the ATPase domains (domains 1 and 2) of RFC. The collar composed of the C-termini of all RFC subunits (domain 3) caps the RFC spiral. (**B**) Model of the interaction between RFC-PCNA complex and DNA. Two cut-away views of the RFC-PCNA complex, showing the tracking of the minor groove in double-stranded DNA by the N-termini of helices α4 and α5 of each RFC subunit (indicated in light yellow). (**C**) Schematic representation of the interactions between RFC, PCNA and DNA. The subunits RFC-A, RFC-B and RFC-C contact the PCNA ring. The helices α4 and α5 of each RFC subunit track the minor groove of DNA. The three PCNA subunits are indicated by three shades of grey in A. RFC subunits are indicated as follows: purple RFC-A, blue RFC-B, red RFC-C, green RFC-D and orange RFC-E. Panels (B) and (C) courtesy of Gregory Bowman and John Kuriyan and reprinted by permission from Macmillan Publishers Ltd (G. D. Bowman, M. O'Donnell and J. Kuriyan, Structural analysis of a eukaryotic sliding DNA clamp-clamp loader complex, *Nature*, 2004, **429**, 724–730), copyright (2004).[56]

RFC-C in the spiral, contacts the intersubunit region of the clamp between the grooves recognised by RFC-A and RFC-C. Interactions between RFC-B and PCNA are more polar, and therefore may be weaker than the hydrophobic interactions of subunits RFC-A and RFC-C with PCNA. All three RFC subunits interact with the clamp through the C-terminal part of helix α4; the N-terminus of the same helix is involved in the interaction with DNA. The spiral arrangement of RFC subunits and their rather limited contact area with the PCNA ring results in the appearance of an extensive and wedge-shaped gap between the clamp and RFC-E subunit, located on the 'top end' of the RFC spiral (Figure 2.4A). The distance between PCNA and RFC-E at the broad end of the wedge is about 18Å. In the presence of DNA, PCNA could adopt a more spiral shape, which should allow opening of the PCNA ring.

2.4.3.2 ATP Utilisation

Studies on ATPase activity of the RFC complex and its core composed of four small subunits revealed that ATP hydrolysis is required only during the last step of the loading process, *i.e.* during release of the PCNA clamp from the clamp loader. The core complex and RFC display similar ATPase activities that are equally stimulated by PCNA. However, ATPγS stabilises binding between the core complex and PCNA, whereas ATP does not, indicating that ATP hydrolysis causes dissociation of the PCNA-core complex. In contrast, the stability of the PCNA-RFC complex is unaffected by ATP hydrolysis.[94,95] The more extensive contact between the RFC-A subunit and the hydrophobic groove in PCNA could suppress the ATPase activity.

The complexity of ATP utilisation during the clamp loading cycle makes measurements of an accurate ATP-binding stoichiometry a rather challenging task. These difficulties could be overcome by using ATPγS, which can replace ATP during the PCNA loading reaction but hinders the release of the loader from the DNA-bound clamp. Using this strategy, it has been shown that the ATP utilisation cycle displays an interesting and striking departure from the prokaryotic mechanism. The T4-phage gp44/62 system and the γ complex from *E. coli* bind saturating numbers of ATP molecules (four and three, respectively) prior to interaction with the clamp.[96–99] In contrast RFC on its own binds only two ATPγS molecules but, in the presence of DNA or PCNA, RFC binds three molecules of ATPγS; in the presence of both PCNA and DNA, RFC binds four molecules of ATPγS. These binding stoichiometries suggest that each step in the reaction pathway of the eukaryotic clamp loader stimulates binding of an additional ATP molecule. Thus, although two ATPs can initially bind to RFC, the remaining two ATP-binding sites are either buried or have an extremely low affinity for ATP. Binding of PCNA to RFC-ATP_2 induces a conformational change which makes one additional ATP-binding site available. Upon binding of DNA to the resulting PCNA-RFC-ATP_3 complex, another conformational change in RFC makes the final ATP-binding site available. Binding of the fourth ATP is required for the loading process to proceed to completion.[90] This scheme not only indicates that clamp loading proceeds *via* an ordered mechanism requiring binding of the clamp to the loader prior to binding of primer/template DNA, but also highlights a marked stepwise regulation of the binding and usage of ATP molecules in the eukaryotic system.

The crystal structure of the RFC-PCNA complex provides additional information about ATP utilisation by RFC. The crystals were obtained in the presence of ATPγS and the RFC complex used in the study contained mutated 'arginine fingers' in four subunits (RFC-B, RFC-C, RFC-D and RFC-E). Substitution of an arginine with a glutamine residue in the conserved SRC (Ser-Arg-Cys) motif of the arginine finger significantly decreased ATPase activity of the complex, thereby increasing the stability of the RFC-PCNA complex. Examination of the crystal structure revealed four nucleotide molecules localised between the interfaces of adjacent subunits in the spiral: RFC-A : B, RFC-B : C, RFC-C : D and RFC-D : E. Surprisingly, a fifth nucleotide molecule (the data did not allow distinction between ATPγS and ADP) was also found to be bound to the RFC-E subunit, which contains a non-functional Walker B ATP hydrolysis motif. The nucleotide bound to RFC-E is packed not against the AAA+ module of RFC-A (a scenario for symmetric AAA+ proteins like *N*-ethylmaleimide sensitive factor (NSF)[100,101]), but it interacts with the C-terminal domain 4 of RFC-A, which does not belong to a AAA+ module. The importance of ATP binding to RFC-E and the dependence on full length RFC-A is unclear. Comparison of the RFC-PCNA structure with the crystal structure of several other AAA+ proteins containing bound nucleotide[100–103] highlights the relatively tight structural arrangements of the amino acids around the four ATP molecules bound to subunits RFC-A through D. This suggests that the crystal structure captured an ATP-hydrolysis competent state of RFC immediately preceding the enzymatic reaction and the release of the clamp.[56]

2.4.3.3 Opening of the PCNA Ring

During the loading process, the PCNA clamp needs to be opened, and once on DNA, closed again. The homology between the RFC-PCNA complex and the *E. coli* $\delta\gamma_3\delta'$-complex-β-clamp loading system was the basis for the wrench–stator–motor loading mechanism proposed for the eukaryotic clamp.[85,104] According to this model, RFC-A is similar to the δ-wrench in the γ-complex of *E. coli* pol III holoenzyme (see Table 2.1), which is thought to open up the β-clamp (analogous to PCNA). The RFC-B, RFC-C and RFC-D subunits bind and hydrolyse ATP, and are similar to the three γ motor subunits. Like its functional orthologue δ′, the RFC-E subunit with an inactive ATP-binding domain is proposed to function as a stator which modulates the interactions between RFC-A and PCNA. This mechanism however, needed to be revisited after the observation that the four member core RFC complex is able to unload the PCNA sliding clamp from DNA, a process which requires opening of the clamp,[105] and after the finding of a spiral arrangement for the RFC subunits. Recent studies on the loading process do not assign any specific function to any particular RFC subunits and suggest that the interaction with the spiral-shaped loader (in its ATP-bound form) could distort the structure of the clamp, so that PCNA itself would adopt spiral-like structure. The perpendicular tension imposed on the PCNA ring could lead to the clamp opening. Sequential hydrolysis of ATP molecules bound to RFC would lead in the first step to the restoration of native PCNA configuration (as observed in the crystal structure of RFC-PCNA) and, upon completion of ATP hydrolysis, to the release of the clamp from RFC.[105–107] This model, however, remains disputed and probably only the crystal structure of an open PCNA ring bound to RFC would provide a definite proof of this mechanism.

In recent years, we have achieved significant progress in our understanding of RFC function. Biochemical analyses have revealed the detailed mechanism of the clamp loading cycle, the order of events, DNA substrate specificity and ATP utilisation during this process, while structural studies have provided atomic detail of the interactions between RFC and the PCNA clamp. These studies also explained how RFC could interact with DNA during the clamp loading process and how ATP hydrolysis could trigger release of the PCNA ring from RFC. The development of new methodologies should soon result in answers to the outstanding questions of the clamp loading cycle, particularly how RFC opens up the PCNA ring and how interaction with DNA leads to the clamp closing around the double helix.

2.5 Outlook and Potential Applications

In recent years it has become apparent that the regulation of the ATPase activity in AAA+ ATPases is key in faithful loading of replication factors. However, little is known about what triggers ATP hydrolysis and what the consequences of the hydrolysis events are. Existing data suggest that ATP hydrolysis drives disassembly rather than assembly of the ORC-Cdc6-MCM complex of the pre-RC; a Cdc6 ATP-hydrolysis mutant blocks release of Cdt1 and an Orc1 ATP hydrolysis mutant blocks release of MCM2-7.[67] Further experiments are needed to clarify the molecular mechanism of MCM loading. MCM ring opening could be achieved by formation of an asymmetric spiral-shaped MCM interaction surface by ORC-Cdc6 and Cdt1 on DNA, which forms a tight complex with the MCM ring. This spiral structure can then proceed to crack the MCM ring structure, by analogy to RFC and PCNA. However, we do not know if the MCM ring is loaded onto dsDNA, or on ssDNA like its bacterial homologue DnaB.[108] If MCMs are loaded onto dsDNA, then this might inactivate their helicase activity if the complex is unable to separate the DNA strands; however, the various models of MCM2-7 helicase activity (Chapter 3) accommodate dsDNA binding without loss of helicase function. The dsDNA bound complex would require activation, probably during S phase. Archaeal MCM, an active helicase, forms double hexamers on DNA[109] (see Chapters 3 and 12). It is therefore important to know the oligomerisation state of MCM on DNA, for example whether hexamers or double hexamers are formed.

Pre-RC formation is dependent on many reactions which are choreographed to load MCM2-7 onto DNA. Blocking MCM loading in human cancer cells by ectopic expression of geminin results in poor MCM loading, gross replication defects, unsuccessful mitosis and consequently p53-independent cell death;[110] oddly, loss of geminin leads to essentially the same phenotypes.[111] In contrast, normal cells react to geminin overexpression by arrest in G1, apparently due to a checkpoint that detects unsuccessful pre-RC formation.[110] Cancer cells probably progress past G1, as they may be unable to detect incomplete pre-RC assembly. This failure might be one requirement in the progression from the pre-malignant to the cancerous state.[112] Indeed, pre-malignant cells frequently arrest at G1 with partially formed pre-RC.[113] This may be due to a misregulated E2F pathway. The E2F pathway is frequently altered in cancers and results in aberrant pre-RC assembly and cell cycle arrest. Once cells acquire mutations in the pre-RC checkpoint, they can break free from this block and progress into S phase.[114] However, incomplete pre-RC assembly will result in incomplete replication and yield massive DNA damage in these cells, which should induce apoptosis or senescence.[115] These checkpoint mutant cells are now rendered sensitive to pre-RC inhibitors (see Chapter 14), as they have lost their ability to detect their pre-RC status. Inhibitors of MCM loading, which can target many reactions during pre-RC formation (Figure 2.3), should therefore be promising targets for anti-cancer therapy. These inhibitors should specifically target pre-RC checkpoint mutant cells, leaving normal cells unharmed.

Acknowledgements

We thank Andrei Chabes, Pippa Clarke, Erika Mancini and Carla Margulies for critical reading and helpful comments on this manuscript. Work in the group of C. Speck is supported by the MRC.

References

1. J. F. Diffley, J. H. Cocker, S. J. Dowell and A. Rowley, Two steps in the assembly of complexes at yeast replication origins *in vivo*, *Cell*, 1994, **78**, 303–316.
2. F. Jacob, S. Brenner and F. Cuzin, On the regulation of DNA replication in bacteria, *Cold Spring Harb. Symp. Quant. Biol.*, 1963, **28**, 329–348.
3. W. Messer, The bacterial replication initiator DnaA. DnaA and oriC, the bacterial mode to initiate DNA replication, *FEMS Microbiol. Rev.*, 2002, **26**, 355–374.
4. E. R. Barry and S. D. Bell, DNA replication in the archaea, *Microbiol. Mol. Biol. Rev.*, 2006, **70**, 876–887.
5. B. Stillman, Origin recognition and the chromosome cycle, *FEBS Lett.*, 2005, **579**, 877–884.

6. R. S. Fuller, B. E. Funnell and A. Kornberg, The DnaA protein complex with the *E. coli* chromosomal replication origin (oriC) and other DNA sites, *Cell*, 1984, **38**, 889–900.
7. D. Bramhill and A. Kornberg, Duplex opening by DnaA protein at novel sequences in initiation of replication at the origin of the *E. coli* chromosome, *Cell*, 1988, **52**, 743–755.
8. C. Speck and W. Messer, Mechanism of origin unwinding: sequential binding of DnaA to double- and single-stranded DNA, *EMBO J.*, 2001, **20**, 1469–1476.
9. J. A. Huberman and A. D. Riggs, On the mechanism of DNA replication in mammalian chromosomes, *J. Mol. Biol.*, 1968, **32**, 327–341.
10. W. L. Fangman and B. J. Brewer, Activation of replication origins within yeast chromosomes, *Annu. Rev. Cell Biol.*, 1991, **7**, 375–402.
11. Y. Marahrens and B. Stillman, A yeast chromosomal origin of DNA replication defined by multiple functional elements, *Science*, 1992, **255**, 817–823.
12. G. M. Wilmes and S. P. Bell, The B2 element of the *Saccharomyces cerevisiae* ARS1 origin of replication requires specific sequences to facilitate pre-RC formation, *Proc. Natl. Acad. Sci. U.S.A.*, 2002, **99**, 101–106.
13. S. P. Bell and B. Stillman, ATP-dependent recognition of eukaryotic origins of DNA replication by a multiprotein complex, *Nature*, 1992, **357**, 128–134.
14. D. G. Lee and S. P. Bell, Architecture of the yeast origin recognition complex bound to origins of DNA replication, *Mol. Cell. Biol.*, 1997, **17**, 7159–7168.
15. H. Rao and B. Stillman, The origin recognition complex interacts with a bipartite DNA binding site within yeast replicators, *Proc. Natl. Acad. Sci. U.S.A.*, 1995, **92**, 2224–2228.
16. K. A. Gavin, M. Hidaka and B. Stillman, Conserved initiator proteins in eukaryotes, *Science*, 1995, **270**, 1667–1671.
17. M. Gossen, D. T. Pak, S. K. Hansen, J. K. Acharya and M. R. Botchan, A *Drosophila* homolog of the yeast origin recognition complex, *Science*, 1995, **270**, 1674–1677.
18. J. L. Li and L. S. Cox, Characterisation of a sexual stage-specific gene encoding ORC1 homologue in the human malaria parasite *Plasmodium falciparum*, *Parasitol. Int.*, 2003, **52**, 41–52.
19. C. Cvetic and J. C. Walter, Eukaryotic origins of DNA replication: could you please be more specific?, *Semin. Cell Dev. Biol.*, 2005, **16**, 343–353.
20. M. G. Clarey, J. P. Erzberger, P. Grob, A. E. Leschziner, J. M. Berger, E. Nogales and M. Botchan, Nucleotide-dependent conformational changes in the DnaA-like core of the origin recognition complex, *Nat. Struct. Mol. Biol.*, 2006, **13**, 684–690.
21. C. Speck, Z. Chen, H. Li and B. Stillman, ATPase-dependent cooperative binding of ORC and Cdc6 to origin DNA, *Nat. Struct. Mol. Biol.*, 2005, **12**, 965–971.

22. A. E. Gorbalenya and E. V. Koonin, One more conserved sequence motif in helicases, *Nucleic Acids Res.*, 1988, **16**, 7734.
23. J. E. Walker, M. Saraste, M. J. Runswick and N. J. Gay, Distantly related sequences in the alpha- and beta-subunits of ATP synthase, myosin, kinases and other ATP-requiring enzymes and a common nucleotide binding fold, *EMBO J*, 1982, **1**, 945–951.
24. A. Seybert and D. B. Wigley, Distinct roles for ATP binding and hydrolysis at individual subunits of an archaeal clamp loader, *EMBO J.*, 2004, **23**, 1360–1371.
25. M. J. Davey, D. Jeruzalmi, J. Kuriyan and M. O'Donnell, Motors and switches: AAA+ machines within the replisome, *Nat. Rev. Mol. Cell Biol.*, 2002, **3**, 826–835.
26. J. P. Erzberger and J. M. Berger, Evolutionary relationships and structural mechanisms of AAA+ proteins, *Annu. Rev. Biophys. Biomol. Struct.*, 2006, **35**, 93–114.
27. L. M. Iyer, D. D. Leipe, E. V. Koonin and L. Aravind, Evolutionary history and higher order classification of AAA+ ATPases, *J. Struct. Biol.*, 2004, **146**, 11–31.
28. J. P. Erzberger, M. M. Pirruccello and J. M. Berger, The structure of bacterial DnaA: implications for general mechanisms underlying DNA replication initiation, *EMBO J.*, 2002, **21**, 4763–4773.
29. J. Liu, C. L. Smith, D. DeRyckere, K. DeAngelis, G. S. Martin and J. M. Berger, Structure and function of Cdc6/Cdc18: implications for origin recognition and checkpoint control, *Mol. Cell*, 2000, **6**, 637–648.
30. M. Gaudier, B. S. Schuwirth, S. L. Westcott and D. B. Wigley, Structural basis of DNA replication origin recognition by an ORC protein, *Science*, 2007, **317**, 1213–1216.
31. E. L. Dueber, J. E. Corn, S. D. Bell and J. M. Berger, Replication origin recognition and deformation by a heterodimeric archaeal Orc1 complex, *Science*, 2007, **317**, 1210–1213.
32. J. Giordano-Coltart, C. Y. Ying, J. Gautier and J. Hurwitz, Studies of the properties of human origin recognition complex and its Walker A motif mutants, *Proc. Natl. Acad. Sci. U.S.A.*, 2005, **102**, 69–74.
33. A. Ranjan and M. Gossen, A structural role for ATP in the formation and stability of the human origin recognition complex, *Proc. Natl. Acad. Sci. U.S.A.*, 2006, **103**, 4864–4869.
34. K. Siddiqui and B. Stillman, ATP-dependent assembly of the human origin recognition complex, *J. Biol. Chem.*, 2007, **282**, 32370–32383.
35. S. Vashee, P. Simancek, M. D. Challberg and T. J. Kelly, Assembly of the human origin recognition complex, *J. Biol. Chem.*, 2001, **276**, 26666–26673.
36. J. Mendez, X. H. Zou-Yang, S. Y. Kim, M. Hidaka, W. P. Tansey and B. Stillman, Human origin recognition complex large subunit is degraded by ubiquitin-mediated proteolysis after initiation of DNA replication, *Mol. Cell*, 2002, **9**, 481–491.

37. J. L. Bowers, J. C. Randell, S. Chen and S. P. Bell, ATP hydrolysis by ORC catalyzes reiterative Mcm2-7 assembly at a defined origin of replication, *Mol. Cell*, 2004, **16**, 967–978.
38. C. J. Li and M. L. DePamphilis, Mammalian Orc1 protein is selectively released from chromatin and ubiquitinated during the S-to-M transition in the cell division cycle, *Mol. Cell. Biol.*, 2002, **22**, 105–116.
39. H. Feng and E. T. Kipreos, Preventing DNA re-replication—divergent safeguards in yeast and metazoa, *Cell Cycle*, 2003, **2**, 431–434.
40. S. G. Prasanth, K. V. Prasanth and B. Stillman, Orc6 involved in DNA replication, chromosome segregation, and cytokinesis, *Science*, 2002, **297**, 1026–1031.
41. S. Chen, M. A. de Vries and S. P. Bell, Orc6 is required for dynamic recruitment of Cdt1 during repeated Mcm2-7 loading, *Genes Dev.*, 2007, **21**, 2897–2907.
42. I. N. Chesnokov, O. N. Chesnokova and M. Botchan, A cytokinetic function of Drosophila ORC6 protein resides in a domain distinct from its replication activity, *Proc. Natl. Acad. Sci. U.S.A.*, 2003, **100**, 9150–9155.
43. D. Remus, E. L. Beall and M. R. Botchan, DNA topology, not DNA sequence, is a critical determinant for *Drosophila* ORC-DNA binding, *EMBO J.*, 2004, **23**, 897–907.
44. R. Y. Chuang, L. Chretien, J. Dai and T. J. Kelly, Purification and characterization of the *Schizosaccharomyces pombe* origin recognition complex: interaction with origin DNA and Cdc18 protein, *J. Biol. Chem.*, 2002, **277**, 16920–16927.
45. M. Gaczynska, P. A. Osmulski, Y. Jiang, J. K. Lee, V. Bermudez and J. Hurwitz, Atomic force microscopic analysis of the binding of the *Schizosaccharomyces pombe* origin recognition complex and the spOrc4 protein with origin DNA, *Proc. Natl. Acad. Sci. U.S.A.*, 2004, **101**, 17952–17957.
46. G. Chen and A. Stenlund, The E1 initiator recognizes multiple overlapping sites in the papillomavirus origin of DNA replication, *J. Virol.*, 2001, **75**, 292–302.
47. R. Y. Chuang and T. J. Kelly, The fission yeast homologue of Orc4p binds to replication origin DNA *via* multiple AT-hooks, *Proc. Natl. Acad. Sci. U.S.A.*, 1999, **96**, 2656–2661.
48. M. Balasov, R. P. Huijbregts and I. Chesnokov, Role of the Orc6 protein in origin recognition complex-dependent DNA binding and replication in *Drosophila melanogaster*, *Mol. Cell. Biol.*, 2007, **27**, 3143–3153.
49. A. W. Thomae, D. Pich, J. Brocher, M. P. Spindler, C. Berens, R. Hock, W. Hammerschmidt and A. Schepers, Interaction between HMGA1a and the origin recognition complex creates site-specific replication origins, *Proc. Natl. Acad. Sci. U.S.A.*, 2008, **105**, 1692–1697.
50. Z. Deng, J. Dheekollu, D. Broccoli, A. Dutta and P. M. Lieberman, The origin recognition complex localizes to telomere repeats and prevents telomere-circle formation, *Curr. Biol.*, 2007, **17**, 1989–1995.

51. J. Zhou, C. Chau, Z. Deng, W. Stedman and P. M. Lieberman, Epigenetic control of replication origins, *Cell Cycle*, 2005, **4**, 889–892.
52. J. Zhou, C. M. Chau, Z. Deng, R. Shiekhattar, M. P. Spindler, A. Schepers and P. M. Lieberman, Cell cycle regulation of chromatin at an origin of DNA replication, *EMBO J.*, 2005, **24**, 1406–1417.
53. E. Danis, K. Brodolin, S. Menut, D. Maiorano, C. Girard-Reydet and M. Mechali, Specification of a DNA replication origin by a transcription complex, *Nat. Cell Biol.*, 2004, **6**, 721–730.
54. C. Girard-Reydet, D. Gregoire, Y. Vassetzky and M. Mechali, DNA replication initiates at domains overlapping with nuclear matrix attachment regions in the xenopus and mouse c-myc promoter, *Gene*, 2004, **332**, 129–138.
55. J. M. Lemaitre, E. Danis, P. Pasero, Y. Vassetzky and M. Mechali, Mitotic remodeling of the replicon and chromosome structure, *Cell*, 2005, **123**, 787–801.
56. G. D. Bowman, M. O'Donnell and J. Kuriyan, Structural analysis of a eukaryotic sliding DNA clamp-clamp loader complex, *Nature*, 2004, **429**, 724–730.
57. S. Ozaki, H. Kawakami, K. Nakamura, N. Fujikawa, W. Kagawa, S. Y. Park, S. Yokoyama, H. Kurumizaka and T. Katayama, A common mechanism for the ATP-DnaA-dependent formation of open complexes at the replication origin, *J. Biol. Chem.*, 2008, **283**, 8351–8362.
58. D. G. Lee, A. M. Makhov, R. D. Klemm, J. D. Griffith and S. P. Bell, Regulation of origin recognition complex conformation and ATPase activity: differential effects of single-stranded and double-stranded DNA binding, *EMBO J.*, 2000, **19**, 4774–4782.
59. J. H. Cocker, S. Piatti, C. Santocanale, K. Nasmyth and J. F. Diffley, An essential role for the Cdc6 protein in forming the pre-replicative complexes of budding yeast, *Nature*, 1996, **379**, 180–182.
60. C. Liang, M. Weinreich and B. Stillman, ORC and Cdc6p interact and determine the frequency of initiation of DNA replication in the genome, *Cell*, 1995, **81**, 667–676.
61. M. Weinreich, C. Liang and B. Stillman, The Cdc6p nucleotide-binding motif is required for loading MCM proteins onto chromatin, *Proc. Natl. Acad. Sci. U.S.A.*, 1999, **96**, 441–446.
62. C. Speck and B. Stillman, Cdc6 ATPase activity regulates ORC·Cdc6 stability and the selection of specific DNA sequences as origins of DNA replication, *J. Biol. Chem.*, 2007, **282**, 11705–11714.
63. K. J. Harvey and J. Newport, Metazoan origin selection: origin recognition complex chromatin binding is regulated by CDC6 recruitment and ATP hydrolysis, *J. Biol. Chem.*, 2003, **278**, 48524–48528.
64. G. Perkins and J. F. Diffley, Nucleotide-dependent prereplicative complex assembly by Cdc6p, a homolog of eukaryotic and prokaryotic clamp-loaders, *Mol. Cell*, 1998, **2**, 23–32.
65. A. Schepers and J. F. Diffley, Mutational analysis of conserved sequence motifs in the budding yeast Cdc6 protein, *J. Mol. Biol.*, 2001, **308**, 597–608.

66. N. Takahashi, S. Tsutsumi, T. Tsuchiya, B. Stillman and T. Mizushima, Functions of sensor 1 and sensor 2 regions of *Saccharomyces cerevisiae* Cdc6p *in vivo* and *in vitro*, *J. Biol. Chem.*, 2002, **277**, 16033–16040.
67. J. C. Randell, J. L. Bowers, H. K. Rodriguez and S. P. Bell, Sequential ATP hydrolysis by Cdc6 and ORC directs loading of the Mcm2-7 helicase, *Mol. Cell*, 2006, **21**, 29–39.
68. P. J. Gillespie, A. Li and J. J. Blow, Reconstitution of licensed replication origins on *Xenopus* sperm nuclei using purified proteins, *BMC Biochem.*, 2001, **2**, 15.
69. Y. Kawasaki, H. D. Kim, A. Kojima, T. Seki and A. Sugino, Reconstitution of *Saccharomyces cerevisiae* prereplicative complex assembly *in vitro*, *Genes Cells*, 2006, **11**, 745–756.
70. U. Sivaprasad, A. Dutta and S. Bell, in *DNA Replication and Human Disease, ed.* M. L. DePamphilis, Cold Spring Harbor Laboratory Press, New York, 2006, pp. 141–154.
71. D. Angelov, V. A. Bondarenko, S. Almagro, H. Menoni, F. Mongelard, F. Hans, F. Mietton, V. M. Studitsky, A. Hamiche, S. Dimitrov and P. Bouvet, Nucleolin is a histone chaperone with FACT-like activity and assists remodeling of nucleosomes, *EMBO J*, 2006, **25**, 1669–1679.
72. S. Tada, A. Li, D. Maiorano, M. Mechali and J. J. Blow, Repression of origin assembly in metaphase depends on inhibition of RLF-B/Cdt1 by geminin, *Nat. Cell Biol.*, 2001, **3**, 107–113.
73. T. Seki and J. F. Diffley, Stepwise assembly of initiation proteins at budding yeast replication origins in vitro, *Proc. Natl. Acad. Sci. U.S.A.*, 2000, **97**, 14115–14120.
74. S. Tanaka and J. F. Diffley, Interdependent nuclear accumulation of budding yeast Cdt1 and Mcm2-7 during G1 phase, *Nat. Cell Biol.*, 2002, **4**, 198–207.
75. T. Asano, M. Makise, M. Takehara and T. Mizushima, Interaction between ORC and Cdt1p of *Saccharomyces cerevisiae*, *FEMS Yeast Res.*, 2007, **7**, 1256–1262.
76. J. Mendez and B. Stillman, Perpetuating the double helix: molecular machines at eukaryotic DNA replication origins, *Bioessays*, 2003, **25**, 1158–1167.
77. J. A. Melo, J. Cohen and D. P. Toczyski, Two checkpoint complexes are independently recruited to sites of DNA damage *in vivo*, *Genes Dev.*, 2001, **15**, 2809–2821.
78. X. V. Gomes, S. L. Gary and P. M. Burgers, Overproduction in *Escherichia coli* and characterization of yeast replication factor C lacking the ligase homology domain, *J. Biol. Chem.*, 2000, **275**, 14541–14549.
79. V. N. Podust, N. Tiwari, S. Stephan and E. Fanning, Replication factor C disengages from proliferating cell nuclear antigen (PCNA) upon sliding clamp formation, and PCNA itself tethers DNA polymerase delta to DNA, *J. Biol. Chem.*, 1998, **273**, 31992–31999.
80. F. Uhlmann, J. Cai, E. Gibbs, M. O'Donnell and J. Hurwitz, Deletion analysis of the large subunit p140 in human replication factor C reveals

regions required for complex formation and replication activities, *J. Biol. Chem.*, 1997, **272**, 10058–10064.

81. G. Cullmann, K. Fien, R. Kobayashi and B. Stillman, Characterization of the five replication factor C genes of *Saccharomyces cerevisiae*, *Mol. Cell. Biol.*, 1995, **15**, 4661–4671.
82. V. Ellison and B. Stillman, Reconstitution of recombinant human replication factor C (RFC) and identification of an RFC subcomplex possessing DNA-dependent ATPase activity, *J. Biol. Chem.*, 1998, **273**, 5979–5987.
83. D. Jeruzalmi, M. O'Donnell and J. Kuriyan, Crystal structure of the processivity clamp loader gamma (gamma) complex of *E. coli* DNA polymerase III, *Cell*, 2001, **106**, 429–441.
84. V. N. Podust, N. Tiwari, R. Ott and E. Fanning, Functional interactions among the subunits of replication factor C potentiate and modulate its ATPase activity, *J. Biol. Chem.*, 1998, **273**, 12935–12942.
85. Y. Shiomi, J. Usukura, Y. Masamura, K. Takeyasu, Y. Nakayama, C. Obuse, H. Yoshikawa and T. Tsurimoto, ATP-dependent structural change of the eukaryotic clamp-loader protein, replication factor, *C, Proc. Natl. Acad. Sci. U.S.A.*, 2000, **97**, 14127–14132.
86. X. Li and P. M. Burgers, Molecular cloning and expression of the *Saccharomyces cerevisiae* RFC3 gene, an essential component of replication factor C, *Proc. Natl. Acad. Sci. U.S.A.*, 1994, **91**, 868–872.
87. Z. Q. Pan, M. Chen and J. Hurwitz, The subunits of activator 1 (replication factor C) carry out multiple functions essential for proliferating-cell nuclear antigen-dependent DNA synthesis, *Proc. Natl. Acad. Sci. U.S.A.*, 1993, **90**, 6–10.
88. F. Bunz, R. Kobayashi and B. Stillman, cDNAs encoding the large subunit of human replication factor C, *Proc. Natl. Acad. Sci. U.S.A.*, 1993, **90**, 11014–11018.
89. B. L. Allen, F. Uhlmann, L. K. Gaur, B. A. Mulder, K. L. Posey, L. B. Jones and S. H. Hardin, DNA recognition properties of the N-terminal DNA binding domain within the large subunit of replication factor C, *Nucleic Acids Res.*, 1998, **26**, 3877–3882.
90. X. V. Gomes, S. L. Schmidt and P. M. Burgers, ATP utilization by yeast replication factor C. II. Multiple stepwise ATP binding events are required to load proliferating cell nuclear antigen onto primed DNA, *J. Biol. Chem.*, 2001, **276**, 34776–34783.
91. R. Fotedar, R. Mossi, P. Fitzgerald, T. Rousselle, G. Maga, H. Brickner, H. Messier, S. Kasibhatla, U. Hubscher and A. Fotedar, A conserved domain of the large subunit of replication factor C binds PCNA and acts like a dominant negative inhibitor of DNA replication in mammalian cells, *EMBO J.*, 1996, **15**, 4423–4433.
92. R. Mossi, Z. O. Jonsson, B. L. Allen, S. H. Hardin and U. Hubscher, Replication factor C interacts with the C-terminal side of proliferating cell nuclear antigen, *J. Biol. Chem.*, 1997, **272**, 1769–1776.

93. C. Venclovas, M. E. Colvin and M. P. Thelen, Molecular modeling-based analysis of interactions in the RFC-dependent clamp-loading process, *Protein Sci.*, 2002, **11**, 2403–2416.
94. K. J. Gerik, S. L. Gary and P. M. Burgers, Overproduction and affinity purification of *Saccharomyces cerevisiae* replication factor C, *J. Biol. Chem.*, 1997, **272**, 1256–1262.
95. X. V. Gomes and P. M. Burgers, ATP utilization by yeast replication factor C. I. ATP-mediated interaction with DNA and with proliferating cell nuclear antigen, *J. Biol. Chem.*, 2001, **276**, 34768–34775.
96. B. Ason, R. Handayani, C. R. Williams, J. G. Bertram, M. M. Hingorani, M. O'Donnell, M. F. Goodman and L. B. Bloom, Mechanism of loading the *Escherichia coli* DNA polymerase III beta sliding clamp on DNA. Bona fide primer/templates preferentially trigger the gamma complex to hydrolyze ATP and load the clamp, *J. Biol. Chem.*, 2003, **278**, 10033–10040.
97. M. M. Hingorani, L. B. Bloom, M. F. Goodman and M. O'Donnell, Division of labor—sequential ATP hydrolysis drives assembly of a DNA polymerase sliding clamp around DNA, *EMBO J.*, 1999, **18**, 5131–5144.
98. D. J. Sexton, B. F. Kaboord, A. J. Berdis, T. E. Carver and S. J. Benkovic, Dissecting the order of bacteriophage T4 DNA polymerase holoenzyme assembly, *Biochemistry*, 1998, **37**, 7749–7756.
99. M. A. Trakselis, A. J. Berdis and S. J. Benkovic, Examination of the role of the clamp-loader and ATP hydrolysis in the formation of the bacteriophage T4 polymerase holoenzyme, *J. Mol. Biol.*, 2003, **326**, 435–451.
100. C. U. Lenzen, D. Steinmann, S. W. Whiteheart and W. I. Weis, Crystal structure of the hexamerization domain of N-ethylmaleimide-sensitive fusion protein, *Cell*, 1998, **94**, 525–536.
101. R. C. Yu, P. I. Hanson, R. Jahn and A. T. Brunger, Structure of the ATP-dependent oligomerization domain of N-ethylmaleimide sensitive factor complexed with ATP, *Nat. Struct. Biol.*, 1998, **5**, 803–811.
102. S. Lee, M. E. Sowa, Y. H. Watanabe, P. B. Sigler, W. Chiu, M. Yoshida and F. T. Tsai, The structure of ClpB: a molecular chaperone that rescues proteins from an aggregated state, *Cell*, 2003, **115**, 229–240.
103. H. Niwa, D. Tsuchiya, H. Makyio, M. Yoshida and K. Morikawa, Hexameric ring structure of the ATPase domain of the membrane-integrated metalloprotease FtsH from *Thermus thermophilus* HB8, *Structure*, 2002, **10**, 1415–1423.
104. D. Jeruzalmi, O. Yurieva, Y. Zhao, M. Young, J. Stewart, M. Hingorani, M. O'Donnell and J. Kuriyan, Mechanism of processivity clamp opening by the delta subunit wrench of the clamp loader complex of *E. coli* DNA polymerase III, *Cell*, 2001, **106**, 417–428.
105. N. Y. Yao, A. Johnson, G. D. Bowman, J. Kuriyan and M. O'Donnell, Mechanism of proliferating cell nuclear antigen clamp opening by replication factor C, *J. Biol. Chem.*, 2006, **281**, 17528–17539.

106. I. Dionne, N. J. Brown, R. Woodgate and S. D. Bell, On the mechanism of loading the PCNA sliding clamp by RFC, *Mol. Microbiol.*, 2008, **68**, 216–222.
107. A. Johnson, N. Y. Yao, G. D. Bowman, J. Kuriyan and M. O'Donnell, The replication factor C clamp loader requires arginine finger sensors to drive DNA binding and proliferating cell nuclear antigen loading, *J. Biol. Chem.*, 2006, **281**, 35531–35543.
108. W. Messer, F. Blaesing, D. Jakimowicz, M. Krause, J. Majka, J. Nardmann, S. Schaper, H. Seitz, C. Speck, C. Weigel, G. Wegrzyn, M. Welzeck and J. Zakrzewska-Czerwinska, Bacterial replication initiator DnaA. Rules for DnaA binding and roles of DnaA in origin unwinding and helicase loading, *Biochimie*, 2001, **83**, 5–12.
109. J. P. Chong, M. K. Hayashi, M. N. Simon, R. M. Xu and B. Stillman, A double-hexamer archaeal minichromosome maintenance protein is an ATP-dependent DNA helicase, *Proc. Natl. Acad. Sci. U.S.A.*, 2000, **97**, 1530–1535.
110. S. Shreeram, A. Sparks, D. P. Lane and J. J. Blow, Cell type-specific responses of human cells to inhibition of replication licensing, *Oncogene*, 2002, **21**, 6624–6632.
111. M. Melixetian, A. Ballabeni, L. Masiero, P. Gasparini, R. Zamponi, J. Bartek, J. Lukas and K. Helin, Loss of geminin induces rereplication in the presence of functional p53, *J. Cell Biol.*, 2004, **165**, 473–482.
112. S. Shreeram and J. J. Blow, The role of the replication licensing system in cell proliferation and cancer, *Prog. Cell Cycle Res.*, 2003, **5**, 287–293.
113. M. A. Gonzalez, S. E. Pinder, G. Callagy, S. L. Vowler, L. S. Morris, K. Bird, J. A. Bell, R. A. Laskey and N. Coleman, Minichromosome maintenance protein 2 is a strong independent prognostic marker in breast cancer, *J. Clin. Oncol.*, 2003, **21**, 4306–4313.
114. E. Lau, T. Tsuji, L. Guo, S. H. Lu and W. Jiang, The role of pre-replicative complex (pre-RC) components in oncogenesis, *FASEB J.*, 2007, **21**, 3786–3794.
115. M. Braig, S. Lee, C. Loddenkemper, C. Rudolph, A. H. Peters, B. Schlegelberger, H. Stein, B. Dorken, T. Jenuwein and C. A. Schmitt, Oncogene-induced senescence as an initial barrier in lymphoma development, *Nature*, 2005, **436**, 660–665.

CHAPTER 3

Ring Structures and Six-fold Symmetry in DNA Replication

LYNNE S. COX[a] AND STEPHEN KEARSEY[b]

[a] Department of Biochemistry, South Parks Road, Oxford, OX1 3QU, UK;
[b] Department of Zoology, South Parks Road, Oxford, OX1 3PS, UK

3.1 Introduction

DNA replication is absolutely dependent upon opening up of the duplex template through the action of DNA-dependent helicases that utilise the energy from ATP hydrolysis to break hydrogen bonds between complementary base pairs (see Chapter 1). Such helicases must be able to bind to DNA at the replication start site or origin, and to act processively with the replisome complex to allow replication fork progression. In order to do this, they possess DNA binding motifs together with protein–protein interaction domains that permit recruitment to replication origins *via* helicase loader proteins (Chapter 2) and form large multimeric complexes, encircling the DNA, to confer high processivity. Though not similar at the amino acid level, structural motifs are highly conserved in replicative helicases in prokaryotes (including bacteriophage), archaea and eukaryotes. The multimeric complexes almost uniformly show six-fold symmetry, with replicative helicases acting as hexamers, double hexamers or dodecamers.

In this chapter we consider the action of the MCM proteins as replicative helicases, and also discuss helicases that are not required directly for fork progression but which act to facilitate replication over unusual DNA structures, or which function in restarting stalled forks. In addition, we consider how sliding clamps mediate processivity of the replicative polymerases; we focus on

Molecular Themes in DNA Replication
Edited by Lynne S. Cox

Published by the Royal Society of Chemistry, www.rsc.org

the eukaryotic paradigm PCNA,[i] but also examine the structurally related 9-1-1 complex. Like the helicases, such clamps show little conservation at the level of primary sequence but structurally are extremely well conserved. Six-fold symmetry is a key motif both in replicative helicases and sliding clamps. The importance of such structural symmetry in relation to helicase and clamp function is discussed in this chapter.

3.2 Replicative Helicases

The archetypal replicative helicases of *E. coli* (DnaB) and phage T4 (gp41) and T7 (gp4) form toroidal structures that encircle the DNA and act processively to open the duplex template ahead of the replisome complex.[1,2] The characteristic six-fold symmetry in such helicases arises from assembly of subunits into hexamers or dodecamers. For example, phage T4 gp41 hexamer comprises a hexagonal trimer of asymmetric dimers, the association of which is driven by binding to either ATP or GTP.[3] These prokaryotic paradigms extend to eukaryotic models, perhaps the best characterised of which is the SV40 large T antigen, which acts as a double hexameric helicase (reviewed in ref. 2); it is the only viral replication protein necessary for replication of the SV40 dsDNA genome in infected mammalian cells, as the remaining replisome components are supplied by the host cell.[4] The main replicative helicase in eukaryotes has for some time been thought to be provided by a complex of MCM proteins;[5] it was only in 2008 that full helicase activity of the MCM2-7 complex was observed[6] (see Section 3.2.4), strongly supporting the view that this is the replicative helicase for bulk DNA synthesis during S phase of eukaryotic cells.

3.2.1 Identification of MCM Proteins

The MCM2-7 proteins are essential for DNA replication and are found in all eukaryotes (reviewed in ref. 5,7–11). MCM proteins were identified from a genetic screen for *Saccharomyces cerevisiae* mutants defective in minichromosome maintenance,[12] although a number of these factors were also identified from cell division cycle (cdc) and related cell cycle screens in both *S. cerevisiae* and *Schizosaccharomyces pombe* (reviewed in ref. 7). Genomic analyses have identified in addition MCM8 and MCM9,[13,14] based on their sequence similarity to MCM2-7 (Figure 3.1). A further factor, MCM10, has little sequence similarity to MCM2-9 (Figure 3.1), but it is also essential for DNA replication (see Sections 3.2.7 and 3.2.8).

3.2.2 Structure and Biochemical Properties of MCM2-7: Analogies with Archaeal MCMs

Detailed structural analysis of the complete eukaryotic MCM2-7 complex has not yet been reported. However by analogy to archaeal MCM homohexameric

[i]Proliferating cell nuclear antigen

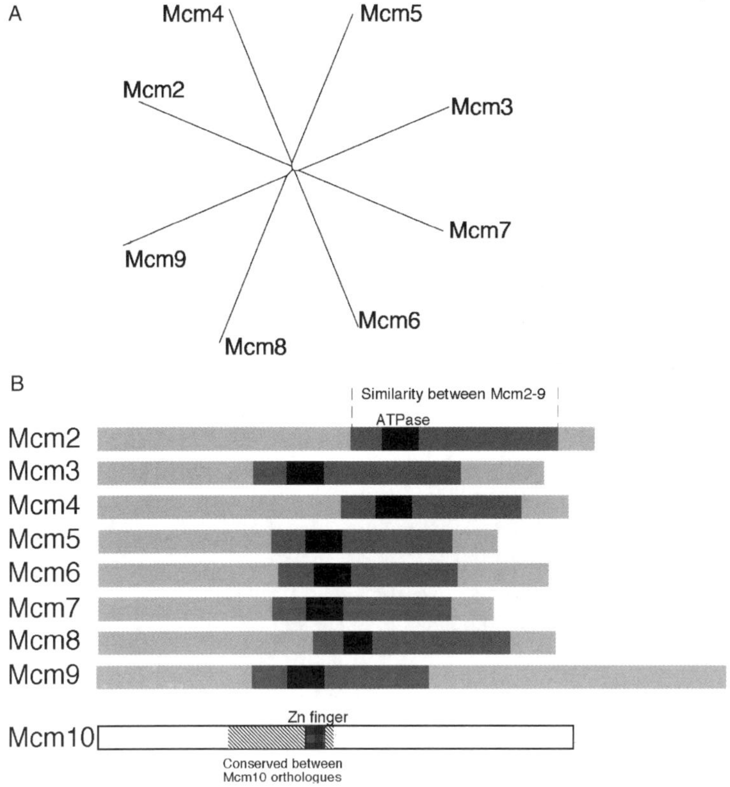

Figure 3.1 **MCM family of proteins.** (**A**) Phylogenetic tree showing that MCM2-9 proteins are approximately equally divergent from one another. The tree was constructed using ClustalX alignment of human MCM2-9 sequences. (**B**) Comparison of MCM2-9 and MCM10 proteins, showing conserved regions and motifs. Adapted from ref. 84 and ref. 221.

complexes[15–19] (Figure 3.2), the eukaryotic MCM2-7 complex is thought to adopt a ring-shaped structure with six fold symmetry, with a central hole sufficiently large to accommodate either single-stranded DNA (ssDNA) or double-stranded (dsDNA). It is still unclear whether the helicase complex exists as a single hexameric unit on DNA as has been observed for the *Sulfolobus solfataricus* (Sso) MCM complex,[18] or rather as a dodecamer as suggested for *Methanobacterium thermoautotrophicum* (Mth) MCM[15] (see also Chapter 12 Figure 12.2), but in either case, it is thought to form a ring either in solution or on DNA.

Loading of the MCM2-7 complex onto DNA is thus likely to involve topological linkage, and ORC and the associated Cdc6 and Cdt1 proteins which are likely to act as helicase loaders, may function by transiently opening a pre-assembled MCM2-7 ring to allow topological linkage with DNA. Energy for this conformational change comes from ATP hydrolysis through the

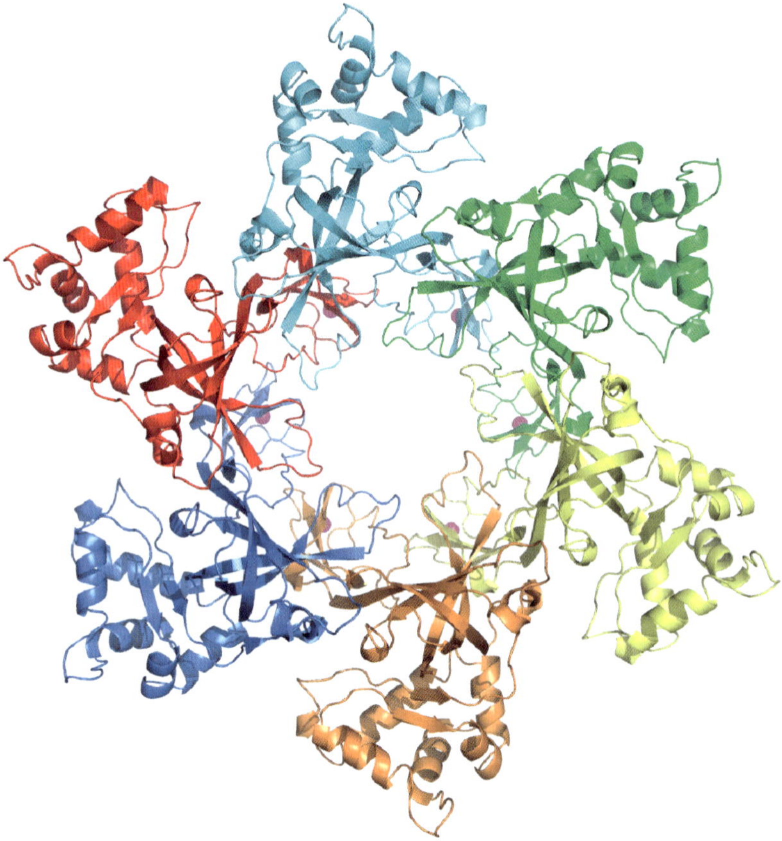

Figure 3.2 **Ring structure of homohexameric archaeal MCM**. The MCM N terminal domain from *Sulfolobus solfataricus* (Sso) is shown, with each subunit depicted in a different colour. Reprinted by permission from the authors and Macmillan Publishers Ltd (W. Liu, B. Pucci, M. Rossi, F. M. Pisani and R. Ladenstein, Structural analysis of the *Sulfolobus solfataricus* MCM protein N-terminal domain, *Nucleic Acids Research*, **36**, 3235–3243), copyright (2008).[18]

AAA+ ATPase activity of several of the helicase loader proteins (Chapter 2). Notably, Mth MCM forms double heptamers in the presence of a nucleotide analogue, but the conformation shifts to double hexamer on association with duplex DNA.[20]

This alteration in MCM structure has been further explored using cryo-electron microscopy of the MCM complex together with DNA. Such studies shows that the DNA wraps around the hexameric MCM complex, involving a significant conformational change in the N-terminal domain of the MCM to expose a helix-turn-helix DNA binding motif.[21] It is possible that this motif is involved in origin melting, prior to activation of the helicase activity of MCM. Additionally, the C-terminal winged helix domain of archaeal MCM may bind to DNA, though this is currently unresolved.[22–24] Single molecule fluorescence resonance energy transfer studies have demonstrated that Sso MCM loads onto 3′ DNA tails but interacts also with the 5′ tail, increasing the stability of the MCM complex bound at the replication fork, and that this directional loading may contribute to substrate selectivity.[25]

3.2.3 AAA+ ATPase Activity in MCM2-7

In addition to the helicase loaders ORC and Cdc6, MCM2-7 proteins are themselves also members of the AAA+ family of ATPases[26] (see Chapter 2), and share a conserved region of $\sim$400 amino acids which contains the consensus ATP binding motif (Figure 3.1). The ATPase motif in several of the MCM2-7 subunits is essential for viability[27] and mutation of the Walker A motif in any one MCM subunit ablates *in vitro* ATPase activity of the complex.[28] Mutational studies further demonstrate that subcomplex MCM4,6,7 possesses canonical ATP binding sites, while MCM2,3,5 acts in a regulatory fashion, though the presence of all six subunits is necessary for full ATPase activity.[28] More recent MCM dimer co-purification studies have shown that the MCM ATPase active sites are formed by a combination of Walker A and B boxes from one MCM subunit together with a catalytically critical arginine finger from an adjacent subunit.[29] The AAA+ ATPase domain appears to be located underneath the C-terminal winged helix (WH) domain.[21] The ATPase function of the MCM2-7 complex does not appear to be required for its chromatin binding (*i.e.* pre-RC formation, Chapters 1 and 2), but it is needed for DNA unwinding during DNA replication.[30] The ATPase activity of MCM from the archaeaon *Thermoplasma acidophilum* is enhanced by binding to Cdc6, probably through inducing a conformational change in the helicase,[31] though this contrasts with the potentially inhibitory effect of Cdc6 on the helicase activity of MCMs reported for other archaeal species.[32–34] MCM2, 4, 6 and 7 subunits also have zinc finger motifs, and mutational studies indicate that this motif is important for helicase function (reviewed in ref. 8). The zinc finger is not necessary for oligomerisation of the archaeal Mth MCM helicase, but it is needed for efficient ssDNA binding, DNA-dependent stimulation of ATPase activity, and helicase activity.[35] Given the strong similarities between archaeal and eukaryotic MCMs, it is likely that the zinc finger motifs in eukaryotic MCM2, 4, 6 and 7 are similarly important in helicase activity.

3.2.4 MCM2-7 Constitute the Replicative Helicase

Eukaryotic MCM2-7 proteins form a heterohexameric MCM2,3,4,5,6,7 complex with conserved helicase domains. Homohexameric archaeal Mth MCM has been shown to have processive helicase activity,[36,37] suggesting that eukaryotic MCM complexes similarly act as processive helicases at the replication fork. However, experimental support has until recently been elusive. Early *in vitro* studies showed that an MCM4,6,7 subcomplex had weak helicase activity,[38–40] consistent with the observation of an MCM2,4,6,7 subcomplex in *Xenopus* egg extracts.[41] More recently, DNA unwinding activity correlating with ATPase activity has been observed in MCM4,7 and MCM4,6,7 complexes, both of which form hexameric rings that can be opened to load onto DNA.[42] Since only the complete MCM2-7 complex is thought to be functional in DNA replication *in vivo*,[43,44] the relevance of subcomplex activity has until recently been unclear.

The question of processive replicative helicase activity of MCM2-7 in higher eukaryotes has now been resolved by the demonstration of helicase activity of isolated MCM2-7.[6] Interestingly, the lack of helicase activity of the full MCM2-7 complex in previous studies is thought to be due to the presence of chloride ions, which are inhibitory to the complete MCM2-7 complex but not to the MCM4,6,7 subcomplex.[6] Moreover, the interface between MCM2 and MCM5 may act as an ATP-dependent 'gate' affecting the ability of the complex to associate with single stranded or duplex DNA; closing of this gate results in helicase activity of MCM2-7 *in vitro*.[6] MCM2-7 is stably associated with the replication fork in a replisome progression complex (RPC) that includes GINS and Cdc45.[45] Indeed, a complex consisting of eukaryotic Cdc45, MCM2-7 and GINS (termed the CMG complex) has been isolated from *Drosophila* and has helicase activity.[46]

3.2.5 Models for MCM2-7 Helicase Action

How the MCM2-7 helicase complex acts to unwind DNA is still the subject of intense research activity. Several models based on DNA entrapment have been proposed to explain the mechanism of DNA unwinding, all of which make the common assumption that the MCM2-7 helicase forms a ring that encloses one or two strands of DNA.[47] In the steric exclusion model, the MCM2-7 hexamer encircles a ssDNA strand and translocates, using steric exclusion to unwind DNA (Figure 3.3A). Other models suggest that DNA unwinding occurs with MCM2-7 translocating along dsDNA. In the rotary pump model, immobilized MCM2-7 complexes located at a distance coordinately rotate DNA in opposite directions causing unwinding of the intervening DNA[48] (Figure 3.3B). Alternatively, two MCM2-7 complexes may be in close proximity and pump dsDNA towards the hexamer interface, resulting in ssDNA being extruded, as proposed for SV40 T antigen helicase[49] (Figure 3.3C). Finally the ploughshare model also envisages translocation of the MCM2-7 complex around dsDNA; here a single hexamer translocates along dsDNA, but coupled to the helicase is an additional protein that acts as a 'ploughshare' to separate the two DNA strands as they emerge from the helicase[47] (Figure 3.3D). GINS and Cdc45 have been proposed to act as the molecular ploughshare.

During this helicase activity, the six-fold symmetry of the MCM2-7 complex probably contributes in several ways. It is important to remember that the DNA template to be unwound adopts a right-handed helical structure; thus simply moving along the DNA (or spooling DNA through a fixed helicase, as in

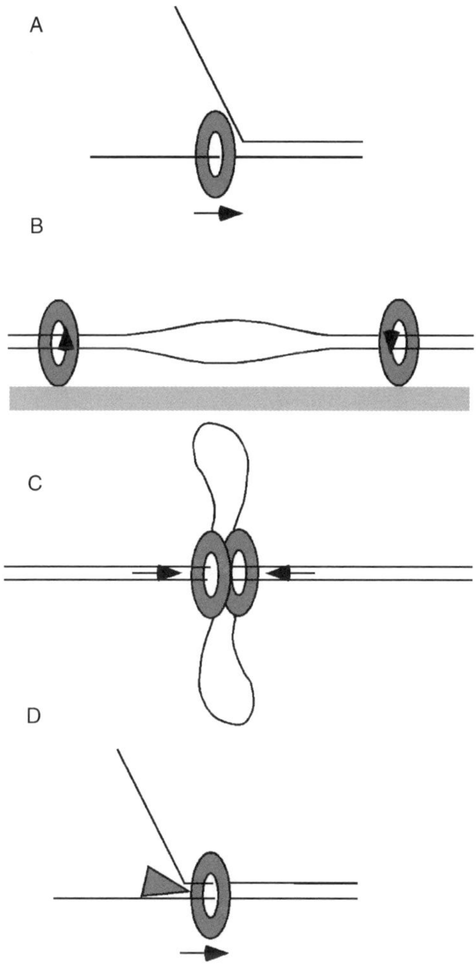

Figure 3.3 **Models for helicase activity of MCM2-7 proteins.** (**A**) Steric exclusion model where translocation of MCM2-7 complex along ssDNA causes displacement of the other non-enclosed strand. (**B**) Rotary pumping model, where dsDNA is rotated in opposite directions by immobilised MCM2-7 complexes, leading to unwinding. (**C**) Pumping of dsDNA into adjacent MCM2-7 hexamers leads to extrusion of ssDNA. (**D**) Ploughshare model where MCM2-7 translocates along dsDNA, with an attached protein (grey wedge) which separates DNA strands as they emerge from the MCM2-7 ring. Adapted from ref. 47.

the rotary pump model) may result in rotation of the helicase complex, exposing different subunits to DNA during the twisting movement. An ordered cycle of ATP hydrolysis by the MCM2-7 complex as it rotates along DNA has been proposed,[28] by analogy with the rotary action of the F1/F0 ATPase. This model[28] (Figure 3.4) is absolutely dependent on the six-fold symmetry of the

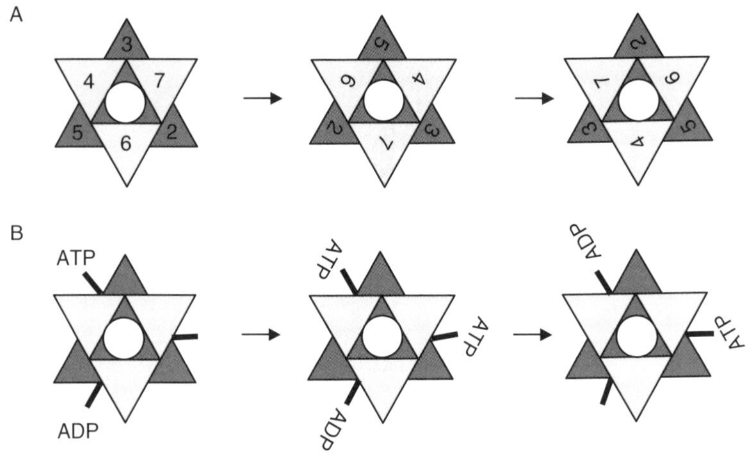

Figure 3.4 Sequential ATP hydrolysis by MCM2-7 heterohexamer. (**A**) The six subunits of MCM heterohexamer are arranged with the 4,6,7 subcomplex forming one triangular structure with the regulatory 2,3,5 subcomplex 'triangle' offset by 60°. Two alternative arrangements are possible for this configuration; only one is shown here. The ATPase actives sites are formed by interaction between adjacent subunits. By analogy with the rotating F1/F0 ATPase, the hexameric MCM complex is proposed to rotate along DNA.[28] (**B**) The rotation of the complex is thought to allow an ordered sequence of ATP hydrolysis such that ATPase active sites can be occupied by ATP, ADP or unoccupied. Figure adapted from ref. 28.

complex. However, it should be emphasised that little is known concerning the mode of action of the MCM2-7 helicase and, given the range of possibilities (Figure 3.3), such models are still hypothetical and await formal experimental confirmation.

Interestingly, the MCM2-7 complex may not simply unwind the duplex template, but might also be involved in stripping off the nucleosomes from the template by virtue of its association with histone chaperones FACT and Asf,[50–52] and histone deacetylases[53] (see Chapter 10 for details).

3.2.6 Regulation of MCM2-7 Helicase

As discussed above, MCM2, 3 and 5 appear to have a regulatory role, with MCM4,6,7 acting catalytically in ATP hydrolysis. The rotary action coupled to ATP hydrolysis is likely to account for helicase activity of MCM2-7. However, such activity must be regulated not only at the intra- and inter-molecular level, but must also be coordinated with other factors at the replication fork, and be responsive to cellular conditions including the appropriate cell cycle phase (as replication occurs only in S phase) and the need to arrest replication in the presence of DNA damage.

3.2.6.1 Cell Cycle Regulation of MCM2-7 Loading at Replication Origins

Binding of MCM2-7 to replication start sites requires a helicase loading complex comprising DNA-bound ORC, Cdt1 and Cdc6. The binding of MCM2-7 to DNA is restricted to late mitosis/G1 phase by several overlapping mechanisms which together ensure a very low probability of licensing during S or G2 phases of the cell cycle, where it would otherwise lead to re-replication of parts of the genome. As there are tens of thousands of origins in the genome and a single relicensing event can lead to genome instability, there is a need for very stringent control both of pre-RC assembly and of origin firing. Regulatory mechanisms vary considerably across evolution (reviewed in ref. 54). For instance, in budding yeast but not other eukaryotes, unbound MCM2-7 are exported from the nucleus at the end of G1 in a step that requires CDK phosphorylation, thus helping to block further pre-RC formation,[55,56] though chromatin dissociation occurs in frogs and mammals[57,58] (reviewed in ref. 9).

Rather than a single MCM2-7 complex at each fork, multiple MCM2-7 complexes are loaded at ORC during licensing,[59] and the stoichiometry of MCMs on DNA is very high (approximately 20-fold excess over replication origins[60–65]). Whilst single MCM2-7 complexes can be loaded onto pre-RCs when Orc4 ATP hydrolysis activity is lost through mutation, reiterative loading is blocked; such mutation is lethal *in vivo*.[59] Perhaps one complex is recruited through direct protein–protein interactions, but ATP hydrolysis by ORC is required for topological linkage of the MCM complexes around the template. Since DNA replication occurs bi-directionally from origins, only two MCM2-7 complexes are theoretically needed per origin; if one complex is sufficient for helicase activity at each fork emanating from the central origin, why are MCMs in such excess? It is possible that multiple MCM2-7 complexes may increase the probability of origin firing, or may facilitate rescue of replication under conditions where replication forks are stalled by DNA damage.[66]

3.2.6.2 Cell Cycle Regulation of MCM Activity

Phosphorylation of the MCM2-7 complex may be important for activation of DNA synthesis, either *via* switching on the helicase function, or by stabilising the complex. The MCM2-7 complex is a substrate for the S phase-specific Cdc7- Dbf4 kinase.[67–70] In *S. cerevisiae*, the requirement for Cdc7 in replication activation can be bypassed by a structural mutation in MCM5.[71,72] Additionally, fission yeast MCM10 (see below) stimulates phosphorylation of MCM2 in the MCM2-7 complex by Dfp1-Hsk1 *in vitro*, the fission yeast orthologue of Cdc7-Dbf4 kinase.[73] These observations suggest activation of MCM helicase activity at the start of S phase upon phosphorylation by Cdc7-Dbf4 kinase or S phase specific CDKs.[74–77] Cdk1-dependent phosphorylation of MCM3 on serine 112 triggers assembly of MCM3 into a complex with the other MCM subunits; the subcomplex prior to MCM3 phosphorylation and binding is highly unstable.[78] Thus phosphorylation-dependent MCM2-7 complex assembly might contribute to the activating role of S phase CDKs for the MCM2-7 helicase. In *S. cerevisiae*, however, CDK activation of DNA replication is more likely to result from Sld2 and Sld3 phosphorylation, which promotes their interaction with Dpb11. It has been suggested that this stabilized complex then recruits GINS and Cdc45 to the pre-RC, which in turn activates the MCM2-7 helicase.[46,79]

3.2.6.3 Inhibiting MCM Activity on DNA Damage

Unlike the activating role of phosphorylation by S phase-specific kinases, checkpoint kinases ATM and ATR also phosphorylate MCM2 and MCM3, respectively, but this results in inhibition of helicase activity[80] (reviewed in ref. 81), with the ATM-phosphorylated form of MCM3 (on Ser-725 and Ser-732) found predominantly soluble in the nucleoplasm rather than chromatin-associated.[82] Furthermore, ATR interaction with MCM7 is important in checkpoint-dependent replication arrest on radiation damage of DNA.[80] Recovery from replication arrest following a replication block caused by dNTP depletion is also mediated through MCM2-7, since MCM4 deletion mutants lacking 84 residues from the C-terminus are slow in recovering from replication checkpoint arrest.[83]

3.2.7 MCM8 and MCM9: AAA+ Helicases

Two additional AAA+ proteins, MCM8 and MCM9, related in sequence to the ubiquitous MCM2-7 family, have been identified (reviewed in ref. 84). These proteins are widely distributed in eukaryotes, but while they are found in vertebrates and plants, they are not present in yeasts, suggesting that they are ancestral but have been lost in some lineages (reviewed in ref. 85). It is possible that they serve some important role in replication specific to multicellular organisms. However, putative MCM8 genes have been identified in the genomes of malarial parasites *Plasmodium vivax*,[86] *Plasmodium falciparum* (Li *et al.* unpublished results, see Chapter 13) and in the slime mould *Dictyostelium discoidum*.[87]

Human MCM8 has been reported to function at an early step in pre-RC formation, being necessary for efficient loading of Cdc6 and MCM2-7 onto chromatin.[88] Similarly, MCM9 functions with ORC in recruiting MCM2-7 to pre-RCs. Depletion of MCM9 from cell-free extracts of *Xenopus* eggs (an *in vitro* system which permits regulated nuclear DNA replication[89]), prevents not only MCM2-7 binding during pre-RC assembly, but also destabilises Cdt1; similarly, Cdt1 depletion results in a decrease in MCM9 association with chromatin and an overall drop in MCM9 levels.[90] Moreover, MCM9 prevents accumulation of the replication inhibitor geminin on chromatin during licensing by associating with Cdt1 and hence antagonizing geminin function.[90] Thus both MCM8 and MCM9 appear to be important in assembling a functional pre-RC. However, MCM8 has also been reported to facilitate the elongation step of DNA replication and assist with recruitment of the single-strand DNA binding protein Replication Protein A (RPA);[91] it may do this by acting as a helicase with ATPase activity, as demonstrated *in vitro*.[91] It is therefore possible that MCM8 is an early component of the pre-RC that functions both in MCM2-7 helicase recruitment and as a helicase in its own right. However, *Drosophila* bearing mutation in the MCM8 orthologue, *rec,* appear normal except for a defect in meiotic recombination.[85] First pass structural modelling (Figure 3.5) shows that both human MCM8 and MCM9 can adopt a typical MCM fold seen in archaeal MCMs (Mth and Sso, respectively), but the structural similarity exists only over a relatively short region of the proteins. Full structural analysis and the functional consequences of these structures remain to be determined experimentally but, like the other MCM proteins, the importance of MCM8 and MCM9 in replication makes them interesting diagnostic and potentially therapeutic targets in cancer.

3.2.8 Hexameric MCM10 Acts in Replication Elongation

MCM10 is also essential for DNA replication,[92–96] and small interfering RNA (siRNA) knockdown of MCM10 in human cells results in incomplete

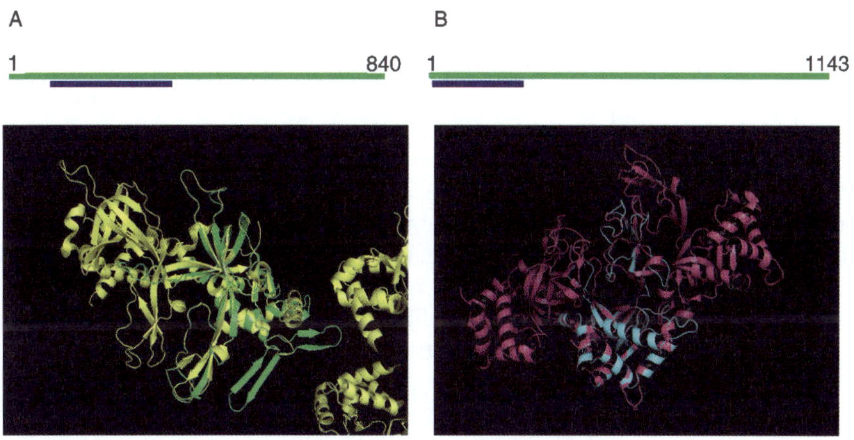

Figure 3.5 **MCM8 and MCM9 adopt a typical MCM fold**. First pass structural predictions were conducted using SWISS-MODEL[222] and MacPyMOL (http://delsci.com/macpymol/). (**A**) Human MCM8 (Swissprot no. Q9UJA3) (green) superimposed on the structure of archaeal Mth MCM from *M. thermoautotrophicum* (PDB accession 1LTL) (yellow). The blue line above the structure (residues 94–363) shows the portion of full length MCM8 (green line) for which valid structural predictions could be made, with 23.8% sequence identity and an E value of 9.4e-41. (**B**) Human MCM9 (Swissprot Q9NXL9) (blue) was modelled onto the structure of archaeal Sso MCM (PDB 2VL6) (magenta). The blue line above the structure (residues 10–270) shows the portion of full length MCM9 (green line) for which valid structural predictions could be made, with 21.9% sequence identity and an E value of 1.13e-37.

replication.[97] MCM10 is unusual among replication proteins in its functional conservation between species; for instance, fission yeast MCM10 can complement a budding yeast MCM10 mutant.[95]

In yeast, MCM10 binds to licensed origins,[97] and in mammalian cells, MCM2-7 are also required at pre-RCs for MCM10 recruitment.[93,94] The presence of MCM10 at licensed origins may then facilitate the chromatin association of other proteins such as Cdc45 and DNA pol α-primase. But although MCM10 interacts with and stabilises the catalytic subunit of DNA polα in budding yeast,[97] in human cells, siRNA-mediated knockdown of MCM10 did not prevent association of Cdc45 or pol α-primase with chromatin;[98] this either reflects species-specific differences in loading of replisome components, or possibly a redundant role of MCM10 in polymerase recruitment. Recruitment of DNA polα by MCM10 requires also the action of CTF4/And1.[99] MCM10 is also reported to possess primase activity by virtue of its C-terminal domain; mutation within this domain abolished primase activity *in vitro* and was lethal *in vivo*.[100] Furthermore, MCM10 associates with PCNA *via* a classical PCNA-interacting peptide or PIP (see Sections 3.4.3 and 3.4.4), mutation of which abolishes cell proliferation in *S. cerevisiae*, suggesting a role for MCM10 in replication elongation.[101]

MCM10 adopts a hexameric ring structure[102] (Figure 3.6), very much like those of active helicases,[1,16] but it lacks helicase domains and shares little primary sequence similarity with MCM2-7 (Figure 3.1). X-ray crystallography at a resolution of 2.6 Å confirms that six MCM10 molecules associate in the complex.[103] The CCCH zinc-finger motif of MCM10 is important for homocomplex formation,[104] and may play a role in DNA contact. It is likely to function as a clamp encircling DNA (the central cavity diameter of 35Å[102] (Figure 3.6) is entirely consist with this idea), acting in the transition between pre-initiation and initiation states at the very start of S phase, and perhaps as a clamp for pol α-primase during elongation. Like the MCM2-7 helicase, the MCM10 hexamer may move with the replication fork,[92,93,105] where it might serve as a central coordinator much like the τ complex in *E. coli*, keeping helicases and polymerases together at the replication fork.[102]

3.3 Non-replisome RecQ Helicases that Contribute to DNA Replication

In addition to the replication fork helicase MCM2-7, further helicases come into play during DNA replication to resolve unusual DNA structures such as hairpin loops, G4 quadruplexes and four-way Holliday junctions. These structures arise naturally within eukaryotic chromosomes as a consequence of DNA sequence (*e.g.* at telomeres or fragile sites), or result from DNA damage and/or replication fork stalling and collapse. The RecQ family of 3′–5′ helicases comprise a large group of proteins highly conserved across evolution in all three domains of life (reviewed in refs. 106,107) that are implicated in DNA

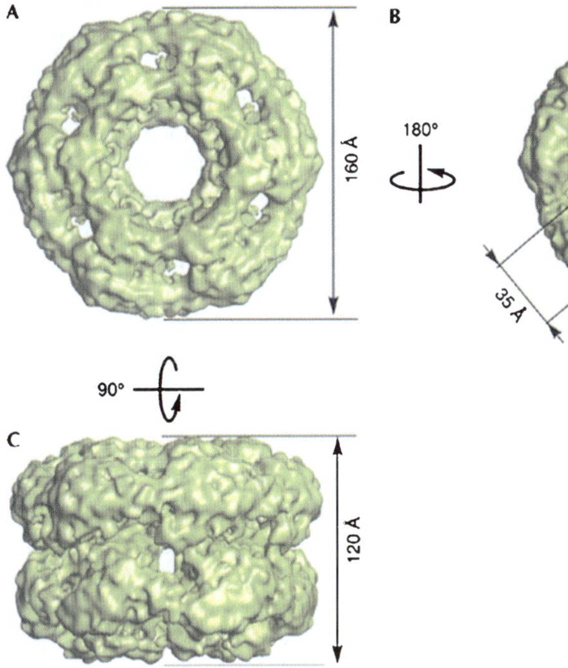

Figure 3.6 **MCM10 ring structure**. Human MCM10 assembles into a homohexamer with a central hole of 35Å and an overall diameter of ~160 Å and height of 120 Å. Surface representations of (**A**) top, (**B**) bottom and (**C**) side of the complex are shown. Reprinted by permission from the authors and Macmillan Publishers Ltd (A. L. Okorokov, A. Waugh, J. Hodgkinson, A. Murthy, H. K. Hong, E. Leo, M. B. Sherman, K. Stoeber, E. V. Orlova and G. H. Williams, Hexameric ring structure of human MCM10 DNA replication factor, *EMBO Reports*, 2007, **8**, 925–930), copyright (2008).[102]

replication, as well as recombination and DNA repair. In humans, three members of the RecQ family are associated with human disease:

- WRN in premature ageing Werner syndrome (WS) (reviewed in ref. 108,109);
- BLM in cancer-prone Bloom syndrome (BS) (reviewed in ref. 110);
- RecQ4 in segmental progeroid Rothmund–Thomson syndrome (reviewed in ref. 111).

3.3.1 WRN

A role for WRN in DNA replication is suggested by several lines of experimental data (reviewed in ref. 108,112). Firstly, the WRN protein has been shown by immunofluorescence to localise at DNA replication sites in both actively replicating and hydroxyurea (HU)-arrested S phase.[113–115] In Werner syndrome patient cells lacking functional WRN, significant DNA replication defects are observed, including an extension of the S phase DNA replication period of the cell cycle[113,116] and abnormal DNA replication fork progression, as shown by combing of individual replicating DNA molecules.[113,117] The replication defect can be corrected by ectopic expression of a Holliday junction resolvase,[118] consistent with *in vitro* assays demonstrating that WRN can resolve Holliday junctions in a non-recombinational way by promoting branch migration.[115] Furthermore, WRN shows low processivity 3′–5′ helicase activity *in vitro*.[119] These data suggest that rather than acting as the processive replication fork helicase, WRN acts instead at stalled replication forks or atypical DNA structures in the genome such as telomeres[120,121] and fragile sites;[122] consistent with this, WRN shows preferences for these types of substrate *in vitro*.[123–125]

WRN also possesses an amino terminal exonuclease domain,[126] which has been suggested to fold into a hexamer on DNA or in association with PCNA (see Section 3.4).[127,128] It is not yet clear whether the helicase or exonuclease predominates in replication, but recent studies in *Arabidopsis*[129] and *Drosophila*,[130,131] where the WRN exonuclease is encoded by loci lacking any helicase-encoding domains, should permit dissection of the role of WRN exonuclease in DNA replication. No structural data are yet available for full length mammalian WRN, but the exonuclease domain has been characterised at high resolution by X-ray crystallography.[132]

Based on the crystal structure of monomeric human WRN exonuclease domain, a hexameric structure was proposed by modelling against a putative DnaQ-family nuclease from *Arabidopsis thaliana* (PDB accession 1VK0);[132] although biochemical studies such as size exclusion gel filtration chromatography or analytical centrifugation have not confirmed multimers greater than trimers.[127,133] A trimer–hexamer equivalent observed by atomic force microscopy is driven towards the hexamer form in the presence of DNA with 3′ recessed ends.[128] The predicted hexamer[132] has an internal cavity of 30Å diameter a depth of 35Å and an outer diameter of 85Å[132] (Figure 3.7), which is remarkably similar both to the Ku70/80 hexamer with which WRN interacts and by which its exonuclease activity is stimulated,[134–136] but also to the sliding clamp PCNA (see Section 3.4). This similarly suggests that the Ku70/80 and WRN exonuclease may stack, permitting efficient protein hand-off to allow seamless processing of DNA at breaks without releasing free DNA ends. In support of this idea, dGMP was co-crystallised with the WRN exonuclease and found to contact tryptophan 145 in the cavity[132] (Figure 3.7) where DNA would pass through from the Ku complex.

Hydrolysis of the phosphodiester bond during WRN exonuclease activity is metal ion dependent, with Mg^{2+} or Mn^{2+} coordinated by acidic residues aspartate 82 and glutamate 84, together with aspartates 143 and 216 which fold to form the active site of human WRN exonuclease.[132] Whether the exonuclease adopts a hexameric conformation in the context of the full-length protein (which possesses both helicase domains and the HRDC and RQC DNA- and

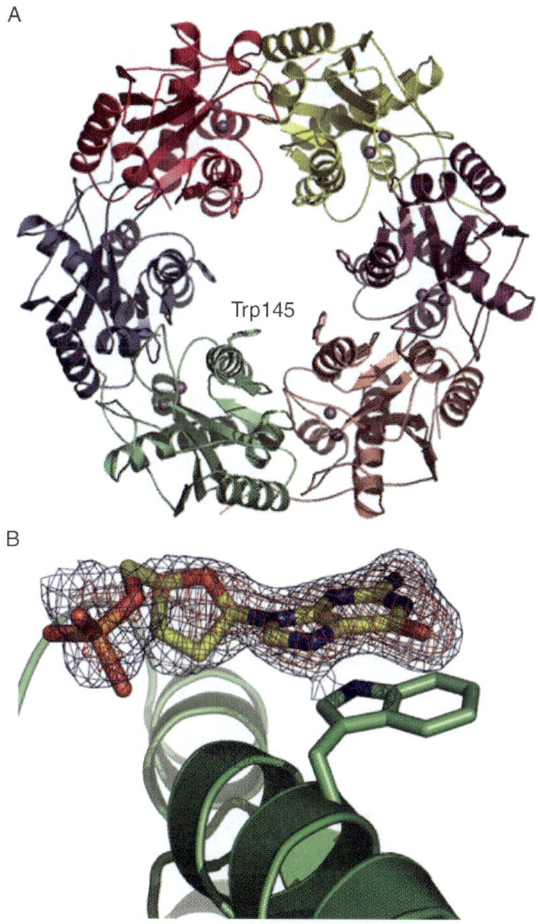

Figure 3.7 **Hexameric human WRN exonuclease domain.** (**A**) The monomeric fold determined from the crystal structure of the amino terminal exonuclease domain of human WRN was modelled as a homohexamer against template *Arabidopsis* DnaQ-family nuclease (PDB accession 1VK0). (**B**) dGMP stacks against tryptophan 145 of WRN exonuclease, consistent with this region of the protein interacting with the DNA substrate passing through the ring. Reprinted by permission from the authors and Macmillan Publishers Ltd (J. J. Perry, S. M. Yannone, L. G. Holden, C. Hitomi, A. Asaithamby, S. Han, P. K. Cooper, D. J. Chen and J. A. Tainer, WRN exonuclease structure and molecular mechanism imply an editing role in DNA end processing, *Nature Structure Molecular Biology*, 2006, **13**, 414–422), copyright (2006).[132]

protein-interaction motifs) awaits confirmation from structural studies of the native protein. However, coordination of helicase and exonuclease activities have been observed *in vitro*,[132,137,138] so it is possible that WRN exonuclease, helicase, HRDC[139] and RQC[140] domains can each adopt independent functional folds joined by flexible linker regions.

3.3.2 BLM

BLM is highly related to WRN within the characteristic RecQ family helicase domain; the yeast homologue is Sgs1. It is important in processing replication intermediates, possibly hemicatenenes and double Holliday junctions or telomeric DNA, as suggested by BLM's substrate preferences *in vitro*.[123,141] (See Chapter 5 for a fuller discussion of the biological functions of Sgs1 considered to be the *S. cerevisiae* homologue of (BLM) in DNA replication, and interactions with replication factors Dna2 and Fen1.)

Electron microscopy analysis shows two alternative conformations of BLM: a tetramer and a hexamer.[142] The hexamers have an outer diameter of ~130 Å with the diameter of the central cavity 35 Å, remarkably similar to that of MCM10,[102] archaeal MCM homohexamers,[15–19] Ku70/80,[143] WRN exonuclease[132] and PCNA.[144,145] This hole is thus likely to be able to accommodate duplex DNA, so like the hexameric MCM2-7 helicase, BLM helicase probably encircles the DNA template. No X-ray crystal structures of full length BLM have been reported to date. However, high resolution structural analyses of the HRDC[146] and RQC[147] domains do not immediately suggest a hexameric fold, as is also the case for WRN HRDC[139] and RQC[140] domains. The HRDC domain of BLM is required for dissolution of double Holliday junctions (DHJs);[141] domain swap experiments would be highly informative in determining if this activity is specific to BLM HRDC, or whether related domains from other RecQ helicases such as WRN can similarly resolve DHJs.

3.4 The Sliding Clamp PCNA

3.4.1 PCNA Structure

PCNA (proliferating cell nuclear antigen) was originally characterised as a protein present at high levels in the nuclei of proliferating cells.[148,149] It is widely used as a prognostic/diagnostic indicator in neoplastic disease.[150] In eukaryotes, it assembles into homotrimers[151] from three identical subunits of ~29 kDa each, forming a toroidal ring lined by positively charged alpha helices that interact with DNA. (Note that a larger alternative form of PCNA has been reported in carrots and the malarial parasite, which may dimerise, akin to the sliding clamp in *E. coli*[152–155]). The diameter of the central hole of the ring is suitable to accommodate duplex DNA,[144–145,156] and can even encompass additional secondary structures such as a four-base stem loop and a 15-mer bubble in the template.[157] Although trimeric, the overall structure shows six-fold pseudosymmetry (Figure 3.8A), and this structure (though not sequence) is highly conserved even in the dimeric β clamp from *E. coli*,[144,152] suggesting that these may have arisen by convergent evolution. The importance of PCNA's symmetry may lie in the protein partners that PCNA associates with during the process of DNA replication; many partners have been characterised (reviewed in ref. 158–160), and this list has recently been extended by a proteomics

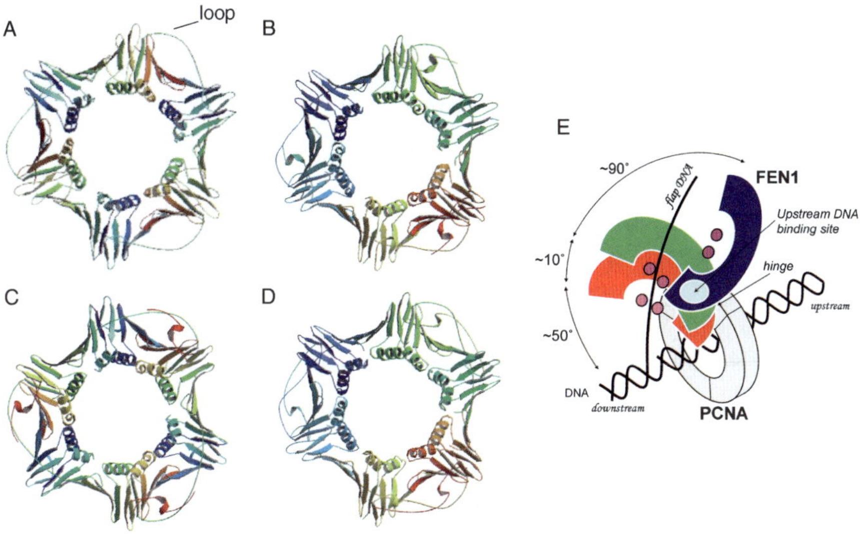

Figure 3.8 **PCNA forms a homotrimeric ring with a hydrophobic binding pocket in the interdomain connector loop.** (**A**) Native human PCNA (PDB 1W60[156]). (**B**) PCNA structure in complex with a peptide derived from amino acids 542–566 of DNA polymerase δ p66 subunit (PDB 1U76[176]). (**C**) Structure of PCNA bound to a peptide representing amino acids 331–350 of human FEN1 (PDB 1U7b[176]). (**D**) PCNA complexed with the PIP region of p21 (PDB 1AXC[145]). Note that all these proteins bind in a hydrophobic pocket formed by the loop, though each makes different contacts within that pocket (see also cover figure). (**E**) Proposed model for how FEN1 swings out onto DNA flaps whilst bound to PCNA. Reprinted by permission from the authors and Macmillan Publishers Ltd (S. Sakurai, K. Kitano, H. Yamaguchi, K. Hamada, K. Okada, K. Fukuda, M. Uchida, E. Ohtsuka, H. Morioka and T. Hakoshima, Structural basis for recruitment of human flap endonuclease 1 to PCNA, *EMBO Journal*, 2004, **24**, 683-693), copyright (2004).[179]

approach to identify proteins that bind to PCNA *in vivo*.[161] Those shown to be important in DNA replication are discussed further below.

3.4.2 PCNA Loading and Interaction with DNA

PCNA is loaded onto the 3′ recessed DNA ends at the primer-template junction of the replication fork by the action of a clamp loader. RFC, a hetero-pentameric AAA+ ATPase, performs this role in normal DNA replication (see Chapter 2), but alternative RFC-like proteins (reviewed in ref. 162) act to load PCNA at stalled forks (Elg1 and ScRad24/hRad17),[163] and to remodel PCNA during establishment of sister chromatid cohesion (Ctf18) so that it can recruit the protein Eco1 important in establishment of cohesion[164,165] (see Chapter 9).

The PCNA ring is essentially a flattened doughnut, so how does it associate with the clamp loader RFC complex, which forms a right-handed spiral in its pentameric form? Molecular dynamics simulations suggest that PCNA can distort significantly, bending and twisting around the sheets, particularly at the interface between subunits but also at the interdomain surface, and that such twisting results in an oscillation between lateral opening and a right-handed spiral conformation.[166] This spiral form of PCNA can readily match the spiral configuration of the RFC clamp loader which is necessary for its ability to recognise DNA at primer-template junctions and to hydrolyse ATP. The binding of RFC to PCNA is regulated at least in part by phosphorylation, with modification of the PCNA-binding domain of RFC by calcium/calmodulin dependent kinase II (CaMKII) leading to a significant decrease in RFC's association with PCNA,[167] which in cells would result in decreased DNA synthesis. Notably, CaMKII cannot phosphorylate RF-C that is already bound to PCNA and DNA.[167]

3.4.3 PCNA Interactions with Polymerases

Once loaded at the fork, PCNA recruits the leading and lagging strand DNA polymerases, polε[168] and polδ[169,170] respectively (reviewed in ref. 171), and ensures their processivity by encircling the template and retaining the polymerases in proximity to their template DNA. DNA polymerase δ subunits can bind into a hydrophobic pocket of PCNA formed in the interdomain connector loop (Figure 3.8B), *via* classical PCNA-interacting peptide (PIP) motifs[158,172–174] (*e.g.* polymerase δ p66 subunit PIP has the sequence KANR**Q**VSI**T**G**FF**QRK[175,176]).

3.4.4 PCNA Partners in Okazaki Fragment Processing

On the lagging strand, displacement of the RNA primer (and possibly the DNA primer synthesised by error-prone DNA polα) by the incoming polymerase δ results in formation of a 5′ flap that is cleaved by the structure specific nuclease FEN1,[177] alone or in combination with other nucleases (see Chapter 5). Interestingly, FEN1 binds *via* a PIP (**Q**GRL**D**D**FF**) to the hydrophobic pocket on PCNA[173](Figure 3.8C); it is likely that this binding displaces DNA polδ[178] and is important for hand-off (Chapter 6). The trimeric nature of PCNA together with a flexible hinge region on FEN1 may be critical in allowing FEN1 to swing into the ideal position for cleavage[179,180] (Figure 3.8E). Once FEN1 has cleaved the 5′ flap, a nick in the phosphodiester backbone remains: this is joined by the action of DNA ligase 1, a further partner of PCNA[181] that again binds to the same hydrophobic pocket and shares a consensus PIP.[182] FEN1 and ligase 1 compete with each other *in vitro* for PCNA binding,[183] and are likely to do so *in vivo*. The importance of the FEN1–PCNA interaction can be seen by the severe replication defects and neonatal mortality of transgenic knock-in mice in which the FEN1–PCNA interaction is disrupted by mutation of critical phenylalanine residues within FEN1 PIP.[184]

The structural basis of PCNA's ability to act as a 'hand-off' platform in Okazaki fragment processing is still the subject of active research. Molecular dynamic modelling of PCNA–DNA contacts suggests that:

- the double-stranded DNA axis tilts with respect to the plane of the PCNA ring;
- basic lysines and arginines lining the central hole of PCNA make direct interactions with the phosphodiester backbone of the DNA.

There is also movement within the PCNA C-terminal domain and linker region involved in protein–protein interactions, which helps to explain how PCNA can act as a platform for coordinated hand-off of DNA intermediates from DNA polymerase to FEN1 to DNA ligase during Okazaki fragment processing.[185] Flexibility within the linker region has been demonstrated by high resolution structural analysis of PCNA in complex with PIPs from three different partners: DNA polδ p66, FEN1 and p21[145,156,176] (Figure 3.8B–D).

PCNA therefore acts as a platform for sequential loading of the various factors necessary for Okazaki fragment processing.[158,186] Association of the C-termini of the three PCNA monomers with Cyclin A-Cdk2 additionally provides a ready means for phosphorylation of Cdk2 targets such as DNA ligase and the large subunit of RFC.[187] These post-translational modifications are likely to be important in regulating binding or function of PCNA's client proteins.

3.4.5 PCNA in Establishing Epigenetic Modifications during Replication

Establishing epigenetic methylation patterns on newly replicated DNA involves fork-associated PCNA acting as a platform for recruiting enzymes that modify either the DNA or chromatin. For example, PCNA interacts with DNA methyltransferase 1 (Dnmtl) in a transient and highly dynamic manner, as observed by photobleaching studies.[188] Modification of nascent chromatin by CAF-1 may also require PCNA-dependent recruitment of CAF-1 to replication sites,[189,190] together with WSTF and HDAC chromatin remodelling factors.[191] PCNA's recruitment of these factors to the replication fork may be important in contributing to epigenetic inheritance (see Chapter 10 for more details).

3.4.6 PCNA on DNA Damage

PCNA also plays an important role in recruiting appropriate factors to deal with replication stress and template damage. For example, binding of PCNA to uracil DNA glycosylase[192] through its classical PIP motif may be important in base excision repair of singly damaged bases. PCNA is subject to modification by different post-translational modifications including monoubiquitination, polyubiquitination and SUMOylation which alter PCNA's interactions with DNA polymerases and other replication components, and can shift replication from processive mode to a DNA repair or translesion synthesis mode (see Chapter 7 for further details). Interaction of PCNA with the translesion DNA polymerases has major functional consequences, reducing misincorporation against an 8oxoG template by 1200-fold, possibly by recruiting polλ or polη in preference to other translesion polymerases.[193] The p53-induced cyclin kinase inhibitor p21 (CDKN1) can outcompete polλ for PCNA binding through interactions at the same hydrophobic pocket on PCNA.[194] Furthermore, p21 blocks the DNA replication activity of PCNA,[172,195] and this inhibition occurs through p21 PIP sequence **QTSMTDFY** binding to the same hydrophobic pocket region on PCNA[172] (Figure 3.8D) as that bound by FEN1,[173] DNA ligase 1,[196] and the RecQ helicase, WRN,[114] that is thought to be recruited to stalled replication forks.[113] This is of particular interest since p53 is rapidly induced upon DNA damage or other cellular stresses, leading to increased levels of p21 (reviewed in ref. 197), thus providing a possible direct mechanism for linking damage to inhibition of processive DNA synthesis and induction of translesion synthesis or stalled fork protection/restart. However, it is not yet clear if the kinetics of p21 induction may be too slow to throw the switch to translesion synthesis or repair in a biologically relevant time frame.

Thus it appears that many of the regulatory steps in DNA replication, particularly the sequential nature of Okazaki fragment processing, translesion synthesis, and the sensing of stalled forks together with recruitment of factors that can resolve structures at stalled forks, are all dependent upon binding to the hydrophobic pocket (or in some cases, the nearby C-terminal region) of PCNA.

3.5 The 9-1-1 Damage Response Sliding Clamp

Cells respond to DNA damage and replication stress by initiating checkpoint pathways that result in arrest of DNA replication and recruitment of factors to repair DNA damage or to restart stalled replication forks. The pathways include DNA damage/replication stress sensors, adaptors, relay proteins and effector kinases (reviewed in ref. 198). As noted in Section 3.4.6, PCNA is important in recruiting factors such as translesion polymerases to sites of damage.

A mechanism has evolved to allow subtle discrimination in recruitment of proteins such as DNA ligase 1 and FEN1 to sites of damage *vs.* sites of Okazaki fragment processing, dependent upon the clamp to which they bind. Thus, whilst FEN1 and DNA ligase 1 bind to PCNA during processive replication and under conditions such as base excision repair, on replication stress and other types of DNA damage they appear to associate instead with a PCNA-like factor that provides a similar loading platform and clamp function. This clamp comprises three separate proteins, Rad9, Rad1 and Hus1 and is known as the 9-1-1 complex (the orthologous proteins in *S. cerevisiae* are ScRad17, ScMec3 and ScDdc1) (reviewed in ref. 199). Structurally, the 9-1-1 complex is trimeric and forms a toroid,[200] as shown by electron microscopy,[201,202] with the N-terminal domains of each of the three proteins important in mediating intermolecular interactions with the C-termini of adjacent proteins in the complex.[203]

Upon replication fork stalling by treatment with aphidicolin, 9-1-1 becomes bound to chromatin.[204] It is loaded by an RFC-like molecule containing the four small subunits of RFC, together with yeast ScRad24[205] or human HsRad17[206] proteins, which are highly similar to the large subunit RFC1.[ii] Such loading occurs only on DNA coated with the ssDNA binding protein RPA, since although the 9-1-1 complex itself can bind naked DNA, the Rad24-RFC clamp loader cannot.[207] This loading ensures that 9-1-1 is loaded only at 5′ DNA junctions,[207] a substrate specificity that is presumably important in repair or fork restart. Once loaded, 9-1-1 recruits the TopBP1 protein, the activation domain of which is necessary to activate ATM/R kinase (Mec1/Ddc2 in yeast). In turn, the activated ATM/R kinase phosphorylates not only the downstream checkpoint kinase Chk1,[208] but also the Rad1 and Hus1 subunits of the 9-1-1 complex, together with the clamp loader Rad24-RFC and the RPA ssDNA binding protein. Such phosphorylation presumably provides some form of feedback mechanism to regulate further 9-1-1 loading and damage processing (Figure 3.9). Rad9 is not simply a sensor of damage, but acts in DNA repair (reviewed in ref. 209), by recruiting repair factors FEN1,[210] DNA polymerase β[211] DNA ligase[212,213] (though not through a classical PIP[214]), the mismatch repair protein MutY DNA glycosylase[215] and translesion DNA polymerases ζ, ε and κ [216,217] Confusingly, whilst loss of the 9-1-1 complex reduces polζ-dependent mutagenesis,[217] no stimulation of polymerase ζ activity by 9-1-1 has been reported.[216]

The structural basis of 9-1-1 action as damage sensor, clamp, loading platform and active player in DNA replication stress and repair still awaits high resolution crystal structures of the trimeric complex, together with molecular dynamics modelling.

3.6 Perspective

Ring-shaped structures with overall six-fold symmetry are widely found in key replication factors—particularly helicases and sliding clamps. Such structures presumably arose to accommodate duplex DNA and permit high processivity either of the intrinsic enzyme activity (helicases) or of associated protein partners (sliding clamps). Current molecular modelling and high-resolution structural analyses have started to uncover how these proteins function and why the six-fold symmetry is so critical to their modes of action. Challenges still remaining include complete structural analysis of full length WRN and BLM, and the 9-1-1 and MCM2-7 complexes, together with determination of the

[ii] Note that the terminology in yeast and humans is potentially confusing in that, in *S. cerevisiae*, ScRad17 is orthologous to human HsRad1, has intrinsic nuclease activity and is an integral component of the Rad17-Mec3-Ddc1 complex loaded onto DNA by ScRad24-RFC; in humans, HsRad17 serves in the role of a large subunit of RFC to load 9-1-1.[206] Here, the loader is designated as either Rad24-RFC or Rad17-RFC according to the name used in the specific paper cited.

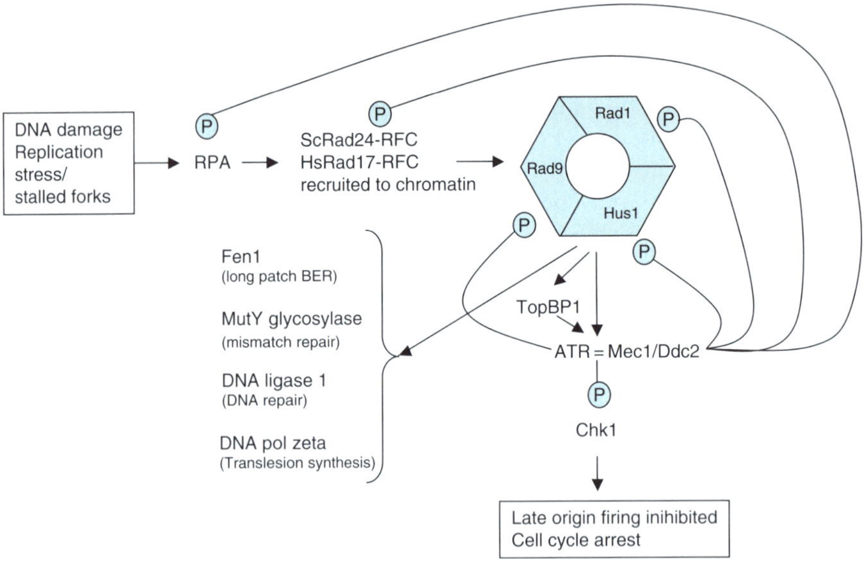

Figure 3.9 **Proposed pathway for 9-1-1 activity in DNA repair and replication stress.** Following DNA damage or replication stress, the PCNA-like heterotrimeric Rad9-Rad1-Hus1 (9-1-1) complex is recruited to chromatin at RPA-coated sites by the action of the clamp loader Rad24-RFC_{2-5} in yeast or Rad17-RFC_{2-5} in humans. Once bound, 9-1-1 recruits TopBP1 that in turn activates the checkpoint kinases ATM/R, equivalent to yeast Mec1/Ddc2. Phosphorylation (P) of Chk1 kinase by ATM/R then triggers replication arrest and DNA repair. ATM/R also phosphorylates components of the 9-1-1 complex itself, plus the recruiting factors RPA (subunits RPA1 and RPA2) and RFC. 9-1-1 associates with DNA repair proteins at the site of damage (see text for details).

in vivo stoichiometries of PCNA client proteins with PCNA under conditions of processive replication, damaged template and stalled forks.

The central role that these helicases and clamps proteins play in DNA replication makes them very useful in diagnosis and prognosis in hyperproliferative disease. MCM protein levels are generally high in rapidly proliferating neoplastic cells[218,219] and are downregulated when cells exit the cell cycle, both in quiescence and senescence,[220] demonstrating their importance in cell proliferation. Indeed, MCM2 and MCM5 are proving useful diagnostic and prognostic indicators in various human cancers.[218,219] Similarly, PCNA has been used widely as a diagnostic marker of hyperproliferative and neoplastic disease[150] (though see Chapter 14 for caveats). Thus MCMs and PCNA appear to be interesting potential targets for pharmaceutical intervention; for example, regulating PCNA's activity to promote repair over replication may be a useful goal in cancer therapy.[156] However, the essential nature of helicases and clamps in DNA replication requires that any pharmacologically active agents are precisely targeted to tumours. Such difficulties may stand in the way of therapeutic exploitation of our current knowledge of these DNA replication proteins.

Acknowledgements

The work in LSC's laboratory is supported by the Biotechnology and Biological Sciences Research Council [(grant numbers BB/E000924/1 and BB/E016995/1)] and SK's lab is supported by CRUK (grant number C814/A9035).

References

1. S. S. Patel and K. M. Picha, Structure and function of hexameric helicases, *Annu. Rev. Biochem.*, 2000, **69**, 651–697.
2. N. Tuteja and R. Tuteja, Prokaryotic and eukaryotic DNA helicases. Essential molecular motor proteins for cellular machinery, *Eur. J. Biochem.*, 2004, **271**, 1835–1848.
3. F. Dong, E. P. Gogol and P. H. von Hippel, The phage T4-coded DNA replication helicase (gp41) forms a hexamer upon activation by nucleoside triphosphate, *J. Biol. Chem.*, 1995, **270**, 7462–7473.
4. Y. Ishimi, A. Claude, P. Bullock and J. Hurwitz, Complete enzymatic synthesis of DNA containing the SV40 origin of replication, *J. Biol. Chem.*, 1988, **263**, 19723–19733.
5. K. Labib and J. F. Diffley, Is the MCM2-7 complex the eukaryotic DNA replication fork helicase?, *Curr. Opin. Genet. Dev.*, 2001, **11**, 64–70.
6. M. L. Bochman and A. Schwacha, The Mcm2-7 complex has *in vitro* helicase activity, *Mol. Cell*, 2008, **31**, 287–293.
7. S. E. Kearsey and K. Labib, MCM proteins: evolution, properties, and role in DNA replication, *Biochim. Biophys. Acta*, 1998, **1398**, 113–136.
8. S. L. Forsburg, Eukaryotic MCM proteins: beyond replication initiation, *Microbiol. Mol. Biol. Rev.*, 2004, **68**, 109–131.
9. B. K. Tye, MCM proteins in DNA replication, *Annu. Rev. Biochem.*, 1999, **68**, 649–686.
10. J. M. Bailis and S. L. Forsburg, MCM proteins: DNA damage, mutagenesis and repair, *Curr. Opin. Genet. Dev.*, 2004, **14**, 17–21.
11. A. Costa and S. Onesti, The MCM complex: (just) a replicative helicase?, *Biochem. Soc. Trans.*, 2008, **36**, 136–140.
12. G. T. Maine, P. Sinha and B. K. Tye, Mutants of *S. cerevisiae* defective in the maintenance of minichromosomes, *Genetics*, 1984, **106**, 365–385.
13. M. Lutzmann, D. Maiorano and M. Mechali, Identification of full genes and proteins of MCM9, a novel, vertebrate-specific member of the MCM2-8 protein family, *Gene*, 2005, **362**, 51–56.
14. D. Gozuacik, M. Chami, D. Lagorce, J. Faivre, Y. Murakami, O. Poch, E. Biermann, R. Knippers, C. Brechot and P. Paterlini-Brechot, Identification and functional characterization of a new member of the human Mcm protein family: hMcm8, *Nucleic Acids Res.*, 2003, **31**, 570–579.
15. R. J. Fletcher, B. E. Bishop, R. P. Leon, R. A. Sclafani, C. M. Ogata and X. S. Chen, The structure and function of MCM from archaeal *M. thermoautotrophicum*, *Nat. Struct. Biol.*, 2003, **10**, 160–167.

16. T. Pape, H. Meka, S. Chen, G. Vicentini, M. van Heel and S. Onesti, Hexameric ring structure of the full-length archaeal MCM protein complex, *EMBO Rep.*, 2003, **4**, 1079–1083.
17. Y. Gomez-Llorente, R. J. Fletcher, X. S. Chen, J. M. Carazo and C. San Martin, Polymorphism and double hexamer structure in the archaeal minichromosome maintenance (MCM) helicase from *Methanobacterium thermoautotrophicum*, *J. Biol. Chem.*, 2005, **280**, 40909–40915.
18. W. Liu, B. Pucci, M. Rossi, F. M. Pisani and R. Ladenstein, Structural analysis of the *Sulfolobus solfataricus* MCM protein N-terminal domain, *Nucleic Acids Res.*, 2008, **36**, 3235–3243.
19. A. Costa, T. Pape, M. van Heel, P. Brick, A. Patwardhan and S. Onesti, Structural basis of the *Methanothermobacter thermautotrophicus* MCM helicase activity, *Nucleic Acids Res.*, 2006, **34**, 5829–5838.
20. A. Costa, T. Pape, M. van Heel, P. Brick, A. Patwardhan and S. Onesti, Structural studies of the archaeal MCM complex in different functional states, *J. Struct. Biol.*, 2006, **156**, 210–219.
21. A. Costa, G. van Duinen, B. Medagli, J. Chong, N. Sakakibara, Z. Kelman, S. K. Nair, A. Patwardhan and S. Onesti, Cryo-electron microscopy reveals a novel DNA-binding site on the MCM helicase, *EMBO J.*, 2008, **27**, 2250–2258.
22. E. R. Jenkinson and J. P. Chong, Minichromosome maintenance helicase activity is controlled by N- and C-terminal motifs and requires the ATPase domain helix-2 insert, *Proc. Natl. Acad. Sci. U.S.A.*, 2006, **103**, 7613–7618.
23. E. R. Barry, A. T. McGeoch, Z. Kelman and S. D. Bell, Archaeal MCM has separable processivity, substrate choice and helicase domains, *Nucleic Acids Res.*, 2007, **35**, 988–998.
24. B. Pucci, M. De Felice, M. Rocco, F. Esposito, M. De Falco, L. Esposito, M. Rossi and F. M. Pisani, Modular organization of the *Sulfolobus solfataricus* mini-chromosome maintenance protein, *J. Biol. Chem.*, 2007, **282**, 12574–12582.
25. E. Rothenberg, M. A. Trakselis, S. D. Bell and T. Ha, MCM forked substrate specificity involves dynamic interaction with the 5′-tail, *J. Biol. Chem.*, 2007, **282**, 34229–34234.
26. E. V. Koonin, A common set of conserved motifs in a vast variety of putative nucleic acid-dependent ATPases including MCM proteins involved in the initiation of eukaryotic DNA replication, *Nucleic Acids Res.*, 1993, **21**, 2541–2547.
27. E. B. Gomez, M. G. Catlett and S. L. Forsburg, Different phenotypes *in vivo* are associated with ATPase motif mutations in *Schizosaccharomyces pombe* minichromosome maintenance proteins, *Genetics*, 2002, **160**, 1305–1318.
28. A. Schwacha and S. P. Bell, Interactions between two catalytically distinct MCM subgroups are essential for coordinated ATP hydrolysis and DNA replication, *Mol. Cell*, 2001, **8**, 1093–1104.

29. M. L. Bochman, S. P. Bell and A. Schwacha, Subunit organization of Mcm2-7 and the unequal role of active sites in ATP hydrolysis and viability, *Mol. Cell. Biol.*, 2008, **28**, 5865–5873.
30. C. Y. Ying and J. Gautier, The ATPase activity of MCM2-7 is dispensable for pre-RC assembly but is required for DNA unwinding, *EMBO J.*, 2005, **24**, 4334–4344.
31. G. T. Haugland, N. Sakakibara, A. L. Pey, C. R. Rollor, N. K. Birkeland and Z. Kelman, *Thermoplasma acidophilum* Cdc6 protein stimulates MCM helicase activity by regulating its ATPase activity, *Nucleic Acids Res.*, 2008, **36**, 5602–5609.
32. J. -H. Shin, B. Grabowski, R. Kasiviswanathan, S. D. Bell and Z. Kelman, Regulation of minichromosome maintenance helicase activity by Cdc6, *J. Biol. Chem.*, 2003, **278**, 38059–38067.
33. R. Kasiviswanathan, J. -H. Shin and Z. Kelman, Interactions between the archaeal Cdc6 and MCM proteins modulate their biochemical properties, *Nucleic Acids Res.*, 2005, **33**, 4940–4950.
34. M. De Felice, L. Esposito, B. Pucci, F. Carpentieri, M. De Falco, M. Rossi and F. M. Pisani, Biochemical characterization of a CDC6-like protein from the Crenarchaeon *Sulfolobus solfataricus*, *J. Biol. Chem.*, 2003, **278**, 46424–46431.
35. A. Poplawski, B. Grabowski, S. E. Long and Z. Kelman, The zinc finger domain of the archaeal minichromosome maintenance protein is required for helicase activity, *J. Biol. Chem.*, 2001, **276**, 49371–49377.
36. Z. Kelman, J. K. Lee and J. Hurwitz, The single minichromosome maintenance protein of *Methanobacterium thermoautotrophicum* DeltaH contains DNA helicase activity, *Proc. Natl. Acad. Sci. U.S.A.*, 1999, **96**, 14783–14788.
37. J. P. Chong, M. K. Hayashi, M. N. Simon, R. M. Xu and B. Stillman, A double-hexamer archaeal minichromosome maintenance protein is an ATP-dependent DNA helicase, *Proc. Natl. Acad. Sci. U.S.A.*, 2000, **97**, 1530–1535.
38. Y. Ishimi, A DNA helicase activity is associated with an MCM4, -6, and -7 protein complex, *J. Biol. Chem.*, 1997, **272**, 24508–24513.
39. J. K. Lee and J. Hurwitz, Isolation and characterization of various complexes of the minichromosome maintenance proteins of *Schizosaccharomyces pombe*, *J. Biol. Chem.*, 2000, **275**, 18871–18878.
40. Z. You, Y. Ishimi, H. Masai and F. Hanaoka, Roles of Mcm7 and Mcm4 subunits in the DNA helicase activity of the mouse Mcm4/6/7 complex, *J. Biol. Chem.*, 2002, **277**, 42471–42479.
41. P. Thommes, Y. Kubota, H. Takisawa and J. J. Blow, The RLF-M component of the replication licensing system forms complexes containing all six MCM/P1 polypeptides, *EMBO J.*, 1997, **16**, 3312–3319.
42. D. M. Kanter, I. Bruck and D. L. Kaplan, Mcm subunits can assemble into two different active unwinding complexes, *J. Biol. Chem.*, 2008, **283**, 31172–31182.

43. T. A. Prokhorova and J. J. Blow, Sequential MCM/P1 subcomplex assembly is required to form a heterohexamer with replication licensing activity, *J. Biol. Chem.*, 2000, **275**, 2491–2498.
44. K. Labib, J. A. Tercero and J. F. Diffley, Uninterrupted MCM2-7 function required for DNA replication fork progression, *Science*, 2000, **288**, 1643–1647.
45. K. Labib and A. Gambus, A key role for the GINS complex at DNA replication forks, *Trends Cell Biol.*, 2007, **17**, 271–278.
46. S. E. Moyer, P. W. Lewis and M. R. Botchan, Isolation of the Cdc45/Mcm2-7/GINS (CMG) complex, a candidate for the eukaryotic DNA replication fork helicase, *Proc. Natl. Acad. Sci. U.S.A.*, 2006, **103**, 10236–10241.
47. T. S. Takahashi, D. B. Wigley and J. C. Walter, Pumps, paradoxes and ploughshares: mechanism of the MCM2-7 DNA helicase, *Trends Biochem. Sci.*, 2005, **30**, 437–444.
48. R. A. Laskey and M. A. Madine, A rotary pumping model for helicase function of MCM proteins at a distance from replication forks, *EMBO Rep.*, 2003, **4**, 26–30.
49. R. Wessel, J. Schweizer and H. Stahl, Simian virus 40 T-antigen DNA helicase is a hexamer which forms a binary complex during bidirectional unwinding from the viral origin of DNA replication, *J. Virol.*, 1992, **66**, 804–815.
50. B. C. Tan, C. T. Chien, S. Hirose and S. C. Lee, Functional cooperation between FACT and MCM helicase facilitates initiation of chromatin DNA replication, *EMBO J.*, 2006, **25**, 3975–3985.
51. A. Gambus, R. C. Jones, A. Sanchez-Diaz, M. Kanemaki, F. van Deursen, R. D. Edmondson and K. Labib, GINS maintains association of Cdc45 with MCM in replisome progression complexes at eukaryotic DNA replication forks, *Nat. Cell Biol.*, 2006, **8**, 358–366.
52. A. Groth, A. Corpet, A. J. Cook, D. Roche, J. Bartek, J. Lukas and G. Almouzni, Regulation of replication fork progression through histone supply and demand, *Science*, 2007, **318**, 1928–1931.
53. Y. Doyon, C. Cayrou, M. Ullah, A. J. Landry, V. Cote, W. Selleck, W. S. Lane, S. Tan, X. J. Yang and J. Cote, ING tumor suppressor proteins are critical regulators of chromatin acetylation required for genome expression and perpetuation, *Mol. Cell*, 2006, **21**, 51–64.
54. S. E. Kearsey and S. Cotterill, Enigmatic variations: divergent modes of regulating eukaryotic DNA replication, *Mol. Cell*, 2003, **12**, 1067–1075.
55. M. E. Liku, V. Q. Nguyen, A. W. Rosales, K. Irie and J. J. Li, CDK phosphorylation of a novel NLS-NES module distributed between two subunits of the Mcm2-7 complex prevents chromosomal rereplication, *Mol. Biol. Cell*, 2005, **16**, 5026–5039.
56. K. Labib, J. F. Diffley and S. E. Kearsey, G1-phase and B-type cyclins exclude the DNA-replication factor Mcm4 from the nucleus, *Nat. Cell Biol.*, 1999, **1**, 415–422.

57. I. T. Todorov, A. Attaran and S. E. Kearsey, BM28, a human member of the MCM2-3-5 family, is displaced from chromatin during DNA replication, *J. Cell Biol.*, 1995, **129**, 1433–1445.
58. T. Krude, C. Musahl, R. A. Laskey and R. Knippers, Human replication proteins hCdc21, hCdc46 and P1Mcm3 bind chromatin uniformly before S-phase and are displaced locally during DNA replication, *J. Cell Sci.*, 1996, **109**, 309–318.
59. J. L. Bowers, J. C. Randell, S. Chen and S. P. Bell, ATP hydrolysis by ORC catalyzes reiterative Mcm2-7 assembly at a defined origin of replication, *Mol. Cell*, 2004, **16**, 967–978.
60. R. Burkhart, D. Schulte, D. Hu, C. Musahl, F. Gohring and R. Knippers, Interactions of human nuclear proteins P1Mcm3 and P1Cdc46, *Eur. J. Biochem.*, 1995, **228**, 431–438.
61. M. Lei, Y. Kawasaki and B. K. Tye, Physical interactions among Mcm proteins and effects of Mcm dosage on DNA replication in *Saccharomyces cerevisiae*, *Mol. Cell. Biol.*, 1996, **16**, 5081–5090.
62. A. Rowles, J. P. Chong, L. Brown, M. Howell, G. I. Evan and J. J. Blow, Interaction between the origin recognition complex and the replication licensing system in *Xenopus*, *Cell*, 1996, **87**, 287–296.
63. S. Donovan, J. Harwood, L. S. Drury and J. F. Diffley, Cdc6p-dependent loading of Mcm proteins onto pre-replicative chromatin in budding yeast, *Proc. Natl. Acad. Sci. U.S.A.*, 1997, **94**, 5611–5616.
64. H. M. Mahbubani, J. P. Chong, S. Chevalier, P. Thommes and J. J. Blow, Cell cycle regulation of the replication licensing system: involvement of a Cdk-dependent inhibitor, *J. Cell Biol.*, 1997, **136**, 125–135.
65. M. C. Edwards, A. V. Tutter, C. Cvetic, C. H. Gilbert, T. A. Prokhorova and J. C. Walter, MCM2-7 complexes bind chromatin in a distributed pattern surrounding the origin recognition complex in *Xenopus* egg extracts, *J. Biol. Chem.*, 2002, **277**, 33049–33057.
66. X. Q. Ge, D. A. Jackson and J. J. Blow, Dormant origins licensed by excess Mcm2 7 are required for human cells to survive replicative stress, *Genes Dev.*, 2007, **21**, 3331–3341.
67. M. Lei, Y. Kawasaki, M. R. Young, M. Kihara, A. Sugino and B. K. Tye, Mcm2 is a target of regulation by Cdc7-Dbf4 during the initiation of DNA synthesis, *Genes Dev.*, 1997, **11**, 3365–3374.
68. W. H. Cho, Y. J. Lee, S. I. Kong, J. Hurwitz and J. K. Lee, CDC7 kinase phosphorylates serine residues adjacent to acidic amino acids in the minichromosome maintenance 2 protein, *Proc. Natl. Acad. Sci. U.S.A.*, 2006, **103**, 11521–11526.
69. H. Masai, E. Matsui, Z. You, Y. Ishimi, K. Tamai and K. Arai, Human Cdc7-related kinase complex. *In vitro* phosphorylation of MCM by concerted actions of Cdks and Cdc7 and that of a criticial threonine residue of Cdc7 bY Cdks, *J. Biol. Chem.*, 2000, **275**, 29042–29052.
70. H. Masai and K. Arai, Cdc7 kinase complex: a key regulator in the initiation of DNA replication, *J. Cell. Physiol.*, 2002, **190**, 287–296.

71. C. F. Hardy, O. Dryga, S. Seematter, P. M. Pahl and R. A. Sclafani, Mcm5/cdc46-bob1 bypasses the requirement for the S phase activator Cdc7p, *Proc. Natl. Acad. Sci. U.S.A.*, 1997, **94**, 3151–3155.
72. M. L. Hoang, R. P. Leon, L. Pessoa-Brandao, S. Hunt, M. K. Raghuraman, W. L. Fangman, B. J. Brewer and R. A. Sclafani, Structural changes in Mcm5 protein bypass Cdc7-Dbf4 function and reduce replication origin efficiency in *Saccharomyces cerevisiae*, *Mol. Cell. Biol.*, 2007, **27**, 7594–7602.
73. J. K. Lee, Y. S. Seo and J. Hurwitz, The Cdc23 (Mcm10) protein is required for the phosphorylation of minichromosome maintenance complex by the Dfp1-Hsk1 kinase, *Proc. Natl. Acad. Sci. U.S.A.*, 2003, **100**, 2334–2339.
74. I. Pereverzeva, E. Whitmire, B. Khan and M. Coue, Distinct phosphoisoforms of the *Xenopus* Mcm4 protein regulate the function of the Mcm complex, *Mol. Cell. Biol.*, 2000, **20**, 3667–3676.
75. Y. Komamura-Kohno, K. Karasawa-Shimizu, T. Saitoh, M. Sato, F. Hanaoka, S. Tanaka and Y. Ishimi, Site-specific phosphorylation of MCM4 during the cell cycle in mammalian cells, *FEBS J.*, 2006, **273**, 1224–1239.
76. M. Coue, S. E. Kearsey and M. Mechali, Chromatin binding, nuclear localization and phosphorylation of *Xenopus* cdc21 are cell-cycle dependent and associated with the control of initiation of DNA replication, *EMBO J.*, 1996, **15**, 1085–1097.
77. A. Montagnoli, B. Valsasina, D. Brotherton, S. Troiani, S. Rainoldi, P. Tenca, A. Molinari and C. Santocanale, Identification of Mcm2 phosphorylation sites by S-phase-regulating kinases, *J. Biol. Chem.*, 2006, **281**, 10281–10290.
78. D. I. Lin, P. Aggarwal and J. A. Diehl, Phosphorylation of MCM3 on Ser-112 regulates its incorporation into the MCM2-7 complex, *Proc. Natl. Acad. Sci. U.S.A.*, 2008, **105**, 8079–8084.
79. S. Tanaka, T. Umemori, K. Hirai, S. Muramatsu, Y. Kamimura and H. Araki, CDK-dependent phosphorylation of Sld2 and Sld3 initiates DNA replication in budding yeast, *Nature*, 2007, **445**, 328–332.
80. D. Cortez, G. Glick and S. J. Elledge, Minichromosome maintenance proteins are direct targets of the ATM and ATR checkpoint kinases, *Proc. Natl. Acad. Sci. U.S.A.*, 2004, **101**, 10078–10083.
81. D. Shechter and J. Gautier, MCM proteins and checkpoint kinases get together at the fork, *Proc. Natl. Acad. Sci. U.S.A.*, 2004, **101**, 10845–10846.
82. Y. Shi, G. E. Dodson, P. S. Mukhopadhyay, N. P. Shanware, A. T. Trinh and R. S. Tibbetts, Identification of carboxyl-terminal MCM3 phosphorylation sites using polyreactive phosphospecific antibodies, *J. Biol. Chem.*, 2007, **282**, 9236–9243.
83. N. Nitani, C. Yadani, H. Yabuuchi, H. Masukata and T. Nakagawa, Mcm4 C-terminal domain of MCM helicase prevents excessive formation of single-stranded DNA at stalled replication forks, *Proc. Natl. Acad. Sci. U.S.A.*, 2008, **105**, 12973–12978.

84. D. Maiorano, M. Lutzmann and M. Mechali, MCM proteins and DNA replication, *Curr. Opin. Cell Biol.*, 2006, **18**, 130–136.
85. H. L. Blanton, S. J. Radford, S. McMahan, H. M. Kearney, J. G. Ibrahim and J. Sekelsky, REC, *Drosophila* MCM8, drives formation of meiotic crossovers, *PLoS Genet.*, 2005, **1**, e40.
86. J. M. Carlton, J. H. Adams, J. C. Silva, S. L. Bidwell, H. Lorenzi, E. Caler, J. Crabtree, S. V. Angiuoli, E. F. Merino, P. Amedeo, Q. Cheng, R. M. Coulson, B. S. Crabb, H. A. Del Portillo, K. Essien, T. V. Feldblyum, C. Fernandez-Becerra, P. R. Gilson, A. H. Gueye, X. Guo, S. Kang'a, T. W. Kooij, M. Korsinczky, E. V. Meyer, V. Nene, I. Paulsen, O. White, S. A. Ralph, Q. Ren, T. J. Sargeant, S. L. Salzberg, C. J. Stoeckert, S. A. Sullivan, M. M. Yamamoto, S. L. Hoffman, J. R. Wortman, M. J. Gardner, M. R. Galinski, J. W. Barnwell and C. M. Fraser-Liggett, Comparative genomics of the neglected human malaria parasite *Plasmodium vivax*, *Nature*, 2008, **455**, 757–763.
87. L. Eichinger, J. A. Pachebat, G. Glockner, M. A. Rajandream, R. Sucgang, M. Berriman, J. Song, R. Olsen, K. Szafranski, Q. Xu, B. Tunggal, S. Kummerfeld, M. Madera, B. A. Konfortov, F. Rivero, A. T. Bankier, R. Lehmann, N. Hamlin, R. Davies, P. Gaudet, P. Fey, K. Pilcher, G. Chen, D. Saunders, E. Sodergren, P. Davis, A. Kerhornou, X. Nie, N. Hall, C. Anjard, L. Hemphill, N. Bason, P. Farbrother, B. Desany, E. Just, T. Morio, R. Rost, C. Churcher, J. Cooper, S. Haydock, N. van Driessche, A. Cronin, I. Goodhead, D. Muzny, T. Mourier, A. Pain, M. Lu, D. Harper, R. Lindsay, H. Hauser, K. James, M. Quiles, M. Madan Babu, T. Saito, C. Buchrieser, A. Wardroper, M. Felder, M. Thangavelu, D. Johnson, A. Knights, H. Loulseged, K. Mungall, K. Oliver, C. Price, M. A. Quail, H. Urushihara, J. Hernandez, E. Rabbinowitsch, D. Steffen, M. Sanders, J. Ma, Y. Kohara, S. Sharp, M. Simmonds, S. Spiegler, A. Tivey, S. Sugano, B. White, D. Walker, J. Woodward, T. Winckler, Y. Tanaka, G. Shaulsky, M. Schleicher, G. Weinstock, A. Rosenthal, E. C. Cox, R. L. Chisholm, R. Gibbs, W. F. Loomis, M. Platzer, R. R. Kay, J. Williams, P. H. Dear, A. A. Noegel, B. Barrell and A. Kuspa, The genome of the social amoeba *Dictyostelium discoideum*, *Nature*, 2005, **435**, 43–57.
88. M. Volkening and I. Hoffmann, Involvement of human MCM8 in prereplication complex assembly by recruiting hcdc6 to chromatin, *Mol. Cell. Biol.*, 2005, **25**, 1560–1568.
89. J. J. Blow and R. A. Laskey, Initiation of DNA replication in nuclei and purified DNA by a cell-free extract of *Xenopus* eggs, *Cell*, 1986, **47**, 577–587.
90. M. Lutzmann and M. Mechali, MCM9 binds Cdt1 and is required for the assembly of prereplication complexes, *Mol. Cell*, 2008, **31**, 190–200.
91. D. Maiorano, O. Cuvier, E. Danis and M. Mechali, MCM8 is an MCM2-7-related protein that functions as a DNA helicase during replication elongation and not initiation, *Cell*, 2005, **120**, 315–328.

92. J. Gregan, K. Lindner, L. Brimage, R. Franklin, M. Namdar, E. A. Hart, S. J. Aves and S. E. Kearsey, Fission yeast Cdc23/Mcm10 functions after pre-replicative complex formation to promote Cdc45 chromatin binding, *Mol. Biol. Cell*, 2003, **14**, 3876–3887.
93. J. A. Wohlschlegel, S. K. Dhar, T. A. Prokhorova, A. Dutta and J. C. Walter, *Xenopus* Mcm10 binds to origins of DNA replication after Mcm2-7 and stimulates origin binding of Cdc45, *Mol. Cell*, 2002, **9**, 233–240.
94. L. Homesley, M. Lei, Y. Kawasaki, S. Sawyer, T. Christensen and B. K. Tye, Mcm10 and the MCM2-7 complex interact to initiate DNA synthesis and to release replication factors from origins, *Genes Dev.*, 2000, **14**, 913–926.
95. S. J. Aves, N. Tongue, A. J. Foster and E. A. Hart, The essential *Schizosaccharomyces pombe* cdc23 DNA replication gene shares structural and functional homology with the *Saccharomyces cerevisiae* DNA43 (MCM10) gene, *Curr. Genet.*, 1998, **34**, 164–171.
96. A. M. Merchant, Y. Kawasaki, Y. Chen, M. Lei and B. K. Tye, A lesion in the DNA replication initiation factor Mcm10 induces pausing of elongation forks through chromosomal replication origins in *Saccharomyces cerevisiae*, *Mol. Cell. Biol.*, 1997, **17**, 3261–3271.
97. R. M. Ricke and A. K. Bielinsky, Mcm10 regulates the stability and chromatin association of DNA polymerase-alpha, *Mol. Cell*, 2004, **16**, 173–185.
98. J. H. Park, S. W. Bang, Y. Jeon, S. Kang and D. S. Hwang, Knockdown of human MCM10 exhibits delayed and incomplete chromosome replication, *Biochem. Biophys. Res. Commun.*, 2008, **365**, 575–582.
99. W. Zhu, C. Ukomadu, S. Jha, T. Senga, S. K. Dhar, J. A. Wohlschlegel, L. K. Nutt, S. Kornbluth and A. Dutta, Mcm10 and And-1/CTF4 recruit DNA polymerase alpha to chromatin for initiation of DNA replication, *Genes Dev.*, 2007, **21**, 2288–2299.
100. K. Fien and J. Hurwitz, Fission yeast Mcm10p contains primase activity, *J. Biol. Chem.*, 2006, **281**, 22248–22260.
101. S. Das-Bradoo, R. M. Ricke and A. K. Bielinsky, Interaction between PCNA and diubiquitinated Mcm10 is essential for cell growth in budding yeast, *Mol. Cell. Biol.*, 2006, **26**, 4806–4817.
102. A. L. Okorokov, A. Waugh, J. Hodgkinson, A. Murthy, H. K. Hong, E. Leo, M. B. Sherman, K. Stoeber, E. V. Orlova and G. H. Williams, Hexameric ring structure of human MCM10 DNA replication factor, *EMBO Rep.*, 2007, **8**, 925–930.
103. N. Y. Jung, W. J. Bae, J. H. Chang, Y. C. Kim and Y. Cho, Cloning, expression, purification, crystallization and preliminary X-ray diffraction analysis of the central zinc-binding domain of the human Mcm10 DNA-replication factor, *Acta Crystallogr. Sect. F: Struct. Biol. Cryst. Commun.*, 2008, **64**, 495–497.
104. C. R. Cook, G. Kung, F. C. Peterson, B. F. Volkman and M. Lei, A novel zinc finger is required for Mcm10 homocomplex assembly, *J. Biol. Chem.*, 2003, **278**, 36051–36058.

105. R. M. Ricke and A. K. Bielinsky, A conserved Hsp10-like domain in Mcm10 is required to stabilize the catalytic subunit of DNA polymerase-alpha in budding yeast, *J. Biol. Chem.*, 2006, **281**, 18414–18425.
106. H. Plchova, F. Hartung and H. Puchta, Biochemical characterization of an exonuclease from Arabidopsis thaliana reveals similarities to the DNA exonuclease of the human Werner syndrome protein, *J. Biol. Chem.*, 2003, **278**, 44128–44138.
107. I. D. Hickson, RecQ helicases: caretakers of the genome, *Nat. Rev. Cancer*, 2003, **3**, 169–178.
108. L. S. Cox and R. G. Faragher, From old organisms to new molecules: integrative biology and therapeutic targets in accelerated human ageing, *Cell. Mol. Life Sci.*, 2007, **64**, 2620–2641.
109. B. A. Kudlow, B. K. Kennedy and R. J. Monnat Jr., Werner and Hutchinson-Gilford progeria syndromes: mechanistic basis of human progeroid diseases, *Nat. Rev. Mol. Cell Biol.*, 2007, **8**, 394–404.
110. I. D. Hickson, S. L. Davies, J. L. Li, N. C. Levitt, P. Mohaghegh, P. S. North and L. Wu, Role of the Bloom's syndrome helicase in maintenance of genome stability, *Biochem. Soc. Trans.*, 2001, **29**, 201–204.
111. V. A. Bohr, Rising from the RecQ-age: the role of human RecQ helicases in genome maintenance, *Trends Biochem. Sci.*, 2008, **33**, 609–620.
112. J. M. Sidorova, Roles of the Werner syndrome RecQ helicase in DNA replication, *DNA Repair (Amst.)*, 2008, **7**, 1776–1786.
113. A. M. Rodriguez-Lopez, D. A. Jackson, F. Iborra and L. S. Cox, Asymmetry of DNA replication fork progression in Werner's syndrome, *Aging Cell*, 2002, **1**, 30–39.
114. A. M. Rodriguez-Lopez, D. A. Jackson, J. O. Nehlin, F. Iborra, A. V. Warren and L. S. Cox, Characterisation of the interaction between WRN, the helicase/exonuclease defective in progeroid Werner's syndrome, and an essential replication factor, PCNA, *Mech. Ageing Dev.*, 2003, **124**, 167–174.
115. A. Constantinou, M. Tarsounas, J. K. Karow, R. M. Brosh, V. A. Bohr, I. D. Hickson and S. C. West, Werner's syndrome protein (WRN) migrates Holliday junctions and co-localizes with RPA upon replication arrest, *EMBO Rep.*, 2000, **1**, 80–84.
116. M. Poot, H. Hoehn, T. M. Runger and G. M. Martin, Impaired S-phase transit of Werner syndrome cells expressed in lymphoblastoid cell lines, *Exp. Cell Res.*, 1992, **202**, 267–273.
117. J. M. Sidorova, N. Li, A. Folch and R. J. Monnat Jr., The RecQ helicase WRN is required for normal replication fork progression after DNA damage or replication fork arrest, *Cell Cycle*, 2008, **7**, 796–807.
118. A. M. Rodriguez-Lopez, M. C. Whitby, C. M. Borer, M. A. Bachler and L. S. Cox, Correction of proliferation and drug sensitivity defects in the progeroid Werner's syndrome by Holliday junction resolution, *Rejuvenation Res.*, 2007, **10**, 27–40.
119. M. D. Gray, J. C. Shen, A. S. Kamath-Loeb, A. Blank, B. L. Sopher, G. M. Martin, J. Oshima and L. A. Loeb, The Werner syndrome protein is a DNA helicase, *Nat. Genet.*, 1997, **17**, 100–103.

120. A. Machwe, L. Xiao and D. K. Orren, TRF2 recruits the Werner syndrome (WRN) exonuclease for processing of telomeric DNA, *Oncogene*, 2004, **23**, 149–156.
121. P. R. Laud, A. S. Multani, S. M. Bailey, L. Wu, J. Ma, C. Kingsley, M. Lebel, S. Pathak, R. A. DePinho and S. Chang, Elevated telomere-telomere recombination in WRN-deficient, telomere dysfunctional cells promotes escape from senescence and engagement of the ALT pathway, *Genes Dev.*, 2005, **19**, 2560–2570.
122. L. M. Pirzio, P. Pichierri, M. Bignami and A. Franchitto, Werner syndrome helicase activity is essential in maintaining fragile site stability, *J. Cell Biol.*, 2008, **180**, 305–314.
123. P. L. Opresko, P. A. Mason, E. R. Podell, M. Lei, I. D. Hickson, T. R. Cech and V. A. Bohr, POT1 stimulates RecQ helicases WRN and BLM to unwind telomeric DNA substrates, *J. Biol. Chem.*, 2005, **280**, 32069–32080.
124. A. Machwe, L. Xiao, J. Groden and D. K. Orren, The Werner and Bloom syndrome proteins catalyze regression of a model replication fork, *Biochemistry*, 2006, **45**, 13939–13946.
125. A. Machwe, L. Xiao, R. G. Lloyd, E. Bolt and D. K. Orren, Replication fork regression *in vitro* by the Werner syndrome protein (WRN): holliday junction formation, the effect of leading arm structure and a potential role for WRN exonuclease activity, *Nucleic Acids Res.*, 2007, **35**, 5729–5747.
126. S. Huang, B. Li, M. D. Gray, J. Oshima, I. S. Mian and J. Campisi, The premature ageing syndrome protein, WRN, is a $3' \rightarrow 5'$ exonuclease, *Nat. Genet.*, 1998, **20**, 114–116.
127. S. Huang, S. Beresten, B. Li, J. Oshima, N. A. Ellis and J. Campisi, Characterization of the human and mouse WRN $3' \rightarrow 5'$ exonuclease, *Nucleic Acids Res.*, 2000, **28**, 2396–2405.
128. Y. Xue, G. C. Ratcliff, H. Wang, P. R. Davis-Searles, M. D. Gray, D. A. Erie and M. R. Redinbo, A minimal exonuclease domain of WRN forms a hexamer on DNA and possesses both $3'$–$5'$ exonuclease and $5'$-protruding strand endonuclease activities, *Biochemistry*, 2002, **41**, 2901–2912.
129. B. Li, N. Conway, S. Navarro and L. Comai, A conserved and species-specific functional interaction between the Werner syndrome-like exonuclease atWEX and the Ku heterodimer in *Arabidopsis*, *Nucleic Acids Res.*, 2005, **33**, 6861–6867.
130. R. D. Saunders, I. Boubriak, D. J. Clancy and L. S. Cox, Identification and characterization of a *Drosophila* ortholog of WRN exonuclease that is required to maintain genome integrity, *Aging Cell*, 2008, **7**, 418–425.
131. I. Boubriak, P. A. Mason, D. J. Clancy, J. Dockray, R. D. Saunders and L. S. Cox, DmWRNexo is a $3'$–$5'$ exonuclease: phenotypic and biochemical characterization of mutants of the *Drosophila* orthologue of human WRN exonuclease, *Biogerontology*, 2009, **10**, 267–277.
132. J. J. Perry, S. M. Yannone, L. G. Holden, C. Hitomi, A. Asaithamby, S. Han, P. K. Cooper, D. J. Chen and J. A. Tainer, WRN exonuclease structure and molecular mechanism imply an editing role in DNA end processing, *Nat. Struct. Mol. Biol.*, 2006, **13**, 414–422.

133. S. Sharma, J. A. Sommers, H. C. Driscoll, L. Uzdilla, T. M. Wilson and R. M. Brosh Jr., The exonucleolytic and endonucleolytic cleavage activities of human exonuclease 1 are stimulated by an interaction with the carboxyl-terminal region of the Werner syndrome protein, *J. Biol. Chem.*, 2003, **278**, 23487–23496.
134. B. Li, S. Navarro, N. Kasahara and L. Comai, Identification and biochemical characterization of a Werner's syndrome protein complex with Ku70/80 and poly(ADP-ribose) polymerase-1, *J. Biol. Chem.*, 2004, **279**, 13659–13667.
135. B. Li and L. Comai, Requirements for the nucleolytic processing of DNA ends by the Werner syndrome protein-Ku70/80 complex, *J. Biol. Chem.*, 2001, **276**, 9896–9902.
136. B. Li and L. Comai, Functional interaction between Ku and the Werner syndrome protein in DNA end processing, *J. Biol. Chem.*, 2000, **275**, 28349–28352.
137. P. L. Opresko, J. P. Laine, R. M. Brosh Jr, M. M. Seidman and V. A. Bohr, Coordinate action of the helicase and 3′ to 5′ exonuclease of Werner syndrome protein, *J. Biol. Chem.*, 2001, **276**, 44677–44687.
138. P. L. Opresko, M. Otterlei, J. Graakjaer, P. Bruheim, L. Dawut, S. Kolvraa, A. May, M. M. Seidman and V. A. Bohr, The Werner syndrome helicase and exonuclease cooperate to resolve telomeric D loops in a manner regulated by TRF1 and TRF2, *Mol. Cell*, 2004, **14**, 763–774.
139. K. Kitano, N. Yoshihara and T. Hakoshima, Crystal structure of the HRDC domain of human Werner syndrome protein, WRN, *J. Biol. Chem.*, 2007, **282**, 2717–2728.
140. J. S. Hu, H. Feng, W. Zeng, G. X. Lin and X. G. Xi, Solution structure of a multifunctional DNA- and protein-binding motif of human Werner syndrome protein, *Proc. Natl. Acad. Sci. U.S.A.*, 2005, **102**, 18379–18384.
141. L. Wu, K. Lung Chan, C. Ralf, D. A. Bernstein, P. L. Garcia, V. A. Bohr, A. Vindigni, P. Janscak, J. L. Keck and I. D. Hickson, The HRDC domain of BLM is required for the dissolution of double Holliday junctions, *EMBO J.*, 2005, **24**, 2679–2687.
142. J. K. Karow, R. H. Newman, P. S. Freemont and I. D. Hickson, Oligomeric ring structure of the Bloom's syndrome helicase, *Curr. Biol.*, 1999, **9**, 597–600.
143. J. R. Walker, R. A. Corpina and J. Goldberg, Structure of the Ku heterodimer bound to DNA and its implications for double-strand break repair, *Nature*, 2001, **412**, 607–614.
144. T. S. Krishna, X. P. Kong, S. Gary, P. M. Burgers and J. Kuriyan, Crystal structure of the eukaryotic DNA polymerase processivity factor PCNA, *Cell*, 1994, **79**, 1233–1243.
145. J. M. Gulbis, Z. Kelman, J. Hurwitz, M. O'Donnell and J. Kuriyan, Structure of the C-terminal region of p21(WAF1/CIP1) complexed with human PCNA, *Cell*, 1996, **87**, 297–306.
146. Z. Liu, M. J. Macias, M. J. Bottomley, G. Stier, J. P. Linge, M. Nilges, P. Bork and M. Sattler, The three-dimensional structure of the HRDC

domain and implications for the Werner and Bloom syndrome proteins, *Structure Fold. Des.*, 1999, **7**, 1557–1566.

147. M. D. Huber, M. L. Duquette, J. C. Shiels and N. Maizels, A conserved G4 DNA binding domain in RecQ family helicases, *J. Mol. Biol.*, 2006, **358**, 1071–1080.
148. J. E. Celis, P. Madsen, S. Nielsen and A. Celis, Nuclear patterns of cyclin (PCNA) antigen distribution subdivide S-phase in cultured cells–some applications of PCNA antibodies, *Leuk. Res.*, 1986, **10**, 237–249.
149. J. E. Celis, P. Madsen, A. Celis, H. V. Nielsen and B. Gesser, Cyclin (PCNA, auxiliary protein of DNA polymerase delta) is a central component of the pathway(s) leading to DNA replication and cell division, *FEBS Lett.*, 1987, **220**, 1–7.
150. O. W. Kamel, D. P. LeBrun, R. E. Davis, G. J. Berry and R. A. Warnke, Growth fraction estimation of malignant lymphomas in formalin-fixed paraffin-embedded tissue using anti-PCNA/Cyclin 19A2. Correlation with Ki-67 labeling, *Am. J. Pathol.*, 1991, **138**, 1471–1477.
151. P. Schurtenberger, S. U. Egelhaaf, R. Hindges, G. Maga, Z. O. Jonsson, R. P. May, O. Glatter and U. Hubscher, The solution structure of functionally active human proliferating cell nuclear antigen determined by small-angle neutron scattering, *J. Mol. Biol.*, 1998, **275**, 123–132.
152. X. -P. Kong, R. Onrsut, M. O'Donnell and J. Kuriyan, Three-dimensional structure of the beta subunit of *E. coli* DNA polymerase III holoenzyme: a sliding DNA clamp, *Cell*, 1992, **69**, 425–437.
153. S. Hata, H. Kouchi, Y. Tanaka, E. Minami, T. Matsumoto, I. Suzuki and J. Hashimoto, Identification of carrot cDNA clones encoding a second putative proliferating cell nuclear antigen, DNA polymerase delta auxiliary protein, *Eur. J. Biochem.*, 1992, **203**, 367–371.
154. J. L. Li, A. V. Warren and L. S. Cox, Identification of a second proliferating cell nuclear antigen in the human malarial pathogen *Plasmodium falciparum*, *Int. J. Parasitol.*, 2002, **32**, 1683–1692.
155. S. Patterson, C. Whittle, C. Robert and D. Chakrabarti, Molecular characterization and expression of an alternate proliferating cell nuclear antigen homologue, PfPCNA2, in *Plasmodium falciparum*, *Biochem. Biophys. Res. Commun.*, 2002, **298**, 371–376.
156. G. Kontopidis, S. Y. Wu, D. I. Zheleva, P. Taylor, C. McInnes, D. P. Lane, P. M. Fischer and M. D. Walkinshaw, Structural and biochemical studies of human proliferating cell nuclear antigen complexes provide a rationale for cyclin association and inhibitor design, *Proc. Natl. Acad. Sci. U.S.A.*, 2005, **102**, 1871–1876.
157. N. Yao, J. Hurwitz and M. O'Donnell, Dynamics of beta and proliferating cell nuclear antigen sliding clamps in traversing DNA secondary structure, *J. Biol. Chem.*, 2000, **275**, 1421–1432.
158. L. S. Cox, Who binds wins: competition for PCNA rings out cell-cycle changes, *Trends Cell Biol.*, 1997, **7**, 493–498.
159. E. Warbrick, The puzzle of PCNA's many partners, *Bioessays*, 2000, **22**, 997–1006.

160. Z. O. Jonsson and U. Hubscher, Proliferating cell nuclear antigen: more than a clamp for DNA polymerases, *Bioessays*, 1997, **19**, 967–975.
161. L. Meslet-Cladiere, C. Norais, J. Kuhn, J. Briffotaux, J. W. Sloostra, E. Ferrari, U. Hubscher, D. Flament and H. Myllykallio, A novel proteomic approach identifies new interaction partners for proliferating cell nuclear antigen, *J. Mol. Biol.*, 2007, **372**, 1137–1148.
162. J. Majka and P. M. Burgers, The PCNA-RFC families of DNA clamps and clamp loaders, *Prog. Nucleic Acid Res. Mol. Biol.*, 2004, **78**, 227–260.
163. P. Kanellis, R. Agyei and D. Durocher, Elg1 forms an alternative PCNA-interacting RFC complex required to maintain genome stability, *Curr. Biol.*, 2003, **13**, 1583–1595.
164. G. O. Bylund and P. M. Burgers, Replication protein A-directed unloading of PCNA by the Ctf18 cohesion establishment complex, *Mol. Cell. Biol.*, 2005, **25**, 5445–5455.
165. J. Kim, K. Robertson, K. J. Mylonas, F. C. Gray, I. Charapitsa and S. A. MacNeill, Contrasting effects of Elg1-RFC and Ctf18-RFC inactivation in the absence of fully functional RFC in fission yeast, *Nucleic Acids Res.*, 2005, **33**, 4078–4089.
166. S. L. Kazmirski, Y. Zhao, G. D. Bowman, M. O'Donnell and J. Kuriyan, Out-of-plane motions in open sliding clamps: molecular dynamics simulations of eukaryotic and archaeal proliferating cell nuclear antigen, *Proc. Natl. Acad. Sci. U.S.A.*, 2005, **102**, 13801–13806.
167. G. Maga, R. Mossi, R. Fischer, M. W. Berchtold and U. Hubscher, Phosphorylation of the PCNA binding domain of the large subunit of replication factor C by Ca2+/calmodulin-dependent protein kinase II inhibits DNA synthesis, *Biochemistry*, 1997, **36**, 5300–5310.
168. G. Maga and U. Hubscher, DNA polymerase ε interacts with proliferating cell nuclear antigen in primer recognition and elongation, *Biochemistry*, 1995, **34**, 891–901.
169. P. M. Burgers, Mammalian cyclin/PCNA (DNA polymerase delta auxiliary protein) stimulates processive DNA synthesis by yeast DNA polymerase III, *Nucleic Acids Res.*, 1988, **16**, 6297–6307.
170. L. D. Langston and M. O'Donnell, DNA polymerase delta is highly processive with PCNA and undergoes collision release upon completing DNA, *J. Biol. Chem.*, 2008, **106**, 6031–6038.
171. P. Garg and P. M. Burgers, DNA polymerases that propagate the eukaryotic DNA replication fork, *Crit. Rev. Biochem. Mol. Biol.*, 2005, **40**, 115–128.
172. E. Warbrick, D. P. Lane, D. M. Glover and L. S. Cox, A small peptide inhibitor of DNA replication defines the site of interaction between the cyclin-dependent kinase inhibitor p21WAF1 and proliferating cell nuclear antigen, *Curr. Biol.*, 1995, **5**, 275–282.
173. E. Warbrick, D. P. Lane, D. M. Glover and L. S. Cox, Homologous regions of Fen1 and $p21^{Cip1}$ compete for binding to the same site on PCNA: a potential mechanism to co-ordinate DNA replication and repair, *Oncogene*, 1997, **14**, 2313–2321.

174. E. Warbrick, PCNA binding through a conserved motif, *Bioessays*, 1998, **20**, 195–199.
175. M. Ducoux, S. Urbach, G. Baldacci, U. Hubscher, S. Koundrioukoff, J. Christensen and P. Hughes, Mediation of proliferating cell nuclear antigen (PCNA)-dependent DNA replication through a conserved p21(Cip1)-like PCNA-binding motif present in the third subunit of human DNA polymerase delta, *J. Biol. Chem.*, 2001, **276**, 49258–49266.
176. J. B. Bruning and Y. Shamoo, Structural and thermodynamic analysis of human PCNA with peptides derived from DNA polymerase-δ p66 subunit and flap endonuclease-1, *Structure*, 2004, **12**, 2209–2219.
177. L. R. Hiraoka, J. J. Harrington, D. S. Gerhard, M. R. Lieber and C. L. Hsieh, Sequence of human FEN-1, a structure-specific endonuclease, and chromosomal localization of the gene (FEN1) in mouse and human, *Genomics*, 1995, **25**, 220–225.
178. X. V. Gomes and P. M. Burgers, Two modes of FEN1 binding to PCNA regulated by DNA, *EMBO J.*, 2000, **19**, 3811–3821.
179. S. Sakurai, K. Kitano, H. Yamaguchi, K. Hamada, K. Okada, K. Fukuda, M. Uchida, E. Ohtsuka, H. Morioka and T. Hakoshima, Structural basis for recruitment of human flap endonuclease 1 to PCNA, *EMBO J.*, 2004, **24**, 683–693.
180. F. Storici, G. Henneke, E. Ferrari, D. A. Gordenin, U. Hubscher and M. A. Resnick, The flexible loop of human FEN1 endonuclease is required for flap cleavage during DNA replication and repair, *EMBO J.*, 2002, **21**, 5930–5942.
181. D. S. Levin, A. E. McKenna, T. A. Motycka, Y. Matsumoto and A. E. Tomkinson, Interaction between PCNA and DNA ligase I is critical for joining of Okazaki fragments and long-patch base-excision repair, *Curr. Biol.*, 2000, **10**, 919–922.
182. E. W. Refsland and D. M. Livingston, Interactions among DNA ligase I, the flap endonuclease and proliferating cell nuclear antigen in the expansion and contraction of CAG repeat tracts in yeast, *Genetics*, 2005, **171**, 923–934.
183. L. A. Henricksen, J. Veeraraghavan, D. R. Chafin and R. A. Bambara, DNA ligase I competes with FEN1 to expand repetitive DNA sequences *in vitro*, *J. Biol. Chem.*, 2002, **277**, 22361–22369.
184. L. Zheng, H. Dai, J. Qiu, Q. Huang and B. Shen, Disruption of the FEN-1/PCNA interaction results in DNA replication defects, pulmonary hypoplasia, pancytopenia, and newborn lethality in mice, *Mol. Cell. Biol.*, 2007, **27**, 3176–3186.
185. I. Ivanov, B. R. Chapados, J. A. McCammon and J. A. Tainer, Proliferating cell nuclear antigen loaded onto double-stranded DNA: dynamics, minor groove interactions and functional implications, *Nucleic Acids Res.*, 2006, **34**, 6023–6033.
186. G. Maga, G. Villani, V. Tillement, M. Stucki, G. A. Locatelli, I. Frouin, S. Spadari and U. Hubscher, Okazaki fragment processing: modulation of the strand displacement activity of DNA polymerase delta by the

concerted action of replication protein A, proliferating cell nuclear antigen, and flap endonuclease-1, *Proc. Natl. Acad. Sci. U.S.A.*, 2001, **98**, 14298–14303.

187. S. Koundrioukoff, Z. O. Jonsson, S. Hasan, R. N. de Jong, P. C. van der Vliet, M. O. Hottiger and U. Hubscher, A direct interaction between proliferating cell nuclear antigen (PCNA) and Cdk2 targets PCNA-interacting proteins for phosphorylation, *J. Biol. Chem.*, 2000, **275**, 22882–22887.
188. L. Schermelleh, A. Haemmer, F. Spada, N. Rosing, D. Meilinger, U. Rothbauer, M. C. Cardoso and H. Leonhardt, Dynamics of Dnmt1 interaction with the replication machinery and its role in post-replicative maintenance of DNA methylation, *Nucleic Acids Res.*, 2007, **35**, 4301–4312.
189. K. Shibahara and B. Stillman, Replication-dependent marking of DNA by PCNA facilitates CAF-1-coupled inheritance of chromatin, *Cell*, 1999, **96**, 575–585.
190. Z. Zhang, K. Shibahara and B. Stillman, PCNA connects DNA replication to epigenetic inheritance in yeast, *Nature*, 2000, **408**, 221–225.
191. R. A. Poot, L. Bozhenok, D. L. van den Berg, S. Steffensen, F. Ferreira, M. Grimaldi, N. Gilbert, J. Ferreira and P. D. Varga-Weisz, The Williams syndrome transcription factor interacts with PCNA to target chromatin remodelling by ISWI to replication foci, *Nat. Cell Biol.*, 2004, **6**, 1236–1244.
192. R. Ko and S. E. Bennett, Physical and functional interaction of human nuclear uracil-DNA glycosylase with proliferating cell nuclear antigen, *DNA Repair (Amst.)*, 2005, **4**, 1421–1431.
193. G. Maga, G. Villani, E. Crespan, U. Wimmer, E. Ferrari, B. Bertocci and U. Hubscher, 8-oxo-guanine bypass by human DNA polymerases in the presence of auxiliary proteins, *Nature*, 2007, **447**, 606–608.
194. G. Maga, G. Blanca, I. Shevelev, I. Frouin, K. Ramadan, S. Spadari, G. Villani and U. Hubscher, The human DNA polymerase lambda interacts with PCNA through a domain important for DNA primer binding and the interaction is inhibited by p21/WAF1/CIP1, *FASEB J.*, 2004, **18**, 1743–1745.
195. M. K. K. Shivji, S. J. Grey, U. P. Strausfeld, R. D. Wood and J. J. Blow, Cip1 inhibits DNA replication but not PCNA-dependent nucleotide excision repair, *Curr. Biol.*, 1994, **4**, 1062–1068.
196. J. M. Pascal, O. V. Tsodikov, G. L. Hura, W. Song, E. A. Cotner, S. Classen, A. E. Tomkinson, J. A. Tainer and T. Ellenberger, A flexible interface between DNA ligase and PCNA supports conformational switching and efficient ligation of DNA, *Mol. Cell*, 2006, **24**, 279–291.
197. L. S. Cox and D. P. Lane, Tumour suppressors, kinases and clamps: how p53 regulates the cell cycle in response to DNA damage, *Bioessays*, 1995, **17**, 501–508.
198. J. M. Bradbury and S. P. Jackson, The complex matter of DNA double-strand break detection, *Biochem. Soc. Trans.*, 2003, **31**, 40–44.

199. E. R. Parrilla-Castellar, S. J. Arlander and L. Karnitz, Dial 9-1-1 for DNA damage: the Rad9-Hus1-Rad1 (9-1-1) clamp complex, *DNA Repair (Amst.)*, 2004, **3**, 1009–1014.
200. Y. Shiomi, A. Shinozaki, D. Nakada, K. Sugimoto, J. Usukura, C. Obuse and T. Tsurimoto, Clamp and clamp loader structures of the human checkpoint protein complexes, Rad9-1-1 and Rad17-RFC, *Genes Cells*, 2002, **7**, 861–868.
201. J. D. Griffith, L. A. Lindsey-Boltz and A. Sancar, Structures of the human Rad17-replication factor C and checkpoint Rad 9-1-1 complexes visualized by glycerol spray/low voltage microscopy, *J. Biol. Chem.*, 2002, **277**, 15233–15236.
202. M. P. Thelen, C. Venclovas and K. Fidelis, A sliding clamp model for the Rad1 family of cell cycle checkpoint proteins, *Cell*, 1999, **96**, 769–770.
203. M. A. Burtelow, P. M. Roos-Mattjus, M. Rauen, J. R. Babendure and L. M. Karnitz, Reconstitution and molecular analysis of the hRad9-hHus1-hRad1 (9-1-1) DNA damage responsive checkpoint complex, *J. Biol. Chem.*, 2001, **276**, 25903–25909.
204. R. E. Jones, J. R. Chapman, C. Puligilla, J. M. Murray, A. M. Car, C. C. Ford and H. D. Lindsay, XRad17 is required for the activation of XChk1 but not XCds1 during checkpoint signaling in *Xenopus*, *Mol. Biol. Cell*, 2003, **14**, 3898–3910.
205. C. M. Green, H. Erdjument-Bromage, P. Tempst and N. F. Lowndes, A novel Rad24 checkpoint protein complex closely related to replication factor C, *Curr. Biol.*, 2000, **10**, 39–42.
206. V. P. Bermudez, L. A. Lindsey-Boltz, A. J. Cesare, Y. Maniwa, J. D. Griffith, J. Hurwitz and A. Sancar, Loading of the human 9-1-1 checkpoint complex onto DNA by the checkpoint clamp loader hRad17-replication factor C complex *in vitro*, *Proc. Natl. Acad. Sci. U.S.A.*, 2003, **100**, 1633–1638.
207. J. Majka, S. K. Binz, M. S. Wold and P. M. Burgers, Replication protein A directs loading of the DNA damage checkpoint clamp to 5′-DNA junctions, *J. Biol. Chem.*, 2006, **281**, 27855–27861.
208. S. Delacroix, J. M. Wagner, M. Kobayashi, K. Yamamoto and L. M. Karnitz, The Rad9-Hus1-Rad1 (9-1-1) clamp activates checkpoint signaling *via* TopBP1, *Genes Dev.*, 2007, **21**, 1472–1477.
209. C. E. Helt, W. Wang, P. C. Keng and R. A. Bambara, Evidence that DNA damage detection machinery participates in DNA repair, *Cell Cycle*, 2005, **4**, 529–532.
210. E. Friedrich-Heineken, M. Toueille, B. Tannler, C. Burki, E. Ferrari, M. O. Hottiger and U. Hubscher, The two DNA clamps Rad9/Rad1/Hus1 complex and proliferating cell nuclear antigen differentially regulate flap endonuclease 1 activity, *J. Mol. Biol.*, 2005, **353**, 980–989.
211. M. Toueille, N. El-Andaloussi, I. Frouin, R. Freire, D. Funk, I. Shevelev, E. Friedrich-Heineken, G. Villani, M. O. Hottiger and U. Hubscher, The human Rad9/Rad1/Hus1 damage sensor clamp interacts with DNA

polymerase beta and increases its DNA substrate utilisation efficiency: implications for DNA repair, *Nucleic Acids Res.*, 2004, **32**, 3316–3324.

212. W. Wang, L. A. Lindsey-Boltz, A. Sancar and R. A. Bambara, Mechanism of stimulation of human DNA ligase I by the Rad9-rad1-Hus1 checkpoint complex, *J. Biol. Chem.*, 2006, **281**, 20865–20872.
213. E. Smirnova, M. Toueille, E. Markkanen and U. Hubscher, The human checkpoint sensor and alternative DNA clamp Rad9-Rad1-Hus1 modulates the activity of DNA ligase I, a component of the long-patch base excision repair machinery, *Biochem. J*, 2005, **389**, 13–17.
214. T. Ellenberger and A. E. Tomkinson, Eukaryotic DNA ligases: structural and functional insights, *Annu. Rev. Biochem.*, 2008, **77**, 313–338.
215. G. Shi, D. Y. Chang, C. C. Cheng, X. Guan, C. Venclovas and A. L. Lu, Physical and functional interactions between MutY glycosylase homologue (MYH) and checkpoint proteins Rad9-Rad1-Hus1, *Biochem. J*, 2006, **400**, 53–62.
216. P. Garg, C. M. Stith, J. Majka and P. M. Burgers, Proliferating cell nuclear antigen promotes translesion synthesis by DNA polymerase zeta, *J. Biol. Chem.*, 2005, **280**, 23446–23450.
217. S. Sabbioneda, B. K. Minesinger, M. Giannattasio, P. Plevani, M. Muzi-Falconi and S. Jinks-Robertson, The 9-1-1 checkpoint clamp physically interacts with polζ and is partially required for spontaneous polζ-dependent mutagenesis in *Saccharomyces cerevisiae*, *J. Biol. Chem.*, 2005, **280**, 38657–38665.
218. G. Mukherjee, B. Muralidhar, U. D. Bafna, R. A. Laskey and N. Coleman, MCM immunocytochemistry as a first line cervical screening test in developing countries: a prospective cohort study in a regional cancer centre in India, *Br. J. Cancer*, 2007, **96**, 1107–1111.
219. C. Scarpini, V. White, B. Muralidhar, A. Patterson, N. Hickey, N. Singh, J. Mullerat, M. Winslet, R. J. Davies, M. L. Phillips, P. Stacey, R. A. Laskey, R. Miller, M. Nathan and N. Coleman, Improved screening for anal neoplasia by immunocytochemical detection of minichromosome maintenance proteins, *Cancer Epidemiol. Biomarkers Prev.*, 2008, **17**, 2855–2864.
220. H. Harada, H. Nakagawa, M. Takaoka, J. Lee, M. Herlyn, J. A. Diehl and A. K. Rustgi, Cleavage of MCM2 licensing protein fosters senescence in human keratinocytes, *Cell Cycle*, 2008, **7**, 3534–3538.
221. M. Izumi, K. Yanagi, T. Mizuno, M. Yokoi, Y. Kawasaki, K. Y. Moon, J. Hurwitz, F. Yatagai and F. Hanaoka, The human homolog of *Saccharomyces cerevisiae* Mcm10 interacts with replication factors and dissociates from nuclease-resistant nuclear structures in G(2) phase, *Nucleic Acids Res.*, 2000, **28**, 4769–4777.
222. T. Schwede, J. Kopp, N. Guex and M. C. Peitsch, SWISS-MODEL: An automated protein homology-modeling server, *Nucleic Acids Res.*, 2003, **31**, 3381–3385.

CHAPTER 4

Mechanisms for High Fidelity DNA Replication

STEPHANIE A. NICK McELHINNY, ZACHARY F. PURSELL AND THOMAS A. KUNKEL

Laboratories of Molecular Genetics and Structural Biology, National Institute of Environmental Health Sciences, NIH, DHHS, Research Triangle Park, NC 27709, USA

4.1 General Organization of the Eukaryotic Replication Fork

The human nuclear genome consists of three billion base pairs of duplex DNA. In order to duplicate this astounding number of base pairs in a matter of minutes to hours while maintaining genomic stability over many generations, replication must be both efficient and highly accurate.[1,2] This tremendous task is complicated by the fact that the two strands of DNA are oriented antiparallel to each other, yet nascent DNA is only polymerized in the 5′ to 3′ direction. Thus, coordinated replication of the two DNA strands is intrinsically asymmetric, with the first (leading strand) thought to be replicated in a continuous manner and the other (lagging strand) replicated shortly thereafter as a series of short Okazaki fragments that are subsequently processed and ligated (Figure 4.1A, B). Replication is still further complicated by the fact that DNA is constantly under chemical and physical attack from normal cellular processes and the external environment. These stresses create a wide variety of structurally distinct DNA lesions that can block replication to varying extents and must be overcome. For all of these reasons, complete and accurate replication

Molecular Themes in DNA Replication
Edited by Lynne S. Cox

Published by the Royal Society of Chemistry, www.rsc.org

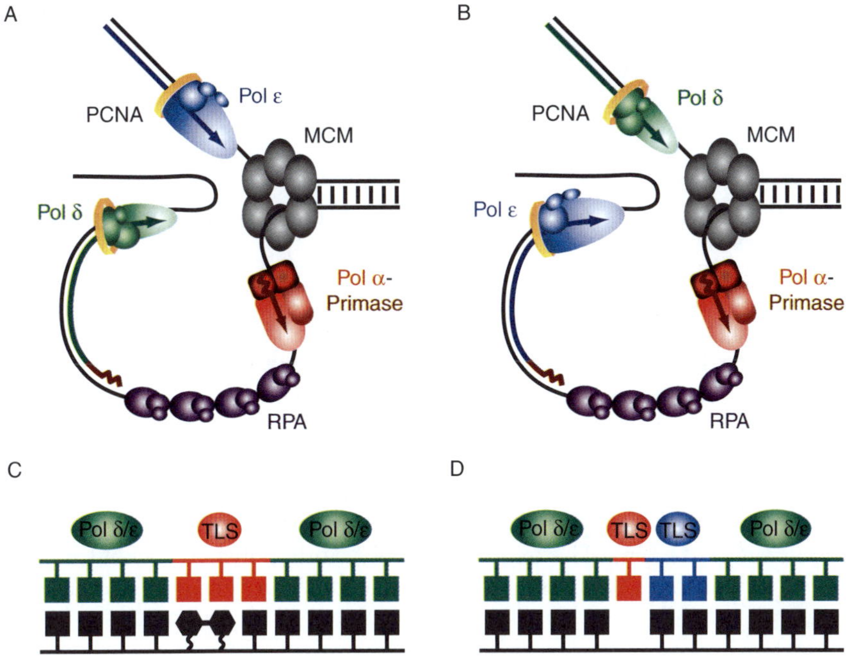

Figure 4.1 **Models of eukaryotic replication.** Models for the replication of undamaged DNA by eukaryotic polymerases in which leading and lagging strand synthesis is performed primarily by (**A**) Pol ε and Pol δ, respectively, or (**B**) Pol δ and Pol ε, respectively. The sliding clamp PCNA is depicted as a gold ring, single-stranded binding protein RPA is purple, and the MCM helicase complex is grey. RNA synthesis on Okazaki fragments by primase (red with black outline) is depicted as a dark red zigzag and DNA synthesis by Pol α (red) as a dark red line. DNA synthesis by Pol δ or Pol ε is shown as a green or blue line, respectively. The subunits of the polymerases, RPA and MCM are shown approximately to scale according to the molecular weight of the human proteins. The images were inspired by and adapted from Figure 1 in ref. 12 and Figure 7 in ref. 13. (**C**) The one-polymerase model of translesion synthesis (TLS) proposes that a single polymerase performs the complete bypass reaction, including insertion opposite the lesion and extension from the non-canonical primer terminal base pair. In the example shown, a replicative polymerase (Pol δ or Pol ε) stalls at a UV-induced thymine dimer—depicted as linked hexagons with damaged bonds (wavy lines) on the template strand (black). A TLS polymerase (e.g. Pol η for a T-T dimer) is recruited and synthesizes across from the lesion and extends the altered primer terminus [translesion synthesis (TLS) shown in red]. The replicative polymerase then returns to continue replication (shown in green). (**D**) The two-polymerase model postulates that, for certain lesions, one TLS polymerase inserts a nucleotide opposite the lesion and a second TLS polymerase extends the distorted primer terminus. In the example shown, a replicative polymerase (Pol δ or Pol ε) stalls at an abasic site (depicted by the absence of a square base). The first TLS polymerase (e.g. Rev1) inserts opposite the abasic site (synthesis shown in red), while a second TLS polymerase (e.g. Pol ζ) then extends the primer (shown in blue) until the replicative polymerase can continue synthesis (shown in green). The template strand is shown in black. See ref. 16–18 for recent reviews; images in panels C and D were adapted from Figure 3 in ref. 91.

of eukaryotic genomes requires the activity of processive and highly accurate polymerases, as well as specialized DNA polymerases that perform lesion bypass (Figure 4.1C, D).

4.2 Eukaryotic Polymerases Involved in Replication

The human genome encodes at least 16 DNA polymerases that are grouped by sequence homology into four families: A, B, X and Y.[3,4] The family X members fill short gaps (Pol β, Pol μ and Pol λ) and insert short non-templated additions (TdT) during DNA repair processes, and are reviewed elsewhere.[5] Family A members include Pol γ, which is responsible for DNA synthesis within mitochondria (reviewed in refs. 6,7; see Chapter 11), Pol ν, which is implicated in interstrand crosslink repair and somatic hypermutation of immunoglobulin genes (for example, see refs. 8,9 and references therein), and Pol θ, which is implicated in somatic hypermutation and translesion synthesis (refs. 10,11 and references therein). The majority of eukaryotic polymerases that participate in replication of normal and damaged DNA are members of families B and Y.

4.2.1 Family B DNA Polymerases

Three members of family B DNA polymerases replicate the vast majority of undamaged DNA in the nuclear genome: Pol α, Pol δ and Pol ε (Figure 4.2A) (reviewed in refs. 12,13). A fourth family B member, Pol ζ, has a specialized role in replicating damaged DNA (see Chapter 7 and refs. 14,15). These enzymes share with all other polymerases a domain (coloured blue in Figure 4.2A) that contains conserved amino acids critical for catalyzing the polymerase reaction. Pol δ and Pol ε, but not Pol α or Pol ζ, also contain a domain (coloured yellow in Figure 4.2A) encoding a 3′ exonuclease that can excise mismatched bases during DNA synthesis to increase replication fidelity. In addition to their catalytic subunits, all four family B enzymes also contain one or more additional subunits that contribute to their functions (Figure 4.2B).

A prime example derives from the fact that all DNA polymerases require a primer strand to initiate DNA synthesis. Thus, DNA replication is initiated at origins of replication and at the beginning of each Okazaki fragment on the lagging strand through *de novo* synthesis of an 8–12 nucleotide RNA primer (Figure 4.1A, B) by the primase subunits of Pol α (Figure 4.2B). Pol α then extends this short RNA chain by adding approximately 20 nucleotides of DNA. These RNA–DNA chains are then used as primers by Pol δ and Pol ε to complete the bulk of DNA synthesis during replication. The division of labour between these two polymerases during leading and lagging strand synthesis (Figure 4.1A, B) has been a question of longstanding interest[12,13] and is considered further below.

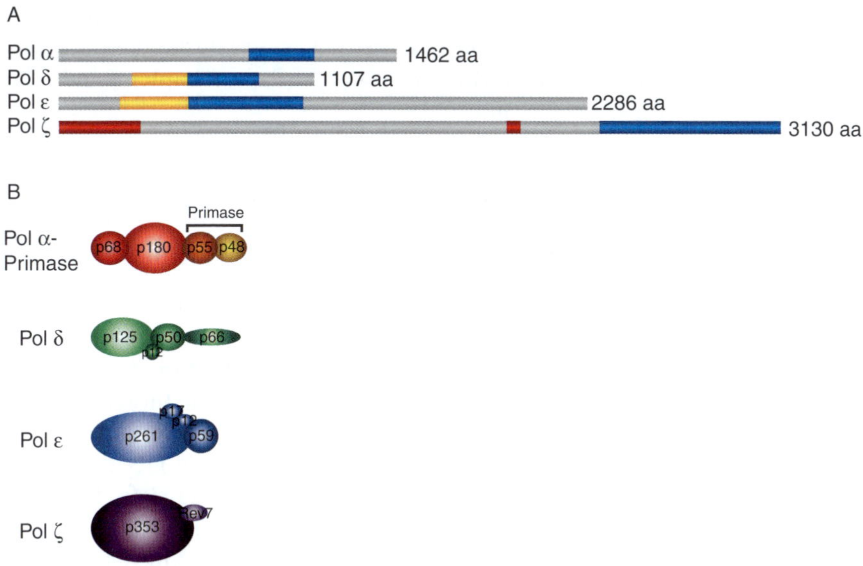

Figure 4.2 **Organization of the B family DNA polymerases.** (**A**) The domain organization of the catalytic subunits of the human B family polymerases is shown with the number of amino acid residues indicated to the right of each polypeptide. Polymerase domains are shown in blue and exonuclease domains in yellow. For Pol ζ, regions of homology with the *S. cerevisiae* Rev3 protein (catalytic subunit of yeast Pol ζ) are shown in red. The image was adapted from Figure 1 in ref. 3. (**B**) Subunit interactions of human B family polymerases are shown as reviewed in refs. 92–94 and to scale according to molecular weight. The image was adapted from Figure 3 in ref. 12.

4.2.2 Family Y DNA Polymerases

Family Y polymerases are implicated in translesion synthesis (TLS), a process by which cells bypass DNA lesions that would otherwise block replication.[16,17] Human family Y members include Pol η, Pol ι, Pol κ and Rev1. As described below (see also Chapter 7), family Y polymerases are relatively inaccurate, such that synthesis by these enzymes is likely to be localized to small patches around lesions to minimize the potential for mutagenic consequences.

There are currently two major models (reviewed in refs. 16–18) to describe the roles of TLS polymerases during replication, depending on the identities of the DNA lesion and the TLS polymerase. In the simplest model (Figure 4.1C), a single translesion polymerase is responsible for bypass, which requires insertion opposite the lesion and extension from the resulting primer terminus. In the ‘two TLS polymerase model’ (Figure 4.1D), one polymerase (*e.g.* Pol η or Pol ι) is proposed to insert a nucleotide opposite a lesion, and a second polymerase (*e.g.* Pol ζ or Pol κ) may then extend the damaged primer-template. Post-translational modification of the proteins involved in TLS, particularly monoubiquitylation of the PCNA[i] clamp (see Chapter 7), is critical for the exchange of replicative and TLS polymerases that occurs during lesion bypass and for the control of TLS. For further details of TLS, see Chapter 7 and other excellent recent reviews on this subject.[16,17]

4.3 Structural Insights into Replication

4.3.1 Structures and Domains of Family B and Y Polymerases

The structural organization of the polymerase domain has been likened to a hand consisting of a palm, fingers, and thumb.[19] This is illustrated in Figure 4.3A, which depicts the structure of bacteriophage RB69 Pol,[20] a family B member that along with ϕ29 Pol[21] currently serve as good structural models for eukaryotic Pol α, Pol δ, Pol ε and Pol ζ, whose crystal structures have not yet been reported. The three subdomains form a cleft that binds DNA, with the polymerase active site located in the palm at the base of the cleft. Family Y polymerases contain an additional 'little finger' subdomain not found in other families [Figure 4.3B, depicting *Sulfolobus solfataricus* (Sso) Dpo4 Pol[22]] that contributes to the TLS capacity of Y family members (reviewed in refs. 23,24).

In addition to their polymerase domains, many polymerases contain other domains that contribute to their functions. For RB69 Pol (Figure 4.3A), Pol δ and Pol ε (Figure 4.2A), this includes a domain harbouring 3′ proofreading exonuclease activity, while Pol α, Pol ζ and family Y polymerases lack intrinsic proofreading activity.

The coding sequences of the catalytic subunits of many DNA polymerases also contain regions that are involved in interactions with their associated subunits (Figure 4.2B) and other replication accessory proteins (*e.g.* PCNA), or regions that can be post-translationally modified (*e.g.* monoubiquitylated) in response to DNA damage (reviewed in refs. 12,13,16,17).

4.3.2 Catalytic Mechanisms of Polymerization and Excision

DNA polymerization is a phosphoryl transfer reaction involving nucleophilic attack by the 3′ hydroxyl of the primer terminus on the α-phosphate of the incoming deoxyribonucleotide triphosphate (dNTP) (see Chapter 1). The products of this reaction are pyrophosphate and a DNA chain increased in length by one nucleotide. The catalytic mechanism[25] is conserved among DNA polymerases regardless of family and begins with binding of a primer-template by the polymerase. The primer terminus to which the dNTP will be added is bound at the polymerase active site, which is comprised of three sequence motifs (A, B and C) that show remarkable structural conservation (Figure 4.3C). Conserved tyrosine residues within the active site are important for activity; yeast strains with a tyrosine to alanine substitution in the SL/MYPS/N motif in region II of Pol δ are inviable and the homologous change in Pol ε

[i] Proliferating cell nuclear antigen

results in formation of microcolonies.[26] The catalytic centre includes three carboxylate residues (one in Motif A and two in Motif C) that, upon dNTP binding, coordinate two divalent metal ions (usually Mg^{2+}). Nucleotide binding results in conformational changes in both the polymerase and the DNA (reviewed in ref. 27). While the specific nature of these changes depends on the polymerase (see, for example, refs. 21,28 and references therein), the dNTP-induced conformational changes result in assembly of an active site with proper geometry for the phosphoryl transfer reaction, which proceeds *via* in-line displacement and results in inversion of the stereochemical configuration of the α-phosphorous atom, as depicted structurally in Figure 4.3D for the family X member Pol λ. Release of pyrophosphate is associated with translocation of the newly formed base pair out of the templating position so that it can serve as the primer terminus of the next cycle of catalysis.[21] (Greater detail on the reaction mechanism can be found in ref. 5 and ref. 29, and references therein.) Exonucleolytic excision of mismatched primer terminal bases proceeds by a mechanism similar to that of polymerization, *via* an in-line displacement involving two metal ions coordinated by three carboxylate residues (reviewed in ref. 30). The transition between polymerization and editing modes requires the primer terminus to travel 40Å between the two active sites and results in a 40° rotation of the helical axis of the DNA duplex.[20]

4.3.3 Replication Machines

The roles played by polymerases *in vivo* are hardly solo performances. Many supporting proteins coordinate with DNA polymerases to achieve efficient and accurate DNA replication (reviewed in refs. 12,13). How these multiple components interact in the context of a 'replication machine' is now being revealed by structural studies of multiprotein replication subassemblies.

An elegant recent example is a cryo-EM structure (Figure 4.4A) of the yeast Pol ε holoenzyme (Figure 4.2B), which revealed a globular catalytic subunit of Pol ε connected to an extended structure consisting of the three additional subunits (Dpb2, Dpb3, and Dpb4).[31] A cleft is apparent in the catalytic subunit that is wide enough to accommodate duplex DNA, and the extended domain formed by the associated subunits faces this cleft in a manner consistent with the reported role of these subunits in DNA interaction. In addition, the dimensions of the holoenzyme complex are consistent with binding a DNA substrate of ~40 nucleotides in length, corresponding to the minimal DNA substrate required for highly processive synthesis by Pol ε. A second example is a structure-based model illustrating how the C-terminus of RB69 Pol interacts with the sliding clamp,[32] a protein that improves the processivity of replicative polymerases (Figure 4.4B; orthologous to eukaryotic PCNA, see Chapters 3 and 7).

These and the structures of other multisubunit replication proteins, such as the single-stranded DNA binding protein complex Replication Protein A (RPA),[33] the RFC clamp loader complex,[34] the GINS complex, which is essential for replication initiation and progression,[35–37] and bacteriophage T7

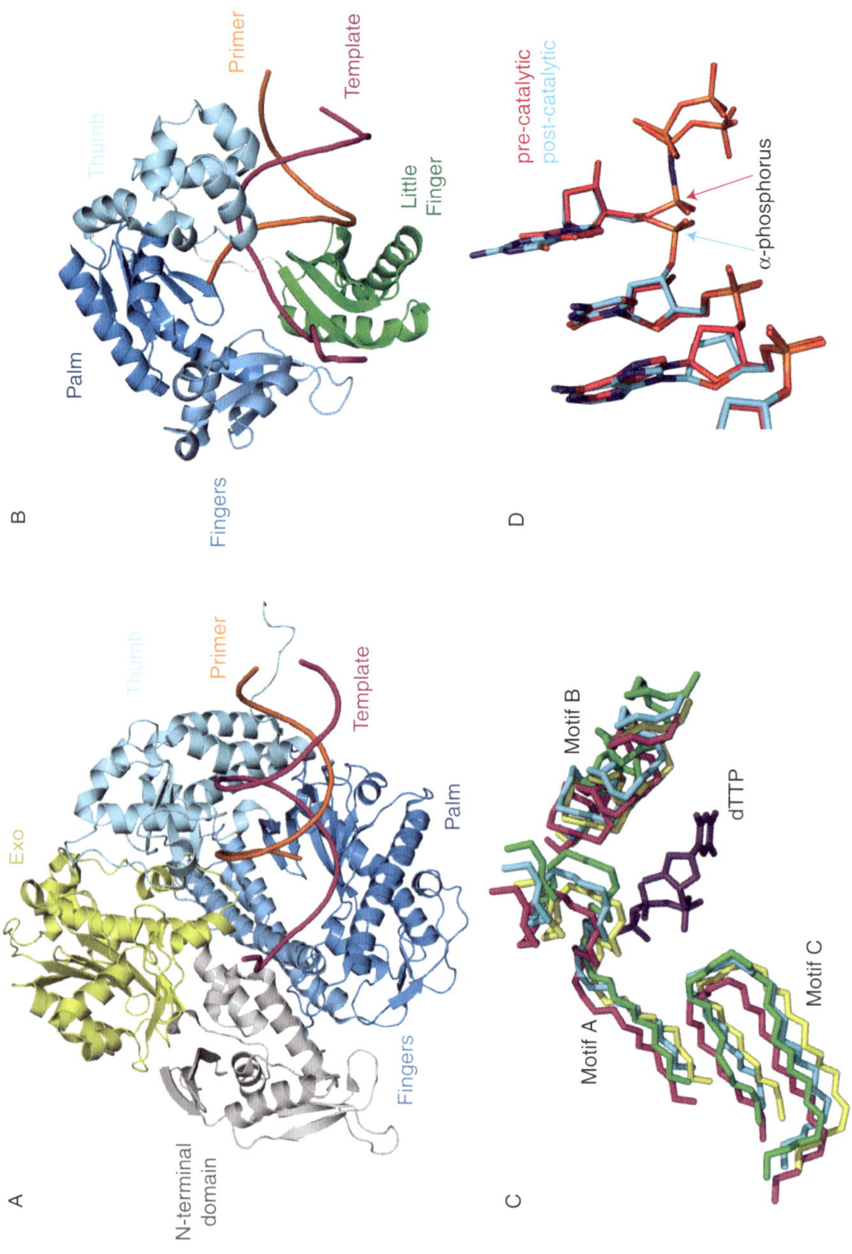
A
N-terminal domain
Exo
Thumb
Primer
Template
Palm
Fingers
B
Palm
Thumb
Primer
Template
Little Finger
Fingers
C
Motif A
Motif B
dTTP
Motif C
D
pre-catalytic
post-catalytic
α-phosphorus

primase-helicase,[38] are highly informative of the mechanism of replication and represent critical steps toward eventually replacing cartoons of replication forks (*e.g.* Figure 4.1) with actual structures of replication machines operating at forks. The potential of this approach is beautifully illustrated by an animated model of a hybrid replisome generated using bacterial and eukaryotic replication protein structures (www.wehi.edu.au/wehi-tv/dna/index.html).

4.4 DNA Replication Fidelity

High fidelity replication of the eukaryotic genome depends on the selectivity of DNA polymerases for inserting correctly paired dNTPs onto properly aligned primer-templates and on the ability of 3′ exonuclease activities to excise polymerization errors during replication. Rare replication errors that escape proofreading can subsequently be corrected by DNA mismatch repair, a subject that is extensively reviewed elsewhere[39–41] (and see Chapter 7). In this section, we describe the error rates of eukaryotic family B and Y members, we consider the main mechanisms by which polymerases avoid or generate these errors, and we discuss the contribution of proofreading and accessory proteins to replication fidelity.

4.4.1 Base Substitution Error Rates

The two most common errors made during DNA synthesis are single base substitutions and single base deletions. The rates at which eukaryotic family B and Y polymerases generate these two types of errors as they copy the *lacZ* coding sequence in M13mp2 DNA *in vitro* are shown in Figure 4.5.

Analysis of the base substitution error rates leads to several interesting conclusions. The three polymerases responsible for the bulk of replication of

---

Figure 4.3 **Structural analysis of the mechanism of polymerization.** The structures of prokaryotic polymerases with homology to eukaryotic DNA polymerases (**A**) RB69 Pol[20] and (**B**) Sso Dpo4 Pol[22] are shown in complex with duplex DNA using coordinates from PDB accession numbers 1IG9 and 1JX4, respectively. In both panels the palm, fingers and thumb subdomains are dark, medium and light blue, respectively, and the primer and template strands are shown in orange and magenta, respectively. The exonuclease domain of RB69 Pol is yellow, and the little finger domain of Dpo4 Pol is green. (**C**) Overlay of Motifs A, B, and C from four B family polymerases: bacteriophage RB69 Pol, green (PDB 1IG9); *Thermococcus gorgonarius* (Tgo) Pol, blue (PDB 1TGO); *Desulfurococcus Tok* (D.Tok) Pol, pink (PDB 1QQC); *Pyrococcus kodakaraensis* (KOD) Pol, yellow (PDB 1WNS). The incoming dTTP from the RB69 Pol structure is shown in purple. The image is adapted from an overlay of A and B family polymerases in ref. 95. (**D**) Overlay of the primer strand and incoming triphosphate from the pre-catalytic structure of Pol λ (magenta; PDB 2PFO) with the primer strand and pyrophosphate from the post-catalytic structure of Pol λ (cyan; PDB 1XSP) illustrating the inversion of the stereochemical configuration of the α-phosphorus atom *via* an in-line displacement mechanism.

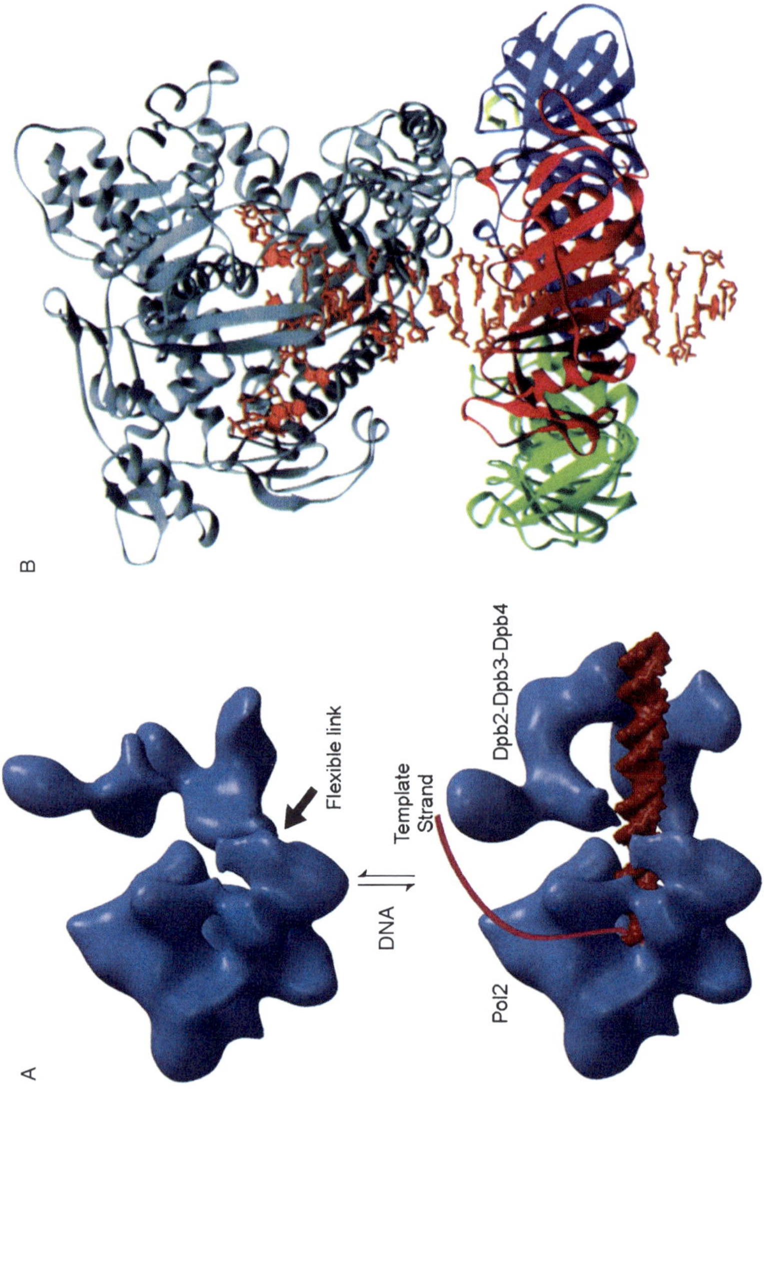
A
Flexible link
DNA
Template
Strand
Pol2
Dpb2-Dpb3-Dpb4
B

the nuclear genome (Pols α, δ and ε) are all more accurate (Figure 4.5A) than the polymerases implicated in TLS (Pols ζ, η and κ; Figure 4.5B). This is true even when the proofreading activities of Pol δ and Pol ε are intentionally inactivated by mutation of their exonuclease active sites in order to determine their inherent nucleotide selectivity in the absence of error correction. The higher fidelity of the major replicative polymerases is consistent with their roles in maintaining genome stability while copying millions of base pairs per genome duplication. Moreover, the lower fidelity of the polymerases implicated in TLS is consistent with their ability to copy damaged templates with distorted helix geometry. The error rates in Figure 4.5 are 'average' rates for errors at many template locations. However, rates for different types of errors can vary widely, in some instances more than 100-fold, depending on the composition of the error (*e.g.* 12 different single base–base mismatches are possible), the sequence context surrounding the error, and the polymerase performing the synthesis (for further details, see refs. 8,27,42,43 and references therein). Potential explanations for these differences in error rates have resulted from investigating the mechanisms responsible for generating base substitutions and insertion–deletion errors ('indels') during DNA synthesis.

4.4.2 Mechanisms Controlling Nucleotide Selectivity

The ability of template-dependent polymerases to select the correct nucleotide for incorporation depends on several parameters. As pointed out by Watson and Crick,[44] this includes base–base hydrogen bonding between correctly paired nucleotides. However this alone does not completely dictate nucleotide selectivity, as many polymerases, including Pol α, Pol δ and Pol ε, have lower base substitution error rates than can be explained simply by free energy differences between correct and incorrect base pairs in aqueous solution (reviewed in ref. 45). It has been proposed that polymerases achieve improved fidelity by partially excluding water from the polymerase active site.[46] This would result in

Figure 4.4 **Replication machines.** (**A**) The cryo-EM structure of yeast Pol ε holoenzyme is shown with a model for the interaction of the polymerase with DNA. This image originally appeared as Figure 9 in ref. 31, and is reprinted with permission from the authors and Macmillan Publishers Ltd (F. J. Asturias, I. K. Cheung, N. Sabouri, O. Chilkova, D. Wepplo and E. Johansson, Structure of *Saccharomyces cerevisiae* DNA polymerase epsilon by cryo-electron microscopy, *Nature Structure Molecular Biology*, 2006, **13**, 35–43.), copyright (2006).[31] (**B**) Model for the interaction of bacteriophage RB69 Pol with the RB69 sliding clamp (PCNA orthologue). RB69 Pol is grey, the DNA is brown, and the subunits of the trimeric sliding clamp are green, red and blue. This image originally appeared as Figure 6 in ref. 32, and is reprinted from: Y. Shamoo and T. A. Steitz, Building a replisome from interacting pieces: sliding clamp complexed to a peptide from DNA polymerase and a polymerase editing complex, *Cell*, **99**, 155–166, copyright (1999) with permission from Elsevier and the authors.[32]

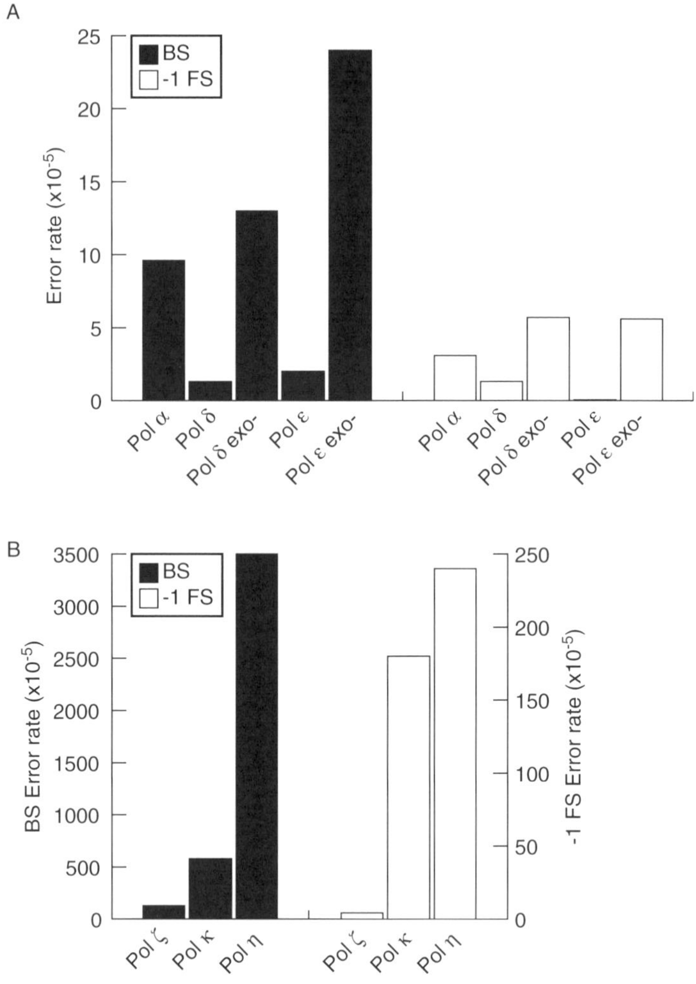

Figure 4.5 **Error rates of B and Y family polymerases for single base substitutions and single base deletions.** (**A**) Error rates of *S. cerevisiae* Pol α, Pol δ, and Pol ε for single base substitution (BS; black bars) and single base frameshift (-1 FS; white bars) errors. (**B**) Error rates of *S. cerevisiae* Pol ζ, human Pol κ and human Pol η. The scale for base substitution error rates is shown on the left and that for -1 frameshifts on the right. Note the difference in scale between panels A and B. The fidelity of each polymerase was determined using the *lacZ*-α forward mutation assay, in which a polymerase performs synthesis along a 407-nucleotide gapped template containing the wild type *lacZ*-α complementation sequence. This substrate allows for detection and calculation of error rates for all 12 base–base mismatches as well as frameshift and deletion errors in a wide variety of sequence contexts, including homonucleotide runs. See ref. 96 for primary references.

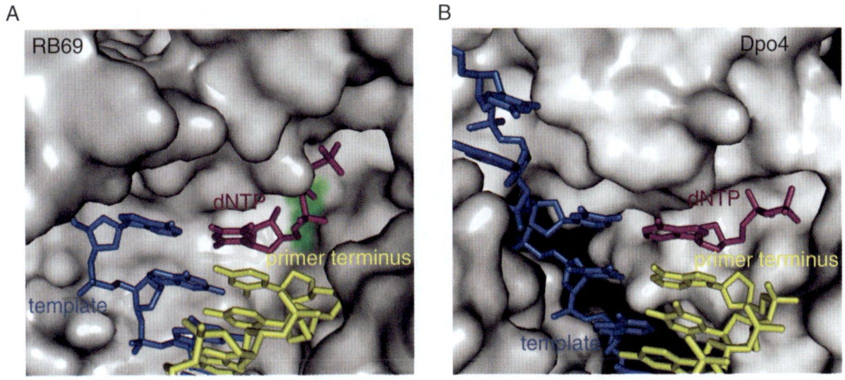

Figure 4.6 **Active sites of high and low fidelity polymerases.** Surface representation of the active sites of (**A**) bacteriophage RB69 Pol and (**B**) archaeal Sso Dpo4 Pol showing the restricted geometry of the RB69 Pol polymerase active site compared with the more open, solvent-accessible active site of Dpo4 Pol (error-prone). The template and primer strands are shown in blue and yellow, respectively, and the incoming deoxyribonucleotide triphosphate is purple. The surface of Leu415 of Motif A in RB69 Pol (discussed in Sections 4.4.5 and 4.5 of the text) is painted green.

increased enthalpy differences and decreased entropy differences, thus amplifying the difference in free energy between a correct and incorrect base pair.

There is also structural selection in favour of correctly paired nucleotides (see, for example, ref. 47); the geometries of accurately paired A-T and C-G base pairs are remarkably similar to each other, but differ from mismatched base pairs. Thus, the aberrant geometry of incorrect base pairs is believed to result in steric clashes with residues in and around the active site of accurate polymerases that disfavour insertion of incorrectly paired nucleotides. For example, the nascent base pair binding pocket of RB69 Pol snugly encloses a correct base pair, but has little additional space for binding of incorrect base pairs with altered geometry (Figure 4.6A). In contrast, the active site of Y family Sso Dpo4 Pol is considerably more spacious, allowing flexibility in the base pair geometry that can be accommodated (Figure 4.6B). The active site of Dpo4 Pol is also more solvent accessible,[22] suggesting that enthalpy–entropy compensation may contribute less to the fidelity of Dpo4 Pol.

Complementing structural studies are a large number of steady state and pre-steady state kinetic studies of incorporation of single dNTPs, probing which step(s) in the polymerization cycle are rate limiting to the nucleotide selectivity of accurate and inaccurate DNA polymerases (refs. 29,48,49 and references therein). These studies reveal the extent to which nucleotide selectivity depends on both the binding affinities and insertion rates of correct versus incorrect dNTPs. Particular emphasis is often placed on rate-limiting conformational changes that precede chemical bond formation. These changes are induced by dNTP binding, and they can differ depending on whether the incoming dNTP is correct or incorrect and also on the polymerase. For example, many DNA polymerases undergo large (20–40°; see Figure 1 of ref. 27), 'open-to-closed' conformational transitions in the relative locations of subdomains (for example, ref. 28 and references therein), while other polymerases do not (for example, ref. 50). Recent studies (ref. 28 and references therein) suggest that the large, open-to-closed conformational changes occur faster than the kinetic step that limits insertion. Nucleotide selectivity may therefore depend on more localized conformational changes (*e.g.* in the positions of critical amino acid side chains and DNA atoms) that occur as dNTP binding induces assembly of an active site with the proper geometry needed to achieve the transition state.[29] A recent study of yeast Pol ζ showed that mutation of lysine 1061 to alanine results in both decreased misinsertion efficiency and decreased mismatch extension efficiency;[51] K1061 is thought to be important in contacting the triphosphate group of an incoming dNTP. (Details of the structural and kinetic basis of nucleotide selectivity during replication are discussed in refs. 22,27–29,44–51.)

4.4.3 Indel Error Rates and Mechanisms for Indel Formation

All DNA polymerases studied to date generate single base insertion/deletion errors (indels) *via* strand misalignments arising during DNA replication (for a recent review see ref. 52). Insertion intermediates contain an unpaired base in the primer strand and are typically generated at lower rates than are deletions, which involve an unpaired base in the template strand (insertions require more hydrogen bonds to be disrupted[52]). The average rates at which eukaryotic family B and Y polymerases generate single base deletion errors as they copy the *lacZ* coding sequence in M13mp2 DNA *in vitro* are shown in Figure 4.5.

As is also the case for base substitutions, the three polymerases responsible for the bulk of replication of the nuclear genome (Pols α, δ and ε) have lower error rates for -1 deletions (Figure 4.5A) than the polymerases implicated in TLS (Pols ζ, η and κ; Figure 4.5B). Again, this is true even for exonuclease-deficient variants of Pol δ and Pol ε.

The higher indel fidelity of the major replicative polymerases is consistent with their roles in maintaining the stability of genomes that contain large amounts of repetitive DNA (see below). This increased fidelity is also consistent with structural information indicating that the family B polymerases make more extensive contacts with the primer-template[20] than do family Y polymerases,[22] (Figure 4.6) and thus are likely to be less tolerant of unpaired bases in the DNA duplex.

The error rates in Figure 4.5 are 'average' rates for deletion errors at many template locations. However, deletion error rates can vary widely not only by polymerase (Figure 4.5) but also depending on sequence context, often being higher in repetitive sequences. Higher indel error rates during copying of repetitive sequences are consistent with the original proposal by Streisinger *et al.*[53] that misaligned bases in repetitive sequences can be stabilized by correct base pairing at the primer terminus. Indeed, recent structural studies show that human Pol λ, which has a particularly high deletion error rate, can bind to a misaligned intermediate containing an unpaired, extrahelical base in the template strand upstream of the polymerase active site while maintaining correct base pairing of the primer terminus and proper geometry for continued synthesis.[54]

An important element of the mechanism of indel formation is the basis of strand slippage during the polymerization cycle. Among several ideas that have been put forth (reviewed in ref. 52), one model proposed that slippage may initiate when an incoming dNTP pairs directly with a base that is adjacent to the templating base, leaving an uncopied base within the polymerase active site that is eventually deleted or added. Structural analysis of *Sso* Dpo4 Pol[22] supports this model, which may be most relevant to that subset of DNA polymerases that can simultaneously accommodate two template nucleotides in the nascent base pair binding pocket, *i.e.* certain family Y polymerases (see Figure 4D in ref. 52).

Another model proposes that slippage may be initiated by dNTP misinsertion followed by relocation of the primer terminus to an adjacent complementary template base (see Figure 5 in ref. 52). This 'misinsertion–primer relocation' model is limited to specific sequence contexts that provide a template base complementary to the misinserted nucleotide just upstream of the primer terminus.

A third idea is melting-misalignment, wherein the duplex DNA frays and strand misalignments occur when the single-stranded primer is transferred between the polymerase and 3′ exonuclease active sites of a proofreading-proficient DNA polymerase.[55] Melting-misalignment is relevant only to the minority of DNA polymerases that have 3′ exonuclease activity such as Pol δ and Pol ε.

A fourth hypothesis states that slippage may initiate as a polymerase associates with or dissociates from the primer-template (reviewed in ref. 56). This may be generally relevant to all polymerases and sequence contexts, as all polymerases make and break many non-covalent contacts with the primer-template during catalytic cycling. Just as for base substitutions, a key step in indel formation may be dNTP-induced conformational changes, particularly those involving movement of the template strand relative to the primer strand in order to achieve the geometry required for catalysis.[57]

4.4.4 Intrinsic Exonucleolytic Proofreading

On average, 3′ exonucleolytic proofreading improves the fidelity with which Pol δ and Pol ε copy undamaged templates by ~ten to a hundred fold. This can be observed *in vitro* by comparing the error rates of wild-type Pol δ and Pol ε with their respective exonuclease-deficient variants (Figure 4.5), and *in vivo* by measuring the increase in mutation rate resulting from exonuclease deficiency in yeast Pol δ and Pol ε (for example, ref. 58) and in mouse Pol δ.[59,60]

Proofreading occurs when the presence of a mismatch at or within a few nucleotides upstream of the primer terminus slows further polymerization. This increases the time available for the primer terminus to fray and generate a single-stranded primer that can transition to the exonuclease active site for excision of the nascent error (Figure 4.7A). This process could occur *via* direct strand transfer without enzyme dissociation (intra-molecular proofreading), or

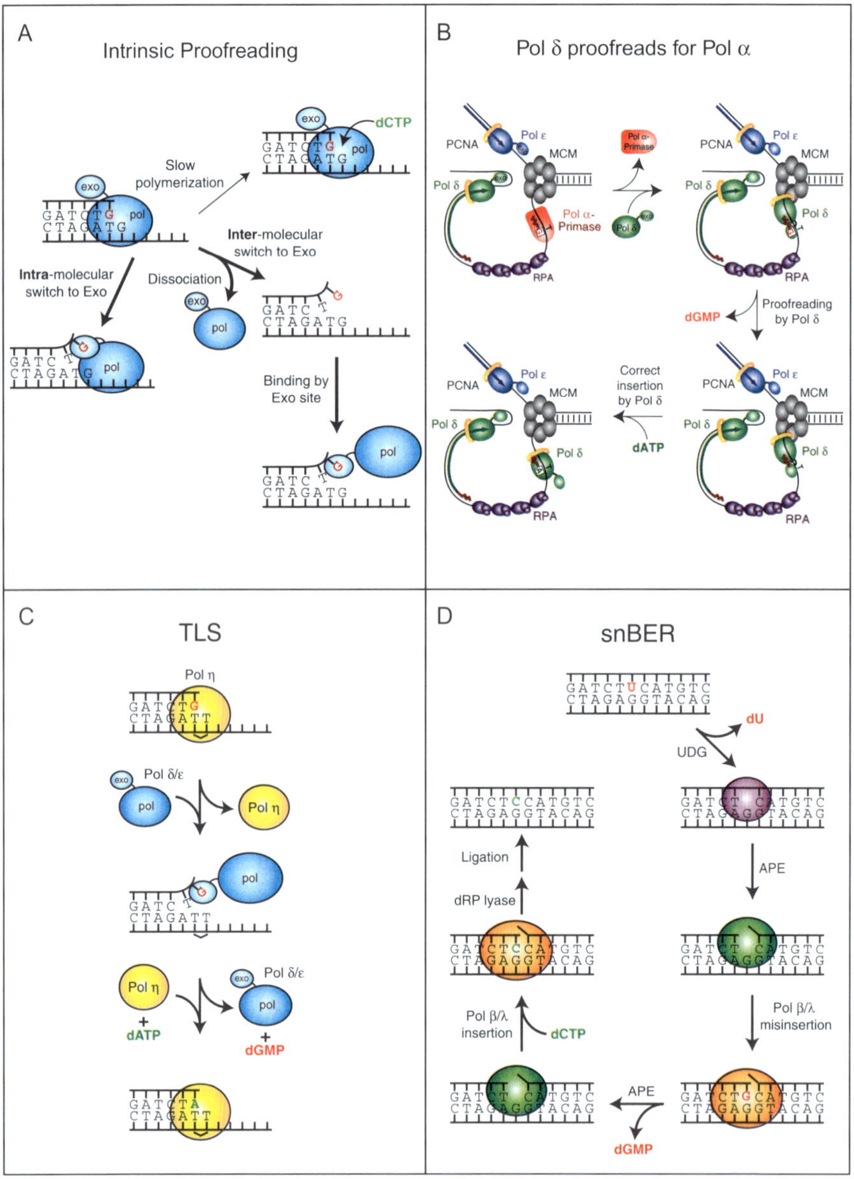

with intervening dissociation (inter-molecular proofreading).[61] The requirement to partition the primer strand between the two active sites can account for substantial differences in the efficiency of proofreading depending on the nature of the mismatch to be extended or proofread. For example, proofreading contributes very little to replication fidelity for indels in long homopolymeric runs[62,63] because the mismatch (*i.e.* the unpaired base) can be separated from the polymerase active site by several correct base pairs that provide a good substrate for further polymerization, whilst reducing fraying and subsequent partitioning to the exonuclease active site.

For this reason, it is mismatch repair, not proofreading, that is the primary error correction mechanism to ensure the genomic stability of simple repetitive sequences. This explains why microsatellite instability is the mutational signature of tumour cells defective in mismatch repair.[64] Similarly, some damaged base–base mismatches are not proofread efficiently. An excellent example is the 8-oxo-G.A mismatch that results from replicating DNA damaged by oxidative stress. The efficiency of proofreading this error by T7 Pol (family A)[65] and RB69 Pol (family B)[66] is low partly because the geometry of an 8-oxo-G.A mismatch is similar to that of a correct base pair.

4.4.5 Extrinsic Proofreading

Pol α lacks an intrinsic exonuclease activity and has a base substitution error rate of $\sim 10^{-4}$ (Figure 4.5), the highest of the three major replicative polymerases. Assuming that Pol α synthesizes ~10 nucleotides of each 250-nucleotide Okazaki fragment, it would replicate ~2% of the human genome and could potentially generate ~12 000 mismatches during each genome replication.

How might these replication errors be corrected? A recent genetic study employing a mutant yeast Pol α allele examined the possibility that an independent 3′ exonuclease might proofread errors made by Pol α. Substituting

Figure 4.7 **Models for exonucleolytic proofreading.** (**A**) Intra-molecular and inter-molecular models for intrinsic proofreading by exonuclease-proficient polymerases. The polymerase domain is depicted as a large oval and the exonuclease domain (exo) as a smaller oval linked to the polymerase domain. (**B**) Extrinsic proofreading during DNA replication. A simplified version of the eukaryotic replication fork is shown using the same colour scheme as in Panel A of Figure 4.1. In the first panel, a dG.T mismatch is generated by Pol α during Okazaki fragment synthesis. Following dissociation of Pol α, the exonuclease domain of Pol δ binds to and recognizes the dG.T mismatch, and then excises the mismatched dG. The polymerase domain of Pol δ can then bind to the primer terminus and insert a correctly paired dA opposite the template T (see ref. 68). (**C**) Extrinsic proofreading by Pol δ or Pol ε of errors generated by TLS polymerases. (**D**) Extrinsic proofreading by apurinic endonuclease (APE) of Pol β or Pol λ errors generated during single nucleotide base excision repair (snBER). This image originally appeared as Figure 1 in ref. 69 and is reproduced with permission (copyright Landes Bioscience).

methionine for a conserved leucine in the Pol α active site (see Figure 4.6) resulted in reduced replication fidelity *in vitro* and *in vivo*,[67] and this mutator phenotype was strongly increased upon inactivation of the proofreading exonuclease of Pol δ, but not that of Pol ε.[68] Among several non-exclusive explanations that were considered for this observation, the results support the hypothesis that the 3′ exonuclease of Pol δ proofreads errors generated by Pol α during initiation of Okazaki fragments (Figure 4.7B). Given that eukaryotes encode many other specialized, naturally proofreading-deficient DNA polymerases (some with even lower fidelity than Pol α), extrinsic proofreading could be relevant to other DNA transactions that control genome stability (reviewed in ref. 69), such as proofreading during translesion DNA synthesis by family Y DNA polymerases[70] (Figure 4.7C), or during base excision repair-related gap-filling by exonuclease-deficient family X polymerases[71] (Figure 4.7D).

4.4.6 Role of Accessory Proteins in DNA Replication Fidelity

Polymerases function in complexes with other proteins that have the potential to modulate replication fidelity. In comparison to the dominant roles of the polymerases themselves in determining nucleotide selectivity and proofreading efficiency, accessory proteins have generally been found to affect replication fidelity by only a few-fold (see ref. 72–74 and references therein). Nonetheless, strong effects of accessory proteins on replication fidelity have been observed in certain cases. For example, the processivity factor thioredoxin greatly reduces the rate of single base insertions by T7 Pol (family A).[75] Another example involves *E. coli* Pol IV, a Y family translesion polymerase that is prone to generate single base deletions. The accessory proteins UmuD and RecA are able to suppress these errors both *in vitro* and *in vivo* through a mechanism involving direct interaction of both proteins with Pol IV.[76] A model of the Pol IV-RecA-UmuD ternary complex suggests that RecA and UmuD act in concert to enclose the relatively open active site of Pol IV, which may prevent the template bulging necessary for single base deletion mutagenesis. A recent example of a robust effect of accessory proteins on eukaryotic replication fidelity involves RPA and PCNA, both of which function at the replication fork (Figure 4.1). While RPA and PCNA have modest (less than five-fold) effects on the base substitution and indel error rates of yeast Pol δ, they significantly decrease the rate of formation of large deletion errors between direct repeat sequences by ≥90-fold.[73] RPA and PCNA may suppress these deletion errors either by preventing fraying of the primer terminus, preventing the repositioning and annealing of the frayed terminus to a downstream repeat sequence, or both (see also Chapter 3).

4.5 Division of Labour during Leading and Lagging Strand Replication

The division of labour between Pol δ and Pol ε in copying the leading and lagging strand templates has remained unclear despite many years of outstanding research (reviewed in, for example, refs. 12,13). Studies to date lead to models suggesting that:

- Pol δ performs the majority of synthesis on both strands with Pol ε responsible for only a modest portion of total synthesis;[77]
- Pol δ replicates the lagging strand and Pol ε replicates the leading strand[12,78] (Figure 4.1A);
- Pol δ replicates the leading strand and Pol ε replicates the lagging strand[13] (Figure 4.1B).

Distinguishing between these and other possible models for the division of labour at the replication fork has been complicated by the fact that Pol δ and Pol ε are essential for normal replication, such that yeast strains lacking the catalytic domains of these polymerases are either dead (Pol δ, Pol ε)[78–81] or very sick (Pol ε).[77,82] However, just as studies of a specific mutator allele of Pol α (L868M) have been informative regarding the possibility of extrinsic proofreading,[68] so too are studies of homologous mutator alleles of Pol δ and Pol ε informative regarding their contributions to leading and lagging strand replication. Analysis of the distribution of base substitution mutations made in yeast by a Pol ε M644G mutator allele has implicated Pol ε in some degree of leading strand DNA replication.[43] Recent examination of the mutational signature of a similar Pol δ L612M mutator allele[42] has determined that the majority of Pol δ synthesis is performed on the lagging strand, thus also clarifying the extent of the contribution of Pol ε to leading strand replication.[83] These results favour a nearly equal, strand-specific division of labour at the eukaryotic replication fork, with Pol ε and Pol δ replicating the majority of the leading and lagging strand templates, respectively (see Figure 4.1A).

4.6 Replication Fidelity and Human Health

The importance of the efficiency and fidelity of DNA synthesis by the family B and Y DNA polymerases during replication is underscored by the adverse effects on human health caused by polymerase dysfunction.

The value of efficient TLS polymerases is perhaps best illustrated by the consequences of Pol η deficiency. In the absence of Pol η, cells are unable to efficiently bypass cyclobutane pyrimidine dimers that result from exposure to ultraviolet (UV) light.[84] This is clinically manifested as Xeroderma Pigmentosum Variant (XPV), a condition characterized by increased susceptibility to sunlight-induced skin cancer.[85,86] XPV patients also exhibit an altered specificity of somatic hypermutation of immunoglobulin genes,[87] indicating that Pol η and other polymerases[88] (including Pol ζ[89]) have significant roles in the development of normal immune responses.

Cancer susceptibility is now also associated with defects in each of the three main determinants of nuclear genome replication fidelity: nucleotide selectivity, proofreading and mismatch repair. The yeast L612M Pol δ mutator allele discussed above has strongly reduced nucleotide selectivity,[42] and mice that are heterozygous for glycine and lysine substitutions of the homologous leucine in the active site of mouse Pol δ have increased genomic instability, cancer susceptibility and decreased lifespan.[90] Similarly, mice defective in the proofreading exonuclease activity of Pol δ,[59,60] and humans and mice defective in mismatch repair of replication errors that escape correction at the fork,[41,64] have decreased lifespan associated with development of several types of cancer. Thus, efficient and accurate replication by DNA polymerases is essential for protection against adverse health outcomes and maintenance of genomic stability.

Acknowledgements

We are grateful to Dr. Katarzyna Bebenek for insightful discussion and comments on the manuscript. This work was supported in part by the Intramural Research Program of the NIH (USA), National Institute of Environmental Health Sciences (T.A.K.).

References

1. A. Kornberg and T. A. Baker, *DNA Replication,* W.H. Freeman & Company, New York, 1992.
2. M. L. DePamphilis, *DNA Replication and Human Disease,* Cold Spring Harbor Laboratory Press, Cold Spring Harbor, 2006.
3. K. Bebenek and T. A. Kunkel, Functions of DNA polymerases, *Adv. Protein Chem.*, 2004, **69**, 137–165.
4. P. V. Shcherbakova and T. A. Kunkel, in *DNA Replication and Human Disease,* ed. M. L. DePamphilis, Cold Spring Harbor Laboratory Press, Cold Spring Harbor, 2006.
5. A. F. Moon, M. Garcia-Diaz, V. K. Batra, W. A. Beard, K. Bebenek, T. A. Kunkel, S. H. Wilson and L. C. Pedersen, The X family portrait: structural insights into biological functions of X family polymerases, *DNA Repair (Amst.)*, 2007, **6**, 1709–1725.
6. M. A. Graziewicz, M. J. Longley and W. C. Copeland, DNA polymerase gamma in mitochondrial DNA replication and repair, *Chem. Rev.*, 2006, **106**, 383–405.
7. L. S. Kaguni, DNA polymerase gamma, the mitochondrial replicase, *Annu. Rev. Biochem.*, 2004, **73**, 293–320.
8. M. E. Arana, K. Takata, M. Garcia-Diaz, R. D. Wood and T. A. Kunkel, A unique error signature for human DNA polymerase nu, *DNA Repair (Amst.)*, 2007, **6**, 213–223.
9. K. Takata, T. Shimizu, S. Iwai and R. D. Wood, Human DNA polymerase N (POLN) is a low fidelity enzyme capable of error-free bypass of 5S-thymine glycol, *J. Biol. Chem.*, 2006, **281**, 23445–23455.

10. K. Masuda, R. Ouchida, M. Hikida, T. Kurosaki, M. Yokoi, C. Masutani, M. Seki, R. D. Wood, F. Hanaoka and O. W. J, DNA polymerases eta and theta function in the same genetic pathway to generate mutations at A/T during somatic hypermutation of Ig genes, *J. Biol. Chem.*, 2007, **282**, 17387–17394.
11. M. Seki and R. D. Wood, DNA polymerase theta (POLQ) can extend from mismatches and from bases opposite a (6-4) photoproduct, *DNA Repair (Amst.)*, 2008, **7**, 119–127.
12. P. Garg and P. Burgers, DNA polymerases that propagate the eukaryotic DNA replication fork, *Crit. Rev. Biochem. Mol. Biol.*, 2005, **40**, 115–128.
13. A. Johnson and M. O'Donnell, Cellular DNA replicases: components and dynamics at the replication fork, *Annu. Rev. Biochem.*, 2005, **74**, 283–315.
14. C. W. Lawrence, Cellular functions of DNA polymerase zeta and Rev1 protein, *Adv. Protein Chem.*, 2004, **69**, 167–203.
15. G. N. Gan, J. P. Wittschieben, B. O. Wittschieben and R. D. Wood, DNA polymerase zeta (pol zeta) in higher eukaryotes, *Cell Res.*, 2008, **18**, 174–183.
16. J. G. Jansen, M. I. Fousteri and N. de Wind, Send in the clamps: control of DNA translesion synthesis in eukaryotes, *Mol. Cell*, 2007, **28**, 522–529.
17. A. R. Lehmann, A. Niimi, T. Ogi, S. Brown, S. Sabbioneda, J. F. Wing, P. L. Kannouche and C. M. Green, Translesion synthesis: Y-family polymerases and the polymerase switch, *DNA Repair (Amst.)*, 2007, **6**, 891–899.
18. S. Prakash and L. Prakash, Translesion DNA synthesis in eukaryotes: a one- or two-polymerase affair, *Genes Dev.*, 2002, **16**, 1872–1883.
19. D. L. Ollis, P. Brick, R. Hamlin, N. G. Xuong and T. A. Steitz, Structure of large fragment of *Escherichia coli* DNA polymerase I complexed with dTMP, *Nature*, 1985, **313**, 762–766.
20. M. C. Franklin, J. Wang and T. A. Steitz, Structure of the replicating complex of a pol alpha family DNA polymerase, *Cell*, 2001, **105**, 657–667.
21. A. J. Berman, S. Kamtekar, J. L. Goodman, J. M. Lazaro, M. de Vega, L. Blanco, M. Salas and T. A. Steitz, Structures of phi29 DNA polymerase complexed with substrate: the mechanism of translocation in B-family polymerases, *EMBO J.*, 2007, **26**, 3494–3505.
22. H. Ling, F. Boudsocq, R. Woodgate and W. Yang, Crystal structure of a Y-family DNA polymerase in action: a mechanism for error-prone and lesion-bypass replication, *Cell*, 2001, **107**, 91–102.
23. E. C. Friedberg, P. L. Fischhaber and C. Kisker, Error-prone DNA polymerases: novel structures and the benefits of infidelity, *Cell*, 2001, **107**, 9–12.
24. W. Yang, Portraits of a Y-family DNA polymerase, *FEBS Lett.*, 2005, **579**, 868–872.

25. T. A. Steitz, S. J. Smerdon, J. Jager and C. M. Joyce, A unified polymerase mechanism for nonhomologous DNA and RNA polymerases, *Science*, 1994, **266**, 2022–2025.
26. Y. I. Pavlov, P. V. Shcherbakova and T. A. Kunkel, *In vivo* consequences of putative active site mutations in yeast DNA polymerases alpha, epsilon, delta, and zeta, *Genetics*, 2001, **159**, 47–64.
27. T. A. Kunkel and K. Bebenek, DNA replication fidelity, *Annu. Rev. Biochem.*, 2000, **69**, 497–529.
28. P. J. Rothwell and G. Waksman, A pre-equilibrium before nucleotide binding limits fingers subdomain closure by Klentaq1, *J. Biol. Chem.*, 2007, **282**, 28884–28892.
29. P. Lin, L. C. Pedersen, V. K. Batra, W. A. Beard, S. H. Wilson and L. G. Pedersen, Energy analysis of chemistry for correct insertion by DNA polymerase beta, *Proc. Natl. Acad. Sci. U.S.A.*, 2006, **103**, 13294–13299.
30. C. M. Joyce and T. A. Steitz, Function and structure relationships in DNA polymerases, *Annu. Rev. Biochem.*, 1994, **63**, 777–822.
31. F. J. Asturias, I. K. Cheung, N. Sabouri, O. Chilkova, D. Wepplo and E. Johansson, Structure of *Saccharomyces cerevisiae* DNA polymerase epsilon by cryo-electron microscopy, *Nat. Struct. Mol. Biol.*, 2006, **13**, 35–43.
32. Y. Shamoo and T. A. Steitz, Building a replisome from interacting pieces: sliding clamp complexed to a peptide from DNA polymerase and a polymerase editing complex, *Cell*, 1999, **99**, 155–166.
33. X. Deng, J. E. Habel, V. Kabaleeswaran, E. H. Snell, M. S. Wold and G. E. Borgstahl, Structure of the full-length human RPA14/32 complex gives insights into the mechanism of DNA binding and complex formation, *J. Mol. Biol.*, 2007, **374**, 865–876.
34. G. D. Bowman, M. O'Donnell and J. Kuriyan, Structural analysis of a eukaryotic sliding DNA clamp-clamp loader complex, *Nature*, 2004, **429**, 724–730.
35. Y. P. Chang, G. Wang, V. Bermudez, J. Hurwitz and X. S. Chen, Crystal structure of the GINS complex and functional insights into its role in DNA replication, *Proc. Natl. Acad. Sci. U.S.A.*, 2007, **104**, 12685–12690.
36. J. M. Choi, H. S. Lim, J. J. Kim, O. K. Song and Y. Cho, Crystal structure of the human GINS complex, *Genes Dev.*, 2007, **21**, 1316–1321.
37. K. Kamada, Y. Kubota, T. Arata, Y. Shindo and F. Hanaoka, Structure of the human GINS complex and its assembly and functional interface in replication initiation, *Nat. Struct. Mol. Biol.*, 2007, **14**, 388–396.
38. E. A. Toth, Y. Li, M. R. Sawaya, Y. Cheng and T. Ellenberger, The crystal structure of the bifunctional primase-helicase of bacteriophage T7, *Mol. Cell*, 2003, **12**, 1113–1123.
39. T. A. Kunkel and D. A. Erie, DNA mismatch repair, *Annu. Rev. Biochem.*, 2005, **74**, 681–710.
40. R. R. Iyer, A. Plucienník, V. Burdett and P. L. Modrich, DNA mismatch repair: functions and mechanisms, *Chem. Rev.*, 2006, **106**, 302–323.

41. G. M. Li, Mechanisms and functions of DNA mismatch repair, *Cell Res.*, 2008, **18**, 85–98.
42. S. A. Nick McElhinny, C. M. Stith, P. M. Burgers and T. A. Kunkel, Inefficient proofreading and biased error rates during inaccurate DNA synthesis by a mutant derivative of *Saccharomyces cerevisiae* DNA polymerase delta, *J. Biol. Chem.*, 2007, **282**, 2324–2332.
43. Z. F. Pursell, I. Isoz, E. B. Lundstrom, E. Johansson and T. A. Kunkel, Yeast DNA polymerase epsilon participates in leading-strand DNA replication, *Science*, 2007, **317**, 127–130.
44. J. D. Watson and F. H. Crick, The structure of DNA, *Cold Spring Harb. Symp. Quant. Biol.*, 1953, **18**, 123–131.
45. L. A. Loeb and T. A. Kunkel, Fidelity of DNA synthesis, *Annu. Rev. Biochem.*, 1982, **51**, 429–457.
46. J. Petruska, L. C. Sowers and M. F. Goodman, Comparison of nucleotide interactions in water, proteins, and vacuum: model for DNA polymerase fidelity, *Proc. Natl. Acad. Sci. U.S.A.*, 1986, **83**, 1559–1562.
47. O. Kennard, in *Nucleic Acids and Molecular Biology,* ed. F. Eckstein and D. M. J. Lilley, Springer-Verlag, Berlin Heidelberg, 1987, pp. 25–52.
48. K. A. Johnson, Conformational coupling in DNA polymerase fidelity, *Annu. Rev. Biochem.*, 1993, **62**, 685–713.
49. C. A. Sucato, T. G. Upton, B. A. Kashemirov, J. Osuna, K. Oertell, W. A. Beard, S. H. Wilson, J. Florian, A. Warshel, C. E. McKenna and M. F. Goodman, DNA Polymerase beta fidelity: halomethylene-modified leaving groups in pre-steady-state kinetic analysis reveal differences at the chemical transition state, *Biochemistry*, 2008, **47**, 870–879.
50. M. Garcia-Diaz, K. Bebenek, J. M. Krahn, T. A. Kunkel and L. C. Pedersen, A closed conformation for the Pol lambda catalytic cycle, *Nat. Struct. Mol. Biol.*, 2005, **12**, 97–98.
51. C. A. Howell, C. M. Kondratick and M. T. Washington, Substitution of a residue contacting the triphosphate moiety of the incoming nucleotide increases the fidelity of yeast DNA polymerase zeta, *Nucleic Acids Res*, 2008, **36**, 1731–1740.
52. M. Garcia-Diaz and T. A. Kunkel, Mechanism of a genetic glissando: structural biology of indel mutations, *Trends Biochem. Sci.*, 2006, **31**, 206–214.
53. G. Streisinger, Y. Okada, J. Emrich, J. Newton, A. Tsugita, E. Terzaghi and M. Inouye, Frameshift mutations and the genetic code, *Cold Spring Harb. Symp. Quant. Biol.*, 1966, **31**, 77–84.
54. M. Garcia-Diaz, K. Bebenek, J. M. Krahn, L. C. Pedersen and T. A. Kunkel, Structural analysis of strand misalignment during DNA synthesis by a human DNA polymerase, *Cell*, 2006, **124**, 331–342.
55. S. Fujii, M. Akiyama, K. Aoki, Y. Sugaya, K. Higuchi, M. Hiraoka, Y. Miki, N. Saitoh, K. Yoshiyama, K. Ihara, M. Seki, E. Ohtsubo and H. Maki, DNA replication errors produced by the replicative apparatus of *Escherichia coli*, *J. Mol. Biol.*, 1999, **289**, 835–850.

56. K. Bebenek and T. A. Kunkel, Streisinger revisited: DNA synthesis errors mediated by substrate misalignments, *Cold Spring Harb. Symp. Quant. Biol.*, 2000, **65**, 81–91.
57. K. Bebenek, M. Garcia-Diaz, M. C. Foley, L. C. Pedersen, T. Schlick and T. A. Kunkel, Substrate-induced DNA strand misalignment during catalytic cycling by DNA polymerase lambda, *EMBO Rep.*, 2008, **9**, 459–464.
58. A. Morrison and A. Sugino, The $3' \rightarrow 5'$ exonucleases of both DNA polymerases delta and epsilon participate in correcting errors of DNA replication in *Saccharomyces cerevisiae*, *Mol. Gen. Genet.*, 1994, **242**, 289–296.
59. R. E. Goldsby, N. A. Lawrence, L. E. Hays, E. A. Olmsted, X. Chen, M. Singh and B. D. Preston, Defective DNA polymerase-delta proofreading causes cancer susceptibility in mice, *Nat. Med.*, 2001, **7**, 638–639.
60. R. E. Goldsby, L. E. Hays, X. Chen, E. A. Olmsted, W. B. Slayton, G. J. Spangrude and B. D. Preston, High incidence of epithelial cancers in mice deficient for DNA polymerase delta proofreading, *Proc. Natl. Acad. Sci. U.S.A.*, 2002, **99**, 15560–15565.
61. C. M. Joyce, How DNA travels between the separate polymerase and 3′–5′ exonuclease sites of DNA polymerase I (Klenow fragment), *J. Biol. Chem.*, 1989, **264**, 10858–10866.
62. L. C. Kroutil, K. Register, K. Bebenek and T. A. Kunkel, Exonucleolytic proofreading during replication of repetitive DNA, *Biochemistry*, 1996, **35**, 1046–1053.
63. H. T. Tran, J. D. Keen, M. Kricker, M. A. Resnick and D. A. Gordenin, Hypermutability of homonucleotide runs in mismatch repair and DNA polymerase proofreading yeast mutants, *Mol. Cell. Biol.*, 1997, **17**, 2859–2865.
64. P. T. Rowley, Inherited susceptibility to colorectal cancer, *Annu. Rev. Med.*, 2005, **56**, 539–554.
65. L. G. Brieba, B. F. Eichman, R. J. Kokoska, S. Doublie, T. A. Kunkel and T. Ellenberger, Structural basis for the dual coding potential of 8-oxoguanosine by a high-fidelity DNA polymerase, *EMBO J.*, 2004, **23**, 3452–3461.
66. E. Freisinger, A. P. Grollman, H. Miller and C. Kisker, Lesion (in)tolerance reveals insights into DNA replication fidelity, *EMBO J.*, 2004, **23**, 1494–1505.
67. A. Niimi, S. Limsirichaikul, S. Yoshida, S. Iwai, C. Masutani, F. Hanaoka, E. T. Kool, Y. Nishiyama and M. Suzuki, Palm mutants in DNA polymerases alpha and eta alter DNA replication fidelity and translesion activity, *Mol. Cell. Biol.*, 2004, **24**, 2734–2746.
68. Y. I. Pavlov, C. Frahm, S. A. Nick McElhinny, A. Niimi, M. Suzuki and T. A. Kunkel, Evidence that errors made by DNA polymerase alpha are corrected by DNA polymerase delta, *Curr. Biol.*, 2006, **16**, 202–207.
69. S. A. Nick McElhinny, Y. I. Pavlov and T. A. Kunkel, Evidence for extrinsic exonucleolytic proofreading, *Cell Cycle*, 2006, **5**, 958–962.

70. S. D. McCulloch, R. J. Kokoska, O. Chilkova, C. M. Welch, E. Johansson, P. M. Burgers and T. A. Kunkel, Enzymatic switching for efficient and accurate translesion DNA replication, *Nucleic Acids Res.*, 2004, **32**, 4665–4675.
71. T. Matsuda, B. J. Vande Berg, K. Bebenek, W. P. Osheroff, S. H. Wilson and T. A. Kunkel, The base substitution fidelity of DNA polymerase beta-dependent single nucleotide base excision repair, *J. Biol. Chem.*, 2003, **278**, 25947–25951.
72. A. Bebenek, G. T. Carver, F. A. Kadyrov, G. E. Kissling and J. W. Drake, Processivity clamp gp45 and ssDNA-binding-protein gp32 modulate the fidelity of bacteriophage RB69 DNA polymerase in a sequence-specific manner, sometimes enhancing and sometimes compromising accuracy, *Genetics*, 2005, **169**, 1815–1824.
73. J. M. Fortune, C. M. Stith, G. E. Kissling, P. M. Burgers and T. A. Kunkel, RPA and PCNA suppress formation of large deletion errors by yeast DNA polymerase delta, *Nucleic Acids Res.*, 2006, **34**, 4335–4341.
74. S. D. McCulloch, A. Wood, P. Garg, P. M. Burgers and T. A. Kunkel, Effects of accessory proteins on the bypass of a cis-syn thymine-thymine dimer by *Saccharomyces cerevisiae* DNA polymerase eta, *Biochemistry*, 2007, **46**, 8888–8896.
75. T. A. Kunkel, S. S. Patel and K. A. Johnson, Error-prone replication of repeated DNA sequences by T7 DNA polymerase in the absence of its processivity subunit, *Proc. Natl. Acad. Sci. U.S.A.*, 1994, **91**, 6830–6834.
76. V. G. Godoy, D. F. Jarosz, S. M. Simon, A. Abyzov, V. Ilyin and G. C. Walker, UmuD and RecA directly modulate the mutagenic potential of the Y family DNA polymerase DinB, *Mol. Cell*, 2007, **28**, 1058–1070.
77. T. Kesti, K. Flick, S. Keranen, J. E. Syvaoja and C. Wittenberg, DNA polymerase epsilon catalytic domains are dispensable for DNA replication, DNA repair, and cell viability, *Mol. Cell*, 1999, **3**, 679–685.
78. A. Morrison, H. Araki, A. B. Clark, R. K. Hamatake and A. Sugino, A third essential DNA polymerase in *S. cerevisiae*, *Cell*, 1990, **62**, 1143–1151.
79. A. Boulet, M. Simon, G. Faye, G. A. Bauer and P. M. Burgers, Structure and function of the *Saccharomyces cerevisiae* CDC2 gene encoding the large subunit of DNA polymerase III, *EMBO J.*, 1989, **8**, 1849–1854.
80. K. C. Sitney, M. E. Budd and J. L. Campbell, DNA polymerase III, a second essential DNA polymerase, is encoded by the *S. cerevisiae* CDC2 gene, *Cell*, 1989, **56**, 599–605.
81. M. E. Budd and J. L. Campbell, DNA polymerases delta and epsilon are required for chromosomal replication in *Saccharomyces cerevisiae*, *Mol. Cell. Biol.*, 1993, **13**, 496–505.

82. R. Dua, D. L. Levy and J. L. Campbell, Analysis of the essential functions of the C-terminal protein/protein interaction domain of *Saccharomyces cerevisiae* pol epsilon and its unexpected ability to support growth in the absence of the DNA polymerase domain, *J. Biol. Chem.*, 1999, **274**, 22283–22288.
83. S. A. Nick McElhinny, D. A. Gordenin, C. M. Stith, P. M. Burgers and T. A. Kunkel, Division of labor at the eukaryotic replication fork, *Mol. Cell*, 2008, **30**, 137–144.
84. C. Masutani, M. Araki, A. Yamada, R. Kusumoto, T. Nogimori, T. Maekawa, S. Iwai and F. Hanaoka, *Xeroderma pigmentosum* variant (XP-V) correcting protein from HeLa cells has a thymine dimer bypass DNA polymerase activity, *EMBO J.*, 1999, **18**, 3491–3501.
85. C. Masutani, R. Kusumoto, A. Yamada, N. Dohmae, M. Yokoi, M. Yuasa, M. Araki, S. Iwai, K. Takio and F. Hanaoka, The XPV (*Xeroderma pigmentosum* variant) gene encodes human DNA polymerase eta, *Nature*, 1999, **399**, 700–704.
86. R. E. Johnson, S. Prakash and L. Prakash, Efficient bypass of a thymine–thymine dimer by yeast DNA polymerase, Poeta, *Science*, 1999, **283**, 1001–1004.
87. X. Zeng, D. B. Winter, C. Kasmer, K. H. Kraemer, A. R. Lehmann and P. J. Gearhart, DNA polymerase eta is an A–T mutator in somatic hypermutation of immunoglobulin variable genes, *Nat. Immunol.*, 2001, **2**, 537–541.
88. M. Diaz and C. Lawrence, An update on the role of translesion synthesis DNA polymerases in Ig hypermutation, *Trends Immunol.*, 2005, **26**, 215–220.
89. M. Diaz, L. K. Verkoczy, M. F. Flajnik and N. R. Klinman, Decreased frequency of somatic hypermutation and impaired affinity maturation but intact germinal center formation in mice expressing antisense RNA to DNA polymerase zeta, *J. Immunol.*, 2001, **167**, 327–335.
90. R. N. Venkatesan, P. M. Treuting, E. D. Fuller, R. E. Goldsby, T. H. Norwood, T. A. Gooley, W. C. Ladiges, B. D. Preston and L. A. Loeb, Mutation at the polymerase active site of mouse DNA polymerase delta increases genomic instability and accelerates tumorigenesis, *Mol. Cell. Biol.*, 2007, **27**, 7669–7682.
91. S. D. McCulloch and T. A. Kunkel, The fidelity of DNA synthesis by eukaryotic replicative and translesion synthesis polymerases, *Cell Res.*, 2008, **18**, 148–161.
92. S. A. MacNeill, G. Baldacci, P. M. Burgers and U. Hubscher, A unified nomenclature for the subunits of eukaryotic DNA polymerase delta, *Trends Biochem. Sci.*, 2001, **26**, 16–17.
93. M. Muzi-Falconi, M. Giannattasio, M. Foiani and P. Plevani, The DNA polymerase alpha–primase complex: multiple functions and interactions, *ScientificWorldJournal*, 2003, **3**, 21–33.
94. H. Pospiech and J. E. Syvaoja, DNA polymerase epsilon—more than a polymerase, *ScientificWorldJournal*, 2003, **3**, 87–104.

95. R. N. Venkatesan, J. J. Hsu, N. A. Lawrence, B. D. Preston and L. A. Loeb, Mutator phenotypes caused by substitution at a conserved motif A residue in eukaryotic DNA polymerase delta, *J. Biol. Chem.*, 2006, **281**, 4486–4494.
96. X. Zhong, P. Garg, C. M. Stith, S. A. Nick McElhinny, G. E. Kissling, P. M. Burgers and T. A. Kunkel, The fidelity of DNA synthesis by yeast DNA polymerase zeta alone and with accessory proteins, *Nucleic Acids Res.*, 2006, **34**, 4731–4742.

CHAPTER 5

Coordination of Nucleases and Helicases during DNA Replication and Double-strand Break Repair

MARTIN E. BUDD,[a] LYNNE S. COX[b] AND JUDITH L. CAMPBELL[a]

[a] Braun Laboratories, California Institute of Technology, Pasadena, CA 91125 USA; [b] Department of Biochemistry, University of Oxford, South Parks Road, Oxford, OX1 3QU, UK

5.1 Introduction

Nucleases and helicases are involved in numerous steps in DNA replication and repair. Nucleases act on intermediates in DNA replication created by DNA polymerases (Chapter 4) and helicases (Chapter 3). They can create substrates for repair as in Okazaki fragment processing (OFP) and homologous recombination. They can also create substrates for activation of a checkpoint response, or participate in downregulation of checkpoints. In the special case of telomere replication, they are also involved in essential processing steps (Chapter 8). Nucleases known to act during DNA replication include Dna2, Rad27, Mre11, Sae2, Exo1, RNaseH, Yen1 and Mus81/Mms4. Of these, Dna2, Exo1 and Mre11 are of particular interest because they have been identified as crucial activities that initiate repair of double-strand breaks (DSBs) by homologous recombination and thus form an intrinsic link between DNA

Molecular Themes in DNA Replication
Edited by Lynne S. Cox

Published by the Royal Society of Chemistry, www.rsc.org

replication and repair of DSBs derived from replication fork failure. The action of the nucleases is coordinated with those of a number of helicases and is discussed here in the context of a network of their interactions that combine to maintain genome integrity during DNA replication.

Fidelity of DNA synthesis is traditionally defined at the level of polymerization (see Chapter 4). Three major mechanisms prevent errors occurring in the nucleotide code: correct insertion; proofreading; and mismatch repair. However, the helicase/nuclease network (Figure 5.1) identifies a much more complex system for maintaining the fidelity of transmission of a complex genome. While the network reflects the multitasking nature of DNA replication proteins in DNA replication and repair of exogenous damage, it more probably evolved as a highly coordinated system for preventing or resolving damage due to aberrant steps in endogenous processes such as DNA replication. The future goal of studies of these enzymes lies in understanding the biochemical underpinnings of the entire system as an ensemble, rather than of the individual components. Such a goal can only be realized using combined genetic and biochemical approaches. This chapter summarizes the individual activities in the context of networks. The activities that are emphasized are highlighted by red circles in Figure 5.1. While the attempt at mechanistic integration is new, analysis of the literature suggests we already have an informative framework to build upon. To usefully limit the discussion, this review focuses mainly on the *Saccharomyces cerevisiae* orthologues of these conserved activities.

5.2 The Role of Nucleases in Okazaki Fragment Processing

Most central to the DNA replication process (because it is a constitutive component) is the role of helicase/nuclease coordination in Okazaki fragment maturation (see also Chapters 1 and 3). Okazaki fragments are initiated by pol α-primase, which synthesizes an RNA/DNA primer of about 10–30 nucleotides. A polymerase switch to pol δ then occurs (see Chapter 6), and pol δ synthesizes the remaining fragment. When pol δ reaches the 5′ RNA terminus of the previously synthesized Okazaki fragments, the RNA is nucleolytically removed and the two DNA fragments are ligated together (Figure 5.2).

5.2.1 FEN1

The prevalent model for Okazaki fragment processing in eukaryotic chromosomes originally included a single, structure-specific nuclease called FEN1, whose role in primer removal was discovered during reconstitution of SV40 DNA replication *in vitro*.[1] FEN1's primary enzymatic function is that of a structure-specific nuclease that binds to a free 5′ end and tracks along the single strand until it reaches a single-strand/duplex junction where it cleaves the single strand endonucleotytically (Figure 5.2 left panel). It is thus ideally suited to cleave a 5′ single-stranded RNA/DNA flap, created by pol δ strand

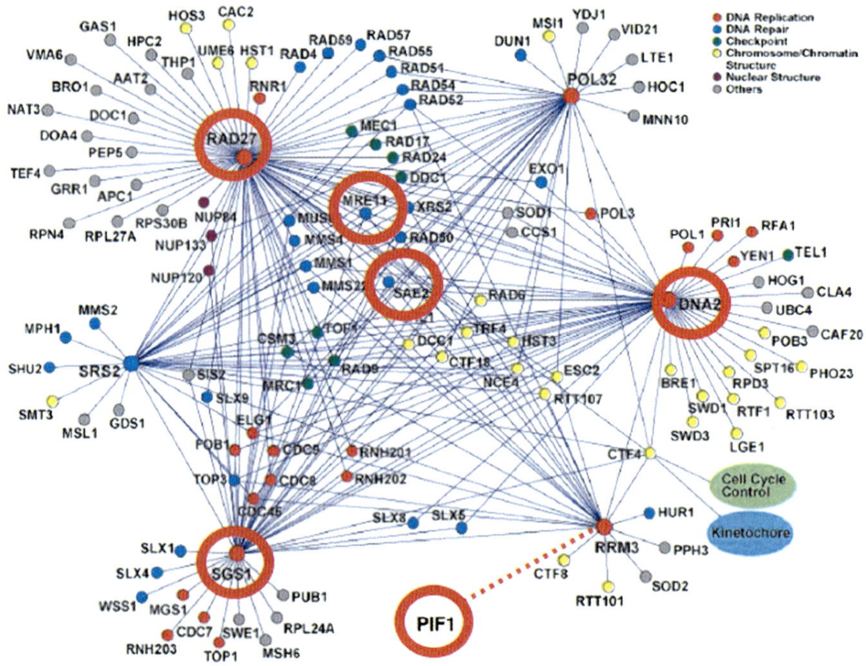

Figure 5.1 **A network for preserving genome stability through the DNA replication fork.** This network was derived from studies in ref. 6; similar but more complex networks are described in refs. 30, 190. *rad27* mutants are synthetically lethal with: helicase mutants *sgs1, srs2, dna2*; replication mutants *pol32, mrc1*; checkpoint mutants *rad9, rad24, rad17, mec1, mrc1, tof1, csm3*; nuclease mutants *exo1, rnh202, mus81/mms4;* repair mutants *rad1, rad50,51,52,54,55,57,59* and *sae2;* cohesion mutants *ctf4, ddc1;* and chromatin mutant *asf1*.[147] The list of synthetically lethal interactions with *rad27* mutants is larger than the list of synthetically lethal interactions of any of the following mutants: *dna2, mre11, sgs1, srs2, rrm3, exo1, sae2, yen1*. The large synthetic lethal list suggests that Rad27 (FEN1) is directly involved in DNA replication and not just repair of replication errors. The absence of *rad27* creates double-strand breaks (DBS) requiring the recombinational repair and checkpoint complexes for viability. The synthetic lethality network of *dna2* mutants does not include the checkpoint mutants *rad9, rad24, rad17, mec1*, nor repair mutants *rad51, 55, 57*, but does include all the genes encoding nucleases involved in OFP: *i.e. rad27, exo1, yen1, rnh202*, repair mutants *rad50, rad52*, and *sae2*.[6] Like *dna2* mutants, *mre11* mutants are not synthetically lethal with checkpoint mutants *rad9*, *rad24*, *mec1*, nor repair mutants *rad51*, *rad52*.[191] The remaining nucleases—Sae2, Exo1, Yen1 and RNaseH—form a synthetic lethal network with many fewer nodes[192] than the Rad27, Dna2 or Mre11 subnetworks, although *dna2* is synthetically lethal with all four.[6] We propose that the high node networks based on *rad27*, *mre11* and *dna2* suggest that these nucleases act first at DNA replication and repair intermediates.

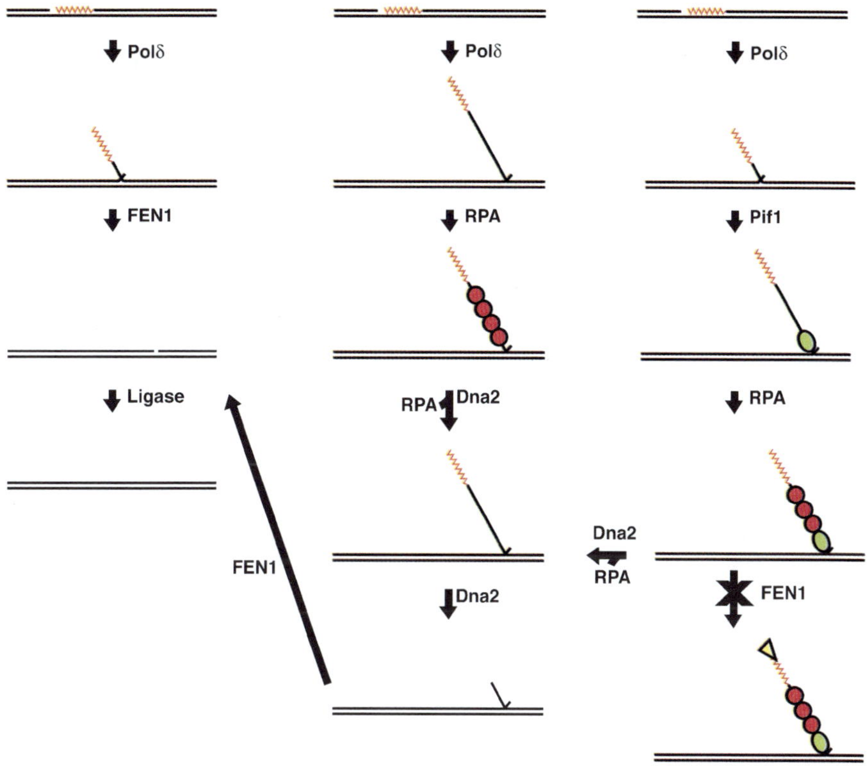

Figure 5.2 **Multiple modes of Okazaki fragment processing (OFP) in eukaryotic cells.** Details are given in the text. The preferred substrate for FEN1 is a double flap as shown: a 5′ flap and a 3′ flap of 1 nucleotide overhang from the upstream fragment.[192,193] An equilibrating flap can occur when both the 5′ flap and the 3′ flap are complementary to the template. Red circles: RPA; green oval, Pif1; yellow triangle, FEN1.

displacement, to form a ligatable nick between adjacent Okazaki fragments (see ref. 2) (Figure 5.2 left panel). In yeast, mutants lacking FEN1 (*i.e. rad27Δ*), are viable but temperature sensitive.[2,3] The viability derives from backup mechanisms, the most direct one consisting of an orthologue of FEN1, the nuclease Exo1 (exonuclease 1). The human counterpart of Exo1 is also a structure-specific nuclease[4] that compensates, albeit inefficiently, for deficiency of both FEN1[5] and Dna2[6] (see Section 5.2.2). There is also an indication that RNaseH can participate in removal of RNA primers in an alternative mode of processing (see Figures 5.1 and 5.2).[6–8] The importance of FEN1 in OFP is highlighted by the finding of very high rates of gross chromosomal rearrangement (GCR) in *rad27Δ* strains, consistent with FEN1 processing the vast majority of Okazaki fragments and of chromosomal breaks occurring in its absence.[2,9]

5.2.2 Dna2

It now appears that the minimal FEN1-mechanism shown in Figure 5.2 (left panel) is not sufficient to maintain a complex genome. With the discovery of Dna2 helicase/nuclease and the fact that Dna2 is an essential protein that interacts with FEN1, the FEN1-alone processing model was expanded to include Dna2.[10–14] Dna2 is an endo/exonuclease that prefers a 5′ flap structure, though it also has limited 3′ to 5′ nuclease activity. The multiple nuclease activities of Dna2 are catalyzed by a single active site homologous to the RecB nuclease of *Escherichia coli*, located in the N-terminal half of the protein.[15–17] Dna2 also has 5′ to 3′ ATP-dependent DNA helicase activity,[10,11,17–19] and the helicase motifs lie in the C-terminal region. Dna2 helicase is a member of a subgroup of superfamily 1 helicases with homology to Upf1 helicase. Interestingly, the other members of the family are RNA helicases involved in modulating translation, while Dna2 is involved in DNA replication and recombination.[20,21] Recently, strand exchange and strand annealing activities have also been observed in Dna2 protein preparations.[22]

The 5′ flap nuclease/helicase activity of Dna2 suggested that it might assist FEN1 in RNA/DNA flap removal. Consistent with this proposal, deletion of *DNA2* results in yeast inviability, and *dna2-1* mutants synthesize only short fragments of DNA under non-permissive conditions.[10] *dna2* mutants are also sensitive to MMS, HU, X-rays and bleomycin.[23] Yeast cells that lack FEN1 and are also mutant for Dna2 (*i.e.* *dna2-1 rad27Δ*) are inviable. Furthermore, *dna2-1* temperature-sensitive growth is suppressed by overexpression of FEN1, supporting the idea of a two-nuclease processing model for OFP (Figure 5.2 centre).[12]

5.2.3 FEN1, Dna2 and RPA Cooperate in OFP

A specific biochemical model for how Dna2 might stimulate FEN1 in OFP arose initially from another genetic observation, namely that *rfa1*-Y29H, a mutation affecting the single-stranded DNA binding protein subunit, Rpa1, was synthetically lethal with *dna2-157* (Cys1255Tyr mutation in Dna2 helicase motif III).[24] Further analysis showed that this behaviour reflects specific interactions between the two proteins. Almost all *rfa1* mutants mapping to the C-terminal DNA binding domain were also synthetically lethal with *dna2-157*. Conversely, almost all *rfa1* mutants in the N-terminal 17 kDa protein–protein interaction domain were viable in combination with *dna2* mutations. Almost all helicase domain *dna2* mutants are synthetically lethal with *rfa-Y29H* (including *dna2-2* (R1253Q), which is completely viable as a single mutant); in contrast, all *dna2* nuclease domain mutants are viable in combination with the *rfa-Y29H* allele, although some are sick due to low expression. Physical association between Dna2 and Replication Protein A (RPA) was also observed, with the C-terminal Dna2 helicase domain interacting primarily with the C-terminal two-thirds of RPA, but interaction between the N-termini of the two proteins also modulates this C–C interaction.[24] Furthermore, RPA stimulates both the helicase and the 5′ to 3′ nuclease activities of Dna2.[24–26] On the other hand, RPA inhibits Dna2-dependent cleavage of 3′ single-stranded regions adjacent to a duplex segment.[27] In terms of physiological function, it is probably significant that, in proteomic experiments involving affinity purification of protein complexes, the Dna2 protein always appears in association with RPA, and *vice versa*.[28–30]

The combined nuclease/helicase activities of Dna2 and its interaction with RPA led to a detailed model of how the two-nuclease OFP processing system might function.[26,31] In this model, DNA pol δ was proposed to displace a 5′ RNA/DNA primer flap longer than 30 nucleotides—a length of DNA that binds RPA efficiently (Figure 5.2 centre). Such a structure had been shown to inhibit FEN1.[32,33] Therefore, it was proposed that the RPA-coated flap first recruited Dna2, which cleaved the flap to within five nucleotides of the branchpoint. Short flaps do not bind RPA tightly and thus Dna2 creates an optimal flap substrate for FEN1. In this model, Dna2, since it was an essential gene, was envisaged as having a role at every Okazaki fragment.

5.2.4 DNA Polymerase δ Exonuclease Activity in OFP

A key element of the two nuclease model is that DNA pol δ needs to displace a flap of 30 nucleotides or longer, because shorter flaps would not stably bind RPA and therefore FEN1 alone would be sufficient. Genetic observations, followed by an investigation of the underlying biochemical principles of the reactions, led to a more complete picture of OFP. Pol3-5DV is a mutant deficient in the 3′ to 5′ proofreading activity of pol δ. While there is no severe defect in OFP in either *rad27Δ* or *pol3-5DV* exonuclease-deficient single mutants, the double mutant shows a severe growth inhibition and a mutator phenotype consistent with a defect in OFP.[34] The major class of mutation that arises is long duplications (up to 100 bp flanked by short direct repeats).[34,35] This suggested a mechanism of OFP requiring both the pol δ 3′ to 5′ exonuclease and FEN1, and tight coupling between them.

Pol δ lacking 3′ to 5′ nuclease activity shows intrinsically increased strand displacement compared to wild-type enzyme, and *in vitro* reconstitution reactions mimicking OFP demonstrated a role for the 3′ to 5′ proofreading activity of pol δ in limiting the extent of strand displacement by pol δ.[36–38] When pol δ extends a primer in the presence of PCNA and FEN1, it usually displaces a downstream blocking DNA oligonucleotide (representing an Okazaki fragment) by only one or two nucleotides, not 30 nucleotides. The strand opening creates a 5′ flap that is coordinately cleaved by FEN1 (nick translation) or which re-anneals to the template creating a 3′ flap that is removed by the 3′ exonuclease of pol δ (idling: dNTP to dNMP turnover is not accompanied by net synthesis or degradation).

The DNA at the junction between Okazaki fragments thus cycles between (i) a ligatable nick formed by synthesis by pol δ and reversal by 3′ exo (idling), and (ii) a ligatable nick formed by limited strand opening giving a 2-3 nucleotide flap cleaved by FEN1 (nick translation). The nick translation reaction can remove RNA, and the idling reaction prevents extensive strand displacement. The latter was demonstrated[38] by showing that:

- only if the 3′ to 5′ exo is defective, does pol δ/PCNA extensively displace the downstream oligonucleotide;
- wild-type pol δ/PCNA only opened the downstream oligo by one or two nucleotides, yielding FEN1-mediated cleavage products of that length when FEN1 was present.

Pol δ and FEN1 also function coordinately to nick translate into a blocking oligonucleotide terminated with RNA—a substrate that mimics an Okazaki fragment. In these reactions, FEN1 is thought to be coordinated through its binding to PCNA,[39] with a hinge region of FEN1 permitting it to swing out to optimise nuclease cleavage[40] (see Figure 3.8E).

5.2.5 Dual Mode OFP

Since the intrinsic properties of pol δ/PCNA and FEN1 thus obviated the need for Dna2, the OFP model has been revised to suggest that there are two alternative modes of processing.[38] The predominant mode of Okazaki fragment processing occurs through short flaps and the one-nuclease (FEN1) mechanism (Figure 5.2 left panel). Dna2 participates only in cases where flaps become long—for instance, under conditions where strand displacement is extensive either because FEN1 is inhibited or due to inhibition of the 3′ exonuclease reaction (Figure 5.2). In keeping with the proposal that extensive strand displacement gave rise to the essential role of Dna2, the double mutant *dna2-1 pol3-01* is inviable and overexpression of *DNA2* suppresses *rad27p pol3-5DV rad51* lethality (*pol3-01* and *pol3-5DV* are 3′ exonuclease-defective mutants of DNA pol δ, *dna2-1* is a Dna2 hypomorph, and *rad27p* encodes a FEN1 defective in PCNA interaction but yeast cells with this mutation grow normally[41]). Note that Rad51 is required for repair of double-strand breaks which are assumed to occur in *rad27p pol2-5DV* mutants.[6,37] Furthermore, deletion of Pol32, a subunit of pol δ that is required for efficient strand displacement, suppresses the phenotype of *dna2-1* and *dna2-2* mutants.[6,42] Finally, further biochemical reconstitution studies defined specific conditions—mutations in the relevant proteins, DNA sequence variation, and reaction conditions—that regulated the proportion of long flaps.[43] In other words, the two-nuclease model is not a constitutive feature of OFP, but occurs only under conditions where flaps become long. This raises the question of what specific conditions lead to the requirement for the additional helicase-nuclease, Dna2. A clue to this question came from genetic analysis that revealed that *pif1Δ* suppresses the lethality of *dna2Δ* and the temperature sensitivity of *dna2-1*.[44]

5.2.6 Pif1 Helicase Regulates OFP

Pif1 is a 5′ to 3′ helicase that has the ability to unwind an RNA/DNA duplex, and is a known inhibitor of telomerase.[45–47] Pif1 also has a mitochondrial form that is required for mitochondrial DNA recombination.[44,47,48] Although not essential in *S. cerevisiae*, its orthologue, Pfh1, is essential in *Schizosaccharomyces pombe*.[49–51] In both organisms, Pif1 shows extensive genetic interactions with DNA pol δ and its subunits.[44,49] *dna2Δ pif1Δ* mutants are temperature-sensitive and sensitive to methyl methanesulfonate (MMS), but, unlike *dna2Δ PIF1* mutants, they are resistant to X-rays. The viability of *dna2Δ pif1Δ* mutants suggests that the presence of functional Pif1 makes Dna2 essential. One way Pif1 helicase could do this is to create a structural intermediate that requires processing by Dna2 helicase–nuclease. Long flaps formed by Pif1 in conjunction with pol δ, a reaction suggested by their genetic interactions, could constitute such a structure. Indeed, RPA does not inhibit FEN1 cleavage of the rare long flaps created by pol δ/PCNA/FEN1 during concerted strand displacement and processing on synthetic substrates. However, RPA does inhibit FEN1 cleavage of long flaps created (more efficiently) by pol δ/PCNA/FEN1 upon addition of Pif1 (Figure 5.2 right panel).[52] Thus, biochemically, Pif1 can direct the OFP reaction toward a mechanism that might require Dna2 to stimulate FEN1. This leads to the further speculation that Pif1 helicase may participate normally in flap generation by pol δ, *i.e.* it could be a polymerase accessory protein. This would be in keeping with *in vitro* kinetic studies where the pol δ/PCNA/FEN1 processing reaction occurs more slowly *in vitro* than the deduced rate of OFP *in vivo*, suggesting that some component present *in vivo* is absent in the reconstituted system;[36,37] Pif1 could be the missing component that yields the physiological rate. To date, however, it has not been investigated as to whether Pif1 increases the rate of processing. Genetic evidence for a Pif1/pol δ interaction *in vivo* stems from the observation that a *dna2Δ pif1Δ* double mutant is temperature-sensitive, but addition of a third mutation, *pol32Δ*, suppresses the defect.[6,44] Since the Pol32 subunit of pol δ enhances the ability of pol δ to strand displace, its deletion (*i.e. pol32Δ*) may reduce strand displacement. In addition, *pifΔ* suppresses the cold sensitivity of a *pol32Δ* strain.[44] Although to date biochemical studies support a model whereby Pif1 creates a substrate for Dna2, it is also possible that absence of Dna2 protein leads to structures that are lethal if Pif1 is present.

The studies summarized here suggest that Pif1 may be a component of the replication fork that is required to regulate the kinetics of OFP, in addition to its well-understood role in regulation of telomere length (see Section 5.6 and Chapter 8). The difference in the degree to which Pif1 is required for viability in *S. cerevisiae* and *S. pombe* may derive from the fact that regulatory mechanisms, even kinetic ones, are less strictly conserved than the underlying machinery of DNA replication. Perhaps the genome of *S. cerevisiae* is not as full of sequences that require the additional helicase for rapid RNA primer removal, or perhaps there are additional backup mechanisms of processing in *S. cerevisae* that are not found in *S. pombe*. It will be interesting to elucidate the differences in future studies.

5.2.7 Flap Processing in OFP

Since FEN1 is a tracking enzyme that requires a free 5′ end to enter a flap and track to the branch point, a mechanism must exist for removing RPA and Dna2 from a flap so that FEN1 is not blocked from reaching its substrate. Detailed analysis of the pairwise sequential actions of RPA, Dna2 and FEN1 support a sequential model for Dna2/Rad27 flap processing.[53] FEN1 itself has been shown to dissociate Dna2 that has been prebound to a DNA flap.[53] This reaction uncovered an interesting new bimodal function for Dna2. Dna2 nuclease requires a free 5′ end for activity (tracking mode).[54] However, in the absence of a 5′ end, Dna2 can bind a flap efficiently, even a flap as short as two nucleotides, though it does not cleave (non-tracking mode) (ref. 53; J.A. Stewart, J. L. Campbell and R.A. Bambara, unpublished data). This would predict that Dna2 should inhibit FEN1 because it would block FEN1 tracking. However, FEN1 can dislodge Dna2 bound in either the tracking or non-tracking modes.

To investigate how Dna2 can cleave an RPA-bound flap, Stewart *et al.*[55] used a nuclease-deficient yeast Dna2 protein to show that Dna2 can dissociate RPA from a flap even in the absence of Dna2-dependent cleavage activity. Furthermore, Dna2 tracking activity was not required for RPA removal although it is needed for Dna2 nucleolytic function, providing further evidence that flap cleavage is not required for dissociation of RPA.[54,55] This activity of Dna2 is probably predicated on specific protein–protein interactions, since yeast Dna2 cannot displace human RPA, perhaps explaining why hRPA inhibits yeast Dna2 cleavage. Furthermore, RPA stimulates Dna2 nuclease preferentially on flaps containing secondary structure by aiding in denaturation of intrastrand structures. In many situations in the cell, RPA must be removed to allow access by other enzymes. Thus, Fanning and coworkers[56,57] have suggested that protein–protein interactions cause a change in conformation of RPA to a weakly bound form that can then be easily displaced by other proteins, allowing access to the DNA of further enzymatic activities (see also Chapter 6). Dna2 may interact with RPA to promote its dissociation from DNA and to allow Dna2 binding and DNA degradation. It is therefore very likely that dissociation of RPA from the DNA is an important activity of Dna2 protein.

5.2.8 Dna2 Helicase Activity in OFP

What is the role of the helicase activity of Dna2 that is essential? Biochemical studies indicate that the Dna2 helicase acts in a coordinated fashion with its nuclease on flaps that have secondary structure, to produce a FEN1-cleavable product.[25,58] Note that Dna2 tracks in the 5′ to 3′ direction and cleaves the strand on which it is tracking. The ability to suppress the lethality of deletion of *DNA2* by deletion of *PIF1* has allowed the coordination to be demonstrated *in vivo* as well. *dna2Δ pif1Δ* strains transformed with plasmids expressing *dna2-helicase-plus; nuclease-minus* (H^+N^-) mutants become inviable, while plasmids expressing *dna2-helicase minus, nuclease minus* (H^-N^-) as well as *dna2- helicase-minus, nuclease-plus* (H^-N^+) plasmids lack toxicity (M. E. Budd and J. L. Campbell, unpublished data). Therefore, Dna2 helicase must be active to generate a requirement for the Dna2 nuclease. Another possible role for the Dna2 helicase/ATPase might be in dislodging RPA (see Section 5.2.7).

5.3 Mismatch Repair in DNA Replication: the Importance of Exo1

Errors occurring during DNA polymerisation include the incorporation of mismatched bases which, if not removed by the nuclease activity of the polymerase itself (see Chapter 4), must be removed following fork passage. Exo1, a $5' \rightarrow 3'$ exonuclease, was initially identified as a mutator in *S. pombe*, suggesting an involvement in mismatch repair.[59] The protein is a member of a family of 5′ to 3′ nucleases including Rad27 (FEN1), Rad2 (excision repair endonuclease), Din7 (a mitochondrial enzyme) and putative nuclease Yen1 (which is synthetically lethal with *dna2*). Exo1 interacts physically with Msh2, a protein that recognizes base mismatches and is required for mismatch repair.[5] Analogous to the situation in *S. pombe*, the mutation rate of *exo1* strains of *S. cerevisiae* was significantly increased over wild type, although it was not as high as in *msh2* strains.[5] The mutation rate of *msh2* and *exo1 msh2* strains was the same, suggesting they function in the same pathway.[5] Addition of Exo1 restored mismatch-dependent, bi-directional excision activity to an *in vitro* human cell extract lacking mismatch repair excision activity.[4] The human mismatch repair protein hMutSa (orthologue of Msh2) activated the exonuclease activity of hExo1 in a mismatch dependent manner.[4] These results suggest that mismatch base repair is the primary function of Exo1 in DNA replication, although Exo1 can also function as a backup nuclease in OFP, DSB processing (see Section 5.4.4) and telomere processing (see Section 5.7). This mismatch repair may make a significant contribution to replication fidelity in much the same way as extrinsic proofreading (see Section 4.4.5).

5.4 Nucleases and Helicases in Double-strand Break Repair

During DNA replication, forks may stall at unusual DNA structures such as hairpin loops, fragile sites, replication slow zones, or at lesions resulting from exposure to DNA damaging agents. In particular, attempts to replicate past a single-strand break on the lagging strand template will result in the formation of a double-strand break (DSB)—a highly mutagenic and lethal lesion if not repaired rapidly in an error-free manner. Replication checkpoints are signalling pathways that detect DNA damage (especially DSBs) and signal, through phosphorylation of adapter (*e.g.* ATM/R) and effector (*e.g.* MRX/N) molecules, to cause replication arrest and trigger DSB repair (reviewed in ref. 60).

The Mre11/Rad50/Xrs2 (MRX) complex plays a critical role in DNA damage checkpoint pathways and in effecting DNA repair (especially of DSBs) during DNA replication, as suggested by the slow growth of *mre11Δ* mutants and synthetic lethality networks such as that shown in Figure 5.1. The MRX complex in yeast (MRN in other systems) is composed of three main protein components: Mre11, Rad50 and Xrs2 (the mammalian orthologue of Xrs2 is Nbs1, mutated in Nijmegen breakage syndrome). Consistent with an important role in DNA replication, repair and genome stability, *mre11*, *rad50*, and *xrs2* mutants are sensitive to X-rays, MMS, ultraviolet (UV) radiation and hydroxyurea (HU),[i] and are slow growing. It has recently become apparent that Sae2 is a nuclease that acts synergistically with Mre11, and *sae2* mutants show phenotypes similar to those of certain *rad50* and *mre11* mutants. Sae2 is therefore discussed in conjunction with Mre11.

5.4.1 Mre11

Mre11 provides the active nuclease of the MRX complex, showing 3′ to 5′ exonuclease activity on double-strand substrates and endonuclease activity on DNA hairpins.[61–63] The Mre11 N-terminus contains five phosphodiesterase motifs also found in *E. coli* sbcD exonuclease.[64,65] The enzyme requires Mn^{2+} for activity rather than Mg^{2+}, since the phosphodiesterase motifs contain four histidine residues which orient Mn^{2+} but will not bind Mg^{2+} [66] (see Figure 5.3A). On nuclease cleavage of DNA, histidine 125 (H125) of Mre11 protonates the leaving 3′ hydroxyl (OH) of the penultimate deoxyribose sugar; consistent with this, mutation of H125 inactivates the nuclease.[66] Four nuclease-deficient *mre11* mutants fail to bind Nbs1 (the mammalian orthologue of Xrs2) in the hMRN complex, suggesting that Nbs1 probably binds near the nuclease active site of Mre11.[67] Biochemically, Xrs2/Nbs1 is unlikely to have catalytic activity, but it probably mediates interaction with other factors important for checkpoint signalling such as BRCA1, perhaps through its forkhead associated domain (FHA) which acts as a phosphopeptide binding motif.[68]

5.4.2 Rad50

The Rad50 component of the MRX/N complex is an ATP binding protein in which the Walker A and Walter B motifs are separated by 600–900 residue heptad repeats which create a coiled coil.[69,70] Rad50 is a member of the Structural Maintenance of Chromosome family, which includes cohesins Smc1 and Smc3 (see Chapter 9), and *E. coli* sbcC. The Rad50 C terminal domain shows homology to ATP binding cassette (ABC) type ATPases, which include the cystic fibrosis transmembrane conductance regulator (CFTR).

Rad50 binding to DNA is dependent on ATP.[71] ATP binds to the P loop in the N-terminus and a conserved region in the C-terminus that performs a

[i] An agent that disrupts nucleotide pool sizes leading to replication fork arrest and induction of the S phase checkpoint.

similar function as region VI in the single subunit helicases, or the arginine finger in hexameric helicases[66,69,72] (see Figure 5.3B). This region is called the signature motif.[73] Mutation of a conserved serine in this signature motif abolishes Rad50 ATP binding. (When the same mutation is made in the CFTR protein, the result is cystic fibrosis.[73]) *rad50* strains with mutations in the Walker A motif or signature motif have the same sensitivity to DNA damaging agents as *rad50Δ* mutants,[73,74] reinforcing the importance of nucleotide binding for Rad50 function. Rad50 ATP binding results in dimerization[69,75] (see Figure 5.3C), and this dimerization leads to enhanced DNA binding by formation of a positively charged surface at the dimer interface. It does this since ATP binding to Rad50 rotates the N- and C-terminal subunits so that the N-terminal DNA binding site of Rad50 aligns with Mre11 bound to the first 40 residues of the coiled coil region in the C-terminus of Rad50 exiting the ATP binding domain[66,76] (see Figures 5.3B and 5.3D). This creates a coupled DNA binding surface.[69]

An additional ramification of Rad50 biochemistry is the demonstration that Rad50 has adenylate kinase activity (*i.e.* it catalyses formation of 2ADP from ATP and AMP). An inhibitor of the adenylate kinase activity of Rad50, Ap_5A, blocks the nonhomologous end joining (NHEJ) reaction mediated by scMre11/Rad50 (scMR), at least in *Xenopus* egg extracts, but has no effect on its ATPase activity nor on the endonuclease activity of hsMRN.[77] (Note that in yeast, Ku-mediated NHEJ is a relatively minor pathway for DSB repair.)

5.4.3 MRX/N Unwinding Creates a Substrate for MRX/N Nuclease Cleavage

Although mammalian MRN does not possess processive helicase activity, a weak ATP-dependent unwinding activity is present.[67] When MRN is assayed in the absence of ATP on an oligonucleotide that forms a hairpin at one end and a protruding single-stranded 3′ region at the other end, cleavage occurs at the hairpin loop at a site where duplex and single-stranded DNA meet. However, in the presence of ATP, numerous cleavages are observed in the 3′ overhang, and these extend into the duplex region in the 3′ to 5′ direction.[67] These cleavages are dependent on the Nbs1 subunit.[67] MRN unwinding activity thus appears to be creating a substrate for its own nuclease. Junctions between single-stranded and double-stranded DNA are hotspots for cutting by yeast MR and MRX/N in the presence of ATP.[62] This cleavage is also dependent on Nbs1, but nuclease cleavage of 3′ overhangs by MRN is inhibited by the addition of RPA or KU.[67]

In *Xenopus* egg extracts, the MRN complex has recently been shown to process DSBs into short single-stranded DNA oligonucleotides of 4–12 bp, which are bound to MRN. The oligonucleotide-bound MRN then activates the checkpoint kinase ATM, as assayed by S1981 phosphorylation.[78] However, this mechanism probably does not function in *S. cerevisiae,* since a nuclease-minus MRX can activate the DNA damage checkpoint, although deactivation is delayed. Moreover, yeast MRX is thought only to activate the DNA damage checkpoint when bound to chromosomes (see below).

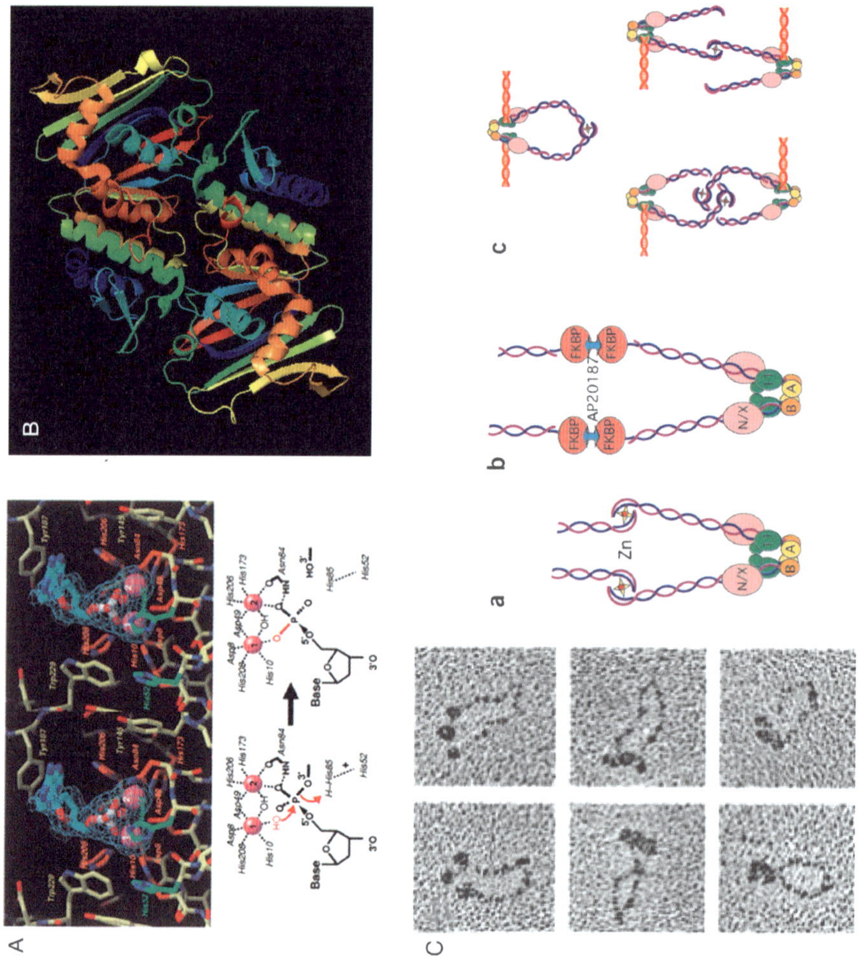
A
B
C
a
b
c
Zn
FKBP
AP20187
N/X

The paradox of the biochemistry of the Mre11 complex is that the enzyme has 3′ to 5′ nuclease activity, but resection of DSBs during double-strand break repair *in vivo* is in the 5′ to 3′ direction. In fact, 3′ ends are stable for up to four hours *in vivo*,[79] reflecting earlier observations of extensive degradation of chromosomes (presumably both 3′ and 5′ ends) at 3.5 hours after X-irradiation of a *rad52* strain.[80] One attractive proposal to reconcile a role for MRN in generating the 3′ ends in DSB repair with its *in vitro* activities is based on the idea that, as MRX unwinds the end of a DSB, the single-stranded DNA forms hairpin loops in the 5′ strand that are cleaved to create a 3′ overhang.[62] Alternatively, other nucleases such as Exo1, Dna2, Sae2 or Rad27 may be associated with MRX and cleave DNA in the 5′ to 3′ direction, revealing a 3′ single strand. Analyzing the phenotypes of Mre11 nuclease defective (*mre11-nd*)

Figure 5.3 **Structural analysis of critical components of the MRN DSB repair complex.** (**A**) Mre11: This crystal structure of *Pyrococcus furiosus* Mre11 shows dimerization of the single active nuclease site of the Mre11 protein bound to Mn^{2+} (pink balls) and to the reaction product, dAMP (red and white stick figure). One active site unites all of the multiple nucleolytic activities of Mre11 in a single mechanism shown schematically below the crystal structure. Reprinted from: K. P. Hopfner, A. Karcher, L. Craig, T. T. Woo, J. P. Carney and J. A. Tainer, Structural biochemistry and interaction architecture of the DNA double-strand break repair Mre11 nuclease and Rad50-ATPase, Cell, **105**, 473–485, copyright (2001) with permission from Elsevier and the authors.[66] (**B**) Rad50 ATPase: ATP-free catalytic domain dimer of Rad50 ABC-ATPase from *Pyrococcus furiosus*. Binding of ATP promotes dimerization, and hydrolysis promotes release of DNA and dimer disassembly. Figure generated in MacPyMOL (http://delsci.com/macpymol/) from pdb data 1US8.[69] (**C**) Electron microscopic imaging of *S. cerevisiae* Rad50/Mre11 dimers demonstrating the coiled-coil and flexible hinge region that allows the two active sites to come together. Reprinted with permission from: D. E. Anderson, K. M. Trujillo, P. Sung and H. P. Erickson, Structure of the Rad50. Mre11 DNA repair complex from *Saccharomyces cerevisiae* by electron microscopy, *J. Biol. Chem.*, 2001, **276**, 37027–37033, copyright (2001) by the American Society for Biochemistry and Molecular Biology.[75] (**D**) Rad50 dimerization: (a) Schematic of the Rad50 zinc binding hook domain at the tip of the coiled-coil. This hook also constitutes the hinge region that allows formation of the Rad50 coiled-coil within the MRX/N complex; (b) An experimental system for dissection the contribution of the hook to dimerization and coiled-coil formation described in ref. 76; (c) Possible intramolecular and intermolecular structures mediated by the hook. Mre11 (green), is shown as a dimer binding between the Rad50 catalytic domains as suggested by electron microscopy; Xrs2/Nbs1 (pink); Rad50 (coil). The yellow A and orange B discs show the location of the Rad50 Walker A and Walker B motifs, respectively. DNA, shown as an orange helix, indicates possible binding modes for DNA. These models suggest how the MRX complex might coordinate the two ends of a DSB. Reprinted by permission from Macmillan Publishers Ltd: M. Lichten, Rad50 connects by hook or by crook, *Nature Structure Molecular Biology*, 2005, **12**, 392–393, copyright (2005).[76]

mutants sheds light on the question whether other nucleases can compensate for Mre11.

Two mutants of particular interest are Mre11 H125N and D56N. These amino acids fall in the conserved phosphodiesterase motif and the mutants have no endonuclease activity *in vitro*;[81] however, they are able to form a complex with Rad50 and Xrs2. Such nuclease-deficient (*nd*) mutants of Mre11 (*i.e. mre11-H125N* and *mre11-D56N*) are much more resistant to X-rays and MMS than strains lacking Mre11 (*i.e. mre11Δ*), although more sensitive than *MRE11* strains,[82,83] and DNA ends at multiple HO-induced breaks are resected at the same rate and extent in *mre11-nd* strains as in *MRN*$^+$.[84] Similarly in meiosis, a residual ATP-dependent DNA unwinding activity of MRX in the *mre11-nd* mutant appears able to create a substrate on which alternative nucleases can act and process DNA ends.[81,85–87] Taken together, these data suggest that other nucleases can compensate for lack of nuclease activity of Mre11, though not for the complete absence of the Mre11 protein.

5.4.4 Dna2 and Exo1 Can Compensate for Mre11 Nuclease in DSB Repair

Exo1 is one nuclease that can compensate for the absence of Mre11. *exo1Δ mre11Δ* and *exo1Δmre11-nd* strains are viable but slightly more sensitive to X-rays than *mre11Δ* and *mre11-nd* mutants, and have a more significant defect in 5′ to 3′ degradation at a DSB (induced by HO endonuclease) than *mre11Δ* strains.[82] Overproduction of Exo1 increases the X-ray survival of an *mre11Δ* and *rad50* strain and the MMS survival of an *mre11Δ* strain, and restores 5′ to 3′ resection in *mre11Δ* strains.[82,88] The DNA damage resistant phenotype of the *mre11-nd exo1Δ* strains suggests that Exo1 is not the only enzyme that can replace Mre11 nuclease activity, but clearly Exo1 can compensate for repair in an *mre11Δ* strain if overproduced. However, a genetic assay that measures actual repair, namely mating type switching after HO cutting, shows that *mre11-H125N exo1Δ* strains are just as proficient in repair as *MRE11 EXO1* strains.[82] This analysis suggests that yet another nuclease must compensate for Mre11 and Exo1 in repair of DSBs. The demonstration that *dna2Δ pif1Δ* mutants are viable has enabled the clear demonstration of a role for Dna2 nuclease in DSB repair. *dna2* point mutants are sensitive to X-rays;[23] however, *dna2Δ pif1Δ* mutants, while sensitive to MMS, are resistant to X-rays. *dna2Δ pif1Δ mre11-nd* are as sensitive to X-rays as *mre11Δ* strains and also grow more slowly than comparable *dna2Δ pif1Δ* mutants. These experiments indicate a requirement for the Mre11 nuclease in X-ray repair if the Dna2 nuclease is missing, and *vice versa*.[89] Remarkably, either Mre11 nuclease or Dna2 nuclease can function in repair, since the *dna2Δ pif1Δ mre11-nd* repair deficiency is complemented by either the *DNA2* or the *MRE11* gene.[89] Further recent studies show that Dna2 is specifically involved in 5′ to 3′ resection of a DSB and that it acts in conjunction with Sgs1 helicase (see Section 5.5) during repair by single-strand annealing of repeats.[90,91] Both of these latter two studies also suggest that Mre11 and Sae2 act earlier than Dna2 and Sgs1 and implicate Exo1 as an alternative to Dna2—at least where extensive resection is required, as in single-strand annealing. Dna2 and a RecQ helicase also function in conjunction with Mre11 in a single-strand annealing reaction in *Xenopus* extracts.[92] Together, Mre11 and Dna2 nucleases seem to perform the major resection required for homology based double-strand break repair.

Mre11 nuclease/helicase interaction differs from that of Dna2 nuclease/helicase. The helicase of Dna2 creates a substrate for the nuclease, which cuts the strand on which it is translocating.[17,58] Dna2 nuclease, however, is the only nuclease that can process substrates created by Dna2 helicase *in vivo*, since Dna2 with defective nuclease supports viability in the absence but not in the presence of its intrinsic helicase (in a *PIF1* deletion, see above).

5.4.5 MRN Recruits Effector Proteins to a DSB

Although Dna2 nuclease can compensate for the absence of Mre11 nuclease activity, Dna2 nuclease cannot compensate for the complete absence of Mre11 protein (*dna2Δ mre11Δ pif1Δ* is inviable).[6] Thus Mre11 has additional functions, perhaps in mediating interaction between the two ends of a DSB.

The relative importance of the various nucleases and helicases in effecting repair at DSBs is suggested by their order of arrival at a DSB and their mutual dependency. These same events initiate a cell cycle checkpoint. The Mre11 nuclease is the first protein detected at a DSB, as measured using real-time observation of fluorescently tagged proteins in living cells,[93] and initiates the DNA damage checkpoint signalling pathway, probably by resecting the break to reveal a single stranded region of DNA which is then bound by RPA, followed by the proteins Tel1, Mec1/Ddc2, Ddc1 and Rad52.[93] The association of the homologous recombination proteins is next: Rad51 depends on Rad52, and the association of Rad55 and Rad57 depends on Rad51.[93] Rad52 associates with breaks after Mre11 has dissociated, suggesting that the binding of Rad52 and Mre11 are mutually exclusive (Figure 5.4).[94] In *mre11-nd* and *sae2Δ* strains, the disappearance of Mre11 foci is delayed; the appearance of Rad52 foci is also significantly delayed.[93] This suggests there is delayed processing of DSBs in *mre11-nd* and *sae2Δ* mutants.

Both Mec1 and Tel1 are homologous S/TQ kinases in the DNA damage signalling pathway. Mec1 is required for DNA damage checkpoint signalling and phosphorylates numerous proteins involved in repair including Rad9, Rad55 and Sae2, while Tel1 functions mainly at telomeres but can compensate for the absence of Mec1 in the DNA damage response if overproduced (*tel1Δ* strains are not sensitive to DNA damaging agents).[95–98] Both Mec1/Ddc2 and Rad52 bind to RPA, and it is their association with RPA that targets these proteins to resected DSBs. Mec1 phosphorylates the mediators Rad9 and Mrc1, and the main effector checkpoint kinase, Rad53, resulting in activation.[95–98]

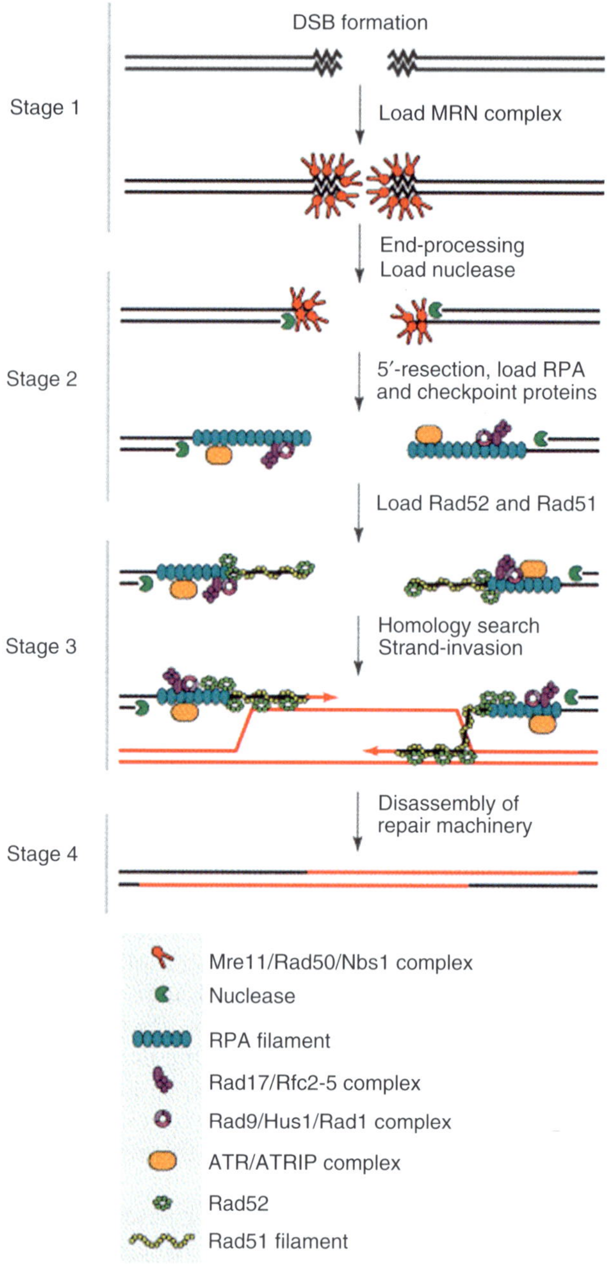

Figure 5.4 **Order of assembly of components at DSBs.** After association of the MRN complex with a DSB, Dna2 and Sgs1 or Exo1 helicases and nucleases may associate and resect 5′ ends, leaving a 3′ single-stranded tail for strand invasion in the strand exchange steps (see text). This stage of DSB repair is represented in the figure by 'nuclease' since the steps have not yet been clearly defined and the specific enzymes involved in specific types of DSB damage remain to be clarified. The remaining stages are described in Section 5.4.5. Reprinted from: M. Lisby and R. Rothstein, DNA damage checkpoint and repair centers, *Current Opinion in Cell Biology*, 2004, **16**, 328–334, copyright (2004), with permission from Elsevier.[94]

In G1 arrested cells, phosphorylation of Rad53 after γ-irradiation is absolutely dependent on MRX.[99] This is sometimes referred to as the MRN-Tel1 checkpoint. Cells of cancer patients with a mutation in either Nbs1 or hMre11 exhibit radiation-resistant DNA synthesis (a phenotype of checkpoint failure) analogous to cells from patients with mutations in ATM (orthologue of Tel1),[100,101] suggesting that the checkpoint role of MRN is conserved from yeast to man. Since RPA recruits Mec1, one expects that Mec1 binding to a DSB would depend on MRX or another resecting nuclease that creates the single-stranded DNA site for RPA binding. In keeping with this suggestion, binding of Mec1 at an HO nuclease-induced break is reduced in an *xrs2Δ* strain, and abolished in an *xrs2 exo1* mutant.[102] After UV treatment (which creates 6-4 photoproducts and pyrimidine dimers that affect usually just one DNA strand of the duplex), MRX is not required for Rad53 phosphorylation, but phosphorylation is nearly blocked in an *xrs2Δ exo1Δ* strain[102] and completely abolished in *mre11Δ exo1Δ* mutants on phleomycin treatment,[102] an agent that leads to DSB formation. Moreover, efficient phosphorylation of Rad53 after low dose (20 mM) treatment of yeast with hydroxyurea is dependent on MRX, although this is not the case after high dose HU treatment (200 mM).[103] DNA replication in the presence of low doses of HU may create DSBs at collapsed replication forks, which require MRX to initiate checkpoint signalling, while high doses of HU may not allow sufficient DNA replication to give rise to DSBs. This idea is supported by the observation that after high dose HU treatment of yeast (100 mM), neither Mre11 foci nor Rad52 foci are observed. However, Mre11 foci are detected in *mec1Δ* strains treated with 100mM HU.[93] Therefore, the absence of DSBs in the presence of high dose HU depends on the presence of a functional Mec1 signalling pathway, suggesting that some HU-induced damage is occurring and that this damage, in the absence of the checkpoint, is converted to DSBs. Mec1 may inhibit a nuclease or other set of events that transforms a stalled fork into a broken fork.

5.4.6 Role of Mre11 and Sae2 in Downregulating the Damage Checkpoint

Mre11 is not only involved in activating the DNA damage checkpoint, but it is also required for checkpoint downregulation, as is Sae2. After an HO-induced DSB, Rad53 phosphorylation persists much longer in a *sae2Δ* mutant than in an *SAE2* strain. The phosphorylation of Rad53 after HO cutting also persists much longer in *rad50S*, *mre11-H125N* and *mre11-D56N* mutants[104] compared with MRX$^+$ strains. The defective dephosphorylation of Rad53 in *sae2Δ*, *rad50S* and *mre11-nd* mutants suggests that processing and repair of DSBs is significantly delayed if the MRX complex has a defect in nuclease activity. Mre11 foci form normally but persist much longer, and the appearance of Rad52 foci is significantly delayed in *sae2Δ* and *mre11-nd* mutants, presumably due to delayed processing of the DSB.[93] Thus, the nuclease activity of Mre11 and Sae2 is not required to activate the checkpoint, but it is required to deactivate it. The ATP-dependent DNA unwinding activity of Rad50 may reveal the single-stranded DNA required to initiate the DNA damage checkpoint in the absence of these nucleases.[105] This suggests a model in which Sae2 is involved in processing the DSB together with Mre11, therefore allowing both repair and checkpoint inactivation to proceed quickly. However, this would be an oversimplification, because overproduction of Sae2 blocks the initiation of the DNA damage checkpoint, Mre11 focus formation and Rad53 phosphorylation after DSB formation (by HO cutting) in spite of the normal generation of resected, single-stranded DNA, the presumed signal for checkpoint activation.[104] A more complete hypothesis to explain these findings is that Sae2 may displace Mre11 from DSBs, and when overproduced, it rapidly disassembles Mre11 bound to DSBs, preventing checkpoint activation. Additionally, the off-rate of MRX from DNA might be significantly increased when Mre11 is acting as a nuclease, explaining the persistence of Mre11-nd foci after DSB induction.

Not only does Sae2 inhibit the DNA damage checkpoint, but Sae2 is also regulated by the DNA damage checkpoint. Sae2 is phosphorylated by Mec1 and Tel1 during S phase and after treatment with HU, bleomycin and MMS;[97] treatments that cause, respectively, replication fork arrest, DSBs or mismatches. This phosphorylation may be involved in activation of Sae2 repair activity, since mutations in the S/TQ putative Mec1/Tel1 phosphorylation sites of Sae2, *sae2(1-9)*, renders cells as sensitive to MMS as does the *sae2Δ* mutation.[97] If Sae2 is activated by Mec1 phosphorylation and if activated Sae2 downregulates the checkpoint, then by phosphorylating Sae2, Mec1 is initiating the downregulation of the DNA damage checkpoint. This model suggests that Mec1 is both an activator and later repressor of the DNA damage checkpoint; it is supported by the observation that, in the *sae2(1-9)* mutant, checkpoint downregulation after UV is significantly delayed. Therefore Mec1 is not acting as a repressor of the checkpoint.[97]

With the DNA damage checkpoint hyperactivated in a *sae2Δ* or *mre11-nd* mutant, one might expect that the complementary Tel1 pathway would be hyperactivated. In keeping with this suggestion, both *sae2* and *rad50S* mutants do suppress the MMS sensitivity of *mec1* mutants in a *TEL1*$^+$ background, and this suppression is absent in a *tel1Δ mec1* mutant.[106] These observations may be explained by the fact that MRX and Tel1 can interact physically, independently of Mec1; Tel1 binds to the C terminus of Xrs2 and is targeted to DSBs by Xrs2.[107]

Xrs2 is thus required for activation of the Tel1 DNA damage checkpoint pathway. To examine the phenotype of disrupting the Xrs2/Tel1 interaction, Mec1 function had to be inactivated, since *MEC1 tel1Δ* strains are resistant to DNA damaging agents. *mec1-81* is a hypomorphic allele of *MEC1* that is sensitive to UV and phleomycin but which, unlike *mec1Δ*, does not result in a proliferative defect in a *tel1Δ* background.[107] *mec1-81 xrs2CTD* (C-terminal deletion that does not bind Tel1) is more sensitive to phleomycin and UV than a *mec1-81* strain, and is as sensitive as a *mec1-81 tel1Δ* strain,[107] showing that *xr2s-CTD* and *tel1* mutants have the same phenotype and may participate in the same pathway. After HO cutting, Tel1 associates with sites near the DSB in a reaction dependent on a physical interaction between Tel1 and the C-terminal domain of Xrs2.[107] The amount of Tel1 associated with a DSB is the same in a *SAE2* and *sae2Δ* strain. Therefore, enhanced Tel1 signalling in a *sae2Δ* strain probably does not result from more Tel1 binding. Instead, an altered MRX/Tel1 physical interaction may be responsible.

5.4.7 Other Nucleases in Checkpoint Regulation

Do other nucleases play a role in checkpoint regulation? For instance, the synthetic lethality of *mre11Δ* and *dna2Δ* suggests that Dna2 nuclease plays a role in activation of Rad53 after DNA damage, but this has not been formally tested. Rad53 is highly phosphorylated in *dna2-1* mutants, even at the permissive temperature and in the absence of exogenous DNA damage, presumably due to activation of the checkpoint by elongated flaps arising from defective OFP.[44] Deletion of *PIF1* suppresses the endogenous Rad53 phosphorylation in the *dna2-1* and *dna2Δ* mutants. Whether there is any defect in Rad53 phosphorylation in *dna2Δ pif1Δ* mutants after induced DNA damage is not known and whether *dna2Δ pif1Δ mre11-nd* mutants, which are viable but X-ray sensitive, show defects in checkpoint signalling is also unknown. *exo1Δ dna2Δ pif1Δ* mutants are inviable, so examining compensatory interactions of Dna2 and Exo1 is difficult. *sae2Δ dna2Δ pif1Δ* is also inviable, so how Dna2 compensates for the absence of Sae2 is also unknown.

5.5 Repair of Stalled Replication Forks

5.5.1 RecQ Helicases and Nucleases

The single RecQ helicase in *S. cerevisiae*, Sgs1, is homologous to Bloom syndrome (BS) helicase (BLM) and Werner syndrome (WS) helicase/nuclease (WRN) in man, mutations of which predispose to cancer and/or premature ageing (see Chapter 3). In both WS and BS, rates of DNA replication are reduced and illegitimate DNA recombination is greatly elevated,[108–110] with high levels of sister chromatid exchange (SCE) and abnormal processing of DNA replication intermediates as defining phenotypes of BS cells[111,112] compared with replication fork asymmetry and premature replicative senescence in WS.[113,114] Like BS cells, *sgs1Δ* mutants have high levels of recombination between chromosomal repeated DNA elements,[115,116] and Sgs1, like BLM and WRN, is a 3′ to 5′ helicase that unwinds DNA forked structures, Holliday junctions, G4 DNA and single-stranded/double-stranded junctions with a 3′ overhang.[117] *Xenopus* egg extracts depleted of BLM have a similar efficiency of DNA replication as undepleted extracts; however the resulting replicated DNA has chromosomal breaks, although an insufficient number to activate the S phase checkpoint.[118] Depletion of the WRN orthologue, FFA1, has a minimal effect on replication rates in *Xenopus* extracts.[119] In yeast, Sgs1 colocalizes with DNA replication forks during S phase,[120] and both BLM and WRN can be observed in human cells at nuclear foci representing sites of DNA replication.[113,121,122]

Suppression of yeast replication-defective nuclease mutants by the human BLM and WRN genes has suggested one mechanism by which BLM and WRN may function in DNA replication. Overproduction of hBLM suppresses the temperature sensitivity of *dna2-1* mutants[123] and partially suppresses the temperature sensitivity of a *dna2Δ pif1Δ rpd3Δ* triple mutant strain (M. E. Budd and J. L. Campbell, unpublished data). Similarly, WRN helicase overexpression also suppresses *dna2-1* temperature sensitivity, and a domain C-terminal to the helicase domain is sufficient for this suppression.[124] Since Rad27 (FEN1) overproduction can also suppress the temperature sensitivity of *dna2-1* cells, the most likely mechanism of BLM/WRN suppression is stimulation of the nuclease activity of FEN1. Consistent with this, biochemical studies have shown that BLM and WRN stimulate the flap endonuclease activity of human FEN1, independently of their helicase activity, but requiring the conserved, winged helix RQC (RecQ C terminal) domain.[125,126] The RQC domain of BLM, which lies just C-terminal of the conserved RecQ helicase domain, is a DNA interaction domain; it binds forked structures and G4 DNA, but not Holliday structures.[127] *sgs1* mutants with deletions into the RQC domain are sensitive to MMS and exhibit hyper-recombination between repeated chromosomal sequences, phenotypes characteristic of OFP defects.[128] BLM with a helicase defect suppresses the growth defect of *dna2-1* mutants, but much less efficiently than wild-type BLM, suggesting that the helicase, in addition to the RQC domain, is required for maximal stimulation of FEN1.[123]

If the RQC domain is sufficient for stimulation of FEN1 nuclease on flaps, then why does helicase-proficient BLM suppress *dna2-1* more effectively than helicase-deficient BLM? The Bambara group showed that ATP is required for maximal BLM stimulation of FEN1 nuclease if the flap has secondary structure (such as a fold back, triplet repeat sequence or bubble),[129] whereas if the flap has no secondary structure, then BLM stimulates FEN1 cleavage as efficiently in the absence of ATP (*i.e.* no helicase activity) as in the presence of ATP (where helicase activity is supported). *In vitro* studies also showed that BLM can stimulate disruption of synthetic recombination-like (strand invasion) intermediates formed by 5′ flaps, such as might arise during aberrant OFP, and thereby indirectly stimulate FEN1 flap cleavage.[130] This reaction occurs either in the presence or absence of ATP. In the presence of ATP, BLM may unwind the strand-switched intermediate, while in the absence of ATP, BLM may use its strand annealing and strand exchange activities to resolve Okazaki fragment intermediates that become involved in aberrant recombination structures, converting them back into FEN1 substrates.[130] The key reaction is conversion of a 5′ flap from a duplex structure to a single-strand required for FEN1 loading and tracking.

sgs1 mutants are synthetically lethal with mutants in genes encoding OFP nucleases, Rad27, Dna2 and RNaseH, and also with the Mus81/Mms4 nuclease. The synthetic lethality of *dna2-2* and *sgs1* is only weakly suppressed by a *rad51* mutation (Rad51 coats DNA to form a large nucleoprotein filament necessary for homologous recombination at Holliday junctions, see Figure 5.4), arguing that most of the lethal damage results from DNA replication defects rather than toxic intermediates potentially formed during recombinational repair of stalled forks. Interestingly, the *dna2-2 sgs1* synthetic lethality/sickness is suppressed by the *fob1* mutation.[131] Fob1 is the rDNA replication fork barrier (RFB) binding protein and is required for replication fork stalling in the ribosomal DNA. Increased fork stalling and breakage is observed at the RFB in both *dna2-2* and in some *sgs1* mutants, and is Fob1 dependent.[132] Increased Holliday junction intermediates are also observed at the RFB in *dna2-2* and *sgs1* mutants,[131] and in human cells deficient in WRN.[113,133] Replication fork stalling may result in more secondary structures at 5′ flaps, and might also allow 5′ flaps to equilibrate to 3′ flaps. Dna2 or Sgs1 helicase may be required to process 3′ flaps, or flaps with secondary structure. Stalled forks may also be converted to DSBs, and Dna2 and Sgs1 may function together in their repair.[90,91]

5.5.2 Sgs1 Resolves Holliday Junctions at Stalled Forks

SGS1 was originally identified as a suppressor of the slow growth phenotype of *top3* mutants.[116] Both BLM and Sgs1 interact functionally and physically with topoisomerase 3, a type I topoisomerase that relaxes negatively supercoiled DNA.[134] The Sgs1/Top3 interaction depends on the N-terminal 158 amino acids of Sgs1.[135] The role of BLM or Sgs1 and Top3 is most probably that of an anti-recombinase that dissolves double Holliday structures.[112,136] Both *sgs1* and *top3* mutants show hyper-recombination in the rDNA and in chromosomal duplicated sites.[115,116] Sgs1 and Top3 function together to disrupt double Holliday structures.[136] Top3 resolves the structures, most probably hemicatenanes, created by Sgs1, that are likely to be toxic if not resolved. Interestingly, the viability of the *top3* mutant depends on Pif1. Overexpression of Pif1 but not Rrm3 suppresses the slow growth phenotype of *top3* mutants.[137] In addition, the *top3Δ pif1Δ* combination is lethal; the lethality is suppressed by *sgs1* mutation, since *sgs1 top3 pif1* yeast cells are viable.[137] Therefore, Pif1 can inefficiently resolve structures created by Sgs1, if Top3 is absent. Since Pif1 is a 5′ to 3′ helicase and Sgs1 is a 3′ to 5′ helicase, they could be working in opposition. Unlike in *S. cerevisae*, *top3Δ* mutants are lethal in *S. pombe*. This lethality is suppressed by deletion of the homologue of the *SGS1* gene, *rqh1*$^+$.[138,139] The Pif1 homologue in *S. pombe*, Pfh1, appears not to suppress *top3* lethality.[51] Therefore, in the case of suppression of *top3* lethality in *S. pombe*, Pfh1 functions like *S. cerevisiae* Rrm3 instead of like Pif1.

Is it possible to separate the Holliday junction dissolving function of Sgs1 and BLM, which causes hyper sister chromatid exchange in BS cells, from their role in DNA replication and OFP? *sgs1(Δ200)* with a deletion of 200 amino acids comprising the conserved HRDC (helicase and RNaseD C-terminal) domain of BLM but retaining the RQC domain, acts like a separation of function mutation in this sense. The HRDC domain is required for BLM and Top3 together to dissolve double Holliday junctions *in vitro*, but is not required for BLM binding to a DNA forked substrate, G4 DNA structures, or for its helicase activity.[140] Thus, it represents a second DNA binding domain in addition to the RQC.

The HRDC domain has also been implicated in the strand annealing and strand exchange activities of BLM.[140] Unlike *sgs1Δ* and *sgs1-RQC* mutants, *sgs1(Δ200)* mutants are resistant to MMS and do not exhibit hyper-recombination at chromosomal duplicated sites, a phenotype putatively attributable to defective replication.[128] However, *sgs1(Δ200)* mutants are defective in a phenotype probably involving the resolution of double Holliday structures, *i.e.* the suppression of the normal growth rate of *top3Δ sgs1Δ* mutants. *top3Δ* strains grow slowly; *top3Δ sgs1Δ* strains grow faster, *top3Δ sgs1Δ* transformed with a *CEN SGS1* plasmid grow slowly, but *top3Δ sgs1Δ* strains transformed with *sgs1(Δ200)* grow like *top3Δ sgs1Δ* strains.[128] One interpretation is that Sgs1(Δ200) is not creating the substrate for Top3, presumably a hemi-catenane derived from a double Holliday structure. In addition, *sgs1(Δ200)*, unlike *sgs1Δ*, is not synthetically lethal with *mms4* (the partner to nuclease Mus81), which may be involved in DNA replication. In addition to the phenotypes of *sgs1(Δ200)*, BLM with a deletion of the HRDC domain complements the temperature sensitivity of a *dna2-1* mutation, while deletion of RQC does not, also suggesting a separation of the DNA replication and Holliday-resolving functions of BLM, and by analogy Sgs1 (L. Liu and J. L. Campbell, unpublished data). Finally, BLM lacking the HRDC domain does not prevent the hyper-sister chromatid recombination phenotype of BS cells.[141]

5.5.3 Mus81 Nuclease in OFP and Stalled Fork Resolution

Mus81/Mms4 is a structure-specific endonuclease that cleaves 3′ flaps.[142] *mus81Δ* is represented in the network shown in Figure 5.1 and is synthetically lethal with *sgs1Δ*, and *sgs1Δ mus81Δ* synthetic lethality is partially suppressed by *rad51Δ*.[143] The triple *sgs1Δ mus81Δ rad51Δ* mutant grows significantly more slowly than *sgs1Δ rad51Δ*. This suggests a complex set of functions in both replication and recombinational repair.

A role for *Mus81* in replication is more directly suggested by the fact that *rnh202*, encoding a subunit of RNaseH2, is synthetically lethal with the *sgs1Δ mus81Δ rad51Δ* triple mutants.[144,145] Mus81/Mms4 thus appears to compensate for the absence of RNaseH removal of RNA/DNA hybrids during OFP. One possible role of Mus81 in OFP could be processing 3′ flaps generated by branch migration of 5′ flaps (equilibration of flaps) at stalled forks whose processing is delayed or impaired. This is consistent with *mus81Δ rad27Δ* synthetic lethality.[146,147] Long 3′ flaps are not expected to form when pol δ is bound to the 3′ OH of the growing Okazaki fragments, *i.e.* without dissociation of pol δ. Such dissociation may occur when forks are stalled, however, such as at the ribosomal RFB, in replication slow zones, or at repeated DNA sequences. If Mus81 and Sgs1 are processing stalled reversed 3′ flaps at the RFB then the triple *mus81Δ sgs1Δ fob1Δ* might be viable. Notably in *S. pombe*,

Mus81 and the RecQ homologue, Rqh1, act to process stalled replication forks prior to their regression into Holliday junctions.[148]

In meiotic cells, Sgs1 functions as an antirecombinase; it is possible that it may do so also in mitotic cells. Sgs1 localizes to and disrupts sites of crossing over.[149] In *sgs1* mutants, the outcome of undisrupted Holliday junctions is joint molecules involving three and four chromosomes.[150] Meiotic cells missing both Sgs1 and Mus81/Mms4 thus accumulate highly elevated levels of joint molecules, which persist through meiosis and are inefficiently processed to crossovers. Expression of either Sgs1 or Mus81/Mms4 during meiosis in the *sgs1 mus81* strain allows dissection of their contribution to crossover formation: Sgs1 prevented formation of joint molecules, whereas Mus81 did not prevent joint molecule formation but allowed resolution of the joint molecules into crossovers.[150–152]

5.5.4 RecQ Proteins Stabilise Stalled Replication Forks

In addition to resolution of structures arising at collapsed or broken replication forks, Sgs1 may be able to prevent fork collapse. A role has been proposed for Sgs1 in the S phase checkpoint, based on observations that Rad53 phosphorylation is slightly defective in an *sgs1* mutant after HU treatment.[120] This is supported by the fact that in *sgs1Δ rad24Δ* strains, phosphorylation of Rad53 after HU treatment is nearly blocked, showing that Sgs1 and Rad24 function in parallel checkpoint pathways (yeast Rad24 is a component of the DNA damage checkpoint pathway important for loading the PCNA-like repair sliding clamp 9-1-1[153] (Rad17-Mec3-Ddc1 in yeast; see Section 3.5).

The role of Sgs1 in Rad53 activation probably involves Sgs1/Rad53 protein–protein interaction and cellular co-localization at the DNA replication fork; human WRN and BLM also co-localise to sites of DNA replication.[121,122,154] However, the sensitivity of the *sgs1Δ* strains to HU is not due to an inability to activate Rad53. Instead, the HU sensitivity may be due to reduced stability of pol ε and pol α at stalled replication forks in *sgs1Δ* mutants.[155] The HRDC domain of Sgs1 is not required for stable maintenance of pol ε at HU-stalled replication forks[156] suggesting that Sgs1 does not need antirecombinase activity to stabilise stalled forks. The enhanced fork arrest observed in *sgs1Δ* strains at the ribosomal replication fork barrier (see Section 5.5.1) might result from the reduced stability of pol ε and pol α at the RFB. Pol ε is also unstable in HU-arrested *mec1* strains, and a combined *mec1 sgs1* mutation increases this instability. These observations on polymerase stability correlate with survival, since double mutant *sgs1 mec1-100* strains show synergistic reduction in survival after HU treatment, compared with single mutant *sgs1* or *mec1-100* strains.[157] The *mec1-100* mutant is resistant to MMS, but does not delay late origin firing in the presence of MMS.[158,159] In a strain with the *mec1-100* mutation, Sgs1 is required to maintain RPA at HU-stalled replication forks. The DNA structures that bind RPA are probably unstable in *sgs1 mec1-100* mutants. Thus Sgs1 may be important in stabilising stalled replication forks, so that restart is rapid and will not require re-licensing or replication from forks emanating from distant origins. In human cells, BLM is implicated in replication fork restart;[160] the finding of WRN at two-thirds of all replication foci[122] and of abnormal fork asymmetry in ~70% of origins in cells lacking WRN[114] supports the importance of RecQ proteins in stabilizing and promoting restart of stalled replication forks.[108,114,133]

Gross chromosomal rearrangements (GCR; *e.g.* deletions, insertions, translocations and telomere additions) are thought to occur at chromosome fragile sites when replication forks stall or fail, and there is a high degree of redundancy in the mechanisms that suppress these rearrangements. Interestingly, the putative replication fork disruptions due to Sgs1 insufficiency correlate with an increased GCR rate in *sgs1, mec1, rad53* and *rad24* strains: *sgs1Δ rad24Δ, sgs1Δ rad53,* and *sgs1Δ mec1* strains show synergistic GCR rates compared with the single mutants.[157,161] The GCR rate of the *sgs1* mutant is similar to that of the *dna2-2* mutant, while that of the *mre11* mutant strain is about 30-fold greater than in the *sgs1* mutant, and the *sgs1 mre11* rate is similar to the single *mre11* mutant.[162]

5.5.5 Implications for Understanding Genome Instability in Human Disease

Does research on BLM, Sgs1, WRN, Dna2, Rad27 and Mus81 provide any answers as to why Bloom and Werner syndrome patients are so highly cancer prone? The answer is unlikely to be simple. One idea is based on the hyper-recombination phenotype of BS and WS cells. Suppose cells are heterozygous for numerous tumour suppressor genes (TSG), in that inactivation of the one active copy of any particular TSG will not promote neoplastic change. However, when many of the TSGs are sequentially or coordinately inactivated, then cancer will almost certainly result. The error rate of DNA replication is 10^{-10} per base pair (Chapter 4), so that TSG inactivation resulting from replication is likely to be rare. However, if BS or WS cells are defective in OFP or restart of stalled replication forks, even if only mildly, then the BLM/WRN mutation might not only raise the overall error rate of replication but also amplify each error very quickly. Thus BLM/WRN mutation would greatly accelerate loss of heterozygosity of the TSGs and lead to high cancer incidence, as is observed in the clinic.[163]

5.6 Nucleases and Helicases in Telomere Maintenance

Telomeres represent chromosomal ends that are, like DSBs, processed by nucleases and helicases, but the processing mechanism is different, unless a telomere becomes a DSB on loss of its protective cap or end binding protein Cdc13.[164] Telomere length is controlled by addition of G/T rich repeats at the ends of chromosomes by telomerase. This is regulated by telomere binding proteins and the OFP processing machinery, which is required to convert the telomerase product into duplex form (see Chapter 8).

5.6.1 Recruitment of Telomerase to the Telomere

In late S phase of *S. cerevisiae*, Est1, a subunit of telomerase, is recruited to telomeres by a pathway requiring Mre11 and Tel1.[165] Tlc1 (telomerase RNA), pol α and pol δ are also required for *de novo* telomere synthesis.[166] This reaction is blocked in cells with a C-terminal deletion mutant of Xrs2 protein that cannot bind Tel1.

Mre11 is thought to create the G/T 3′ telomere overhang strands that are present constitutively throughout the cell cycle, which lengthen in late S phase[167] (the specific function of which in telomere maintenance is not clear), since *mre11Δ* strains have reduced levels of single-stranded GT overhangs[166] and telomeres that are 150 bp shorter than in *MRE11* strains.[168] A role for Mre11 nuclease in telomere addition is further supported by the finding that the X-ray resistant *mre11* mutant, *mre11-3* (H125L/D126V), which is presumed to be nuclease deficient, has normal length telomeres but is blocked in *de novo* telomere synthesis in G2 arrested cells.[169,170] However, the amount of single-stranded GT overhang DNA and overall telomere lengths of *mre11-D56N* and *mre11-H125N,* strains deficient specifically in the nuclease activity, are the same as in wild-type *MRE11* strains.[165] The 5′ to 3′ nuclease digestion that is mediated by MRX and observed in late S phase may therefore not be required for telomere synthesis, but for some other aspect of telomere maintenance. This leaves open the role of Mre11 nuclease in telomere biosynthesis,[171] especially as Mre11 does not appear to be the sole nuclease capable of either creation of the single-strand G/T overhangs in late S phase or telomere length regulation. Interestingly, 5′ to 3′ nuclease digestion of the 5′ C/A strand is not observed during *de novo* telomere synthesis.[170]

5.6.2 Dna2, Exo1, and Sgs1 in Telomere Processing

The apparently normal telomeres in nuclease-deficient Mre11 mutants may arise, at least in part, from the existence of a nuclease that compensates for Mre11 nuclease but not its other functions; the identity of this putative nuclease is unknown. ExoI is a candidate since it plays roles at telomeres under some, but not all, conditions. *exo1Δ* causes no change in telomere length nor does overproduction of Exo1 lengthen telomeres in a *mre11Δ* strain.[82] Therefore Exo1 is not sufficient to compensate for Mre11 in telomere maintenance, and presumably another nuclease compensates. However, when telomeres are uncapped by Cdc13 inactivation, then Exo1 is the major player in 5′ to 3′ degradation.[172,173] In this case the dysfunctional telomere may resemble a DSB, where Exo1 is thought to play a role, at least in extensive resection.

A role also for Dna2 at telomeres is suggested by several lines of evidence. Firstly, *dna2-2* mutants are defective in *de novo* telomere synthesis:[174] the gross chromosomal rearrangement (GCR) rate of the *dna2-2* strain is about 20-fold greater than wild type. Secondly, telomeres of *dna2Δ pif1Δ* mutants are shorter than *pif1Δ* mutants at 30 °C.[44] Moreover, Dna2 protein associates with telomeres during G1, moves to bulk chromosomal DNA during S phase, and then re-associates with telomeres during G2,[174] but is released from telomeres in the presence of DNA damage.

In this context, it is of note that Dna2 not only binds with high affinity to G4 DNA (found at telomeres) but that it can also unwind such structures via its helicase activity.[27] Similarly, Mre11 binds G4 DNA, and shows weak nuclease activity on this template.[175] The G4 binding activity of Dna2 and Mre11 may be involved in telomere synthesis, but their relative importance may depend on Pif1, and any possible G4 binding/unwinding by this helicase. For example, the GCR rate of a *dna2Δ pif1-m2* strain is about half that of the *pif1-m2* strain, defective in nuclear Pif1, suggesting the Dna2 is required for telomere addition.[44] Interestingly, *dna2-2 est2* strains die shortly after telomerase loss, with survivors relying on telomere GT recombination (Type II) rather than subtelomeric Y′ repeat recombination (Type I) for their survival.[44,174] The significance of these observations is the subject of active study.

Telomere maintenance in the absence of telomerase, which occurs through a recombinational pathway between telomeric GT repeats (Type II), also requires the action of the Sgs1 helicase;[176,177] Sgs1 has G4 binding and unwinding activity, and in telomerase-negative (*est2*) cells, absence of Sgs1 leads to rapid telomere-driven senescence.[178] It is of note that again Mre11 is important, since both Mre11 and Sgs1 are required for the recombination between telomere GT repeats observed in telomerase-minus cells.

5.6.3 Preferential Elongation of Short Telomeres

Short telomeres can activate the DNA damage checkpoint, resulting in cellular senescence or apoptosis. Probably to avoid this, short telomeres are preferentially elongated; yeast telomeres of length 300 bp have an 8% chance of being elongated per cell cycle, while a decrease in length to 100 bp increases the frequency to 46%.[179] In *S. cerevisiae,* telomere length is regulated by the Rap1/Rif1/Rif2 complex. Increased numbers of telomere GT repeats (which depends on Mre11—see Section 5.6.2) result in increased Rap1 binding. Rap1 then binds Rif1 and Rif2, which inhibit elongation by telomerase.[180]

To examine the mechanism of preferential elongation of short telomeres, a chromosome was engineered in which a short telomere can be created by site-specific recombination.[181] Shortening of the telomere resulted in enhanced binding of telomerase components Est1 and Est2, but not the capping protein Cdc13. Cdc13 binding to telomeres is similar in wild type, *mre11Δ* and *tel1Δ* yeast cells,[182] suggesting that Cdc13 recruitment is independent of both Mre11 and Tel1. By contrast, increased binding of Est1 and Est2 to telomeres has also been observed in S phase, and this enhanced binding is dependent on Tel1.[183]

A major role of MRX in telomere elongation is in recruiting Tel1 to telomeres in late S phase; *tel1Δ* strains have short telomeres. This can be corrected by galactose-induced overproduction of Tel1 in a *tel1Δ* strain, resulting in Mre11-dependent Rad53 phosphorylation which persists for 12 hours and occurs concurrently with telomere elongation,[184] supporting the importance of Tel1 in telomere elongation. Rad53 phosphorylation lasts much longer when Tel1 is overproduced in a *sae2Δ* or *rad50S* mutant;[97] however the telomeres remain short, perhaps reflecting the requirement for Mre11 nuclease in the *de novo* generation of telomeres after a DSB. The time course of Rad53 phosphorylation upon Tel1 overproduction correlates with MRX binding to telomere ends in wild type, *sae2Δ* and *rad50S* strains. The enhanced binding of Tel1 and Est2 to short telomeres is dependent on MRX, and occurs with reduced Rif2 but not Rif1 binding. Overproduction of Rif2 at the same time as overproduction of Tel1 inhibited both MRX binding to telomere ends and Rad53 phosphorylation, however, suggesting that the Rif2 'counting mechanism' involves inhibition of MRX binding to the telomere.[180,184] Creating a strain with a 90 bp telomere activated MRX-dependent Rad53 phosphorylation over a 24 hour time period, correlating MRX telomere binding and telomere elongation with Rad53 phosphorylation.[184] These data show that in yeast, as in mammalian cells, short telomeres can activate the DNA damage checkpoint, although it is unlikely that telomeres as short as 100 bp occur naturally in wild-type *S. cerevisiae*.

5.6.4 Inhibition of Telomerase by Helicases

Pif1 helicase inhibits telomerase and *pif1Δ* strains have long telomeres.[45,46] The telomere lengths in *MRE11pif1Δ* and *mre11Δ pif1Δ* strains are closer in size than in *MRE11* and *mre11Δ* strains[185] (M. E. Budd and J. L. Campbell, unpublished data), which could suggest that an additional role of Mre11 in telomere synthesis is to suppress Pif1 inhibition of telomerase, and that the nuclease active site of Mre11 might be required for this suppression. This idea is testable by examining *de novo* telomere synthesis in *pifΔ1* and *mre11Δ pif1Δ* and in *mre11-nd pif1Δ* strains, and is supported by data from gross chromosomal rearrangement (GCR) assays developed by Kolodner and collaborators. The GCR selection involves marking a chromosome with *CAN1* and *URA3*, and assaying for chromosomal breakage and repair by quantifying canavanine resistant (*canR*) and 5-FOA resistant colonies.[9] Survivors are characterized by DNA sequencing, which reveals that repair can occur by telomere addition, micro-homology joining or translocation, as well as determining the frequency of each event. As there is no telomere seed sequence at the presumed DSBs initiating the rearrangements, telomere addition is likely to be inefficient in these assays. In wild-type cells, GCR survivors (*canR* and FOA resistant) almost always show telomere additions.[186] *mre11* mutation increases the GCR rate by 600-fold, and only a minority (30%) of the GCR survivors had telomere additions.[187] *mre11Δ pif1Δ* strains, on the other hand, have about a four-fold increased GCR rate relative to *mre11Δ*, but the survivors all showed telomere additions.[185] The *mre11-H125N* mutation increases the GCR rate by 150-fold compared with wild type, and none of the GCR survivors in the *mre11-H125N* strain had telomere additions, showing a requirement for a nuclease active site in telomere additions. By contrast, the GCR survivors of the *mre11-H125N pif1-m2* mutant are all telomere additions. Thus, inactivation of Pif1 eliminates the requirement for the Mre11 nuclease in telomere addition-type GCRs. These results suggest an interaction of the Mre11 nuclease with Pif1 helicase at telomeres.

5.7 Perspective

S. cerevisiae does not require all the nucleases and helicases involved in genome stability for survival because the size of the genome is small and not all damaged DNA replication forks need to be repaired correctly for survival, since replication is possible from a distal fork in a reasonable time frame. However, the ability to process telomeres and repair endogenous DNA damage resulting from stalled forks is compromised when such factors are missing. In order to increase the size of the genome ten, a hundred or even a thousand fold (as in the human genome), orthologues of *MRE11, SAE2, SGS1, PIF1, RAD27, DNA2, EXO1* and *MUS81/MMS4* all become essential to correct the inevitable damage (*e.g.* single- and double-strand breaks and partially processed Okazaki fragments) created by a moving DNA replication fork. An estimate of the number of endogenous DSBs created per cell per division in human cells is between 12 and 50;[188,189] none of these should be left unrepaired, otherwise cell death or neoplastic change leading to cancer remain alternative fates. By utilising a network of interacting helicases and nucleases to maintain genome stability, fungi anticipated the evolution of large complex genomes coding for long-lived complex organisms. An extremely useful outcome of this is the ability to study replication proteins in the simple yeast system, with the confidence that their roles will be highly conserved in higher organisms.

References

1. S. Waga, G. Bauer and B. Stillman, Reconstitution of complete SV40 DNA replication with purified replication factors, *J. Biol. Chem.*, 1994, **269**, 10923–10934.
2. R. E. Johnson, K. K. Gopala, L. Prakash and S. Prakash, Requirement for the yeast *RTH1* 5′ to 3′ exonuclease for the stability of simple repetitive DNA, *Science*, 1995, **269**, 238–240.
3. M. S. Reagan, C. Pittenger, W. Siede and E. C. Friedberg, Characterization of a mutant strain of *Saccharomyces cerevisiae* with a deletion of the *RAD27* gene, a structural homolog of the *RAD2* nucleotide excision-repair gene, *J. Bacteriol.*, 1995, **177**, 364–371.

4. J. Genschel, L. R. Bazemore and P. Modrich, Human exonuclease I is required for 5′ and 3′ mismatch repair, *J. Biol. Chem.*, 2002, **277**, 13302–13311.
5. D. X. Tishkoff, A. L. Boerger, P. Bertrand, N. Filosi, G. M. Gaida, M. Kane and R. D. Kolodner, Identification and characterization of *Saccharomyuces cerevisiae* EXO1, a gene encoding an exonuclease that interacts with MSH2, *Proc. Natl. Acad. Sci. U.S.A.*, 1997, **94**, 7487–7492.
6. M. E. Budd, A. H. Tong, P. Polaczek, X. Peng, C. Boone and J. L. Campbell, A network of multi-tasking proteins at the DNA replication fork preserves genome stability, *PLoS Genet.*, 2005, **1**, 634–650.
7. J. Qiu, Y. Qian, P. Frank, U. Wintersberger and B. Shen, *Saccharomyces cerevisiae* RNaseH(35) functions in RNA primer removal during lagging-strand DNA synthesis, most efficiently in cooperation with Rad27 nuclease, *Mol. Cell. Biol.*, 1999, **19**, 8361–8371.
8. J. J. Turchi, L. Huang, R. S. Murante, Y. Kim and R. A. Bambara, Enzymatic completion of mammalian lagging-strand DNA replication, *Proc. Natl. Acad. Sci. U.S.A.*, 1994, **91**, 9803–9807.
9. C. Chen and R. D. Kolodner, Gross chromosomal rearrangements in *Saccharomyces cerevisiae* replication and recombination defective mutants, *Nature Genet.*, 1999, **23**, 81–85.
10. M. E. Budd and J. L. Campbell, A new yeast gene required for DNA replication encodes a protein with homology to DNA helicases, *Proc. Natl. Acad. Sci. U.S.A.*, 1995, **92**, 7642–7646.
11. M. E. Budd, W.-C. Choe and J. L. Campbell, DNA2 encodes a DNA helicase essential for replication of eukaryotic chromosomes, *J. Biol. Chem.*, 1995, **270**, 26766–26769.
12. M. E. Budd and J. L. Campbell, A yeast replicative helicase, Dna2 helicase, interacts with yeast FEN-1 nuclease in carrying out its essential function, *Mol. Cell. Biol.*, 1997, **17**, 2136–2142.
13. T. Formosa and T. Nitiss, Dna2 mutants reveal interactions with DNA polymerase alpha and Ctf4, a Pol α accessory factor, and show that full DNA2 helicase activity is not essential for growth, *Genetics*, 1999, **151**, 1459–1470.
14. D. F. Fiorentino and G. R. Crabtree, Characterization of *Saccharomyces cerevisiae dna2* mutants suggests a role for the helicase late in S phase, *Mol. Biol. Cell*, 1997, **8**, 2519–2537.
15. L. Aravind, D. R. Walker and E. V. Koonin, Conserved domains in DNA repair proteins and evolution of repair systems, *Nucleic Acids Res.*, 1999, **27**, 1223–1242.
16. S.-H. Bae, J.-A. Kim, E. Choi, K.-H. Lee, H.-Y. Kang, H.-D. Kim, J.-H. Kim, K.-H. Bae, Y. Cho, C. Park and Y.-S. Seo, Tripartite structure of *Saccharomyces cerevisiae* Dna2 helicase/endonuclease, *Nucleic Acids Res.*, 2001, **29**, 3069–3079.
17. M. E. Budd, W.-C. Choe and J. L. Campbell, The nuclease activity of the yeast Dna2 protein, which is related to the RecB-like nucleases, is essential *in vivo*, *J. Biol. Chem.*, 2000, **275**, 16518–16529.

18. K.-H. Lee, D. W. Kim, S.-H. Bae, J.-A. Kim, G.-H. Ryu, Y.-N. Kwon, K.-A. Kim, H.-S. Koo and Y.-S. Seo, The endonuclease activity of the yeast Dna2 enzyme is essential *in vivo*, *Nucleic Acids Res.*, 2000, **28**, 2873–2881.
19. S.-H. Bae, E. Choi, K. Lee, J. Park, S. Lee and Y. Seo, Dna2 of *Saccharomyces cerevisiae* possesses a single-stranded DNA-specific endonuclease activity that is able to act on double-stranded DNA in the presence of ATP, *J. Biol. Chem.*, 1998, **273**, 26880–26890.
20. K. Czaplinski, N. Majlesi, T. Banerjee and S. W. Peltz, Mtt1 is a Upf1-like helicase that interacts with the translation termination factors and whose overexpression can modulate termination efficiency, *RNA*, 2000, **6**, 730–743.
21. Z. Cheng, D. Muhlrad, M. K. Lim, R. Parker and H. Song, Structural and functional insights into the human Upf1 helicase core, *EMBO J.*, 2007, **26**, 253–264.
22. T. Masuda-Sasa, P. Polaczek and J. L. Campbell, Single strand annealing and ATP-independent strand exchange activities of yeast and human DNA2: possible role in Okazaki fragment maturation, *J. Biol. Chem.*, 2006, **281**, 38555–38564.
23. M. E. Budd and J. L. Campbell, The pattern of sensitivity of yeast *dna2* mutants to DNA damaging agents suggests a role in DSB and post-replication repair pathways, *Mutat. Res.*, 2000, **459**, 173–186.
24. K.-H. Bae, H.-S. Kim, S.-H. Bae, H.-Y. Kang, S. Brill and Y.-S. Seo, Bimodal interaction between replication-protein A and Dna2 is critical for Dna2 function both *in vivo* and *in vitro*, *Nucleic Acids Res.*, 2003, **31**, 3006–3015.
25. S.-H. Bae, D. W. Kim, J. Kim, J.-H. Kim, D.-H. Kim, H.-D. Kim, H.-Y. Kang and Y.-S. Seo, Coupling of DNA helicase and endonuclease activities of yeast Dna2 facilitates Okazaki fragment processing, *J. Biol. Chem.*, 2002, **277**, 26632–26641.
26. S. H. Bae, K.-H. Bae, J. A. Kim and Y. S. Seo, RPA governs endonuclease switching during processing of Okazaki fragments in eukaryotes, *Nature*, 2001, **412**, 456–461.
27. T. Masuda-Sasa, P. Polaczek, X. P. Peng, L. Chen and J. L. Campbell, Processing of G4 DNA by Dna2 Helicase/nuclease and RPA provides insights into the mechanism of Dna2/RPA substrate recognition, *J. Biol. Chem.*, 2008.
28. Y. Ho, A. Gruhler, A. Heilbut, G. D. Bader, L. Moore, S.-L. Adams, A. Millar, P. Taylor, K. Bennett, K. Boutilier, L. Yang, C. Wolting, I. Donaldson, S. Schandorff, J. Shewnarane, M. Vo, J. Taggart, M. Goudreault, B. Muskat, C. Alfarano, D. Dewar, Z. Lin, K. Michalickova, A. R. Willem, H. Sassin, P. A. Nielsen, K. J. Rasmussen, J. R. Andersen, L. E. Johansen, L. H. Hansen, H. Jespersen, A. Podtelejnikov, E. Nielsen, J. Crawford, V. Poulsen, B. D. Sørensen, J. Matthiesen, C. R. Hendrickson, F. Gleeson, T. Pawson, M. F. Moran, D. Durocher, M. Mann,

C. W. V. Hogue, D. Figeys and M. Tyers, Systematic identification of protein complexes in *Saccharomyces cerevisiae* by mass spectrometry, *Nature*, 2002, **415**, 180–182.

29. G. Butland, J. M. Peregrin-Alvarez, J.-L. Li, W. Yang, X. Yang, V. Canadien, A. Starostine, D. Richards, B. Beattie, N. Krogan, M. Davey, J. Parkinson, J. F. Greenblatt and A. Emili, Interaction network containing conserved and essential protein complexes in *Escherichia coli*, *Nature*, 2005, **433**, 531–537.
30. N. J. Krogan, G. Cagney, H. Yu, G. Zhong, X. Guo, A. Ignatchenko, J. Li, S. Pu, N. Datta, A. P. Tikuisis, T. Punna, J. M. Peregrin-Alvarez, M. Shales, X. Zhang, M. Davey, M. D. Robinson, A. Paccanaro, J. E. Bray, A. Sheung, B. Beattie, D. P. Richards, V. Canadien, A. Lalev, F. Mena, P. Wong, A. Starostine, M. M. Canete, J. Vlasblom, S. Wu, C. Orsi, S. R. Collins, S. Chandran, R. Haw, J. J. Rilstone, K. Gandi, N. J. Thompson, G. Musso, P. St Onge, S. Ghanny, M. H. Lam, G. Butland, A. M. Altaf-Ul, S. Kanaya, A. Shilatifard, E. O'shea, J. S. Weissman, C. J. Ingles, T. R. Hughes, J. Parkinson, M. Gerstein, S. J. Wodak, A. Emili and J. F. Greenblatt, Global landscape of protein complexes in the yeast *Saccharomyces cerevisiae*, *Nature*, 2006, **440**, 637–643.
31. S.-H. Bae and Y.-S. Seo, Characterization of the enzymatic properties of the yeast Dna2 helicase/endonuclease suggests a new model for Okazaki fragment processing, *J. Biol. Chem.*, 2000, **275**, 38022–38031.
32. R. S. Murante, L. Huang, J. J. Turchi and R. A. Bambara, The calf 5′ to 3′ exonuclease is also an endonuclease with both activities dependent on primers annealed upstream of the point of cleavage, *J. Biol. Chem.*, 1994, **269**, 1191–1196.
33. R. S. Murante, J. A. Rumbaugh, C. J. Barnes, J. R. Norton and R. A. Bambara, Calf RTH-1 nuclease can remove the initiator RNAs of Okazaki fragments by endonuclease activity, *J. Biol. Chem.*, 1996, **271**, 25888–25897.
34. Y. H. Jin, R. Obert, P. M. J. Burgers, T. A. Kunkel, M. A. Resnick and D. A. Gordenin, The 3′–5′ exonuclease of DNA polymerase δ can substitute for the 5′ flap endonuclease Rad27/Fen1 in processing Okazaki fragments and preventing genome instability, *Proc. Natl. Acad. Sci. U.S.A.*, 2001, **98**, 5122–5127.
35. D. X. Tishkoff, N. Filosi, G. M. Gaida and R. D. Kolodner, A novel mutation avoidance mechanism dependent on *Saccharomyces cerevisiae* RAD27 is distinct from DNA mismatch repair, *Cell*, 1997, **88**, 253–263.
36. R. Ayyagari, X. V. Gomes, D. A. Gordenin and P. M. J. Burgers, Okazaki fragment maturation in yeast. I. Distribution of functions between Fen1 and Dna2, *J. Biol. Chem.*, 2003, **278**, 1618–1625.
37. Y. H. Jin, R. Ayyagari, M. A. Resnick, D. A. Gordenin and P. M. J. Burgers, Okazaki fragment maturation in yeast. II. Cooperation between the polymerase and 3′ to 5′ exonuclease activities of Pol δ in the creation of a ligatable nick, *J. Biol. Chem.*, 2003, **278**, 1626–1633.

38. P. Garg, C. M. Stith, N. Sabouri, E. Johansson and P. M. Burgers, Idling by DNA polymerase δ maintains a ligatable nick during lagging-strand DNA replication, *Genes Dev.*, 2004, **18**, 2764–2773.
39. E. Warbrick, D. P. Lane, D. M. Glover and L. S. Cox, Homologous regions of Fen1 and p21^{Cip1} compete for binding to the same site on PCNA: A potential mechanism to coordinate DNA replication and repair, *Oncogene*, 1997, **14**, 2313–2321.
40. S. Sakurai, K. Kitano, H. Yamaguchi, K. Hamada, K. Okada, K. Fukuda, M. Uchida, E. Ohtsuka, H. Morioka and T. Hakoshima, Structural basis for recruitment of human flap endonuclease 1 to PCNA, *EMBO J.*, 2004, **24**, 683–693.
41. Y. Xie, Y. Liu, J. L. Argueso, L. A. Henricksen, H.-I. Kao, R. A. Bambara and E. Alani, Identification of *rad27* mutations that confer differential defects in mutation avoidance, repeat tract instability, and flap cleavage, *Mol. Cell. Biol.*, 2001, **21**, 4889–4899.
42. E. Johansson, P. Garg and P. M. J. Burgers, The Pol32 subunit of DNA polymerase δ contains separable domains for processive replication and proliferating cell nuclear antigen (PCNA) binding, *J. Biol. Chem.*, 2004, **279**, 1907–1915.
43. M. L. Rossi and R. A. Bambara, Reconstituted Okazaki fragment processing indicates two pathways of primer removal, *J. Biol. Chem.*, 2006, **281**, 26051–26061.
44. M. E. Budd, C. C. Reis, S. Smith, K. Myung and J. L. Campbell, Evidence suggesting that Pif1 helicase functions in DNA replication with the Dna2 helicase/nuclease and DNA polymerase δ, *Mol. Cell. Biol.*, 2006, **26**, 2490–2500.
45. J.-B. Boule, L. R. Vega and V. A. Zakian, The yeast Pif1p helicase removes telomerase from telomeric DNA, *Nature*, 2005, **438**, 57–61.
46. V. P. Schulz and V. A. Zakian, The *Saccharomyces PIF1* DNA helicase inhibits telomere elongation and de novo telomere formation, *Cell*, 1994, **76**, 145–155.
47. J.-Q. Zhou, E. K. Monson, S.-C. Teng, V. P. Schulz and V. A. Zakian, Pif1p helicase, a catalytic inhibitor of telomerase in yeast, *Science*, 2000, **289**, 771–774.
48. A. Lahaye, H. Stahl, D. Thine-Sempoux and F. Foury, PIF1: a DNA helicase in yeast mitochondria, *EMBO J.*, 1991, **10**, 997–1007.
49. G.-H. Ryu, H. Tanaka, D.-H. Kim, J.-H. Kim, S.-H. Bae, Y.-N. Kwon, J. S. Rhee, S. A. MacNeill and Y.-S. Seo, Genetic and biochemical analyses of Pfh1 DNA helicase function in fission yeast, *Nucleic Acids Res.*, 2004, **32**, 4205–4216.
50. H. Tanaka, G.-H. Ryu, Y.-S. Seo, K. Tanaka, H. Okayama, S. A. MacNeill and Y. Yuasa, The fission yeast pfh1+ gene encodes an essential 5′ to 3′ DNA helicase required for the completion of S-phase, *Nucleic Acids Res.*, 2002, **30**, 4728–4739.
51. J.-Q. Zhou, H. Qi, V. P. Schulz, M. K. Mateyak, E. K. Monson and V. A. Zakian, *Schizosaccharomyces pombe* pfh1+ encodes an essential 5′ to 3′

DNA helicase that is a member of the PIF1 subfamily of DNA helicases, *Mol. Biol. Cell*, 2002, **13**, 2180–2191.
52. M. L. Rossi, J. E. Pike, W. Wang, P. M. Burgers, J. L. Campbell and R. A. Bambara, Pif1 helicase directs eukaryotic Okazaki fragments toward the two-nuclease cleavage pathway for primer removal, *J. Biol. Chem.*, 2008, **283**, 27483–27493.
53. J. A. Stewart, J. L. Campbell and R. A. Bambara, Flap endonuclease disengages Dna2 helicase/nuclease from Okazaki fragment flaps, *J. Biol. Chem.*, 2006, **281**, 38565–38572.
54. H.-I. Kao, J. L. Campbell and R. A. Bambara, Dna2p helicase/nuclease is a tracking protein, like FEN1, for flap cleavage during Okazaki fragment maturation, *J. Biol. Chem.*, 2004, **279**, 50840–50849.
55. J. A. Stewart, A. S. Miller, J. L. Campbell and R. A. Bambara, Dynamic removal of replication protein A by Dna2 facilitates primer cleavage during Okazaki fragment processing in *Saccharomyces cerevisiae*, *J. Biol. Chem.*, 2008, **283**, 31356–31365.
56. A. I. Arunkumar, V. Klimovich, X. Jiang, R. D. Ott, L. Mizoue, E. Fanning and W. J. Chazin, Insights into hRPA32 C-terminal domain-mediated assembly of the simian virus 40 replisome, *Nat. Struct. Mol. Biol.*, 2005, **12**, 332–339.
57. E. Fanning, V. Klimovich and A. R. Nager, A dynamic model for replication protein A (RPA) function in DNA processing pathways, *Nucleic Acids Res*, 2006, **34**, 4126–4137.
58. H. I. Kao, J. Veeraraghavan, P. Polaczek, J. L. Campbell and R. A. Bambara, On the roles of *Saccharomyces cerevisiae* Dna2p and FEN1 in Okazaki fragment processing, *J. Biol. Chem.*, 2004, **279**, 15014–15024.
59. P. Szankasi and G. R. Smith, A role for exonuclease I from *S. pombe* in mutation avoidance and mismatch correction, *Science*, 1995, **267**, 1166–1169.
60. J. M. Bradbury and S. P. Jackson, The complex matter of DNA double-strand break detection, *Biochem. Soc. Trans.*, 2003, **31**, 40–44.
61. T. T. Paull and M. Gellert, The 3′ to 5′ exonuclease activity of Mre11 facilitates repair of DNA double-strand breaks, *Mol. Cell*, 1998, **1**, 969–979.
62. K. M. Trujillo and P. Sung, DNA structure-specific nuclease activities in the *Saccharomyces cerevisiae* Rad50*Mre11 complex, *J. Biol. Chem.*, 2001, **276**, 35458–35464.
63. K. M. Trujillo, S. S. Yuan, E. Y. Lee and P. Sung, Nuclease activities in a complex of human recombination and DNA repair factors Rad50, Mre11, and p95, *J. Biol. Chem.*, 1998, **273**, 21447–21450.
64. D. A. Bressan, H. A. Olivares, B. E. Nelms and J. H. Petrini, Alteration of N-terminal phosphoesterase signature motifs inactivates *Saccharomyces cerevisiae* Mre11, *Genetics*, 1998, **150**, 591–600.
65. S. Wilson, M. Tavassoli and F. Z. Watts, *Schizosaccharomyces pombe* rad32 protein: a phosphoprotein with an essential phosphoesterase motif

required for repair of DNA double strand breaks, *Nucleic Acids Res.*, 1998, **26**, 5261–5269.

66. K. P. Hopfner, A. Karcher, L. Craig, T. T. Woo, J. P. Carney and J. A. Tainer, Structural biochemistry and interaction architecture of the DNA double-strand break repair Mre11 nuclease and Rad50-ATPase, *Cell*, 2001, **105**, 473–485.
67. T. T. Paull and M. Gellert, Nbs1 potentiates ATP-driven DNA unwinding and endonuclease cleavage by the Mre11/Rad50 complex, *Genes Dev.*, 1999, **13**, 1276–1288.
68. D. D'Amours and S. P. Jackson, The Mre11 complex: at the crossroads of dna repair and checkpoint signalling, *Nat. Rev. Mol. Cell Biol.*, 2002, **3**, 317–327.
69. K. P. Hopfner, A. Karcher, D. S. Shin, L. Craig, L. M. Arthur, J. P. Carney and J. A. Tainer, Structural biology of Rad50 ATPase: ATP-driven conformational control in DNA double-strand break repair and the ABC-ATPase superfamily, *Cell*, 2000, **101**, 789–800.
70. E. Alani, S. Subbiah and N. Kleckner, The yeast RAD50 gene encodes a predicted 153-kD protein containing a purine nucleotide-binding domain and two large heptad-repeat regions, *Genetics*, 1989, **122**, 47–57.
71. W. E. Raymond and N. Kleckner, RAD50 protein of *S. cerevisiae* exhibits ATP-dependent DNA binding, *Nucleic Acids Res.*, 1993, **21**, 3851–3856.
72. K. P. Hopfner, L. Craig, G. Moncalian, R. A. Zinkel, T. Usui, B. A. Owen, A. Karcher, B. Henderson, J. L. Bodmer, C. T. McMurray, J. P. Carney, J. H. Petrini and J. A. Tainer, The Rad50 zinc-hook is a structure joining Mre11 complexes in DNA recombination and repair, *Nature*, 2002, **418**, 562–566.
73. G. Moncalian, B. Lengsfeld, V. Bhaskara, K. P. Hopfner, A. Karcher, E. Alden, J. A. Tainer and T. T. Paull, The Rad50 signature motif: essential to ATP binding and biological function, *J. Mol. Biol.*, 2004, **335**, 937–951.
74. L. Chen, K. M. Trujillo, S. Van Komen, D. H. Roh, L. Krejci, L. K. Lewis, M. A. Resnick, P. Sung and A. E. Tomkinson, Effect of amino acid substitutions in the Rad50 ATP binding domain on DNA double strand break repair in yeast, *J. Biol. Chem.*, 2005, **280**, 2620–2627.
75. D. E. Anderson, K. M. Trujillo, P. Sung and H. P. Erickson, Structure of the Rad50.Mre11 DNA repair complex from *Saccharomyces cerevisiae* by electron microscopy, *J. Biol. Chem.*, 2001, **276**, 37027–37033.
76. M. Lichten, Rad50 connects by hook or by crook, *Nat. Struct. Mol. Biol.*, 2005, **12**, 392–393.
77. V. Bhaskara, A. Dupre, B. Lengsfeld, B. B. Hopkins, A. Chan, J. H. Lee, X. Zhang, J. Gautier, V. Zakian and T. T. Paull, Rad50 adenylate kinase activity regulates DNA tethering by Mre11/Rad50 complexes, *Mol. Cell*, 2007, **25**, 647–661.
78. A. Jazayeri, A. Balestrini, E. Garner, J. E. Haber and V. Costanzo, Mre11-Rad50-Nbs1-dependent processing of DNA breaks generates

oligonucleotides that stimulate ATM activity, *EMBO J.*, 2008, **27**, 1953–1962.

79. C. Zierhut and J. F. Diffley, Break dosage, cell cycle stage and DNA replication influence DNA double strand break response, *EMBO J.*, 2008, **27**, 1875–1885.
80. K. Ho, Induction of DNA double strand breaks by X-rays in a radiosensitive strain of the yeast *Saccharomyces cerevisiae*, *Mutat. Res.*, 1975, **30**, 377–334.
81. S. Moreau, J. R. Ferguson and L. S. Symington, The nuclease activity of Mre11 is required for meiosis but not for mating type switching, end joining, or telomere maintenance, *Mol. Cell. Biol.*, 1999, **19**, 556–566.
82. S. Moreau, E. A. Morgan and L. S. Symington, Overlapping functions of the *Saccharomyces cerevisiae* Mre11, Exo1 and Rad27 nucleases in DNA metabolism, *Genetics*, 2001, **159**, 1423–1433.
83. S. E. Lee, D. A. Bressan, J. H. Petrini and J. E. Haber, Complementation between N-terminal *Saccharomyces cerevisiae* mre11 alleles in DNA repair and telomere length maintenance, *DNA Repair (Amst.)*, 2002, **1**, 27–40.
84. B. Llorente and L. S. Symington, The Mre11 nuclease is not required for 5′ to 3′ resection at multiple HO-induced double-strand breaks, *Mol. Cell. Biol.*, 2004, **24**, 9682–9694.
85. S. Keeney, C. N. Giroux and N. Kleckner, Meiosis-specific DNA double-strand breaks are catalyzed by Spo11, a member of a widely conserved protein family, *Cell*, 1997, **88**, 375–384.
86. E. Alani, R. Padmore and N. Kleckner, Analysis of wild-type and rad50 mutants of yeast suggests an intimate relationship between meiotic chromosome synapsis and recombination, *Cell*, 1990, **61**, 419–436.
87. T. Usui, T. Ohta, H. Oshiumi, J. Tomizawa, H. Ogawa and T. Ogawa, Complex formation and functional versatility of Mre11 of budding yeast in recombination, *Cell*, 1998, **95**, 705–716.
88. L. K. Lewis, G. Karthikeyan, J. W. Westmoreland and M. A. Resnick, Differential suppression of DNA repair deficiencies of yeast *rad50, mre11* and *xrs2* mutants by *EXO1* and *TLC1* (the RNA component of telomerase), *Genetics*, 2002, **160**, 49–62.
89. M. E. Budd and J. L. Campbell, Interplay of Mre11 nuclease with Dna2 plus Sgs1 in Rad51-dependent recombinational repair, *PLoS ONE*, 2009, **4**, e4267.
90. Z. Zhu, W.-H. Chung, E. Y. Shi, S. E. Lee and G. Ira, Sgs1 helicase and two nucleases Dna2 and Exo1 resect DNA double-strand break ends, *Cell*, 2008, **134**, 981–994.
91. E. P. Mimitou and L. S. Symington, Sae2, Exo1 and Sgs1 collaborate in DNA double-strand break processing, *Nature*, 2008, **455**, 770–774.
92. S. Liao, T. Toczylowski and H. Yan, Identification of the *Xenopus* DNA2 protein as a major nuclease for the $5' \rightarrow 3'$ strand-specific processing of DNA ends, *Nucleic Acids Res.*, 2008, **36**, 6091–6100.

93. M. Lisby, J. H. Barlow, R. C. Burgess and R. Rothstein, Choreography of the DNA damage response: spatiotemporal relationships among checkpoint and repair proteins, *Cell*, 2004, **118**, 699–713.
94. M. Lisby and R. Rothstein, DNA damage checkpoint and repair centers, *Curr. Opin. Cell Biol.*, 2004, **16**, 328–334.
95. A. Emili, MEC1-dependent phosphorylation of Rad9p in response to DNA damage, *Mol. Cell*, 1998, **2**, 183–189.
96. J. E. Vialard, C. S. Gilbert, C. M. Green and N. F. Lowndes, The budding yeast Rad9 checkpoint protein is subjected to Mec1/Tel1-dependent hyperphosphorylation and interacts with Rad53 after DNA damage, *EMBO J.*, 1998, **17**, 5679–5688.
97. E. Baroni, V. Viscardi, H. Cartagena-Lirola, G. Lucchini and M. P. Longhese, The functions if budding yeast Sae2 in the DNA damage response require Mec1- and Tel1-dependent phosphorylation, *Mol. Cell. Biol.*, 2004, **24**, 4151–4165.
98. J. L. Ma, S. J. Lee, J. K. Duong and D. F. Stern, Activation of the checkpoint kinase Rad53 by the phosphatidyl inositol kinase-like kinase Mec1, *J. Biol. Chem.*, 2006, **281**, 3954–3963.
99. N. F. Lowndes and J. R. Murguia, Sensing and responding to DNA damage, *Curr. Opin. Genet. Dev.*, 2000, **10**, 17–25.
100. J. P. Carney, R. S. Maser, H. Olivares, E. M. Davis, M. Le Beau, J. R. Yates, L. Hays, W. F. Morgan and J. H. Petrini, The HMRE11/HRAD50 protein complex and Nijmegen breakage syndrome-linkage of double-strand break repair to the cellular DNA-damage response, *Cell*, 1998, **93**, 477–486.
101. G. S. Stewart, R. S. Maser, T. Stankovic, D. A. Bressan, M. I. Kaplan, N. G. Jaspers, A. Raams, P. J. Byrd, J. H. Petrini and A. M. Taylor, The DNA double-strand break repair gene hMRE11 is mutated in individuals with an ataxia-telangiectasia-like disorder, *Cell*, 1999, **99**, 577–587.
102. D. Nakada, Y. Hirano and K. Sugimoto, Requirement of the Mre11 complex and exonuclease 1 for activation of the Mec1 signaling pathway, *Mol. Cell. Biol.*, 2004, **24**, 10016–10025.
103. D. D'Amours and S. P. Jackson, The yeast Xrs2 complex functions in S phase checkpoint regulation, *Genes Dev.*, 2001, **15**, 2238–2249.
104. M. Clerici, D. Mantiero, G. Lucchini and M. P. Longhese, The *Saccharomyces cerevisiae* Sae2 protein negatively regulates DNA damage checkpoint signalling, *EMBO Rep.*, 2006, **7**, 212–218.
105. J. H. Lee and T. T. Paull, ATM activation by DNA double-strand breaks through the Mre11-Rad50-Nbs1 complex, *Science*, 2005, **308**, 551–554.
106. T. Usui, H. Ogawa and J. H. Petrini, A DNA damage response pathway controlled by Tel1 and the Mre11 complex, *Mol. Cell*, 2001, **7**, 1255–1266.
107. D. Nakada, K. Matsumoto and K. Sugimoto, ATM-related Tel1 associates with double-strand breaks through an Xrs2-dependent mechanism, *Genes Dev.*, 2003, **17**, 1957–1962.

108. C. Z. Bachrati and I. D. Hickson, RecQ helicases: guardian angels of the DNA replication fork, *Chromosoma*, 2008, **117**, 219–233.
109. B. A. Kudlow, B. K. Kennedy and R. J. Monnat Jr., Werner and Hutchinson–Gilford progeria syndromes: mechanistic basis of human progeroid diseases, *Nat. Rev. Mol. Cell. Biol.*, 2007, **8**, 394–404.
110. L. S. Cox and R. G. Faragher, From old organisms to new molecules: integrative biology and therapeutic targets in accelerated human ageing, *Cell. Mol. Life Sci.*, 2007, **64**, 2620–2641.
111. R. Hand and J. German, A retarded rate of DNA chain growth in Bloom's syndrome, *Proc. Natl. Acad. Sci. U.S.A.*, 1975, **72**, 758–762.
112. U. Lonn, S. Lonn, U. Nylen, G. Winblad and J. German, An abnormal profile of DNA replication intermediates in Bloom's syndrome, *Cancer Res.*, 1990, **50**, 3141–3145.
113. A. M. Rodriguez-Lopez, D. A. Jackson, F. Iborra and L. S. Cox, Asymmetry of DNA replication fork progression in Werner's syndrome, *Aging Cell*, 2002, **1**, 30–39.
114. J. M. Sidorova, N. Li, A. Folch and R. J. Monnat Jr, The RecQ helicase WRN is required for normal replication fork progression after DNA damage or replication fork arrest, *Cell Cycle*, 2008, **7**, 796–806.
115. S. Gangloff, J. P. McDonald, C. Bendixen, L. Arthur and R. Rothstein, The yeast type I topoisomerase TOP3 interacts with SGS1, a DNA helicase, *Mol. Cell. Biol.*, 1994, **14**, 8391–8398.
116. P. M. Watt, I. D. Hickson, R. H. Borts and E. J. Louis, *SGS1*, a homologue of the Bloom's and Werner's syndrome genes, is required for maintenance of genome stability in *Saccharomyces cerevisiae*, *Genetics*, 1996, **144**, 935–945.
117. I. D. Hickson, RecQ helicases: caretakers of the genome, *Nat. Rev. Cancer*, 2003, **3**, 169–178.
118. W. Li, S.-M. Kim, J.-K. Lee and W. G. Dunphy, Absence of BLM leads to accumulation of chromosomal DNA breaks during both unperturbed and disrupted S phases, *J. Cell Biol.*, 2004, **165**, 801–812.
119. C.-Y. Chen, J. Graham and H. Yan, Evidence for a replication function of FFA-1, the *Xenopus* orthologue of Werner syndrome protein, *J. Cell Biol.*, 2001, **152**, 985–996.
120. C. Frei and S. M. Gasser, The yeast Sgs1p helicase acts upstream of Rad53p in the DNA replication checkpoint and colocalizes with Rad53p in S-phase-specific foci, *Genes Dev.*, 2000, **14**, 81–96.
121. A. Constantinou, A. A. Davies and S. C. West, Branch migration and Holliday junction resolution catalyzed by activities from mammalian cells, *Cell*, 2001, **104**, 259–268.
122. A. M. Rodriguez-Lopez, D. A. Jackson, J. O. Nehlin, F. Iborra, A. V. Warren and L. S. Cox, Characterisation of the interaction between WRN, the helicase/exonuclease defective in progeroid Werner's syndrome, and an essential replication factor, PCNA, *Mech. Ageing Dev.*, 2003, **124**, 167–174.

123. O. Imamura and J. L. Campbell, The human Bloom syndrome gene suppresses the DNA replication and repair defects of yeast *dna2* mutants, *Proc. Natl. Acad. Sci. U.S.A.*, 2003, **100**, 8193–8198.
124. S. Sharma, J. A. Sommers and R. M. Brosh Jr, *In vivo* function of the conserved non-catalytic domain of Werner syndrome helicase in DNA replication, *Hum. Mol. Genet.*, 2004, **13**, 2247–2261.
125. J. Brosh, R. M. C. von Kobbe, J. A. Sommers, P. Karmakar, P. L. Opresko, J. Piotrowski, I. Dianova, G. L. Dianov and V. A. Bohr, Werner syndrome protein interacts with human flap endonuclease 1 and stimulates its cleavage activity, *EMBO J.*, 2001, **20**, 5791–5801.
126. S. Sharma, J. A. Sommers, L. Wu, V. A. Bohr, I. D. Hickson and R. M. Brosh Jr, Stimulation of flap endonuclease-1 by the Bloom's syndrome protein, *J. Biol. Chem.*, 2004, **279**, 9847–9856.
127. M. D. Huber, M. L. Duquette, J. C. Shiels and N. Maizels, A conserved G4 DNA binding domain in RecQ family helicases, *J. Mol. Biol.*, 2006, **358**, 1071–1080.
128. J. R. Mullen, V. Kaliraman and S. J. Brill, Bipartite structure of the SGS1 DNA helicase in *Saccharomyces cerevisiae*, *Genetics*, 2000, **154**, 1101–1114.
129. W. Wang and R. A. Bambara, Human Bloom protein stimulates flap endonuclease 1 activity by resolving DNA secondary structure, *J. Biol. Chem.*, 2005, **280**, 5391–5399.
130. J. D. Bartos, W. Wang, J. E. Pike and R. A. Bambara, Mechanisms by which Bloom protein can disrupt recombination intermediates of Okazaki fragment maturation, *J. Biol. Chem.*, 2006, **281**, 32227–32239.
131. T. Weitao, M. Budd and J. L. Campbell, Evidence that yeast SGS1, DNA2, SRS2, and FOB1 interact to maintain rDNA stability, *Mutat. Res.*, 2003, **532**, 157–172.
132. L. L. Mays Hoopes, M. Budd, W. Choe, T. Weitao and J. L. Campbell, Mutations in DNA replication genes reduce yeast life span, *Mol. Cell. Biol.*, 2002, **22**, 4136–4146.
133. A. M. Rodriguez-Lopez, M. C. Whitby, C. M. Borer, M. A. Bachler and L. S. Cox, Correction of proliferation and drug sensitivity defects in the progeroid Werner's syndrome by Holliday junction resolution, *Rejuvenation Res.*, 2007, **10**, 27–40.
134. R. A. Kim and J. C. Wang, Identification of the yeast TOP3 gene product as a single strand-specific DNA topoisomerase, *J. Biol. Chem.*, 1992, **267**, 17178–17185.
135. W. M. Fricke, V. Kaliraman and S. J. Brill, Mapping the DNA topoisomerase III binding domain of the Sgs1 DNA helicase, *J. Biol. Chem.*, 2001, **276**, 8848–8855.
136. L. Wu and I. D. Hickson, The Bloom's syndrome helicase suppresses crossing over during homologous recombination, *Nature*, 2003, **426**, 870–874.

137. M. Wagner, G. Price and R. Rothstein, The absence of Top3 reveals an interaction between the Sgs1 and Pif1 DNA helicases in *Saccharomyces cerevisiae*, *Genetics*, 2006, **174**, 555–573.
138. A. Goodwin, S. W. Wang, T. Toda, C. Norbury and I. D. Hickson, Topoisomerase III is essential for accurate nuclear division in *Schizosaccharomyces pombe*, *Nucleic Acids Res.*, 1999, **27**, 4050–4058.
139. M. Maftahi, C. S. Han, L. D. Langston, J. C. Hope, N. Zigouras and G. A. Freyer, The top3(+) gene is essential in *Schizosaccharomyces pombe* and the lethality associated with its loss is caused by Rad12 helicase activity, *Nucleic Acids Res.*, 1999, **27**, 4715–4724.
140. L. Wu, K. L. Chan, C. Ralf, D. A. Bernstein, P. L. Garcia, V. A. Bohr, A. Vindigni, P. Janscak, J. L. Keck and I. D. Hickson, The HRDC domain of BLM is required for the dissolution of double Holliday junctions, *EMBO J.*, 2005, **24**, 2679–2687.
141. V. Yankiwski, J. Noonan and N. F. Neff, The C-terminal domain of the Bloom syndrome DNA helicase is essential for genomic stability, *BMC Cell Biol.*, 2001, **2**, 11.
142. S. A. Bastin-Shanower, W. M. Fricke, J. R. Mullen and S. J. Brill, The mechanism of Mus81-Mms4 cleavage site selection distinguishes it from the homologous endonuclease Rad1-Rad10, *Mol. Cell. Biol.*, 2003, **23**, 3487–3496.
143. F. Fabre, A. Chan, W.-D. Heyer and S. Gangloff, Alternate pathways involving Sgs1/Top3, Mus81/ Mms4, and Srs2 prevent formation of toxic recombination intermediates from single-stranded gaps created by DNA replication, *Proc. Natl. Acad. Sci. U.S.A.*, 2002, **99**, 16887–16892.
144. M. Ii, T. Ii and S. J. Brill, Mus81 functions in the quality control of replication forks at the rDNA and is involved in the maintenance of rDNA repeat number in *Saccharomyces cerevisiae*, *Mutat. Res.*, 2007, **625**, 1–19.
145. M. Ii and S. J. Brill, Roles of SGS1, MUS81, and RAD51 in the repair of lagging-strand replication defects in *Saccharomyces cerevisiae*, *Curr. Genet.*, 2005, **48**, 213–225.
146. A. H. Y. Tong, M. Evangelista, A. B. Parsons Xu, G. D. Bader, N. Page, M. Robinson, S. Raghibizadeh, C. W. V. Hogue, H. Bussey, B. Andrews, M. Tyers and C. Boone, Systematic genetic analysis with ordered arrays of yeast deletion mutants, *Science*, 2001, **294**, 2364–2368.
147. S. Loeillet, B. Palancade, M. Cartron, A. Thierry, G.-F. Fichard, B. Dujon, V. Doye and A. Nicolas, Genetic network interactions among replication, repair and nuclear pore deficiencies in yeast, *DNA Repair (Amst.)*, 2005, **4**, 459–468.
148. C. L. Doe, J. S. Ahn, J. Dixon and M. C. Whitby, Mus81-Eme1 and Rqh1 involvement in processing stalled and collapsed replication forks, *J. Biol. Chem.*, 2002, **277**, 32753–32759.

149. B. Rockmill, J. C. Fung, S. S. Branda and G. S. Roeder, The Sgs1 helicase regulates chromosome synapsis and meiotic crossing over, *Curr. Biol.*, 2003, **13**, 1954–1962.
150. S. D. Oh, J. P. Lao, P. Y. Hwang, A. F. Taylor, G. R. Smith and N. Hunter, BLM ortholog, Sgs1, prevents aberrant crossing-over by suppressing formation of multichromatid joint molecules, *Cell*, 2007, **130**, 259–272.
151. S. D. Oh, J. P. Lao, A. F. Taylor, G. R. Smith and N. Hunter, RecQ helicase, Sgs1, and XPF family endonuclease, Mus81-Mms4, resolve aberrant joint molecules during meiotic recombination, *Mol. Cell*, 2008, **31**, 324–336.
152. L. Jessop and M. Lichten, Mus81/Mms4 endonuclease and Sgs1 helicase collaborate to ensure proper recombination intermediate metabolism during meiosis, *Mol. Cell*, 2008, **31**, 313–323.
153. E. R. Parrilla-Castellar, S. J. Arlander and L. Karnitz, Dial 9-1-1 for DNA damage: the Rad9-Hus1-Rad1 (9-1-1) clamp complex, *DNA Repair (Amst.)*, 2004, **3**, 1009–1014.
154. S. Sengupta, S. P. Linke, R. Pedeux, Q. Yang, J. Farnsworth, S. H. Garfield, K. Valerie, J. W. Shay, N. A. Ellis, B. Wasylyk and C. C. Harris, BLM helicase-dependent transport of p53 to sites of stalled DNA replication forks modulates homologous recombination, *EMBO J.*, 2003, **22**, 1210–1222.
155. J. A. Cobb, L. Bjergbaek, K. Shimada, C. Frei and S. M. Gasser, DNA polymerase stabilization at stalled replication forks requires Mec1 and the RecQ helicase Sgs1, *EMBO J.*, 2003, **22**, 4325–4336.
156. L. Bjergbaek, J. A. Cobb, M. Tsai-Pflugfelder and S. M. Gasser, Mechanistically distinct roles for Sgs1p in checkpoint activation and replication fork maintenance, *EMBO J.*, 2005, **24**, 405–417.
157. J. A. Cobb, T. Schleker, V. Rojas, L. Bjergbaek, J. A. Tercero and S. M. Gasser, Replisome instability, fork collapse, and gross chromosomal rearrangements arise synergistically from Mec1 kinase and RecQ helicase mutations, *Genes Dev.*, 2005, **19**, 3055–3069.
158. J. A. Tercero, M. P. Longhese and J. F. Diffley, A central role for DNA replication forks in checkpoint activation and response, *Mol. Cell*, 2003, **11**, 1323–1336.
159. V. Paciotti, M. Clerici, M. Scotti, G. Lucchini and M. P. Longhese, Characterization of mec1 kinase-deficient mutants and of new hypomorphic mec1 alleles impairing subsets of the DNA damage response pathway, *Mol. Cell. Biol.*, 2001, **21**, 3913–3925.
160. S. L. Davies, P. S. North and I. D. Hickson, Role for BLM in replication-fork restart and suppression of origin firing after replicative stress, *Nat. Struct. Mol. Biol.*, 2007, **14**, 677–679.
161. L. Bjergbaek, J. A. Cobb and S. M. Gasser, RecQ helicases and genome stability: lessons from model organisms and human disease, *Swiss Med. Wkly*, 2002, **132**, 433–442.
162. K. Myung and R. D. Kolodner, Suppression of genome instability by redundant S-phase checkpoint pathways in *Saccharomyces cerevisiae*, *Proc. Natl. Acad. Sci. U.S.A.*, 2002, **99**, 4500–4507.

163. P. Mohaghegh and I. D. Hickson, DNA helicase deficiencies associated with cancer predisposition and premature ageing disorders, *Hum. Mol. Genet.*, 2001, **10**, 741–746.
164. B. Garvik, M. Carson and L. Hartwell, Single-stranded DNA arising at telomeres in cdc13 mutants may constitute a specific signal for the RAD9 checkpoint, *Mol. Cell. Biol.*, 1995, **15**, 6128–6138 [published erratum appears in *Mol. Cell. Biol.*, 1996, **16**, 457].
165. H. Takata, Y. Tanaka and A. Matsuura, Late S phase-specific recruitment of Mre11 complex triggers hierarchical assembly of telomere replication proteins in Saccharomyces cerevisiae, *Mol. Cell*, 2005, **17**, 573–583.
166. S. J. Diede and D. E. Gottschling, Exonuclease activity is required for sequence addition and Cdc13p loading at a de novo telomere, *Curr. Biol.*, 2001, **11**, 1336–1340.
167. M. Larrivee, C. LeBel and R. J. Wellinger, The generation of proper constitutive G-tails on yeast telomeres is dependent on the MRX complex, *Genes Dev.*, 2004, **18**, 1391–1396.
168. S. H. Askree, T. Yehuda, S. Smolikov, R. Gurevich, J. Hawk, C. Coker, A. Krauskopf, M. Kupiec and M. J. McEachern, A genome-wide screen for *Saccharomyces cerevisiae* deletion mutants that affect telomere length, *Proc. Natl. Acad. Sci. U.S.A.*, 2004, **101**, 8658–8663.
169. C. J. Frank, M. Hyde and C. W. Greider, Regulation of telomere elongation by the cyclin-dependent kinase CDK1, *Mol. Cell*, 2006, **24**, 423–432.
170. S. G. Diede and D. E. Gottschling, Telomerase-mediated telomere addition *in vivo* requires DNA primase and DNA polymerase α and δ, *Cell*, 1999, **99**, 723–733.
171. R. J. Wellinger, K. Ethier, P. Labrecque and V. A. Zakian, Evidence for a new step in telomere maintenance, *Cell*, 1996, **85**, 423–433.
172. L. Maringele and D. Lydall, EXO1-dependent single-stranded DNA at telomeres activates subsets of DNA damage and spindle checkpoint pathways in budding yeast yku70Delta mutants, *Genes Dev.*, 2002, **16**, 1919–1933.
173. A. A. Bertuch and V. Lundblad, EXO1 contributes to telomere maintenance in both telomerase-proficient and telomerase-deficient *Saccharomyces cerevisiae*, *Genetics*, 2004, **166**, 1651–1659.
174. W. Choe, M. Budd, O. Imamura, L. Hoopes and J. L. Campbell, Dynamic localization of an Okazaki fragment processing protein suggests a novel role in telomere replication, *Mol. Cell. Biol.*, 2002, **22**, 2002–2017.
175. G. Ghosal and K. Muniyappa, *Saccharomyces cerevisiae* Mre11 is a high-affinity G4 DNA-binding protein and a G-rich DNA-specific endonuclease: implications for replication of telomeric DNA, *Nucleic Acids Res.*, 2005, **33**, 4692–4703.
176. H. Cohen and D. A. Sinclair, Recombination-mediated lengthening of terminal telomeric repeats requires the Sgs1 DNA helicase, *Proc. Natl. Acad. Sci. U.S.A.*, 2001, **98**, 3174–3179.

177. F. B. Johnson, R. A. Marciniak, M. McVey, S. A. Stewart, W. C. Hahn and L. Guarente, The *Saccharomyces cerevisiae* WRN homolog Sgs1p participates in telomere maintenance in cells lacking telomerase, *EMBO J.*, 2001, **20**, 905–913.
178. M. D. Huber, D. C. Lee and N. Maizels, G4 DNA unwinding by BLM and Sgs1p: Substrate specificity and substrate-specific inhibition, *Nucleic Acids Res.*, 2002, **30**, 3954–3961.
179. M. T. Teixeira, M. Arneric, P. Sperisen and J. Lingner, Telomere length homeostasis is achieved via a switch between telomerase-extendible and nonextendible states, *Cell*, 2004, **117**, 323–335.
180. D. L. Levy and E. H. Blackburn, Counting of Rif1p and Rif2p on *Saccharomyces cerevisiae* telomeres regulates telomere length, *Mol. Cell. Biol.*, 2004, **24**, 10857–10867.
181. A. Bianchi and D. Shore, Early replication of short telomeres in budding yeast, *Cell*, 2007, **128**, 1051–1062.
182. Y. Tsukamoto, A. K. Taggart and V. A. Zakian, The role of the Mre11-Rad50-Xrs2 complex in telomerase-mediated lengthening of *Saccharomyces cerevisiae* telomeres, *Curr. Biol.*, 2001, **11**, 1328–1335.
183. M. Sabourin, C. T. Tuzon and V. A. Zakian, Telomerase and Tel1p preferentially associate with short telomeres in *S. cerevisiae*, *Mol. Cell*, 2007, **27**, 550–561.
184. V. Viscardi, D. Bonetti, H. Cartagena-Lirola, G. Lucchini and M. P. Longhese, MRX-dependent DNA damage response to short telomeres, *Mol. Biol. Cell*, 2007, **18**, 3047–3058.
185. S. Smith, A. Gupta, R. D. Kolodner and K. Myung, Suppression of gross chromosomal rearrangements by the multiple functions of the Mre11-Rad50-Xrs2 complex in *Saccharomyces cerevisiae*, *DNA Repair (Amst.)*, 2005, **4**, 606–617.
186. K. Myung, A. Datta and R. D. Kolodner, Suppression of spontaneous chromosomal rearrangements by S phase checkpoint functions in *Saccharomyces cerevisiae*, *Cell*, 2001, **104**, 397–408.
187. K. Myung, C. Chen and R. D. Kolodner, Multiple pathways cooperate in the suppression of genome instability in *Saccharomyces cerevisiae*, *Nature*, 2001, **411**, 1073–1076.
188. J. M. Pennington and S. M. Rosenberg, Spontaneous DNA breakage in single living *Escherichia coli* cells, *Nat. Genet.*, 2007, **39**, 797–802.
189. M. M. Vilenchik and A. G. Knudson, Endogenous DNA double-strand breaks: production, fidelity of repair, and induction of cancer, *Proc. Natl. Acad. Sci. U.S.A.*, 2003, **100**, 12871–12876.
190. X. Pan, P. Ye, D. S. Yuan, X. Wang, J. S. Bader and J. D. Boeke, A DNA integrity network in the yeast *Saccharomyces cerevisiae*, *Cell*, 2006, **124**, 1069–1081.
191. A. H. Y. Tong, G. Lesage, G. D. Bader, H. Ding, H. Xu, X. Xin, J. Young, G. F. Berriz, R. L. Brost, M. Chang, Y. Q. Chen, X. Cheng, G. Chua, H. Friesen, D. S. Goldberg, J. Haynes, C. Humphries, G. He, S. Hussein, L. Ke, N. Krogan, Z. Li, J. N. Levinson, H. Lu, P. Menard,

C. Munyana, A. B. Parsons, O. Ryan, R. Tonikian, T. Roberts, A.-M. Sdicu, J. Shapiro, B. Sheikh, B. Suter, S. L. Wong, L. V. Zhang, H. Zhu, C. G. Burd, S. Munro, C. Sander, J. Rine, J. Greenblatt, M. Peter, A. Bretscher, G. Bell, F. P. Roth, G. W. Brown, B. Andrews, H. Bussey and C. Boone, Global mapping of the yeast genetic interaction network, *Science*, 2004, **303**, 808–813.

192. H. I. Kao, L. A. Henricksen, Y. Liu and R. A. Bambara, Cleavage specificity of *Saccharomyces cerevisiae* flap endonuclease 1 suggests a double-flap structure as the cellular substrate, *J. Biol. Chem.*, 2002, **277**, 14379–14389.
193. G. Henneke, E. Friedrich-Heineken and U. Hubscher, Flap endonuclease 1: A novel tumour suppresser protein, *Trends Biochem. Sci.*, 2003, **28**, 384–390.

CHAPTER 6

Molecular Hand-off Mechanisms in DNA Replication

ELLEN FANNING,[a] XIAOHUA JIANG,[†a] KUN ZHAO[a] AND WALTER J. CHAZIN[b]

[a] Department of Biological Sciences, Vanderbilt University, Nashville, TN, USA; [b] Departments of Biochemistry and Chemistry, Center for Structural Biology, Vanderbilt University, Nashville, TN, USA

6.1 Introduction

DNA replication is a highly complex process that ensures the faithful duplication of the nucleotide sequence, its modifications, assembly into chromatin, and preparation for segregation into daughter cells. Early studies led investigators to postulate that a pre-assembled multiprotein complex, termed a replisome, would perform this task efficiently, much as ribosomes accurately and rapidly translate mRNA into proteins. However, there is little evidence for pre-assembled DNA processing machinery in replication, repair or recombination. Rather, evidence accumulated over the past decade indicates that the processing of DNA proceeds by intricately choreographed entry and exit of individual proteins, which has been termed 'trading places' on DNA or molecular 'hand-off'.[1–4] The critical implication of molecular 'hand-off' is that the protein complexes required at each step are successively remodeled to enable the next step in the process to be performed. Remodeling of the protein complexes is a key characteristic, providing the flexibility to accommodate lesions in the DNA template, multiple levels of regulation, and the recycling of commonly used proteins such as the

† Present address: Department of Biochemistry, Vanderbilt University, Nashville, TN, USA

Molecular Themes in DNA Replication
Edited by Lynne S. Cox

Published by the Royal Society of Chemistry, www.rsc.org

major single-stranded DNA (ssDNA) binding proteins [SSB in prokaryotes and Replication Protein A (RPA) in eukaryotes] or the processivity clamp for DNA polymerases [β clamp in *E. coli* and proliferating cell nuclear antigen (PCNA) in eukaryotes; see also Chapter 3].

Another hallmark of the DNA replication process is the major role of modular proteins, *i.e.* polypeptides composed of multiple folded domains that are connected by flexible linkers. These modular proteins interact physically with one another through contacts in one or more domains or, in some cases, in linkers between domains. The interactions are surprisingly weak when examined individually (Kd in the μM–mM range). We have developed a framework to explain how these properties facilitate molecular 'hand-off'.[5] In this model, the weakness of the interactions is in fact critical to the dynamic remodeling of protein complexes during replication. The requisite overall affinity required to enable effective interaction between proteins derives from the linking together of multiple weak binding modules, or in some cases, through mutual interactions with a third protein. The key feature of this model is that, while providing overall high affinity, multiple weak interactions provide the system with a dynamic character. In particular, the linking of weak interactions provides a mechanism to enable rapid reduction in affinity as only one of the contributing (weak) interactions need be blocked in order to rapidly drop the overall affinity by several orders of magnitude, thereby enabling dissociation from the complex (Figure 6.1).

Two key proteins serve as common platforms in hand-off reactions during DNA replication, as well as other DNA processing pathways:

- a single-strand DNA-binding protein (SSB, T7 gene 2.5 protein, T4 gene 32 protein or RPA);
- a toroidal clamp protein (β clamp or PCNA).

Both classes of protein interact with a rapidly growing number of eukaryotic DNA processing proteins.[6–8] PCNA has two known interaction sites that are targeted by multiple proteins (see Chapter 3), whereas RPA has at least three. Interestingly, these interacting surfaces are often specific for cognate proteins from the same or related species; for example, mammalian proteins cannot substitute for yeast proteins. The interaction partners of RPA and PCNA often have enzymatic activity and use the interaction to facilitate their exchange on DNA. In a sense, RPA and PCNA serve as the lubricants and gears that allow replication machine assemblies to operate smoothly and efficiently. Moreover, they serve a similar function in other DNA processing pathways and therefore act as switch points in crosstalk among replication, repair, recombination and DNA damage signaling pathways.

In this chapter, we discuss protein hand-off mechanisms at key steps in the progression of DNA processing in model replication systems. We have selected examples to highlight fundamental principles of hand-off mechanisms in DNA replication applicable to eukaryotic DNA replication. The chapter concludes with a brief discussion of possible avenues for chemical modulation of hand-off pathways through small molecules.

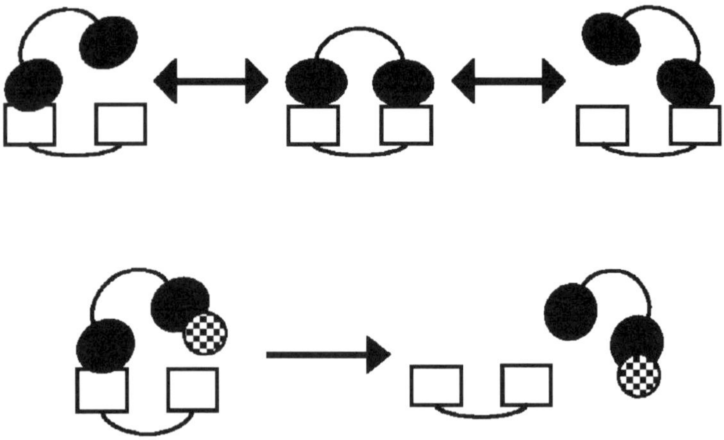

Figure 6.1 **A framework for molecular hand-off between modular proteins.** The upper panel depicts weak interactions between two domains of the modular proteins (open rectangles and filled ovals). The overall affinity between the two proteins is significantly greater because of the linkage between domains in A and B. This interaction can be readily disrupted through introduction of a third protein (checked circle) that binds to only one of the two interacting domains (lower panel), rapidly lowering the overall binding affinity and facilitating dissociation of the modular proteins. Reprinted with permission of American Society for Biochemistry and Molecular Biology from: M. E. Stauffer and W. J. Chazin, Structural mechanisms of DNA replication, repair, and recombination, *Journal of Biological Chemistry,* **279**, 30915–30918, copyright (2004);[5] permission conveyed through Copyright Clearance Center, Inc.

6.2 Primase to Polymerase Switching: Prokaryotic Paradigms

Since replicative DNA polymerases cannot initiate synthesis *de novo* (due to the energetic requirements described in Chapter 1), a specialized RNA polymerase (primase) is required to synthesize a short RNA primer that is then extended by a processive replicative DNA polymerase. The shift from RNA synthesis to DNA synthesis at the replication fork requires a switch from a primase enzyme to a DNA polymerase, and coordination of DNA polymerase loading with fork progress. Association of the polymerase with the helicase that is unwinding the fork may be necessary to obtain this level of coordination.

6.2.1 Lessons from *E. coli*

Primase to polymerase switching in *E. coli* is simpler and more thoroughly investigated than in eukaryotes. Primase (DnaG), associated with the replicative helicase (DnaB), first synthesizes a short RNA oligonucleotide. Acting in a manner analogous to RFC, the pentameric γ complex of the replicative DNA polymerase (Pol III) binds ATP, triggering a conformational change that strengthens its association with β clamp, opening an interface between the two β clamp subunits. ATP hydrolysis by the γ complex allows the clamp to close around the primer–template junction on DNA and displace primase.[9–12] (This tightly parallels the situation in eukaryotes where ATP hydrolysis by RFC is required to load the PCNA clamp onto duplex DNA—see Chapters 2 and 3 and below.) Of note, recent studies of the structure of β clamp associated with DNA reveal that the β clamp binds to the primer template junction, which passes through the ring at a 22° angle.[13] This interaction is proposed to facilitate closure of the clamp around DNA and displacement of the clamp-loader, to retain the clamp at the primer–template junction until the polymerase core re-binds to the clamp and the primer junction, and possibly to facilitate binding of a second polymerase molecule, such as DNA polymerase IV, to the other subunit of the clamp.[13,14]

Thus, with the assistance of the γ complex, primase *hands off* the primer–template junction to the leading strand replicative DNA polymerase.[2] The DNA polymerase core enzyme (αθε) bound to the clamp extends the 3′ OH primer terminus to generate a leading strand daughter DNA complementary to the 3′–5′ parental strand (Figure 6.2). Flexible linkers in the τ subunits (orange) of the γ complex form contacts with the helicase DnaB, greatly stimulating its

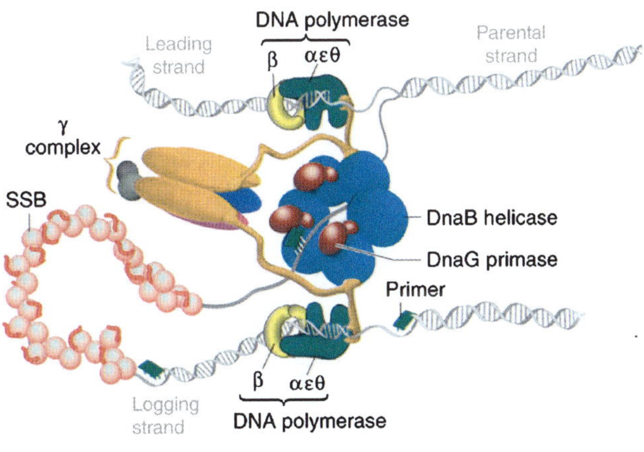

Figure 6.2 **The *E. coli* replisome at a replication fork.** DNA polymerase III consists of three subassemblies: the core (green); the processivity β clamp (yellow); and the pentameric γ clamp-loader complex (orange, blue, purple). Two additional subunits of the gamma complex, chi (χ) and psi (ψ) (grey) are not essential for clamp loading on naked DNA, but interact with SSB-ssDNA complex to facilitate clamp loading. Reprinted by permission from Macmillan Publishers Ltd: N. A. Tanner, S. M. Hamdan, S. Jergic, P. M. Schaeffer, N. E. Dixon and A. M. van Oijen, Single-molecule studies of fork dynamics in Escherichia coli DNA replication, *Nature Structure Molecular Biology*, 2008, **15**, 170–176, copyright (2008).[15] For dynamic representations of the bacterial replisome in operation, view the movie by Drew Berry, WEHI (www.wehi.edu.au/wehi-tv/dna/index.html).

rate of translocation and coupling helicase and DNA polymerase action (Figure 6.2).

Coordinate replication of the two antiparallel parental strands at each fork requires a more complex discontinuous mechanism to synthesize the lagging strand daughter DNA. Single-stranded parental DNA accumulating behind the translocating DnaB helicase loops out and is coated by SSB, limiting intra-strand base pairing and protecting against nuclease action. The χ subunit of the clamp-loader complex interacts with SSB-ssDNA to facilitate β clamp loading at primed sites, followed by primase displacement, extension of the primer by DNA polymerase, and SSB displacement to generate a lagging strand daughter DNA fragment. Extension of such Okazaki fragments ceases when the polymerase 'catches up' to the 5′ end of a previously synthesized fragment. At that point, the DNA polymerase core dissociates from the β clamp, leaving it behind, recycles to the next primed site on the loop, loads a new clamp, and repeats the primer extension cycle.

Interestingly, docking of two or three primase molecules to DnaB helicase triggers a pause in leading strand polymerase progression.[15–17] Leading strand synthesis could thus be regulated dynamically in coordination with priming of the lagging strand. This mechanism is thought to contribute to coupling of leading and lagging strand synthesis, in that primase released from DnaB after primer synthesis would release the brake on the leading strand polymerase. It is tempting to speculate that the DnaB-τ interaction that modulates DNA polymerase progression may play a role in regulation of leading strand synthesis by DnaB-primase complexes.[15] In bacteriophage T7 replication reactions reconstituted with purified proteins, primase also applies the brakes to the leading strand polymerase during primer synthesis, suggesting that this coordinating mechanism may be conserved, at least among prokaryotic replicases.[18]

6.2.2 Hand-off Facilitates Replication Fork Restart After Fork Collapse

In theory, a circular bacterial chromosome of ~4.6 Mbp could be replicated by two forks diverging from a single origin, but lesions in the template DNA often result in fork stalling, collapse and disassembly. To complete replication, bacteria have evolved additional pathways to restart forks through assembly of a new primosome (reviewed in ref. 19). Extensive genetic, biochemical, and most recently structural investigation of the proteins involved in restart pathways, provides fundamental insight into the molecular mechanisms of DNA hand-off from one protein to another.

To allow restart proteins to bind to ssDNA at the fork, SSB that coats the template must be partially cleared, perhaps through the SSB-binding activity of PriA helicase.[6] PriA interacts with DNA at a fork (Figure 6.3A) or at a D-loop structure, which causes the protein to change conformation, exposing a weak binding site for the PriB protein.[20] PriB, a dimer, stimulates the helicase activity of PriA and forms a ternary complex with it on ssDNA (Figure 6.3A). PriB, an

OB-fold protein (see Section 6.3), can also bind to ssDNA and DnaT through contacts that partially overlap on the surface of PriB (Figure 6.3B). Competition among the binding partners for these sites on PriB suggests that the pathway progresses through DnaT-mediated hand-off of ssDNA from PriB to the DnaB-DnaC complex (Figure 6.3A). A comparison of the three binding sites on PriB illustrates the structural basis for the ordered competition, in that the binding site sizes increase from PriA to DnaT. DnaT-mediated hand-off to DnaB-DnaC through a mechanism that is not yet fully understood leads to loading of DnaB on ssDNA, enabling primer synthesis and reassembly of the replisome as in Figure 6.2. Several features of this pathway are also found in hand-off pathways in eukaryotic replication:

- modular or multi-subunit proteins;
- conformational changes that expose new interacting surface patches on the proteins;
- ternary complexes of two or more proteins with ssDNA;
- a structural fold in PriB that is capable of binding either ssDNA or another protein, facilitating close contacts for hand-off.

Restart pathways are conserved among the Gram$^+$ and Gram$^-$ bacteria,[21] but whether similar pathways exist in eukaryotes is not known.

6.3 Structural Basis for RPA Action in Eukaryotic Hand-off

Given the central role of SSBs and RPA in providing a platform for molecular hand-off, a brief consideration of their structures and the manner in which they bind DNA and other DNA processing proteins provides insight into their functions in molecular hand-off.

Both *E. coli* SSB and RPA are comprised of oligonucleotide–oligosaccharide binding (OB)-fold domains (reviewed in ref. 6–7,22–24). *E. coli* SSB is composed of a single OB-fold domain that is responsible for binding DNA and a short amphipathic C-terminal tail that interacts with at least 14 other DNA processing proteins in hand-off reactions.[6] SSB functions as a homotetramer with multiple ssDNA binding modes and cooperativity between multiple tetramers when binding large regions of ssDNA.

RPA is a much more complex protein than SSB. It is comprised of three subunits: RPA70, RPA32, and RPA14, named for their apparent molecular weights (Figure 6.4A). RPA contains six OB-fold domains, although only four (A–D) are involved in binding ssDNA. RPA70 consists of four OB-fold domains (N, A, B and C) tethered together by flexible linkers of different lengths. RPA14 is a single OB-fold domain. RPA32 has a central OB-fold domain (RPA32D) with a disordered N-terminal domain (RPA32N) and a C-terminal winged helix domain (RPA32C) tethered by a long flexible linker.[25]

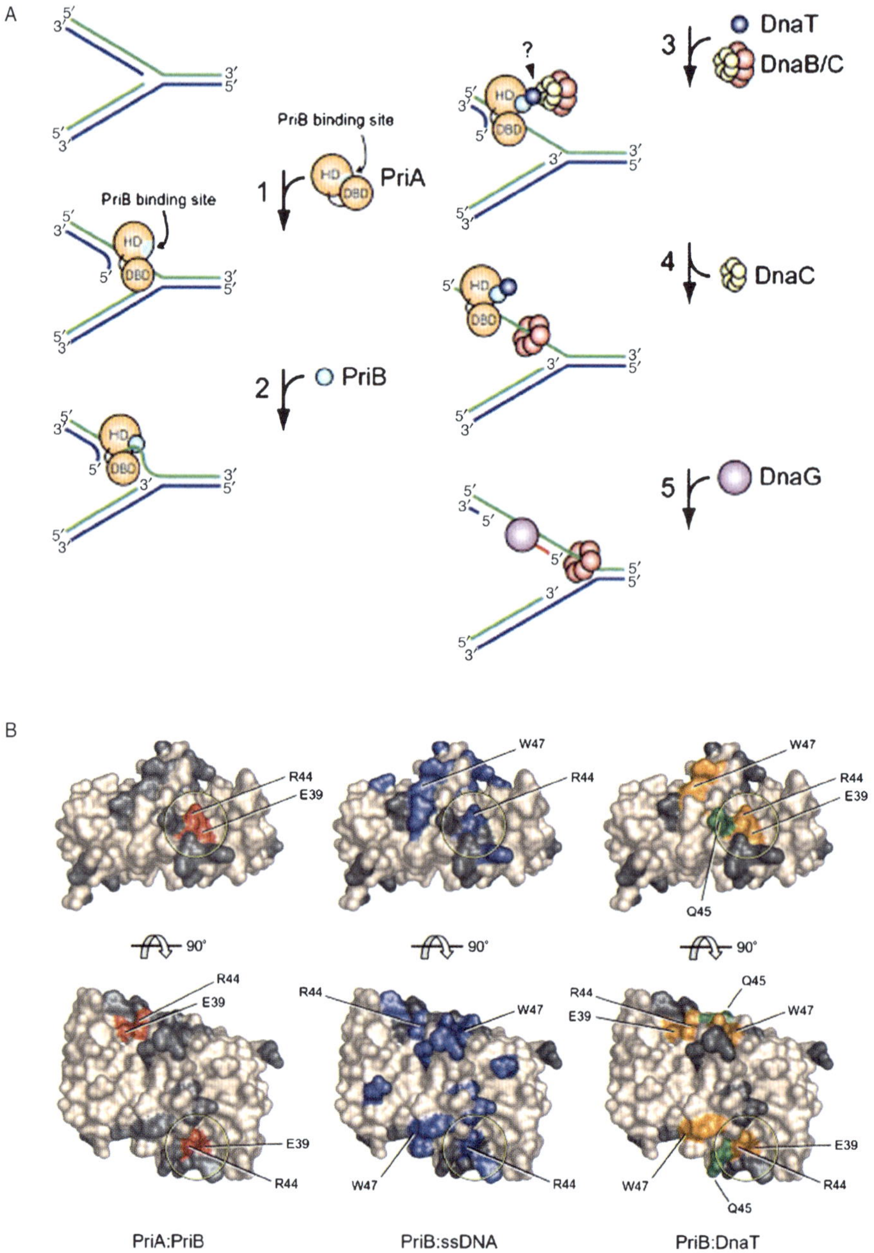
A
5′
3′
3′
5′
PriB binding site
HD
DBD
PriA
1
2
PriB
3
DnaT
DnaB/C
?
4
DnaC
5
DnaG
B
R44
E39
W47
Q45
90°
PriA:PriB
PriB:ssDNA
PriB:DnaT

The three subunits are stably associated through a three-helix bundle formed by RPA70C, RPA32D and RPA14.

RPA binds to proteins through multiple binding regions (Figure 6.4A, red bars). For example, RPA32C has been shown to interact physically through the same surface with at least four different proteins involved in replication, repair and recombination (UNG2, XPA, RAD52 and SV40 large T antigen).[4,26] Each protein that recognizes this surface makes slightly different contacts with it. Similarly, RPA70N binds a variety of proteins in the same basic cleft, including the p53 tumor suppressor, RecQ helicases, the ATR protein kinase-interacting protein ATRIP, Mre11 nuclease, Rad18 ubiquitin ligase (which modifies PCNA for damage bypass synthesis), the Rad9 repair clamp subunit, and the Rad17 clamp loader.[25,27–33] Remarkably, all of these proteins participate primarily in DNA damage signaling in response to stalled or collapsed replication forks (see also Chapters 3, 5 and 7), suggesting that they have similar functions in hand-off pathways that follow RPA binding to troubled forks. Future work to elucidate the mechanisms through which these interactions facilitate replication fork recovery will be of great interest.

RPA binds ssDNA sequentially with 5′-3′ polarity and decreasing affinity from A to D. When fully engaged, RPA occludes 28–30 nucleotides and binds with high affinity (Kd $\sim$ 10^{-9} M).[7,34] However, each individual OB-fold binds with modest affinity, *e.g.* a maximum of 1.7 μM for OB-fold A.[35] The RPA70A and RPA70B domains function in tandem because there is only a short linker between them. Consequently, their binding to ssDNA is strongly coupled and the affinity of RPA70AB (Kd $\sim$0.1 μM) is substantially higher than that of the isolated domains. RPA binds ssDNA in three modes (Figure 6.4B), which are associated with binding by the AB, ABC and ABCD domains and the occlusion of $\sim$10, $\sim$20 and $\sim$30 nucleotides.[36,37] Unfortunately, RPA structural information is restricted to domain fragments, so the changes in the organization of the various domains have yet to be determined. Insights into the consequences

Figure 6.3 **Molecular hand-off in a replication fork restart pathway.** (**A**) PriA, a helicase with 3′–5′ polarity, binds to parental ssDNA at a collapsed fork, unmasking a binding site for PriB, a homodimeric OB-fold protein (step 1). PriB association with PriA stabilizes the complex at the fork (step 2), creating a ternary complex that recruits DnaT through interactions with PriB. The complex with DnaT is proposed to hand-off the ssDNA from PriB to the hexameric replicative helicase DnaB and its loading protein DnaC (step 3). DnaC loads DnaB on ssDNA (step 4) and recruits DnaG primase (step 5), forming a primosome that can reassemble a complete replisome with DNA Pol III. (**B**) Two views of the structure of PriB dimer and its partially overlapping interaction surfaces; PriA (red), ssDNA (blue), and DnaT (orange, green) are shown. Mutations in the indicated PriB residues either diminish or enhance (Q45) PriB interaction with its binding partner. Reprinted from: M. Lopper, R. Boonsombat, S. J. Sandler and J. L. Keck, A hand-off mechanism for primosome assembly in replication restart, *Molecular Cell*, **26**, 781–793, copyright (2007), with permission from Elsevier.[70]

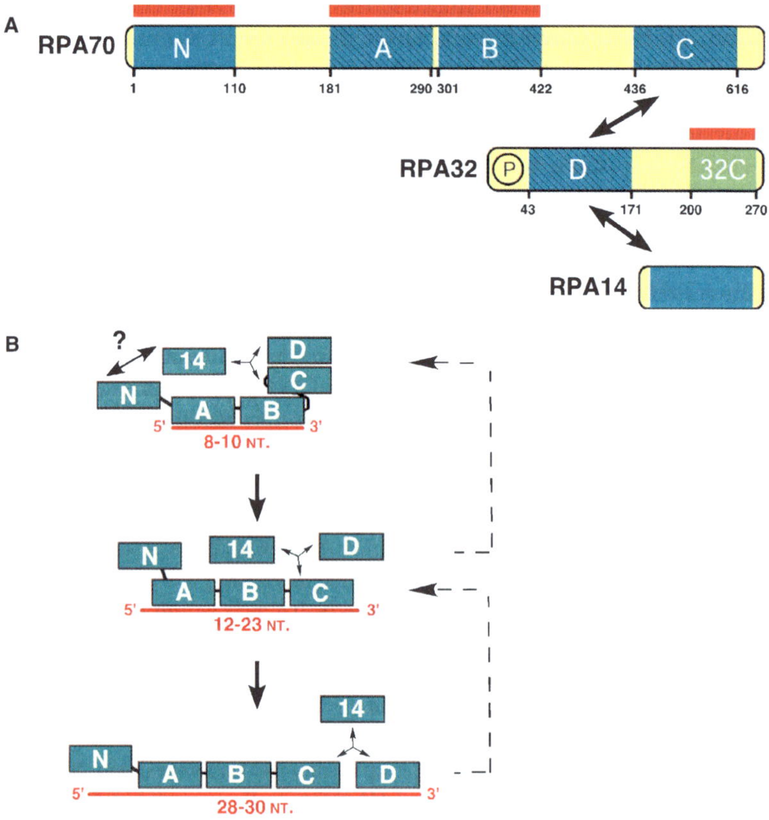

Figure 6.4 **Protein guided remodeling of RPA ssDNA-binding mode in molecular hand-off.** (**A**) Structured domains of the three RPA subunits are indicated in blue (OB-folds) and green (winged helix-loop helix) and amino acid residue numbers are indicated below each subunit. The OB-folds in RPA14 and RPA70N are also termed OB-folds E and F. Flexible regions that link the domains are coloured yellow. Black arrows indicate regions that interact to form the stable heterotrimer. Hatched OB-folds have ssDNA-binding activity. Red bars denote protein-binding regions. (**B**) Sequential 5′ to 3′ binding of OB-folds occludes three different lengths of ssDNA, correlating with an increase in ssDNA-binding affinity (filled arrows). Proteins that bind to RPA are postulated to remodel its ssDNA-binding mode to a weaker one as depicted by dashed arrows (see text for details). The accompanying changes in RPA structure are hypothetical, as the quaternary structure of the trimeric protein is not known. Reprinted by permission from Macmillan Publishers Ltd: E. Fanning, V. Klimovich and A. R. Nager, A dynamic model for replication protein A (RPA) function in DNA processing pathways, *Nucleic Acids Research*, 2006, **34**, 4126–4137), copyright (2006).[7]

of ssDNA binding have been determined from X-ray crystal structures and NMR, which showed that the two domains are structurally independent in the absence of DNA, but oriented in a specific manner when ssDNA is bound.[35,37,38]

The RPA70AB tandem domains are a critical functional nexus in RPA, directly coupling interactions with ssDNA and other DNA processing proteins. For example, the linker between the A and B domains presents an acidic binding site for a basic surface of the SV40 T antigen origin DNA-binding domain (OBD).[39] In the case of T antigen, a ternary complex of RPA70AB bound to both a 10-nucleotide ssDNA and OBD forms. However, in the presence of longer ssDNA, the complex is not stable, RPA adopts a more extended binding mode, and T antigen dissociates.[39] In other words, proteins that bind to RPA may remodel the ssDNA-binding mode of RPA, thereby facilitating their own binding to RPA, the release of ssDNA, or hand-off to another protein[26] (discussed in Section 6.4). Figure 6.5 proposes a model that may serve as a general paradigm for protein-guided remodeling of the ssDNA-binding mode of RPA, which may underlie its frequent participation in hand-off reactions.

6.4 Eukaryotic Primase to Polymerase Switching

In eukaryotes, the DNA polymerase alpha-primase complex (pol α-primase) carries out both primer synthesis and the switch to DNA synthesis on RPA-ssDNA *via* a molecular hand-off. Details on the mechanism for this step in replication are beginning to emerge from studies of SV40 DNA replication.[26] As in the bacterial system, a hexameric DNA helicase (in the SV40 viral system, a dodecamer of large T antigen) mediates the hand-off of ssDNA from RPA-ssDNA to pol α-primase.[1] In normally replicating eukaryotic cells, this hand-off may be mediated by the GINS-Cdc45-Mcm2-7 (GCM) complex[40] (see Chapter 3). Biochemical and structural mapping of interactions between the RPA-binding domain of T antigen (residues 131–259) and the T antigen-binding domains of RPA (RPA70AB and RPA32C) revealed multiple weak interactions,[26,39] consistent with the requirements for molecular hand-off. Functional analysis of mutant proteins underscored the importance of T antigen interactions with both domains of RPA for primer synthesis by pol α-primase and suggested that contacts of the T antigen hexamer with both RPA domains weakened the ssDNA-binding mode of RPA (Figure 6.5). DNA pol α-primase contacts with the helicase domain of T antigen are also critical for primer synthesis on RPA-ssDNA.[41,42] Following synthesis of the RNA primer ($\sim$8–12 nucleotides), a counting mechanism intrinsic to the pol α-primase complex shifts the primer internally from the primase active site in the p48 subunit to the polymerase active site in the p180 subunit in the same complex.[43–45] The polymerase activity extends the primer to generate an RNA-DNA primer of $\sim$30 nucleotides and the polymerase-primase remains bound to RPA-ssDNA at the primed site until displaced by the next hand-off to the clamp-loader RFC.[1]

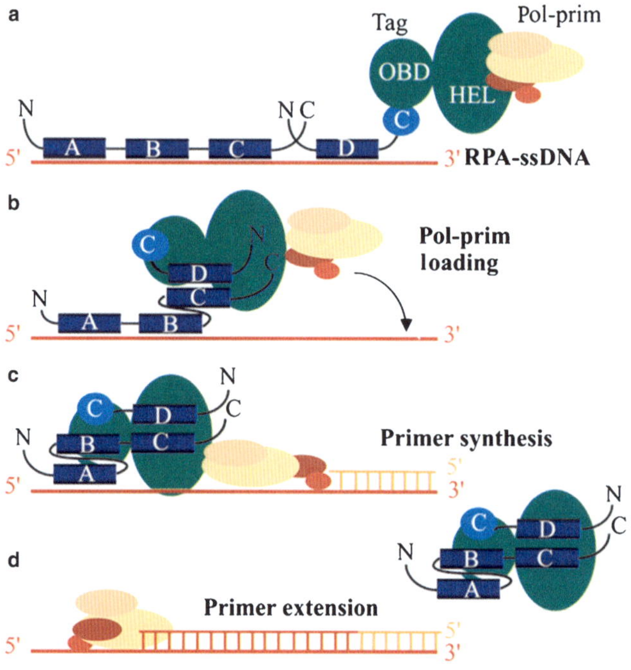

Figure 6.5 **Working model for ssDNA hand-off from RPA to DNA polymerase α-primase (pol-prim) mediated by the SV40 helicase T antigen (Tag).** (**a**) The Tag hexamer binds through its origin-binding domain (OBD) to RPA32C and RPA70AB domains, remodeling the RPA-ssDNA complex into a weaker binding mode. Pol-prim is anchored through the N-termini of two subunits (p180 and p68) to two distinct binding sites in the helicase domains of Tag (**b**). RPA remodeling by Tag is proposed to expose ssDNA for pol-prim binding in a concerted hand-off reaction (**c**), leading to RNA–DNA primer synthesis and release of Tag and RPA (**d**). Reprinted by permission from Macmillan Publishers Ltd: A. I. Arunkumar, V. Klimovich, X. Jiang, R. D. Ott, L. Mizoue, E. Fanning and W. J. Chazin, Insights into hRPA32 C-terminal domain-mediated assembly of the simian virus 40 replisome, *Nature Structure Molecular Biology*, 2005, 12, 332–339, copyright (2005).[26]

6.5 Molecular Hand-off in Loading the Processive Replicative DNA Polymerases

Eukaryotic DNA replication is similar to *E. coli* and SV40 DNA replication, but complicated through the use of at least three different DNA replication fork polymerases (α, δ and ε) and a greater variety of mechanisms to bypass or rescue forks stalled by lesions (see also Chapters 4 and 7). Eukaryotic DNA replication uses different DNA polymerases for leading and lagging strand synthesis (pol ε and pol δ, respectively[46]), which are more processive and error-free than pol α-primase.[47] Pol δ is proposed to be solely a lagging strand

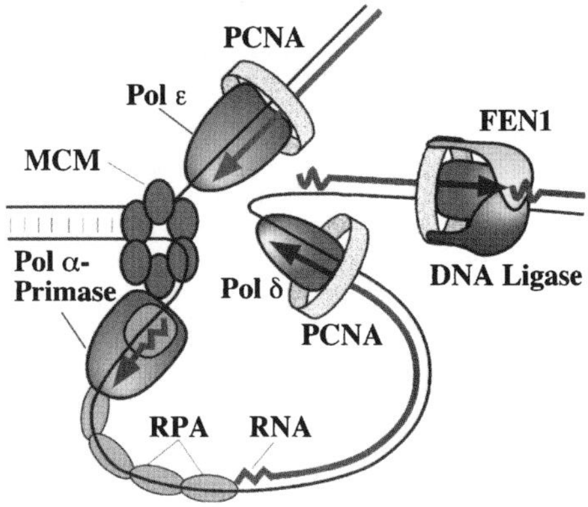

Figure 6.6 **A minimal replisome at a eukaryotic fork.** MCM helicase complex includes additional essential subunits GINS and Cdc45 that are not shown here.[53] The clamp loader RFC and topoisomerases are also not shown. The reconstitution of a simple replication using these proteins substituted the SV40 T antigen helicase for MCM helicase.[56] Figure from: P. Garg and P. M. Burgers, DNA polymerases that propagate the eukaryotic DNA replication fork, *Critical Reviews in Biochemistry and Molecular Biology*, 2005, **40**, 115–128, publisher Taylor & Francis Ltd (www.tandf.co.uk/journals); reprinted by permission of the publisher.[46]

polymerase[48] (see Chapter 4). Recent genetic studies provide clear evidence that pol ε is primarily involved in leading strand DNA synthesis, whereas pol δ acts primarily in lagging strand replication[49,50] (Figure 6.6). Pol ε is loaded onto licensed chromosomal replication origins before the GINS-Cdc45-Mcm2-7 complex unwinds double stranded DNA,[51–53] and it does not interact with PCNA in solution.[54] Thus the eukaryotic primase–polymerase switch on the leading strand differs from that on the lagging strand.

Biochemical reconstitution experiments of eukaryotic DNA replication derive mostly from studies of the relatively simple SV40 DNA replication model with proteins isolated from human cells or recombinant human proteins[55] (Figure 6.7). The viral system uses pol α-primase and pol δ for leading and lagging strand synthesis, both *in vitro* and *in vivo*.[56] After pol α-primase synthesizes a 30 nucleotide RNA–DNA primer, it remains bound to the primer–template junction through interaction with RPA bound to the template strand.[1] Pol α-primase binds through its primase subunits to RPA70[55,57,58] as well as to RPA32.[59] However, the domains and contact surfaces that are needed for interactions between these proteins have not yet been mapped.

Once polymerization by pol α-primase has arrested, in part through its tight association with RPA, it is necessary to switch polymerases to DNA pol δ, a high fidelity enzyme with proofreading capacity (see Chapter 4). However, pol δ does not bind directly to DNA but is held in proximity to the template by binding to the sliding clamp protein PCNA (see Chapter 3). As PCNA exists as a closed toroidal trimer, its ring must be broken open in order to load it onto the primer template junction, a reaction catalyzed by the AAA+ ATPase

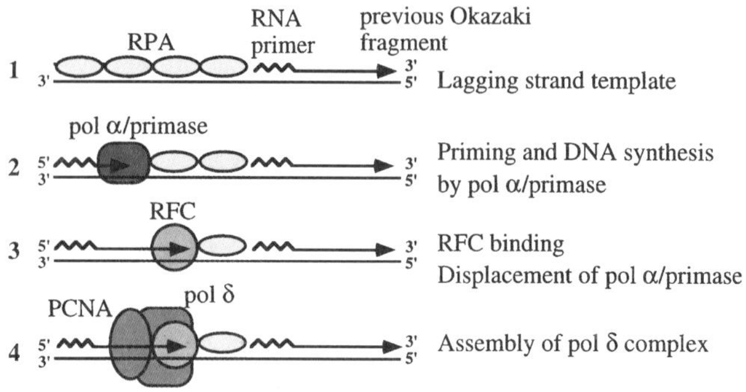

Figure 6.7 **The eukaryotic DNA polymerase switch: hand-off of primed DNA from polymerase alpha-primase to polymerase delta**. The mechanism of elongation of a primed DNA template on the leading strand or for each Okazaki fragment on the lagging strand was elucidated in reactions reconstituted with purified proteins. The 5′ flap displaced when polymerase δ completes an Okazaki fragment is not shown (see Chapters 3 and 5). Modified with permission from Annual Reviews (www.annualreviews.org): S. Waga and B. Stillman, The DNA replication fork in eukaryotic cells, *Annual Review of Biochemistry,* 1998, **67**, 721–751, copyright (1998).[56]

clamp loader, RFC (see Chapter 2). Three subunits of the RFC clamp loader (140, 40 and 38) interact with RPA70,[1] thus localizing the functional heteropentameric RFC clamp loader at the primer–template junction.

Although RFC has four ATP binding sites, only two ATP molecules initially bind to RFC in solution (reviewed in ref. 60). When PCNA binds to RFC-ATP_2, RFC undergoes conformational changes, which makes additional ATP binding site available.[60] The PCNA-RFC-ATP_3 complex interacts with RPA70 through RFC140, 40 and 38 subunits. This competitive interaction displaces pol α-primase from the primer. RFC binding to DNA induces another conformational change in RFC, exposing the final ATP binding site.[61–63] Once all four ATPs are bound to RFC, DNA stimulates ATP hydrolysis and RFC closes the PCNA ring around DNA (see Chapter 2). The observation that multiple PCNA clamps can be loaded at a primer–template junction coated with bacterial SSB rather than RPA suggests that interaction of RFC with RPA also plays a role in limiting the loading reaction to a single cycle.[1]

The complex of RFC-PCNA-RPA at the primer–template junction not only positions the primer properly for extension by the incoming polymerase (Figure 6.8c, d), but also selects the incoming polymerase through a second set of hand-off reactions mediated by interactions of pol δ with RPA, RFC and PCNA.[1,60] Pol δ appears to bind to RPA70, but it is not known to which domain it binds or whether these contacts compete directly with RPA contacts made by the primase subunits of pol δ-primase. Pol δ interacts with RFC through the RFC40 subunit. This subunit is also responsible for displacing pol α-primase in the hand-off reaction that loads PCNA on DNA, suggesting that the two polymerases may compete for RFC40. Lastly, pol δ and RFC compete

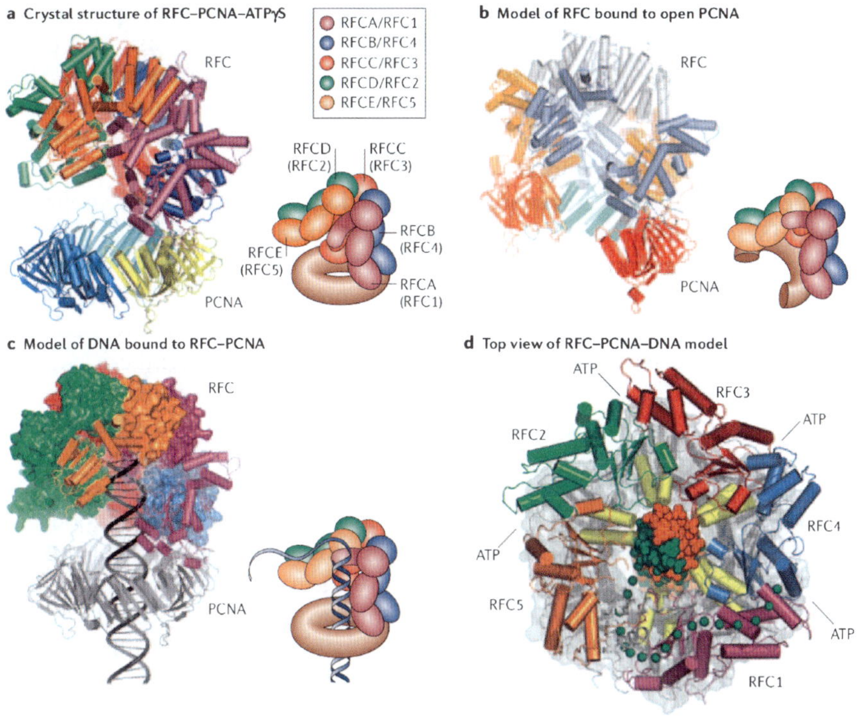

Figure 6.8 **Molecular mechanisms of clamp opening and loading at a primer-template junction.** (**a**) Crystal structure of *S. cerevisiae* RFC-PCNA-ATPγS (left) and a colour-coded model illustrating the interaction of PCNA with three subunits of RFC in a spiral conformation.[78] Additional contact with the other two subunits is involved in opening PCNA. (**b**) Model of PCNA opening in a right-handed spiral prior to loading on DNA in an open ring conformation that is thought to match the RFC-ATP spiral.[79] (**c**, **d**) Side and top views of a colour-coded RFC in complex with closed PCNA (grey) loaded on DNA. The primer-template (blue in **c**, orange and green, respectively, in **d**) is positioned for hand-off to pol δ.[78] RPA is not shown. Reprinted by permission from Macmillan Publishers Ltd: C. Indiani and M. O'Donnell, The replication clamp-loading machine at work in the three domains of life, *Nature Reviews in Molecular and Cellular Biology*, 2006, **7**, 751–761, copyright (2006).[9]

for PCNA, implying that they both bind to the same face of the ring.[1,60] However, RFC is retained in the holoenzyme after pol δ loading, apparently through its interaction with the RPA-ssDNA template. The mechanism of RFC retention is not fully understood. Perhaps RFC has an additional binding site in the holoenzyme complex or on DNA. Alternatively, RFC may undergo cycles of association and dissociation with RPA on the lagging strand template, ensuring a high local concentration of RFC for PCNA loading to initiate new Okazaki fragments.[21]

Pol δ has several binding sites for PCNA and their relative strengths depend on whether PCNA is loaded on DNA.[64] Pol δ in fission yeast and higher eukaryotes has four subunits, whereas pol δ in *S. cerevisiae* consists of three subunits named for the genes that encode them:

- Pol3, the catalytic subunit;
- Pol31, essential for activity and for bridging Pol3 to the third subunit;
- Pol32, not required for viability, but important for robust growth.

All three subunits bind to PCNA independently. Pol3 and Pol31 bind in the hydrophobic pocket of the interdomain connector loop of PCNA loaded on DNA, and these interactions are sufficient for processive synthesis under a variety of conditions. Interestingly, Pol32 carries a conserved PCNA interacting peptide (PIP)[65,66] (see also Chapter 3) at its extreme C-terminus, which binds to the hydrophobic pocket of PCNA that is *not* loaded on DNA. During processive DNA synthesis by the Pol δ–PCNA holoenzyme, this PIP interacts with the C-terminus of PCNA instead of the hydrophobic pocket.[64] Whether the PIP of the Pol32 subunit plays a role in recruitment of pol δ to PCNA at a primer–template junction is not clear, but other roles are conceivable as well. Pol32 subunit interacts physically with the p180 catalytic subunit of pol α-primase,[64] suggesting the possibility that hand-off exchanges between pol α-primase and pol δ during lagging strand synthesis might utilize the Pol32 PIP-PCNA interaction.[67] Similarly, the interaction of Pol32 with the PCNA C-terminus might facilitate the exchange of pol δ for translesion polymerases when a progressing fork encounters a lesion in the template.

6.6 Dynamic Polymerase Exchange at the Replication Fork

The possibility of dynamic exchange of DNA polymerases at the fork is well illustrated in phage T4 DNA replication. The processivity of wild-type T4 gp43 DNA polymerase can be disrupted by a catalytically inactive D408N mutant polymerase, with inhibition kinetics that suggest a dynamic exchange in which the mutant polymerase replaces the catalytically active form.[68] A similar polymerase exchange has also been observed in the T7 replication system.[69] Furthermore, a recent study using a variant *E. coli* clamp loader assembled *in vitro* demonstrated that three DNA pol III core complexes could be loaded onto a primed template.[70] Moreover, leading strand extension by the trimeric replicase was at least as fast as that conducted by dimeric pol III core enzyme, whereas Okazaki fragments on the lagging strand were slightly shorter with the trimeric replicase, suggesting very efficient use of RNA primers by the trimeric core enzyme.[70] In addition, the *E. coli* translesion polymerase Pol IV can replace pol III on the sliding clamp during replication *in vitro*.[14]

Although the exchange of high-fidelity DNA polymerases for a translesion polymerase is crucial to bypass DNA lesions that stall replication forks (Chapters 4 and 7), how these exchanges are coordinated mechanistically is not yet understood in detail.[71–73] Possible triggers for polymerase dissociation or exchange on a 3′ DNA terminus may be completion of an Okazaki fragment, polymerase stalling, or direct or indirect interactions between a stalled replicative polymerase and an error-prone polymerase.[14,74] Like the Pol III to Pol IV exchange on the β clamp in *E. coli*,[14] PCNA may be involved in polymerase switching in eukaryotes (see Chapter 7). Taken together, these data suggest an interchangeability of DNA polymerases that, despite their high processivity, exchange frequently with free polymerases at a replication fork.

6.7 Perspective

Dynamic protein interactions are important for many aspects of cell life. If small molecules could target these interactions, they could have potential for therapeutic applications. Several forays in this direction are beginning to reveal the utility of this approach.[75] For example, the ubiquitin ligase HDM2 is responsible for the normally short half-life of the p53 tumor suppressor, but the interface between HDM2 and p53 can be disrupted by small molecules that appear to have potential in cancer therapy (reviewed in ref. 76).

Recent efforts to develop small molecules to target DNA processing pathways guided by hand-off mechanisms have demonstrated the feasibility of this approach.[77] As noted above, the *E. coli* β clamp has two binding pockets for the five *E. coli* DNA polymerases, allowing their exchange in the replisome at the fork. Using peptides from the C-terminal clamp-binding motif of DNA Pol III, a library of small molecules was screened to identify compounds that could disrupt the interaction. Interestingly, one of these proved capable of selectively targeting Pol III but not Pol IV clamp binding. Significantly, some small molecules identified in the *E. coli*-based screen also disrupted the Pol III-clamp interaction of *Streptococcus pyogenes*, but did not inhibit the interactions among yeast PCNA, RFC, and pol δ, suggesting a novel potential avenue for antibiotic development.[77]

The principles of molecular hand-off and dynamic assembly/disassembly of multi-protein complexes provide a highly effective framework for interpreting biochemical, structural and functional studies of DNA replication machinery. As our understanding grows, this will set the stage for directly targeting DNA processing to manipulate growth and survival of cells *in vivo*.

Acknowledgements

We thank L. S. Cox for invaluable editorial advice, and J. Keck and members of the Chazin and Fanning labs for stimulating discussions. We sincerely apologize to colleagues whose publications could not be cited here due to space constraints. The support of NIH (GM52948 to EF, GM65484 to WJC, P30

CA68485 to the Vanderbilt-Ingram Cancer Center, and P50 ES00267 to the Center for Molecular Toxicology) and Vanderbilt University is gratefully acknowledged.

References

1. A. Yuzhakov, Z. Kelman, J. Hurwitz and M. O'Donnell, Multiple competition reactions for RPA order the assembly of the DNA polymerase delta holoenzyme, *EMBO J.*, 1999, **18**, 6189–6199.
2. A. Yuzhakov, Z. Kelman and M. O'Donnell, Trading places on DNA—a three-point switch underlies primer hand-off from primase to the replicative DNA polymerase, *Cell*, 1999, **96**, 153–163.
3. J. A. Tainer and E. C. Friedberg, Dancing with the elephants: envisioning the structural biology of DNA repair pathways, *Mutat. Res.*, 2000, **460**, 139–141.
4. G. Mer, A. Bochkarev, R. Gupta, E. Bochkareva, L. Frappier, C. J. Ingles, A. M. Edwards and W. J. Chazin, Structural basis for the recognition of DNA repair proteins UNG2, XPA, and RAD52 by replication factor RPA, *Cell*, 2000, **103**, 449–456.
5. M. E. Stauffer and W. J. Chazin, Structural mechanisms of DNA replication, repair, and recombination, *J. Biol. Chem.*, 2004, **279**, 30915–30918.
6. R. D. Shereda, A. G. Kozlov, T. M. Lohman, M. M. Cox and J. L. Keck, SSB as an organizer/mobilizer of genome maintenance complexes, *Crit. Rev. Biochem. Mol. Biol.*, 2008, **43**, 289–318.
7. E. Fanning, V. Klimovich and A. R. Nager, A dynamic model for replication protein A (RPA) function in DNA processing pathways, *Nucleic Acids Res.*, 2006, **34**, 4126–4137.
8. G. L. Moldovan, B. Pfander and S. Jentsch, PCNA, the maestro of the replication fork, *Cell*, 2007, **129**, 665–679.
9. C. Indiani and M. O'Donnell, The replication clamp-loading machine at work in the three domains of life, *Nat. Rev. Mol. Cell Biol.*, 2006, **7**, 751–761.
10. M. J. Davey, D. Jeruzalmi, J. Kuriyan and M. O'Donnell, Motors and switches: AAA+ machines within the replisome, *Nat. Rev. Mol. Cell Biol.*, 2002, **3**, 826–835.
11. D. Jeruzalmi, O. Yurieva, Y. Zhao, M. Young, J. Stewart, M. Hingorani, M. O'Donnell and J. Kuriyan, Mechanism of processivity clamp opening by the delta subunit wrench of the clamp loader complex of *E. coli* DNA polymerase III, *Cell*, 2001, **106**, 417–428.
12. D. Jeruzalmi, M. O'Donnell and J. Kuriyan, Crystal structure of the processivity clamp loader gamma (gamma) complex of *E. coli* DNA polymerase III, *Cell*, 2001, **106**, 429–441.
13. R. E. Georgescu, S. S. Kim, O. Yurieva, J. Kuriyan, X. P. Kong and M. O'Donnell, Structure of a sliding clamp on DNA, *Cell*, 2008, **132**, 43–54.

14. A. Furukohri, M. F. Goodman and H. Maki, A dynamic polymerase exchange with *Escherichia coli* DNA polymerase IV replacing DNA polymerase III on the sliding clamp, *J. Biol. Chem.*, 2008, **283**, 11260–11269.
15. N. A. Tanner, S. M. Hamdan, S. Jergic, P. M. Schaeffer, N. E. Dixon and A. M. van Oijen, Single-molecule studies of fork dynamics in *Escherichia coli* DNA replication, *Nat. Struct. Mol. Biol.*, 2008, **15**, 170–176.
16. J. E. Corn, J. G. Pelton and J. M. Berger, Identification of a DNA primase template tracking site redefines the geometry of primer synthesis, *Nat. Struct. Mol. Biol.*, 2008, **15**, 163–169.
17. G. Wang, M. G. Klein, E. Tokonzaba, Y. Zhang, L. G. Holden and X. S. Chen, The structure of a DnaB-family replicative helicase and its interactions with primase, *Nat. Struct. Mol. Biol.*, 2008, **15**, 94–100.
18. J. B. Lee, R. K. Hite, S. M. Hamdan, X. S. Xie, C. C. Richardson and A. M. van Oijen, DNA primase acts as a molecular brake in DNA replication, *Nature*, 2006, **439**, 621–624.
19. R. C. Heller and K. J. Marians, Replisome assembly and the direct restart of stalled replication forks, *Nat. Rev. Mol. Cell Biol.*, 2006, **7**, 932–943.
20. M. Lopper, R. Boonsombat, S. J. Sandler and J. L. Keck, A hand-off mechanism for primosome assembly in replication restart, *Mol. Cell*, 2007, **26**, 781–793.
21. F. Lecointe, C. Serena, M. Velten, A. Costes, S. McGovern, J. C. Meile, J. Errington, S. D. Ehrlich, P. Noirot and P. Polard, Anticipating chromosomal replication fork arrest: SSB targets repair DNA helicases to active forks, *EMBO J.*, 2007, **26**, 4239–4251.
22. D. L. Theobald, R. M. Mitton-Fry and D. S. Wuttke, Nucleic acid recognition by OB-fold proteins, *Annu. Rev. Biophys. Biomol. Struct.*, 2003, **32**, 115–133.
23. S. K. Binz, A. M. Sheehan and M. S. Wold, Replication protein A phosphorylation and the cellular response to DNA damage, *DNA Repair (Amst.)*, 2004, **3**, 1015–1024.
24. A. Bochkarev and E. Bochkareva, From RPA to BRCA2: lessons from single-stranded DNA binding by the Oβ-fold, *Curr. Opin. Struct. Biol.*, 2004, **14**, 36–42.
25. E. Bochkareva, L. Kaustov, A. Ayed, G. S. Yi, Y. Lu, A. Pineda-Lucena, J. C. Liao, A. L. Okorokov, J. Milner, C. H. Arrowsmith and A. Bochkarev, Single-stranded DNA mimicry in the p53 transactivation domain interaction with replication protein A, *Proc. Natl. Acad. Sci. U.S.A.*, 2005, **102**, 15412–15417.
26. A. I. Arunkumar, V. Klimovich, X. Jiang, R. D. Ott, L. Mizoue, E. Fanning and W. J. Chazin, Insights into hRPA32 C-terminal domain-mediated assembly of the simian virus 40 replisome, *Nat. Struct. Mol. Biol.*, 2005, **12**, 332–339.
27. H. L. Ball, J. S. Myers and D. Cortez, ATRIP binding to replication protein A-single-stranded DNA promotes ATR-ATRIP localization but is

dispensable for Chk1 phosphorylation, *Mol. Biol. Cell*, 2005, **16**, 2372–2381.

28. H. L. Ball, M. R. Ehrhardt, D. A. Mordes, G. G. Glick, W. J. Chazin and D. Cortez, Function of a conserved checkpoint recruitment domain in ATRIP proteins, *Mol. Cell. Biol.*, 2007, **27**, 3367–3377.
29. J. Majka, S. K. Binz, M. S. Wold and P. M. Burgers, Replication protein A directs loading of the DNA damage checkpoint clamp to 5′-DNA junctions, *J. Biol. Chem.*, 2006, **281**, 27855–27861.
30. A. A. Davies, D. Huttner, Y. Daigaku, S. Chen and H. D. Ulrich, Activation of ubiquitin-dependent DNA damage bypass is mediated by replication protein A, *Mol. Cell*, 2008, **29**, 625–636.
31. X. Xu, S. Vaithiyalingam, G. G. Glick, D. A. Mordes, W. J. Chazin and D. Cortez, The basic cleft of RPA70N binds multiple checkpoint proteins, including RAD9, to regulate ATR signaling, *Mol. Cell. Biol.*, 2008, **28**, 7345–7353.
32. L. Zou, D. Liu and S. J. Elledge, Replication protein A-mediated recruitment and activation of Rad17 complexes, *Proc. Natl. Acad. Sci. U.S.A.*, 2003, **100**, 13827–13832.
33. K. Umezu, N. Sugawara, C. Chen, J. E. Haber and R. D. Kolodner, Genetic analysis of yeast RPA1 reveals its multiple functions in DNA metabolism, *Genetics*, 1998, **148**, 989–1005.
34. I. M. Wyka, K. Dhar, S. K. Binz and M. S. Wold, Replication protein A interactions with DNA: differential binding of the core domains and analysis of the DNA interaction surface, *Biochemistry*, 2003, **42**, 12909–12918.
35. A. I. Arunkumar, M. E. Stauffer, E. Bochkareva, A. Bochkarev and W. J. Chazin, Independent and coordinated functions of replication protein A tandem high affinity single-stranded DNA binding domains, *J. Biol. Chem.*, 2003, **278**, 41077–41082.
36. S. A. Bastin-Shanower and S. J. Brill, Functional analysis of the four DNA binding domains of replication protein A: the role of RPA2 in ssDNA binding, *J. Biol. Chem.*, 2001, **276**, 36446–36453.
37. E. Bochkareva, V. Belegu, S. Korolev and A. Bochkarev, Structure of the major single-stranded DNA-binding domain of replication protein A suggests a dynamic mechanism for DNA binding, *EMBO J.*, 2001, **20**, 612–618.
38. A. Bochkarev, R. A. Pfuetzner, A. M. Edwards and L. Frappier, Structure of the single-stranded-DNA-binding domain of replication protein A bound to DNA, *Nature*, 1997, **385**, 176–181.
39. X. Jiang, V. Klimovich, A. I. Arunkumar, E. B. Hysinger, Y. Wang, R. D. Ott, G. D. Guler, B. Weiner, W. J. Chazin and E. Fanning, Structural mechanism of RPA loading on DNA during activation of a simple prereplication complex, *EMBO J.*, 2006, **25**, 5516–5526.
40. K. Labib and A. Gambus, A key role for the GINS complex at DNA replication forks, *Trends Cell Biol.*, 2007, **17**, 271–278.

41. R. D. Ott, C. Rehfuess, V. N. Podust, J. E. Clark and E. Fanning, Role of the p68 subunit of human DNA polymerase alpha-primase in simian virus 40 DNA replication, *Mol. Cell. Biol.*, 2002, **22**, 5669–5678.
42. H. Huang, B. Weiner, H. Zhang, W. J. Chazin and E. Fanning, unpublished data.
43. W. C. Copeland and T. S. Wang, Enzymatic characterization of the individual mammalian primase subunits reveals a biphasic mechanism for initiation of DNA replication, *J. Biol. Chem.*, 1993, **268**, 26179–26189.
44. B. Arezi, B. W. Kirk, W. C. Copeland and R. D. Kuchta, Interactions of DNA with human DNA primase monitored with photoactivatable cross-linking agents: implications for the role of the p58 subunit, *Biochemistry*, 1999, **38**, 12899–12907.
45. L. K. Zerbe and R. D. Kuchta, The p58 subunit of human DNA primase is important for primer initiation, elongation, and counting, *Biochemistry*, 2002, **41**, 4891–4900.
46. P. Garg and P. M. Burgers, DNA polymerases that propagate the eukaryotic DNA replication fork, *Crit. Rev. Biochem. Mol. Biol.*, 2005, **40**, 115–128.
47. Y. I. Pavlov, P. V. Shcherbakova and I. B. Rogozin, Roles of DNA polymerases in replication, repair, and recombination in eukaryotes, *Int. Rev. Cytol.*, 2006, **255**, 41–132.
48. Y. H. Jin, R. Obert, P. M. Burgers, T. A. Kunkel, M. A. Resnick and D. A. Gordenin, The 3′-->5′ exonuclease of DNA polymerase delta can substitute for the 5′ flap endonuclease Rad27/Fen1 in processing Okazaki fragments and preventing genome instability, *Proc. Natl. Acad. Sci. U.S.A.*, 2001, **98**, 5122–5127.
49. Z. F. Pursell, I. Isoz, E. B. Lundstrom, E. Johansson and T. A. Kunkel, Yeast DNA polymerase epsilon participates in leading-strand DNA replication, *Science*, 2007, **317**, 127–130.
50. S. A. Nick McElhinny, D. A. Gordenin, C. M. Stith, P. M. J. Burgers and T. A. Kunkel, Division of labor at the eukaryotic replication fork, *Mol. Cell*, 2008, **30**, 137–144.
51. M. L. Bochman, S. P. Bell and A. Schwacha, Subunit organization of Mcm2-7 and the unequal role of active sites in ATP hydrolysis and viability, *Mol. Cell. Biol.*, 2008, **28**, 5865–5873.
52. M. L. Bochman and A. Schwacha, The Mcm2-7 complex has *in vitro* helicase activity, *Mol. Cell*, 2008, **31**, 287–293.
53. S. E. Moyer, P. W. Lewis and M. R. Botchan, Isolation of the Cdc45/Mcm2-7/GINS (CMG) complex, a candidate for the eukaryotic DNA replication fork helicase, *Proc. Natl. Acad. Sci. U.S.A.*, 2006, **103**, 10236–10241.
54. O. Chilkova, P. Stenlund, I. Isoz, C. M. Stith, P. Grabowski, E. B. Lundstrom, P. M. Burgers and E. Johansson, The eukaryotic leading and lagging strand DNA polymerases are loaded onto primer-ends via separate mechanisms but have comparable processivity in the presence of PCNA, *Nucleic Acids Res.*, 2007, **35**, 6588–6597.

55. T. Tsurimoto, T. Melendy and B. Stillman, Sequential initiation of lagging and leading strand synthesis by two different polymerase complexes at the SV40 DNA replication origin, *Nature*, 1990, **346**, 534–539.
56. S. Waga and B. Stillman, The DNA replication fork in eukaryotic cells, *Annu. Rev. Biochem.*, 1998, **67**, 721–751.
57. I. Dornreiter, L. F. Erdile, I. U. Gilbert, D. von Winkler, T. J. Kelly and E. Fanning, Interaction of DNA polymerase alpha-primase with cellular replication protein A and SV40 T antigen, *EMBO J.*, 1992, **11**, 769–776.
58. K. A. Braun, Y. Lao, Z. He, C. J. Ingles and M. S. Wold, Role of protein–protein interactions in the function of replication protein A (RPA): RPA modulates the activity of DNA polymerase alpha by multiple mechanisms, *Biochemistry*, 1997, **36**, 8443–8454.
59. G. Mass, T. Nethanel and G. Kaufmann, The middle subunit of replication protein A contacts growing RNA–DNA primers in replicating simian virus 40 chromosomes, *Mol. Cell. Biol.*, 1998, **18**, 6399–6407.
60. J. Majka and P. M. Burgers, The PCNA-RFC families of DNA clamps and clamp loaders, *Prog. Nucleic Acid Res. Mol. Biol.*, 2004, **78**, 227–260.
61. S. L. Kazmirski, M. Podobnik, T. F. Weitze, M. O'Donnell and J. Kuriyan, Structural analysis of the inactive state of the *Escherichia coli* DNA polymerase clamp-loader complex, *Proc. Natl. Acad. Sci. U.S.A.*, 2004, **101**, 16750–16755.
62. X. V. Gomes, S. L. Schmidt and P. M. Burgers, ATP utilization by yeast replication factor C. II. Multiple stepwise ATP binding events are required to load proliferating cell nuclear antigen onto primed DNA, *J. Biol. Chem.*, 2001, **276**, 34776–34783.
63. X. V. Gomes and P. M. Burgers, ATP utilization by yeast replication factor C. I. ATP-mediated interaction with DNA and with proliferating cell nuclear antigen, *J. Biol. Chem.*, 2001, **276**, 34768–34775.
64. E. Johansson, P. Garg and P. M. Burgers, The Pol32 subunit of DNA polymerase delta contains separable domains for processive replication and proliferating cell nuclear antigen (PCNA) binding, *J. Biol. Chem.*, 2004, **279**, 1907–1915.
65. L. S. Cox, Who binds wins: competition for PCNA rings out cell-cycle changes, *Trends Cell Biol.*, 1997, **7**, 493–498.
66. E. Warbrick, PCNA binding through a conserved motif, *Bioessays*, 1998, **20**, 195–199.
67. Y. Masuda, M. Suzuki, J. Piao, Y. Gu, T. Tsurimoto and K. Kamiya, Dynamics of human replication factors in the elongation phase of DNA replication, *Nucleic Acids Res.*, 2007, **35**, 6904–6916.
68. J. Yang, Z. Zhuang, R. M. Roccasecca, M. A. Trakselis and S. J. Benkovic, The dynamic processivity of the T4 DNA polymerase during replication, *Proc. Natl. Acad. Sci. U.S.A.*, 2004, **101**, 8289–8294.
69. D. E. Johnson, M. Takahashi, S. M. Hamdan, S. J. Lee and C. C. Richardson, Exchange of DNA polymerases at the replication fork of bacteriophage T7, *Proc. Natl. Acad. Sci. U.S.A.*, 2007, **104**, 5312–5317.

70. P. McInerney, A. Johnson, F. Katz and M. O'Donnell, Characterization of a triple DNA polymerase replisome, *Mol. Cell*, 2007, **27**, 527–538.
71. A. R. Lehmann, A. Niimi, T. Ogi, S. Brown, S. Sabbioneda, J. F. Wing, P. L. Kannouche and C. M. Green, Translesion synthesis: Y-family polymerases and the polymerase switch, *DNA Repair (Amst.)*, 2007, **6**, 891–899.
72. D. F. Jarosz, P. J. Beuning, S. E. Cohen and G. C. Walker, Y-family DNA polymerases in *Escherichia coli*, *Trends Microbiol.*, 2007, **15**, 70–77.
73. E. C. Friedberg, A. R. Lehmann and R. P. Fuchs, Trading places: how do DNA polymerases switch during translesion DNA synthesis?, *Mol. Cell*, 2005, **18**, 499–505.
74. L. D. Langston and M. O'Donnell, DNA polymerase delta is highly processive with proliferating cell nuclear antigen and undergoes collision release upon completing DNA, *J. Biol. Chem.*, 2008, **283**, 29522–29531.
75. J. A. Wells and C. L. McClendon, Reaching for high-hanging fruit in drug discovery at protein–protein interfaces, *Nature*, 2007, **450**, 1001–1009.
76. S. Patel and M. R. Player, Small-molecule inhibitors of the p53-HDM2 interaction for the treatment of cancer, *Expert Opin. Investig. Drugs*, 2008, **17**, 1865–1882.
77. R. E. Georgescu, O. Yurieva, S. S. Kim, J. Kuriyan, X. P. Kong and M. O'Donnell, Structure of a small-molecule inhibitor of a DNA polymerase sliding clamp, *Proc. Natl. Acad. Sci. U.S.A.*, 2008, **105**, 11116–11121.
78. G. D. Bowman, M. O'Donnell and J. Kuriyan, Structural analysis of a eukaryotic sliding DNA clamp–clamp loader complex, *Nature*, 2004, **429**, 724–730.
79. S. L. Kazmirski, Y. Zhao, G. D. Bowman, M. O'Donnell and J. Kuriyan, Out-of-plane motions in open sliding clamps: molecular dynamics simulations of eukaryotic and archaeal proliferating cell nuclear antigen, *Proc. Natl. Acad. Sci. U.S.A.*, 2005, **102**, 13801–13806.

CHAPTER 7

Coping with DNA Damage and Replication Stress

HELLE D. ULRICH

Cancer Research UK London Research Institute, Clare Hall Laboratories, Blanche Lane, South Mimms, Hertfordshire, EN6 3LD, UK

7.1 Introduction

Faithful transmission of genetic information from generation to generation requires the accurate duplication of a full genome within the time frame of an S phase. This accomplishment places high demands on both the precision and speed of the enzymes involved. Mutation rates of less than 10^{-10} per base pair and generation, as commonly observed in eukaryotic cells, result from a combination of polymerase fidelity, exonucleolytic proofreading, accessory replisome proteins and a correction of misinsertions by the mismatch repair system[1] (see also Chapter 4). However, optimal performance by the replication machinery also depends heavily on an undisturbed template. The replicative DNA polymerases in particular are streamlined for accurate and processive polymerisation of native, chemically intact DNA and exhibit little tolerance to damaged or distorted templates.[2,3]

As a reactive molecule in an aqueous environment, DNA is susceptible to a variety of attacks and injuries.[4] An estimated one million lesions, most of which interfere with DNA replication, occur spontaneously in every human cell per day.[5,6] Exogenous sources of DNA damage, such as chemicals and radiation, add to the challenges faced by the replication machinery in a natural environment.[7] As a consequence, cells have developed a range of measures to cope with the inevitable fragility of their genomes.

Molecular Themes in DNA Replication
Edited by Lynne S. Cox

Published by the Royal Society of Chemistry, www.rsc.org

This chapter describes the mechanisms by which eukaryotic cells sense and overcome DNA damage during replication. Beginning with a general summary of DNA repair and the damage response operational during S phase, it discusses the DNA replication checkpoint, triggered specifically by conditions that block the progression of replication forks (see also Chapter 5).

This section highlights the basic concepts of checkpoint signalling and explains how the sensing of replication stress prepares the cell to take appropriate measures to overcome its problems. It is followed by a description of pathways contributing to the actual bypass of DNA lesions, including translesion synthesis by damage-tolerant DNA polymerases as well as recombination-dependent replication restart mechanisms and their consequences for mutagenesis and overall genome stability. Mammalian nomenclature is used throughout unless specific examples from lower eukaryotes are described. By illustrating the manifold and intimate connections between core components of the replication machinery and damage-specific signalling and repair factors, the chapter attempts to give an impression of the seamless transitions between DNA replication and damage processing.

7.2 The Sensing of DNA Damage and Replication Stress

An essential factor in the defence against DNA damage is its perception. Recognition of a lesion is the initial step for its removal, but equally important is its role as a signal to effect global changes in cellular physiology in preparation for DNA repair. Recognition of DNA damage is complicated on one hand by the necessity to identify very subtle changes to the DNA that may not cause overall perturbances, and on the other hand by the diversity of non-physiological structures that cellular repair systems have to deal with.

Endogenous DNA damage affects the bases as well as the sugar–phosphate backbone and involves mostly hydrolytic reactions such as deaminations or base loss, but also methylation and oxidation.[4–6] In addition, exogenous chemicals can form a variety of adducts or DNA crosslinks. Ultraviolet (UV) radiation causes the crosslinking of adjacent pyrimidines, most often in the form of cyclobutane pyrimidine dimers (CPDs) or (6-4) photoproducts.[8] Ionising radiation results in various types of base modifications as well as single- and double-strand breaks. All of these lesions are processed by dedicated repair systems, whose in-depth description is covered by several recent books.[9,10]

The following section provides a brief summary of how repair pathways recognise damaged DNA within and outside of S phase, thereby preventing negative effects on DNA replication. These repair systems are important for minimising the use of damaged DNA as a template not only for replication, but also for transcription, as RNA polymerases face the same problems as replicative DNA polymerases on a damaged template. Replication-specific aspects of DNA damage sensing are then discussed in more detail.

7.2.1 Sensing of DNA Damage

7.2.1.1 Sensing of DNA Damage by Repair Systems

Eukaryotes harbour a number of systems for the direct reversal of DNA damage. For example, photolyases are able to resolve UV-induced CPDs in a light-dependent reaction,[11] and several enzymes exist for the removal of methyl groups from DNA.[12] More commonly, however, lesions are eliminated by excision of the damaged stretch and resynthesis based on the information encoded by the undamaged complementary strand. This task is performed by two pathways: base excision repair (BER) and nucleotide excision repair (NER). They operate in a mostly cell cycle-independent manner and serve as defence mechanisms for virtually all types of DNA damage excluding strand breaks.

BER involves lesion-specific recognition of damaged bases or abasic (AP) sites by a set of dedicated enzymes, mono-functional glycosylases and bi-functional glycosylases/AP lyases, which cleave the glycosidic bond between the damaged base and the sugar. In contrast, global genomic NER (GG-NER) recognises structural distortions rather than specific lesions. It is therefore considered the most universal repair system.[13,14] NER results in the excision of a short patch of DNA on one strand, whose resynthesis requires components of the general DNA replication machinery, PCNA, RFC, Replication Protein A (RPA) and Polymerase δ (see Chapter 1). In a specialised sub-pathway of NER, transcription-coupled repair (TCR), damage recognition is mediated by a stalled RNA Polymerase II complex, which targets repair to the transcribed strand of actively expressed genes.[15,16]

DNA single-strand breaks, involving only one of the two strands, are common lesions that may disturb transcription and—if not repaired—can be converted to more dangerous double strand breaks (DSBs) by DNA replication.[17] They are repaired by ligation in a pathway similar to BER, with the major challenge being the 'cleaning up' of un-ligatable termini.[18,19]

DNA double strand breaks are potentially the most dangerous types of lesions during S phase, as they endanger not only DNA replication and gene expression, but also the accurate segregation of chromosomes during cell division. Their processing requires fundamentally different mechanisms from those of damaged bases or nucleotides. Two distinct pathways of repair, non-homologous end-joining (NHEJ) and homologous recombination (HR), are available. Both have been reviewed in detail elsewhere[9–10,20–25] (see also Chapter 5).

Briefly, NHEJ involves the direct ligation of blunt or compatible termini by a DNA ligase, aided by several additional factors that contribute to recognition, processing and alignment of the ends. While NHEJ is an error-prone pathway that usually results in small deletions around the break site, HR can accurately restore the original sequence given the presence of a homologous region elsewhere in the genome. However, HR can also result in gross chromosomal rearrangements (GCR) by means of genetic cross-over. Repair of a DSB by HR is initiated by a resection of the 5′ ends at the break, resulting in long 3′ overhangs covered initially by the RPA complex, but eventually replaced by the recombinogenic Rad51 filament capable of invading a homologous duplex DNA (see Figure 5.4). As outlined below, the structures of these repair intermediates have profound implications for the perception of the DSB as a damage signal by the checkpoint response.

7.2.1.2 DNA Damage Checkpoints

Checkpoints were originally defined as control points in the cell cycle at which the fulfilment of a certain constraint was monitored as a condition for progression to the next stage.[26] Today, the term is often used more loosely to describe the global physiological response of a cell to conditions of stress that impinge on cell cycle progression. Characteristic for a checkpoint response, the signal emanating from DNA damage usually promotes a cell cycle arrest or delay that is believed to provide time for repair.[27] It has become clear, however, that the DNA damage response is responsible for much more than just a cell cycle arrest; in fact, the signal transduction pathway triggered by DNA damage appears to have a direct impact on lesion processing by activating several repair pathways.

Among the different types of damage, DSBs have been studied in greatest detail as a trigger of the DNA damage checkpoint: two signalling pathways with distinct sensors, transducers and effectors have been revealed[28,29] (Figure 7.1A). They are initiated by two sensor protein kinases:

- ATM, named 'ataxia telangiectasia mutated' after an associated hereditary disease;
- ATR, named 'atm and Rad3 related', referring to its *Schizosaccharomyces pombe* homologue (see also Chapter 5).

ATM is rapidly activated by DSBs through direct association with the DNA termini, accompanied by autophosphorylation and dissociation of its dimeric form into monomers.[30] A multifunctional protein complex, MRN, consisting of the subunits Mre11, Rad50 and Nbs1[31] (see Chapter 5), which is involved in NHEJ as well as HR and is among the earliest factors recruited to DSBs,[32] is required for full ATM activation at DSBs in mammals and yeast.[33–36] Mre11 and Nbs1 are phosphorylated in an ATM-dependent manner; the latter was in turn found to mediate damage-dependent inhibition of DNA replication.[37] The MRN complex, *via* the nuclease activity of Mre11, is also required for the resection of DSBs in preparation for HR.[38,39] Following this resection, the second sensor kinase, ATR, is activated at the single-stranded DNA (ssDNA) overhangs adjacent to the break. The kinase is recruited to DNA by its interaction partner, ATRIP, which directly interacts with the RPA complex,[40] although an RPA-independent affinity of ATR for ssDNA has also been observed, and in some circumstances ATR activation may not require

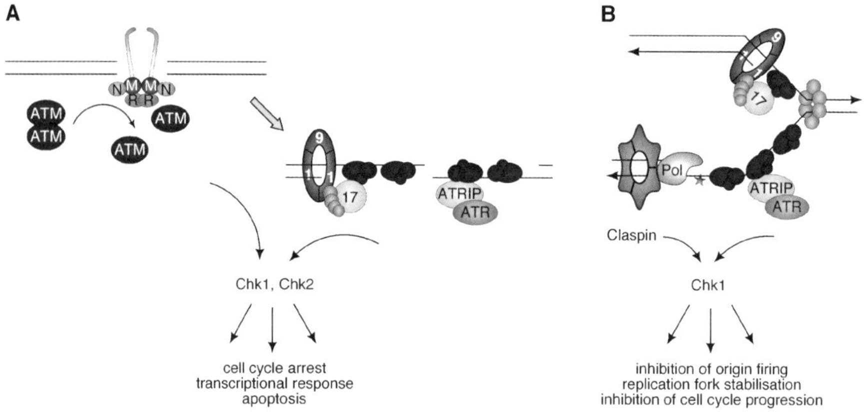

Figure 7.1 **Sensing of DNA damage and replication stress via the checkpoint response.** (**A**) The response to double-strand breaks in DNA (DSBs) is mediated by the two sensor kinases ATM and ATR. ATM activation involves the MRN complex (comprising Mre11, Rad50 and Nbs1), which associates directly with the break. Following MRN-dependent processing of the break ends, RPA-covered ssDNA contributes to the activation of ATR and the 9-1-1 checkpoint clamp by means of direct interactions with the ATRIP protein and the 9-1-1-specific clamp loader (Rad17). (**B**) The replication checkpoint senses problems with replication fork progression (indicated by an asterisk) by means of RPA-covered ssDNA, which is exposed due to the continuing movement of the replicative helicase in the absence of polymerisation. In addition to ATR and the 9-1-1 clamp, replisome components like Claspin (Mrc1) contribute to checkpoint signalling (see Chapter 9).

RPA.[41,42] RPA-coated ssDNA resulting from resection of a DSB also serves as a signal for the recruitment of a third damage sensor—the PCNA-like, heterotrimeric 9-1-1 complex, assembled from the subunits Rad9, Rad1 and Hus1[43,44] (see also Chapter 3). Like PCNA, the 9-1-1 complex is loaded onto DNA by an RFC-like loading factor, comprising a complex of Rad17 (Rad24 in *Saccharomyces cerevisiae*) in association with Rfc2-5.[43] Direct interactions between Rad17 and RPA have been found to be required for recruitment to ssDNA.[45]

Although the sensor kinases and the 9-1-1 clamp associate with damaged DNA independently,[46,47] functional interactions between the factors are important for the checkpoint signal. The 9-1-1 clamp, itself a substrate of the ATR kinase, is required for ATR activation in yeast.[48] In mammalian cells, both ATM and ATR physically interact with and phosphorylate Rad17 in response to DNA damage.[49] Although the mechanism of action of the 9-1-1 clamp is not yet understood, it may facilitate the selection or recruitment of substrates for the checkpoint kinases.

Much less is known about the mechanisms by which other types of damage elicit the checkpoint response, but the same principles as described above for the DSB, *i.e.* recognition of strand breaks by ATM and regions of ssDNA by ATR, aided by the 9-1-1 complex, appear to apply. NER tracts can easily be envisaged as inducers of ATR activity due to the formation of a ssDNA gap and the participation of RPA; physical interactions between NER components and the 9-1-1 clamp suggest a requirement for processing in checkpoint activation by this type of lesion.[50] Surprisingly, however, ATR was shown to preferentially bind to UV-damaged DNA in the absence of RPA or ATRIP,[51] raising the possibility that bulky lesions could be recognised directly by the sensor kinase.[52]

These pathways of checkpoint activation exemplify how a plethora of lesions can elicit a global signal of DNA damage that is transmitted by means of mediator proteins and downstream kinases, and promotes an adaptation of the cell to damage conditions. This involves a transcriptional response as well as cell cycle arrests at G1/S, intra-S and G2/M. The signal transducers are shared by ATM and ATR to some degree, and their targets vary with the cell cycle stage.

While the pathways described above are highly conserved from mammalian cells to simple unicellular eukaryotes, multicellular organisms have incorporated an additional layer of control into their DNA damage signalling repertoire—the choice to undergo controlled cell death *via* apoptosis.[53] This pathway may reflect the overall benefit for the organism to eliminate individual cells that have either accumulated unrepairable damage or undergone GCR that may lead to genomic instability, thus reducing the carcinogenic potential of DNA damage.

A key player in the mammalian response to genotoxic stress is the tumour suppressor protein p53.[54] This extensively studied transcription factor, also known as a 'guardian of the genome',[55] critically affects the choice between DNA repair, cell cycle progression and apoptosis by regulating the activity of numerous target genes involved in various metabolic pathways.[54,56] Consistent with its central position as a cellular gatekeeper, p53 is subject to various post-translational modifications, such as phosphorylation, acetylation, methylation, ubiquitylation, SUMOylation and neddylation, which collectively control its activity, stability and intracellular localisation.[54,57]

7.2.2 Sensing of Replication Stress

The previous section covered the recognition of DNA damage by repair and checkpoint systems in a largely cell cycle-independent manner. Although these pathways are important during S phase, replicating DNA presents additional problems in terms of damage recognition. On one hand, some lesions appear to be silent to the checkpoint until they are replicated, and some replication problems that warrant a checkpoint response may in turn arise in the absence of any structural damage to the DNA. Hence, although the signalling pathways for damage and replication stress overlap to a large extent and follow similar concepts, the cell appears to differentiate between the two phenomena and to take specific measures for the protection of replication forks in S phase.

7.2.2.1 Lesions that Cause Polymerase Blocks

Possibly the most sensitive damage sensors during S phase are the replicative DNA polymerases themselves. As mentioned above (see also Chapter 4), their active sites are streamlined for processive DNA synthesis and do not tolerate abnormal template structures.[2] In addition to obvious lesions 'overlooked' by the NER system, subtle flaws in the DNA can therefore result in a replication block without eliciting a checkpoint response outside of S phase. For example, the alkylating agent methyl methanesulfonate (MMS) causes replication-blocking adducts that elicit a checkpoint response only during DNA replication.[58] A similar situation may apply to other small lesions that are usually removed by BER. Even secondary structures in the single-stranded template, which would form only after strand separation by the helicase, could block progression of the polymerases. However, it is important to note that not all lesions detectable by repair systems would also cause polymerase stalling. For example, 8oxo-dG is normally removed by a dedicated glycosylase, but is efficiently used as a template by replicative polymerases, which tend to misinsert an A opposite this structure. Thus, 8oxo-dG is a mutagenic, but not a replication-blocking, lesion.[59]

Polymerase stalling due to a lesion in the template strand usually causes an asymmetry in the replication fork, as the replicative helicase initially continues to unwind the parental DNA, while DNA synthesis on the other strand proceeds uninhibited. As a consequence, RPA-covered ssDNA accumulates in front of the blocked polymerase[60] and triggers an ATR- and 9-1-1-dependent checkpoint response[61–64] as outlined in Section 7.1 (Figure 7.1B). Importantly, not only lesions, but also other conditions that inhibit DNA synthesis such as nucleotide depletion or polymerase inhibitors like aphidicolin, generate ssDNA at the replication fork due to an uncoupling between polymerase and helicase movement. In this case, replisome progression is significantly slowed down, although not completely stalled. Thus, by means of polymerase stalling, a variety of lesions and replication problems are converted to a common intermediate, RPA-covered ssDNA, as an initiator of the checkpoint response.

In many ways, the ensuing pathway resembles damage signalling in terms of the effectors involved, but the participation of some integral components of the replication machinery as signal transducers indicates S phase-specific features particularly in response to those types of replication stress that do not involve DNA damage;[65] in budding yeast, the non-essential replisome components Mrc1 (the homologue of mammalian Claspin) and Tof1 were identified as required for checkpoint signalling specifically after nucleotide depletion.[66–68] They appear to be responsible for the coupling of DNA unwinding or replisome translocation to the site of DNA synthesis and may serve as replication-specific mediators of the checkpoint response.[67] In their absence, a damage-specific mediator, yeast Rad9 (functionally related to human 53BP1) would transduce a genuine damage signal.[66]

7.2.2.2 *Other Types of Lesions*

Whereas lesions that specifically block the activity of replicative polymerases cause an accumulation of ssDNA at the replication fork, other conditions stall replication by affecting progression of the entire replisome. For example, lesions affecting both DNA strands, such as inter-strand cross-links (ICLs) caused by bifunctional agents like cisplatin, psoralen or nitrogen mustards, would impede the unwinding of the parental strands by the replicative DNA helicase. As a consequence, uncoupling of helicase and polymerase movement does not occur. Similarly, tight association of a protein with the DNA can inhibit unwinding. For example, an array of transcriptional repressors has been shown to block replication fork progression in a bacterial system.[69] In eukaryotes, several natural replication fork barriers, enforced by site-specific association of a protein, are known to control the direction of replication fork movement; for example in the rDNA or, in fission yeast, the mating type locus.[70,71] Depending on the genomic context, these types of replication problems sometimes appear to result in a transient pausing rather than a stalling of replication, and they do not cause a checkpoint response, presumably because of the lack of ssDNA accumulation.[72,73]

Covalent protein-DNA adducts, caused by topoisomerase inhibitors like camptothecin, act as replication-specific DNA-damaging agents, as collisions with the replisome result in the formation of DSBs.[74] Again, studies in budding yeast have shown that elevated levels of ssDNA do not accumulate and the replication checkpoint is not activated under these conditions.[75,76] However, in the absence of RecQ homologues, replication forks stall at high frequency (see also Chapters 3 and 5).[77,78] Finally, unrepaired single-stranded DNA breaks, which can either be caused directly by damaging agents or arise from BER processes, are converted to DSBs in a replication-dependent manner.[79]

7.2.2.3 *Consequences of the Replication Checkpoint*

The vulnerability of DNA replication in the absence of the appropriate checkpoint factors is highlighted by the observations that defects in various components of the ATR-dependent checkpoint strongly sensitise cells to replication stress-induced chromosome fragmentation,[80–82] and that the stability of so-called fragile sites and replication slow zones, supposedly areas difficult to replicate in mammals and yeast, is markedly decreased in the absence of ATR or in response to replication stress.[81,83–86]

Hence, the physiological response to replication stress is particularly aimed at the protection and stabilisation of replication forks.[61–63] Strictly speaking, the cell cycle does not undergo a complete arrest, because even in the presence of MMS, replication forks progress slowly and replication is eventually completed.[87] However, inhibition of replication initiation at late-firing origins causes an overall slow-down of S phase, thereby also preventing the generation of additional sources of fork instability and thus indicating a cross-talk between early and late origins mediated by the checkpoint protein Rad53[88] (see Chapters 1 and 5).

Mrc1/Claspin and Tof1 contribute to the stabilisation of the replisome at the site of DNA synthesis and may limit excessive helicase-dependent unwinding,[67] and the association of polymerases α and ε with the stalled forks is protected.[89] Although the phosphorylation targets relevant to these checkpoint functions are largely unknown, their effects are genetically separable. For example, a mutant of the budding yeast ATR homologue, *mec1-100*, fails to inhibit late origin firing in response to replication stress, but is capable of stabilising stalled replication forks.[90]

The stabilising actions of the replication checkpoint on replication forks allow a differentiation between fork stalling and fork collapse. The latter is usually defined by a loss of replisome components such as the polymerases and the replicative helicase. As discussed in detail below, eukaryotic cells employ damage tolerance mechanisms for the resolution and restart of stalled replication forks, whereas replication fork collapse requires more elaborate measures, usually associated with HR.[61,71] Hence, the replication checkpoint protects the cell against GCR and global genome instability. Accordingly, fork collapse upon failure of the replication checkpoint results in a genuine damage signal that triggers a checkpoint response involving the ATM-dependent reaction to DSBs. Notably, the vulnerability to replication fork collapse appears to be unique to eukaryotes, as in bacteria, the replisome is routinely lost upon fork stalling and HR appears to be the method of choice for reloading of the replisome and replication restart.[91–93]

7.3 DNA Damage Tolerance Mechanisms

Complementary to the strategies for stabilising a stalled replication fork, cells need to take appropriate measures to resolve the problem that caused the stalling in order to be able to complete DNA replication.

Two fundamentally different mechanisms allowing the replisome to overcome DNA damage are described in this section. On one hand, the replisome can be forced to progress through the lesion by exchange of the polymerase, and on the other hand, use of the damaged DNA can be avoided altogether by switching to an alternative template (Figure 7.2). As they do not actually remove lesions, these mechanisms are generally called 'DNA damage tolerance' in order to distinguish them from repair mechanisms. However, the term post-replication repair (PRR) has also been used.

7.3.1 Location and Timing of Damage Bypass

The concept of post-replication repair stems from early observations in *E. coli* of persistent daughter strand gaps in an excision repair-deficient background,[94]

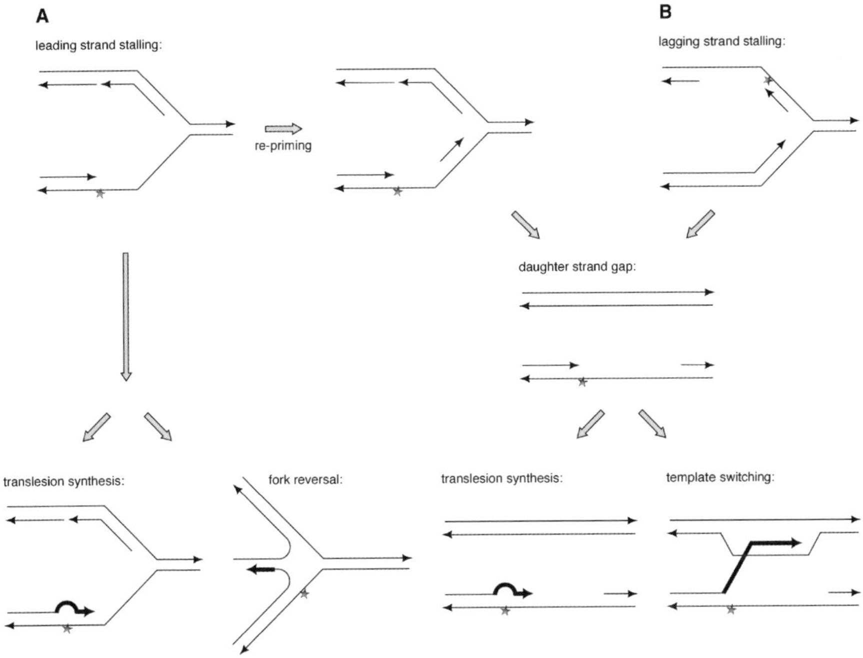

Figure 7.2 **Pathways of DNA damage tolerance.** (**A**) Replication fork stalling on the leading strand can be resolved by translesion synthesis or may result in replication fork reversal, leading to a 'chicken foot' structure. Initiation of DNA synthesis by re-priming downstream of the lesion is also a possibility and results in a daughter strand gap. (**B**) Stalling on the lagging strand likewise results in a daughter strand gap, which can be filled by translesion synthesis or template switching in a post-replicative manner.

which indicated a system dedicated to the filling of these gaps in a post-replicative manner. Similar observations were made in mammals and yeast.[95–97] Even before the molecular details of the relevant pathway were elucidated, however, the misleading nature of its name was noted, as it was realised that damage is not removed (*i.e.* repaired), but bypassed, and rather than acting strictly after completion of S phase, its function was deemed to be equally relevant during DNA synthesis.[98]

The confusion about the name of the pathway illustrates a problem directly related to the mechanism of damage bypass, regarding the location and timing of the event. As outlined in Figure 7.2B, polymerase stalling on the lagging strand would result in a small daughter strand gap due to re-initiation of the following Okazaki fragment. On the leading strand, however, the question arises whether the lesion is bypassed directly at the fork or whether discontinuous DNA synthesis can occur here as well, again leaving a daughter strand gap to be filled behind the fork at a later time (Figure 7.2A). In the bacterial system, there is good evidence for re-priming and post-replicative gap filling,[99] and electron micrographs of stalled replication intermediates in yeast indicate that re-priming is possible in eukaryotes as well.[60] Moreover, experiments using the mammalian SV40 replication system indicated that gap filling, rather than replication fork progression, was limiting for completion of DNA synthesis on UV-damaged DNA, suggesting that replisome movement is not significantly inhibited by lesions.[100] These observations do not exclude a contribution of both replication-associated and post-replicative damage bypass.

7.3.2 Translesion Synthesis

Translesion synthesis (TLS) is defined as the replicative processing of a damaged template. Depending on the nature of the lesion, bypass may be accurate in the sense that the damaged nucleotide is correctly recognised and the complementary one is used for polymerisation, or error-prone, with an inappropriate nucleotide being inserted opposite the lesion.

Of course, in an extreme scenario where the lesion is an abasic [apurinic/apyrimidinic (AP)] site, all information regarding the correct nucleotide is lost, and many lesions affecting the base may obscure or even change its coding capacity.[59] Consequently, TLS is often mutagenic. Its effects have long been known as damage-induced mutagenesis, and in *E. coli*, are termed the SOS response.[101,102] Genetic analysis in yeast and *E. coli* identified several factors responsible for this process without elucidating their enzymatic activities.[103–106] Working models postulated a modification of the properties of the replicative polymerase by these translesion-specific factors,[107] but when one of the genes, *REV3*, was cloned, its sequence was found to resemble that of a DNA polymerase.[108] It has now become clear that TLS in pro- and eukaryotes is mediated directly by a range of non-essential, damage-tolerant DNA polymerases dedicated to lesion bypass.[109–112]

Over the past decade, a wealth of information about these enzymes has been gathered, and although each exhibits some unique characteristics, a number of common properties have been defined (see also Chapter 4). Compared with the replicative polymerases, their accuracy on undamaged templates (with error rates of 10^{-2} to 10^{-4}) is usually low.[112] They lack 3′–5′ exonucleolytic activity and tend to be much less processive. Despite their low accuracy, damage-tolerant polymerases do not act completely unselectively, but exhibit distinct preferences and mutation spectra.

The following section describes the structural features and properties of each of the eukaryotic TLS polymerases before discussing their cooperation on damaged DNA.

7.3.3 Structures and Properties of Damage-tolerant DNA Polymerases

With the notable exception of polymerase ζ, a member of the B-family (see Figure 4.2), all TLS enzymes belong to a highly conserved sub-family of DNA polymerases, termed the Y-family.[113] This class of polymerases shares a

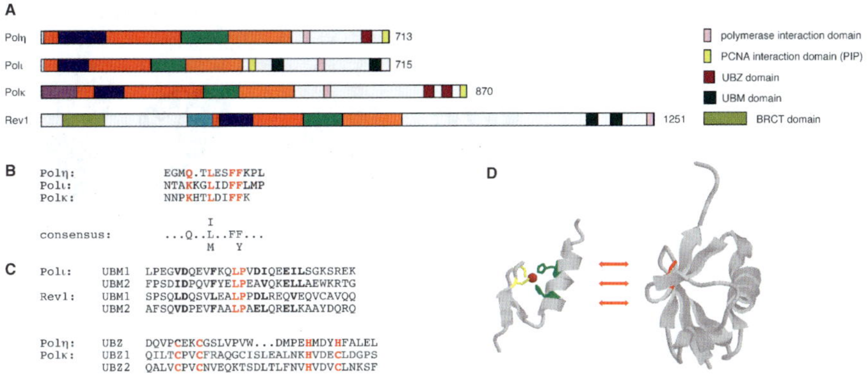

Figure 7.3 **The Y-family of damage tolerant DNA polymerases.** (**A**) Domain structures of the human Y-family DNA polymerases. Colour coding within the catalytic domain: red (palm), blue (fingers), green (thumb), orange (PAD). N-terminal extensions in Polκ and Rev1 are shown in purple and cyan. The scheme was adapted from ref. 123. (**B**) Alignment of the PCNA-interaction peptides (PIP) in human TLS polymerases. Conserved residues are marked in red. (**C**) Alignment of the ubiquitin binding UBM and UBZ domains. Strongly conserved residues are marked in red, less conserved ones in bold. (**D**) Cartoon models of the Polη UBZ domain[206] and ubiquitin. The zinc-coordinating cysteine and histidine residues of the UBZ domain are indicated in yellow and green. Isoleucine 44 of ubiquitin, which marks the interaction surface for most UBDs, is highlighted in red.

catalytic core, characterised by five signature motifs, which is extended by unique C-terminal and in some cases N-terminal domains harbouring protein–protein interaction motifs (Figure 7.3). X-ray structures of two archaeal and the polymerase domains of all the eukaryotic Y-family polymerases have been solved;[114–122] their overall architecture conforms to the 'right-handed' structure of replicative polymerases, comprising palm, thumb and finger domains (Figure 7.4). The active site is situated within the palm domain in a cleft formed by thumb and fingers. While the palm closely resembles that of replicative polymerases, both the fingers and the thumb are notably shorter in the TLS enzymes, leaving the active site wider and more solvent-accessible (Figure 7.5).

A domain unique to the Y-family, termed the 'little finger' or 'polymerase-associated domain' (PAD) is situated at the C-terminus of the catalytic core. Whereas in replicative polymerases, the fingers are responsible for clamping down on the replicating base pair in opposition to the thumb, this role is fulfilled by the PAD in the Y-family enzymes.[115,120] Accordingly, this domain undergoes the largest conformational change upon substrate binding. Whereas replicative polymerases achieve their high fidelity by an inability of the finger domain to close the active site in the presence of a mispaired dNTP or a damaged template nucleotide, the spacious cleft of some Y-family enzymes accommodates both mismatches and distorted templates. Yet, their selectivity

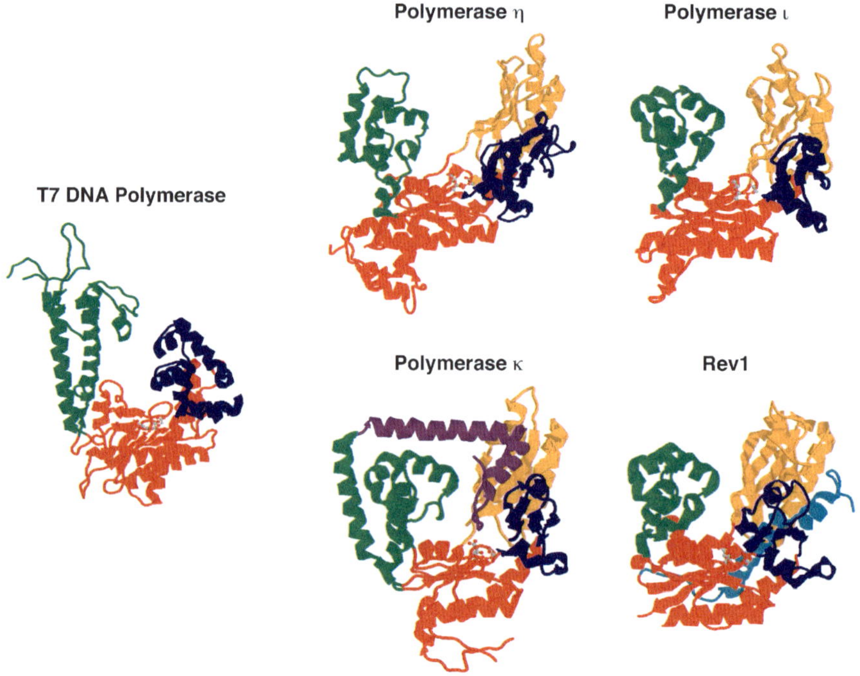

Figure 7.4 **Cartoon models of the catalytic domains of Y-family polymerases** Polη[120] and Rev1[117] from *S. cerevisiae* and human Polι[118] and Polκ[116] The catalytic domain of T7 DNA polymerase, a replicative enzyme, is shown for comparison.[207] Colour coding is according to Figure 7.3. Three conserved acidic residues important for catalytic activity indicate the location of the active site and are situated at the bottom of the cleft.

depends not only on Watson–Crick pairing of the incoming dNTP with the template nucleotide, but also on hydrogen bonding between substrates and enzyme, resulting in unique preferences and TLS efficiencies amongst the different polymerases.[111–112,123]

7.3.3.1 Polymerase η (Polη)

The gene encoding the yeast homologue of Polη, *RAD30*, was cloned before its biochemical function was established, though a contribution to error-free bypass of UV-induced lesions was noted.[124,125] The discovery that the protein was a genuine DNA polymerase capable of correctly bypassing a T–T cyclobutane pyrimidine dimer (CPD)[126] opened the door to a mechanistic understanding of TLS and the identification of further Y-family polymerases.

Cloning and identification of the human homologue also gave insight into a hereditary disease named Xeroderma pigmentosum variant[127] (XP-V), which is associated with Polη deficiency and results in photosensitivity and a strong predisposition for skin cancer.[126,128,129] Consistent with its function in

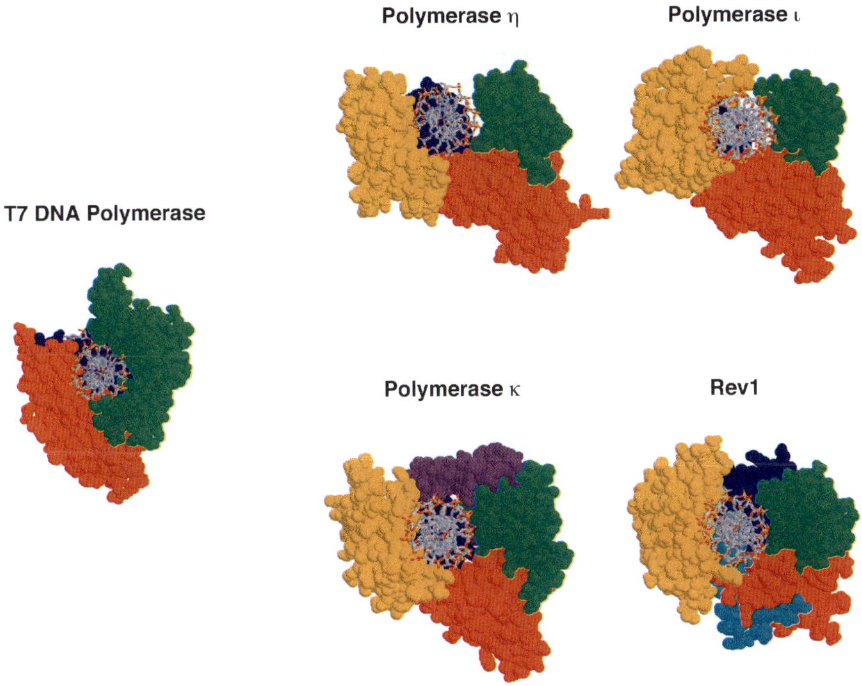

Figure 7.5 **DNA binding by the Y-family polymerases.** Space-filling models are shown for the same group of enzymes as in Figure 7.4, using the same colour code, but rotated by 180°. The DNA substrate in the active site is shown as a stick model. In the case of Polη, a structure with a cisplatin adduct in the active site was used to generate the image.[114]

tolerance to UV light, Polη appears to be adapted particularly for the bypass of CPDs. Although its overall fidelity *in vitro* is low, with an error frequency of 10^{-2} to 10^{-3}, it correctly inserts two As opposite T–T CPDs. In contrast, (6-4) photoproducts are processed inefficiently and inaccurately. With these characteristics, Polη is complementary to the NER system, which recognises (6-4) photoproducts much better than CPDs. The enzyme accurately bypasses 8oxo-dG and tolerates strongly distorting lesions such as cisplatin adducts; however, it does not extend mismatched primer termini well.

The shape of Polη is consistent with these properties. From the X-ray structure of the apo-enzyme (Figure 7.4), its active site was predicted to be wide enough to accommodate both nucleotides of a T–T cyclobutane pyrimidine dimer,[120] and the structure of the polymerase in complex with a template carrying an intrastrand cisplatin adduct, 1,2-d(GpG), has confirmed this notion[114] (Figure 7.5).

The structure also shows how the lesion is processed: by means of hydrogen bonding with the enzyme-bound dCTP, the template-primer complex is pulled towards the incoming nucleotide. Rotation of the template DNA into the active site then allows the chemical step to take place. While insertion of the first nucleotide opposite the adduct is efficient and accurate, the second step is less efficient and more promiscuous due to the rigid structure of the cross-link.

7.3.3.2 Polymerase ι (Polι)

Although related in sequence, Polι differs from Polη in its catalytic properties, in that its fidelity is strongly biased towards purines in the template strand. Opposite dA, the enzyme exhibits error frequencies of 10^{-3} to 10^{-5}. A dG is processed somewhat less efficiently and accurately, but pyrimidines (C and T) are used very inefficiently with misincorporation rates up to 90%.

Polι does not bypass CPDs, but can process abasic sites, N^2 adducts of G and the 3′ nucleotide of (6-4) photoproducts. Its preference for template purines can be explained by the X-ray structure in complex with a primed template and an incoming nucleotide:[118] the template adenine (A) is rotated into the *syn* conformation and is recognised by the dTTP *via* Hoogsteen basepairing to its back side (Figure 7.6). The *in vivo* function of Polι, which is absent in lower eukaryotes, is still somewhat unclear, but there is evidence that it might serve as a backup for other TLS polymerases.

7.3.3.3 Polymerase κ (Polκ)

Polκ, which is found in higher eukaryotes and in fission yeast, but not in *S. cerevisiae*, is most closely related to the *E. coli* TLS polymerase DinB and the archaeal Dpo4 and Dbh, as it shares several sequence motifs within the PAD domain.[123] A unique feature of Polκ, however, is a catalytically important N-terminal α-helical extension that encircles the DNA substrate and constrains the active site[116,121] (Figures 7.4 and 7.5). Whereas in Polη, Polι and Dpo4, the PAD is close to the fingers and binds to duplex DNA, this domain is separated from the core in Polκ. The resulting gap close to the active site is likely to

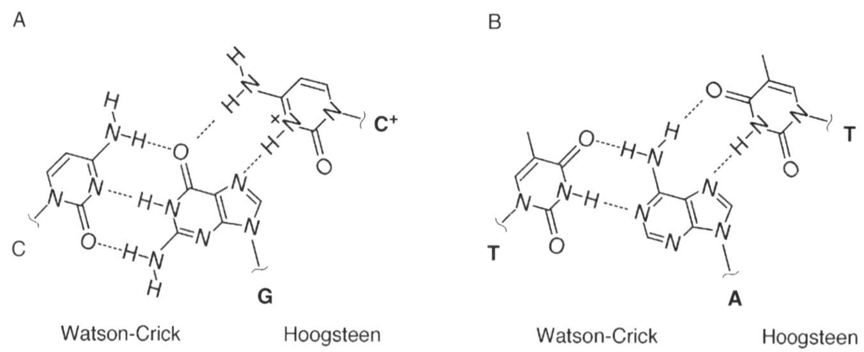

Figure 7.6 **Hydrogen-bonding in Watson–Crick and Hoogsteen base pairs. (A)** G–C base pairs. (**B**) A–T base pairs. Note that in order to engage in Hoogsteen base pairing in the context of a duplex DNA, the purine base needs to rotate to the *syn* conformation.

accommodate bulky lesions well, possibly explaining the enzyme's preference for large hydrophobic adducts at the N^2 position of dG, such as benz[*a*]pyrene diol epoxide (BPDE).

Although Polκ is relatively accurate, with error frequencies of 10^{-3} to 10^{-4}, it can generate -1 frameshifts (see Figure 4.5), possibly by looping out a template base, and is capable of extending mismatched termini. *In vivo* the enzyme might be important for the bypass of adducts caused by steroid hormones like oestrogen.

7.3.3.4 Rev1

The *REV1* gene from *S. cerevisiae* emerged from one of the earliest screens for factors involved in UV-induced mutagenesis in eukaryotes.[105,106] Of all the Y-family members, it was the first to which a biochemical function could be assigned.[130] Yet, its properties are unique among the TLS enzymes, as it can hardly be called a polymerase: opposite a template dG and also an AP site, Rev1 inserts a single dC. Other template nucleotides are accepted with 10^2 to 10^3-fold reduced efficiency. Hence, it acts as a polymerase only on a poly-dG template.

Again, the X-ray structure of the enzyme in a ternary complex with DNA and dCTP provided an explanation for this unusual selectivity and revealed a most surprising mechanism of substrate recognition.[117] An N-terminal extension from the thumb domain reaches across the palm into the gap between the PAD and the fingers, thereby forcing the template dG to flip out of the helix and hydrogen-bond to a site within the protein (Figures 7.4 and 7.5). At the same time, the incoming dCTP bonds to an arginine residue of the enzyme rather than the template. Hence, the preference to pair dC with dG is not due to base complementarity, but is intrinsic to the enzyme. Intriguingly, the dC-transferase activity of Rev1 appears to be largely dispensable for its function *in vivo*.[131] Instead, an N-terminal BRCT domain as well as multiple protein interaction motifs in the C-terminal part of the protein are essential for function, possibly indicating a structural rather than a catalytic role for Rev1 in TLS.

7.3.3.5 Polymerase ζ (Polζ)

Among the TLS polymerases, Rev3, the catalytic subunit of Polζ, is the only member of the B-family. But unlike other members of this family, it lacks a proofreading 3′-5′ exonuclease activity.[132] Yet, with an error frequency of $\sim 10^{-4}$, it is rather more accurate than the other TLS enzymes. The holoenzyme comprises the subunits Rev3 and Rev7. Its capacity to use damaged bases as a template is poor, but it very efficiently extends mismatched or distorted primer termini resulting from a variety of lesions. Whereas deletion of *REV3* in yeast has little consequence beyond an effect on damage-induced mutagenesis

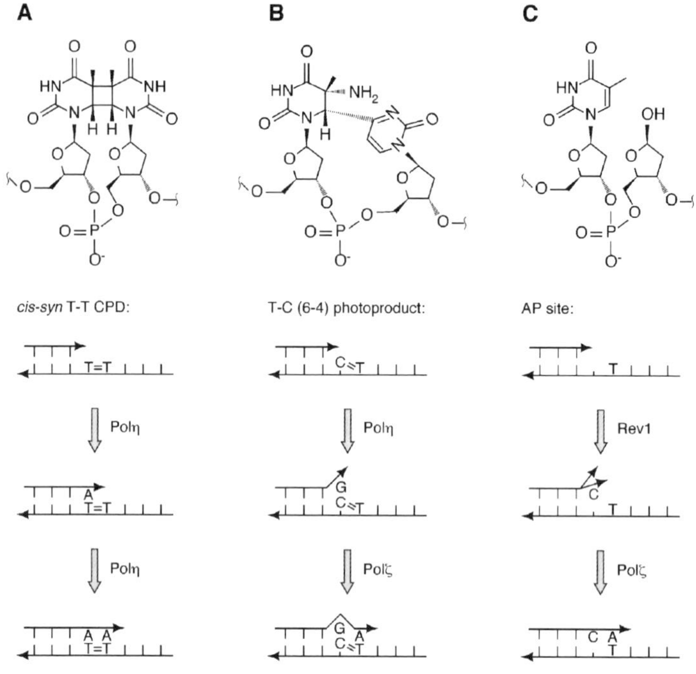

Figure 7.7 **Examples of lesion bypass by damage-tolerant DNA polymerases.** (**A**) A *cis–syn* T–T CPD is bypassed by Polη in an error-free manner. (**B**) The T–C (6-4) photoproduct is a strongly distorting lesion. Polη correctly inserts G opposite the 3′ dC, but cannot extend from the distorted primer terminus. Polζ correctly bypasses the 5′ dT, resulting in error-free TLS. (**C**) An abasic (AP) site is bypassed by Rev1, which inserts a single dC. The mismatched primer terminus can be extended by Polζ, resulting in (mostly) error-prone TLS.

(see below), it causes early embryonic lethality in mice, which makes it difficult to assess its function in higher eukaryotes.[133]

7.3.4 Cooperation between the Polymerases in TLS

As outlined above, Polη, Polι and Rev1 can insert nucleotides opposite lesions, but they poorly extend mismatched primer termini (with the notable exception of a T–T CPD by Polη; see Figure 7.7A). Instead, other TLS polymerases, primarily Polζ and possibly also Polκ, can extend from distorted primers. These characteristics result in cooperation between different TLS polymerases on many lesions (Figure 7.7B, C). Physical interactions of Rev1 with Polη, Polι, Polκ and the Rev7 subunit of Polζ (Figure 7.3A) emphasise the importance of joint activity and may also explain the non-catalytic role of Rev1.

Due to the properties of the damage-tolerant enzymes, TLS has varied and sometimes unpredictable consequences for damage-induced mutagenesis. For example, error-free TLS of CPDs requires Polη. In its absence, the same lesion is processed inaccurately by other TLS enzymes, thus explaining the hypermutability of Polη-deficient XP-V cells.[134] In contrast, loss of Rev1 or Polζ results in a nearly complete loss of damage-induced mutagenesis,[105,106] indicating that the two enzymes are jointly responsible for the bulk of error-prone TLS. A contribution of Polζ and Rev1 to the processing of endogenous lesions is suggested by a reduction in spontaneous mutation rates in their absence; Polζ even appears to promote mutagenesis when replication problems result not from DNA damage but a defect in the replicative polymerases.[135,136] However, in cooperation with Polη on T–C or C–C (6-4) photoproducts, Polζ promotes error-free TLS[111] (Figure 7.7B).

By means of error-prone activity, Polη and Rev1 were shown to contribute to the generation of somatic mutations in immunoglobulin genes.[137–139] It should also be noted that several functions unrelated to TLS have recently been proposed for damage-tolerant polymerases. For example, Polκ has been implicated in NER,[140] while Polη was shown to affect gene conversion rates in chicken B cells and to be particularly suited for the extension of D loops *in vitro*, possibly suggesting a role in HR.[141,142]

7.3.5 Error-free Post-replication Repair

As an alternative to error-prone TLS, it is often possible for the cell to avoid the use of a damaged template for polymerisation, requiring the switch to an alternative template. This can in principle be accomplished with the use of homologous recombination (HR) factors by strand invasion either at the replication fork or in a post-replicative manner. Alternatively, regression of the replication fork and annealing of the two newly synthesised strands in the shape of a 'chicken foot' would bring the stalled primer terminus in contact with an undamaged template (the complementary nascent strand), and resolution of the structure by branch migration after primer extension would result in a gene conversion event without involving HR or cross-over (Figure 7.2A). Although in eukaryotes the chicken foot structure has primarily been observed in checkpoint-deficient cells[143] (and may therefore represent an aberrant or non-productive intermediate), it is possible that all the mechanisms described above contribute to error-free damage avoidance. Genetic evidence in yeast suggests contributions from two independent systems, HR and a mechanistically poorly understood pathway termed error-free post-replication repair (PRR), which results in gene conversion, but does not make use of the general HR machinery.[144] A template switch to the complementary nascent strand (as shown in Figure 7.2B) may account for such non-recombinogenic gene conversion.

7.3.6 Regulation of DNA Damage Tolerance

DNA damage tolerance is a highly regulated process. On the one hand, TLS polymerases need to be recruited to the site of a lesion and take the place of the stalled replicative enzyme. On the other hand, their activity must be strictly limited to situations where it is beneficial, in order to prevent their unrestricted mutagenic action on undamaged DNA. Access of HR factors to the regions of ssDNA at stalled replication forks may also need to be controlled in order to allow less invasive methods of damage bypass.

There is evidence for a regulation of yeast TLS polymerases at multiple levels: *REV1* transcription peaks at G2/M, possibly indicating a post-replicative role of the enzyme.[145] A contribution of the checkpoint is suggested by an ATR (Mec1)-dependent phosphorylation of Rev1.[146] Finally, Polη appears to be controlled by proteasomal degradation under non-damage conditions.[147]

However, by far the most important regulatory influence on damage bypass is exerted by interactions with the replicative clamp protein, PCNA. With the exception of Rev1, all the TLS enzymes are stimulated in their activity by PCNA, even though the clamp has no effect on their processivity.[148,149] Polη, Polκ and Polι harbour conserved PCNA-interaction peptide (PIP) motifs within their C-terminal extensions (Figure 7.3A, B; see also Chapter 3).[150,151] Given the trimeric nature of PCNA, the 'toolbelt' model of TLS proposes that association with PCNA promotes the simultaneous recruitment of multiple damage-tolerant polymerases to replication forks even before they gain access to the site of polymerisation.[110] Interactions between the TLS enzymes might also contribute to their activation; for example, interaction with Polη is required to target Polι to subnuclear foci in response to replication fork stalling.[152] A fine-tuning of the protein–protein interactions at stalled replication forks, however, is achieved by multiple post-translational modifications of PCNA by members of the ubiquitin family (Figure 7.8), as outlined below.

7.3.6.1 PCNA Ubiquitylation by the RAD6 Pathway

The group of genes responsible for DNA damage tolerance is commonly known as the *RAD6* pathway.[98,153] Epistasis analysis in budding yeast has identified separate branches mediating either error-prone TLS or error-free damage avoidance (Figure 7.9A). Two members, *RAD6* and *RAD18*, are required for both sub-pathways, while the others are more specific in their actions (Figure 7.9A). The TLS-specific genes were found to encode all the damage-tolerant polymerases, whereas the other members of the *RAD6* pathway belong to two classes of enzymes involved in post-translational protein modification by the small protein ubiquitin. Ubiquitylation is best known as a signal for degradation of proteins by the 26S proteasome;[154] however, the modifier also contributes to a wide range of other cellular functions by means of changing the properties of its substrate proteins.[155] Whereas the proteasome usually requires polyubiquitin chains where one moiety is attached to the next *via* lysine (K) 48 of ubiquitin, alternative signals are conveyed either by monoubiquitin or by chains of other geometries.[156] As shown in Figure 7.8,

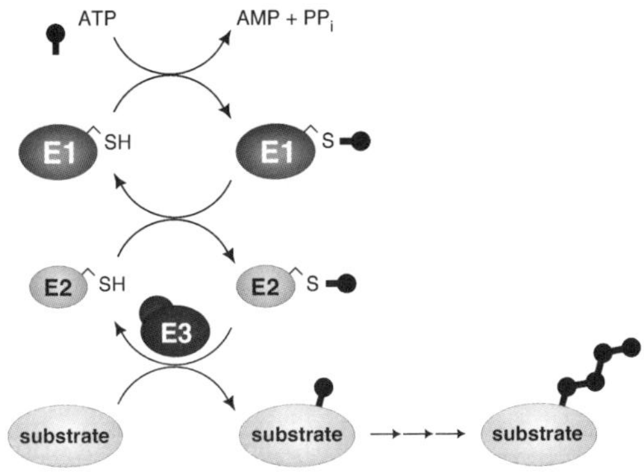

Figure 7.8 The ubiquitin system of post-translational protein modification. Ubiquitin (black symbol) is activated in an ATP-dependent reaction by ubiquitin-activating enzyme (E1). A catalytic cysteine in the enzyme undergoes a thioester linkage with the C-terminal glycine of ubiquitin. Ubiquitin is then transferred to a cysteine in an ubiquitin-conjugating enzyme (E2) and finally forms an isopeptide linkage with a lysine residue on the substrate protein. Selectivity and rate enhancement are provided by an ubiquitin protein ligase (E3). Repeated conjugation of ubiquitin results in poly-ubiquitin chains.

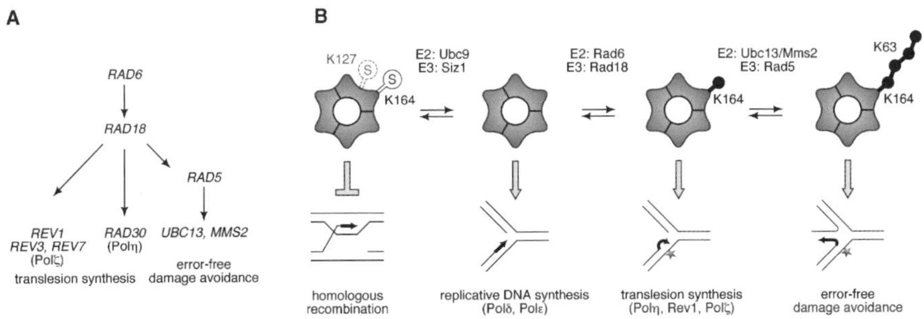

Figure 7.9 The *RAD6* pathway of postreplicative DNA repair. (**A**) Genetic relationships between the components of the *RAD6* pathway. (**B**) PCNA modifications during DNA replication and damage bypass. Trimeric PCNA (ring symbol) is modified by mono- or poly-ubiquitylation (black symbols) (via K63) at K164 in response to DNA damage. In *S. pombe* during S phase, PCNA is SUMOylated (S) at K164 and to a minor extent on K127 in a damage-independent manner. E2 and E3 enzymes for the reactions are shown, and the consequences of the modifications are illustrated, as described in detail in the text.

ubiquitin conjugation requires activation of its C-terminus by formation of a high-energy thioester bond with ubiquitin-activating (E1) enzyme, its transfer to an ubiquitin-conjugating enzyme (E2), and finally the attachment to a lysine of the target protein *via* an isopeptide linkage, usually aided by an ubiquitin protein ligase (E3) that confers selectivity to the reaction. The resulting ubiquitin conjugates are often recognised by downstream effector proteins *via* specific ubiquitin-binding domains (UBDs).[157]

The *RAD6* pathway comprises two E2 enzymes, the multifunctional Rad6 protein[158] and a heterodimeric complex of Ubc13 and Mms2.[159] The latter synthesises ubiquitin chains linked exclusively *via* K63 of ubiquitin. The E2 enzymes were found to interact genetically and physically with the RING finger E3s, Rad18 and Rad5, which target the E2s to chromatin.[160,161] Their single ubiquitylation substrate relevant for DNA damage tolerance was identified to be PCNA;[162] in a first step, Rad6 in conjunction with Rad18 attaches a single ubiquitin moiety to K164 of PCNA. By means of Rad5, Ubc13 and Mms2, this is extended to a K63-linked polyubiquitin chain (Figure 7.9B). Monoubiquitylation of PCNA at K164 by Rad6 and Rad18 is highly conserved among eukaryotes,[162–169] and the same appears to hold true for Rad5-dependent polyubiquitylation, although the relative abundance of mono- versus polyubiquitylated PCNA varies from species to species.[170–173] Interestingly, monoubiquitylation of PCNA in the chicken DT40 cell line was found to depend only partially on Rad18.[168]

7.3.6.2 Consequences of PCNA Ubiquitylation

Polyubiquitylation of PCNA is a prerequisite for error-free damage avoidance. However, this notion is based purely on genetic evidence and the mechanistic details of how this modification conveys damage tolerance remain completely unknown. Although *in vitro*, K63-linked polyubiquitylation can target a model substrate to the proteasome,[174] there is no evidence for degradation of polyubiquitylated PCNA *in vivo*. A genetic link between the *RAD6* pathway and the 26S proteasome has been noted,[175] but this may be caused by an altered free ubiquitin pool. In addition, the Rad5 protein exhibits a DNA-dependent ATPase and helicase activity that might be consistent with a function in replication fork regression;[176] however, inactivation of this activity is of little consequence for PCNA ubiquitylation or sensitivity to replication fork-stalling lesions.[177]

In contrast, the function of PCNA monoubiquitylation is much better understood. Epistasis analysis in budding yeast first indicated that this modification is required for TLS,[178] and it has now become clear that it is indeed a trigger of the switch between replicative and damage-tolerant polymerases at stalled replication forks.[110,179] All the eukaryotic Y-family polymerases harbour UBDs (UBZ and UBM) within their C-terminal extensions (Figure 7.3A, C, D) that promote their preferential interaction with the monoubiquitylated form of PCNA, their recruitment into damage-induced repair foci[180–182] and their catalytic activity on damaged templates.[183] Even linear fusions of ubiquitin to PCNA were found to support TLS *in vivo*.[184] Dependence of Polζ on PCNA ubiquitylation probably results from its cooperation with Rev1.[178]

7.3.6.3 Activation and Downregulation of PCNA-dependent Damage Tolerance

Given the dependence of damage bypass on PCNA, the targeting of TLS activity to sites of stalled replication forks is achieved by locally controlling ubiquitylation of the PCNA clamp. The modification is induced on chromatin-bound PCNA in a replication-dependent manner by conditions that cause polymerase blocks without inhibiting progression of the helicase, such as nucleotide depletion, alkylation, oxidative and UV damage, but not double-strand breaks or camptothecin.[75,164,165,185–187] In *S. pombe*, some damage-independent modification is observed in S phase, possibly indicating a higher incidence of spontaneous replication problems.[185] In G1, PCNA ubiquitylation can be triggered by DNA interstrand crosslinks in a repair pathway involving a combination of NER and TLS.[188] Overall, the activation pattern is consistent with a dependence on ssDNA accumulating at the site of a stalled replication intermediate (Figure 7.1B). A requirement for the ssDNA-binding RPA complex has been noted in yeast and mammals.[75,164] In fact, Rad18 interacts directly with the large and the medium subunits of RPA and can be recruited to ssDNA in this way.[75]

These observations suggest that PCNA ubiquitylation is triggered by a signal identical to that responsible for the replication checkpoint: RPA-covered ssDNA. However, the function of RPA in activating damage bypass is separable from its contribution to replication and checkpoint activation, and PCNA ubiquitylation was found to occur independently of checkpoint signalling in yeast and vertebrates.[75,165,185] The situation is not entirely clear in mammalian cells, where an influence of ATR as well as the cell cycle inhibitor p21 has been suggested.[164,189,190] In addition, physical interactions between TLS polymerases and the 9-1-1 checkpoint clamp have also been noted, possibly indicating a PCNA-independent recruitment mechanism.[191,192]

Control over lesion bypass appears to be exerted not only by activation, but also by downregulation of PCNA ubiquitylation. In higher eukaryotes, an ubiquitin-specific isopeptidase, Usp1, de-ubiquitylates PCNA under non-damage conditions.[168,193] Intriguingly, Usp1 itself appears to be controlled by damage-induced degradation, thus allowing the upregulation of PCNA ubiquitylation in response to DNA damage.[193] It is currently unclear, however, whether Usp1 serves primarily to counteract spurious modification of PCNA in undamaged cells, or rather to remove the modification after lesion bypass is completed.

7.3.6.4 PCNA SUMOylation in S. cerevisiae

In some organisms, PCNA undergoes an additional post-translational modification that affects the processing of replication intermediates. In a damage-independent fashion, budding yeast PCNA is subject to conjugation by the small ubiquitin-related modifier SUMO.[162] Attachment of SUMO follows a mechanism similar to that of ubiquitin, mediated by a distinct set of E1, E2 and

E3 enzymes, and consequences for the modified target proteins are at least as diverse as in the case of ubiquitin.[194] SUMO, like ubiquitin, is attached primarily to K164, but to some extent also to K127. Again the modifications serve to recruit a downstream effector protein, in this case the helicase Srs2 (see Chapter 5), which harbours a conserved SUMO interacting motif (SIM) in its C-terminus and preferentially interacts with the SUMOylated forms of PCNA.[195,196] Srs2 is a 3′–5′ helicase capable of disrupting Rad51 filaments *in vitro*.[197,198] Accordingly, the protein acts in a mostly antirecombinogenic manner *in vivo*, although its functions appear to be rather diverse.[199] By means of protein–protein interactions, S phase-specific SUMOylation of PCNA recruits the Srs2 helicase to replication forks,[195] where it prevents the formation of spontaneous cross-over events.[200] The importance of PCNA SUMOylation for genome stability becomes apparent in mutants of the *RAD6* pathway. Here, the damage sensitivity resulting from defects in PCNA ubiquitylation is suppressed by an increased activity of the HR pathway in *srs2* mutants, indicating that PCNA SUMOylation, *via* Srs2, normally controls the choice of pathways for the processing of stalled replication intermediates by minimising the use of HR. At the same time, the system influences the activity of Polζ in spontaneous mutagenesis.[178] In addition to Srs2 recruitment, SUMOylation might fulfil an additional function by blocking access of other proteins to PCNA. For example, SUMO appears to compete with the PCNA binding of the protein Eco1, involved in the establishment of cohesion in *S. cerevisiae*[201] (see Chapter 9). Hence, abolition of PCNA SUMOylation partially rescues the phenotype of an *eco1* mutant in an Srs2-independent manner. It is currently unclear, however, how relevant this effect is in a wild-type background. An equivalent system is not yet known in other organisms; whereas PCNA SUMOylation—with unknown consequences—has been detected in chicken DT40 cells and *Xenopus laevis* egg extracts,[163,167] a SUMO-interacting Srs2 homologue appears to be absent, and in other species, such as mammals or *S. pombe*, PCNA may not even undergo SUMOylation.[162,185]

7.4 Replication Restart by Homologous Recombination

The mechanisms described above allow the processing and restart of stalled replication forks in a locally restricted fashion by manipulating core components of the replisome and changing its composition. However, this 'non-invasive' strategy is useful only in situations where fork collapse (*i.e.* dissociation of the replisome from the DNA) can be prevented. Lesions that irreversibly block helicase movement or result in DSBs or other pathological structures at the fork will eventually result in replication fork collapse.

Under these conditions, replication restart will usually require HR, a pathway routinely used in prokaryotes, but also predominant in eukaryotic checkpoint mutants that fail to stabilise replication factors upon fork stalling.[61,71] The sub-pathway of HR employed for this purpose is called break-induced replication (BIR) and involves a non-reciprocal recombination event, initiated by the invasion of a single 3′ end emanating from the broken fork into an intact duplex (Figure 7.10), thus forming a D-loop.[202–204] This is followed by the establishment of unidirectional DNA synthesis, which may involve repeated cycles of invasion and dissolution. It is currently unclear whether BIR results in conservative or semi-conservative replication, but models consistent with both outcomes have been proposed (Figure 7.10). The pathway, studied in detail in budding and fission yeast, involves most of the recombination proteins and a poorly defined subset of the factors involved in standard replication as well as helicases of the RecQ family and nucleases (see also Chapter 5).[71,204]

Unlike TLS or error-free damage avoidance, replication fork restart by BIR bears the risk of causing GCR and genome instability on a larger scale. On one hand, the pathway is prone to cause loss of heterozygosity (LOH) due to the extended gene conversion tracts generated in the process. On the other hand, chromosomal aberrations may result from invasion of the break end into

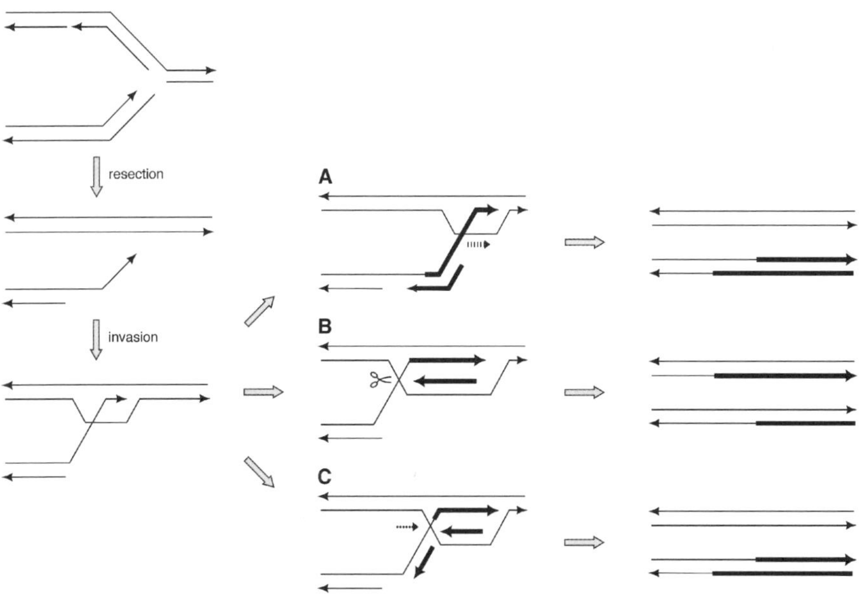

Figure 7.10 **Replication restart by break-induced replication.** A collapsed replication fork involving a DSB is processed by resection of the 5′ strand of the broken end. The resulting 3′ overhang subsequently invades the duplex to form a D-loop. Several alternative hypotheses have been proposed for the processing of this intermediate. (**A**) Propagation of the D-loop along the duplex (dashed arrow) and initiation of lagging strand synthesis on the displaced strand would result in conservative replication. (**B**) Establishment of a replication fork within the D-loop and processing of the resulting Holliday junction would lead to semi-conservative replication. (**C**) Establishment of a replication fork within the D-loop followed by branch migration (dotted arrow) would result in conservative replication. Newly synthesised DNA is indicated in bold. This scheme was adapted from references 203 and 204.

repetitive sequences on the same or even a different chromosome. Both phenomena have been associated with the development of malignancies.[205] This danger illustrates once more the importance of the replication checkpoint for preventing replication fork breakdown and allowing damage tolerance pathways.

7.5 Outlook

This chapter has attempted to give an overview of some of the endogenous and exogenous factors that endanger DNA replication, and the measures that the cell takes to cope with replication problems. Whereas the basic transactions in response to lesions and replication stress are described on the level of DNA, much remains to be learned about the damage response in the context of the living cell. For example, topological problems in the DNA as sources of replication stress or physiological aspects such as nucleotide levels or ongoing transcription are not considered here.

The influence of DNA ends, together with the effect of nucleosomes and higher-order chromatin on DNA replication are discussed elsewhere (Chapters 8 and 10, respectively). It is expected that all these factors influence the fidelity and efficiency with which eukaryotic cells pass on their genetic information.

Acknowledgements

Work in the author's lab is supported by Cancer Research UK.

References

1. T. A. Kunkel and K. Bebenek, DNA replication fidelity, *Annu. Rev. Biochem.*, 2000, **69**, 497–529.
2. M. Hogg, S. S. Wallace and S. Doublie, Bumps in the road: how replicative DNA polymerases see DNA damage, *Curr. Opin. Struct. Biol.*, 2005, **15**, 86–93.
3. E. T. Kool, Active site tightness and substrate fit in DNA replication, *Annu. Rev. Biochem.*, 2002, **71**, 191–219.
4. T. Lindahl, Instability and decay of the primary structure of DNA, *Nature*, 1993, **362**, 709–715.
5. G. P. Holmquist, Endogenous lesions, S-phase-independent spontaneous mutations, and evolutionary strategies for base excision repair, *Mutat. Res.*, 1998, **400**, 59–68.
6. T. A. Kunkel, The high cost of living. American Association for Cancer Research Special Conference: endogenous sources of mutations, Fort Myers, Florida, USA, 11–15 November 1998, *Trends Genet.*, 1999, **15**, 93–94.
7. P. C. Hanawalt, Genomic instability: environmental invasion and the enemies within, *Mutat. Res.*, 1998, **400**, 117–125.

8. J. Cadet, E. Sage and T. Douki, Ultraviolet radiation-mediated damage to cellular DNA, *Mutat. Res.*, 2005, **571**, 3–17.
9. E. C. Friedberg, G. C. Walker, W. Siede, R. D. Wood, R. A. Schultz and T. Ellenberger, *DNA Repair and Mutagenesis,* ASM Press, Washington, DC, 2005.
10. W. Siede, Y. W. Kow and P. W. Doetsch, *DNA Damage Recognition,* Taylor & Francis, New York, 2006.
11. T. Carell, L. T. Burgdorf, L. M. Kundu and M. Cichon, The mechanism of action of DNA photolyases, *Curr. Opin. Chem. Biol.*, 2001, **5**, 491–498.
12. B. Sedgwick, Repairing DNA-methylation damage, *Nat. Rev. Mol. Cell Biol.*, 2004, **5**, 148–157.
13. R. Dip, U. Camenisch and H. Naegeli, Mechanisms of DNA damage recognition and strand discrimination in human nucleotide excision repair, *DNA Repair (Amst.)*, 2004, **3**, 1409–1423.
14. S. C. Shuck, E. A. Short and J. J. Turchi, Eukaryotic nucleotide excision repair: from understanding mechanisms to influencing biology, *Cell Res.*, 2008, **18**, 64–72.
15. J. P. Laine and J. M. Egly, When transcription and repair meet: a complex system, *Trends Genet.*, 2006, **22**, 430–436.
16. A. Sarasin and A. Stary, New insights for understanding the transcription-coupled repair pathway, *DNA Repair (Amst.)*, 2007, **6**, 265–269.
17. A. Kuzminov, Single-strand interruptions in replicating chromosomes cause double-strand breaks, *Proc. Natl. Acad. Sci. U.S.A.*, 2001, **98**, 8241–8246.
18. K. W. Caldecott, Mammalian single-strand break repair: mechanisms and links with chromatin, *DNA Repair (Amst.)*, 2007, **6**, 443–453.
19. P. J. McKinnon and K. W. Caldecott, DNA strand break repair and human genetic disease, *Annu. Rev. Genomics Hum. Genet.*, 2007, **8**, 37–55.
20. Y. Aylon and M. Kupiec, New insights into the mechanism of homologous recombination in yeast, *Mutat. Res.*, 2004, **566**, 231–248.
21. S. Burma, B. P. Chen and D. J. Chen, Role of non-homologous end joining (NHEJ) in maintaining genomic integrity, *DNA Repair (Amst.)*, 2006, **5**, 1042–1048.
22. J. E. Haber, DNA recombination: the replication connection, *Trends Biochem. Sci.*, 1999, **24**, 271–275.
23. M. L. Hefferin and A. E. Tomkinson, Mechanism of DNA double-strand break repair by non-homologous end joining, *DNA Repair (Amst.)*, 2005, **4**, 639–648.
24. X. Li and W. D. Heyer, Homologous recombination in DNA repair and DNA damage tolerance, *Cell Res.*, 2008, **18**, 99–113.
25. M. R. Lieber, Y. Ma, U. Pannicke and K. Schwarz, Mechanism and regulation of human non-homologous DNA end-joining, *Nat. Rev. Mol. Cell Biol.*, 2003, **4**, 712–720.
26. L. H. Hartwell and T. A. Weinert, Checkpoints: controls that ensure the order of cell cycle events, *Science*, 1989, **246**, 629–634.

27. K. A. Nyberg, R. J. Michelson, C. W. Putnam and T. A. Weinert, Toward maintaining the genome: DNA damage and replication checkpoints, *Annu. Rev. Genet.*, 2002, **36**, 617–656.
28. J. C. Harrison and J. E. Haber, Surviving the breakup: the DNA damage checkpoint, *Annu. Rev. Genet.*, 2006, **40**, 209–235.
29. K. K. Khanna, M. F. Lavin, S. P. Jackson and T. D. Mulhern, ATM, a central controller of cellular responses to DNA damage, *Cell Death Differ.*, 2001, **8**, 1052–1065.
30. C. J. Bakkenist and M. B. Kastan, DNA damage activates ATM through intermolecular autophosphorylation and dimer dissociation, *Nature*, 2003, **421**, 499–506.
31. N. Assenmacher and K. P. Hopfner, MRE11/RAD50/NBS1: complex activities, *Chromosoma*, 2004, **113**, 157–166.
32. M. Lisby, J. H. Barlow, R. C. Burgess and R. Rothstein, Choreography of the DNA damage response: spatiotemporal relationships among checkpoint and repair proteins, *Cell*, 2004, **118**, 699–713.
33. J. H. Lee and T. T. Paull, Activation and regulation of ATM kinase activity in response to DNA double-strand breaks, *Oncogene*, 2007, **26**, 7741–7748.
34. D. Nakada, T. Shimomura, K. Matsumoto and K. Sugimoto, The ATM-related Tel1 protein of *Saccharomyces cerevisiae* controls a checkpoint response following phleomycin treatment, *Nucleic Acids Res.*, 2003, **31**, 1715–1724.
35. T. Uziel, Y. Lerenthal, L. Moyal, Y. Andegeko, L. Mittelman and Y. Shiloh, Requirement of the MRN complex for ATM activation by DNA damage, *EMBO J.*, 2003, **22**, 5612–5621.
36. M. van den Bosch, R. T. Bree and N. F. Lowndes, The MRN complex: coordinating and mediating the response to broken chromosomes, *EMBO Rep.*, 2003, **4**, 844–849.
37. J. Falck, J. H. Petrini, B. R. Williams, J. Lukas and J. Bartek, The DNA damage-dependent intra-S phase checkpoint is regulated by parallel pathways, *Nat. Genet.*, 2002, **30**, 290–294.
38. A. Jazayeri, J. Falck, C. Lukas, J. Bartek, G. C. Smith, J. Lukas and S. P. Jackson, ATM- and cell cycle-dependent regulation of ATR in response to DNA double-strand breaks, *Nat. Cell Biol.*, 2006, **8**, 37–45.
39. S. E. Lee, J. K. Moore, A. Holmes, K. Umezu, R. D. Kolodner and J. E. Haber, *Saccharomyces* Ku70, Mre11/Rad50 and RPA proteins regulate adaptation to G2/M arrest after DNA damage, *Cell*, 1998, **94**, 399–409.
40. L. Zou and S. J. Elledge, Sensing DNA damage through ATRIP recognition of RPA-ssDNA complexes, *Science*, 2003, **300**, 1542–1548.
41. S. M. Kim, A. Kumagai, J. Lee and W. G. Dunphy, Phosphorylation of Chk1 by ATM- and Rad3-related (ATR) in *Xenopus* egg extracts requires binding of ATRIP to ATR but not the stable DNA-binding or coiled-coil domains of ATRIP, *J. Biol. Chem.*, 2005, **280**, 38355–38364.

42. K. Unsal-Kacmaz and A. Sancar, Quaternary structure of ATR and effects of ATRIP and replication protein A on its DNA binding and kinase activities, *Mol. Cell. Biol.*, 2004, **24**, 1292–1300.
43. J. Majka and P. M. Burgers, The PCNA-RFC families of DNA clamps and clamp loaders, *Prog. Nucleic Acid Res. Mol. Biol.*, 2004, **78**, 227–260.
44. E. R. Parrilla-Castellar, S. J. Arlander and L. Karnitz, Dial 9-1-1 for DNA damage: the Rad9-Hus1-Rad1 (9-1-1) clamp complex, *DNA Repair (Amst.)*, 2004, **3**, 1009–1014.
45. L. Zou, D. Liu and S. J. Elledge, Replication protein A-mediated recruitment and activation of Rad17 complexes, *Proc. Natl. Acad. Sci. U.S.A.*, 2003, **100**, 13827–13832.
46. T. Kondo, T. Wakayama, T. Naiki, K. Matsumoto and K. Sugimoto, Recruitment of Mec1 and Ddc1 checkpoint proteins to double-strand breaks through distinct mechanisms, *Science*, 2001, **294**, 867–870.
47. J. A. Melo, J. Cohen and D. P. Toczyski, Two checkpoint complexes are independently recruited to sites of DNA damage *in vivo*, *Genes Dev.*, 2001, **15**, 2809–2821.
48. J. Majka, A. Niedziela-Majka and P. M. Burgers, The checkpoint clamp activates Mec1 kinase during initiation of the DNA damage checkpoint, *Mol. Cell*, 2006, **24**, 891–901.
49. S. Bao, R. S. Tibbetts, K. M. Brumbaugh, Y. Fang, D. A. Richardson, A. Ali, S. M. Chen, R. T. Abraham and X. F. Wang, ATR/ATM-mediated phosphorylation of human Rad17 is required for genotoxic stress responses, *Nature*, 2001, **411**, 969–974.
50. M. Giannattasio, F. Lazzaro, M. P. Longhese, P. Plevani and M. Muzi-Falconi, Physical and functional interactions between nucleotide excision repair and DNA damage checkpoint, *EMBO J.*, 2004, **23**, 429–438.
51. K. Unsal-Kacmaz, A. M. Makhov, J. D. Griffith and A. Sancar, Preferential binding of ATR protein to UV-damaged DNA, *Proc. Natl. Acad. Sci. U.S.A.*, 2002, **99**, 6673–6678.
52. G. Jiang and A. Sancar, Recruitment of DNA damage checkpoint proteins to damage in transcribed and nontranscribed sequences, *Mol. Cell. Biol.*, 2006, **26**, 39–49.
53. J. Bartek and J. Lukas, DNA damage checkpoints: from initiation to recovery or adaptation, *Curr. Opin. Cell Biol.*, 2007, **19**, 238–245.
54. K. H. Vousden and D. P. Lane, p53 in health and disease, *Nat. Rev. Mol. Cell Biol.*, 2007, **8**, 275–283.
55. D. P. Lane, Cancer. p53, guardian of the genome, *Nature*, 1992, **358**, 15–16.
56. G. M. Wahl and A. M. Carr, The evolution of diverse biological responses to DNA damage: insights from yeast and p53, *Nat. Cell Biol.*, 2001, **3**, E277–286.
57. R. J. Sims 3rd and D. Reinberg, Is there a code embedded in proteins that is based on post-translational modifications?, *Nat. Rev. Mol. Cell Biol.*, 2008, **9**, 815–820.

58. J. A. Tercero, M. P. Longhese and J. F. Diffley, A central role for DNA replication forks in checkpoint activation and response, *Mol. Cell*, 2003, **11**, 1323–1336.
59. S. Boiteux, L. Gellon and N. Guibourt, Repair of 8-oxoguanine in Saccharomyces cerevisiae: interplay of DNA repair and replication mechanisms, *Free Radic. Biol. Med.*, 2002, **32**, 1244–1253.
60. M. Lopes, M. Foiani and J. M. Sogo, Multiple mechanisms control chromosome integrity after replication fork uncoupling and restart at irreparable UV lesions, *Mol. Cell*, 2006, **21**, 15–27.
61. D. Branzei and M. Foiani, The DNA damage response during DNA replication, *Curr. Opin. Cell Biol.*, 2005, **17**, 568–575.
62. J. A. Cobb, K. Shimada and S. M. Gasser, Redundancy, insult-specific sensors and thresholds: unlocking the S-phase checkpoint response, *Curr. Opin. Genet. Dev.*, 2004, **14**, 292–300.
63. R. D. Paulsen and K. A. Cimprich, The ATR pathway: fine-tuning the fork, *DNA Repair (Amst.)*, 2007, **6**, 953–966.
64. D. Shechter, V. Costanzo and J. Gautier, Regulation of DNA replication by ATR: signaling in response to DNA intermediates, *DNA Repair (Amst.)*, 2004, **3**, 901–908.
65. P. Zegerman and J. F. Diffley, Lessons in how to hold a fork, *Nat. Struct. Biol.*, 2003, **10**, 778–779.
66. A. A. Alcasabas, A. J. Osborn, J. Bachant, F. Hu, P. J. Werler, K. Bousset, K. Furuya, J. F. Diffley, A. M. Carr and S. J. Elledge, Mrc1 transduces signals of DNA replication stress to activate Rad53, *Nat. Cell Biol.*, 2001, **3**, 958–965.
67. Y. Katou, Y. Kanoh, M. Bando, H. Noguchi, H. Tanaka, T. Ashikari, K. Sugimoto and K. Shirahige, S-phase checkpoint proteins Tof1 and Mrc1 form a stable replication-pausing complex, *Nature*, 2003, **424**, 1078–1083.
68. A. J. Osborn and S. J. Elledge, Mrc1 is a replication fork component whose phosphorylation in response to DNA replication stress activates Rad53, *Genes Dev.*, 2003, **17**, 1755–1767.
69. C. Possoz, S. R. Filipe, I. Grainge and D. J. Sherratt, Tracking of controlled *Escherichia coli* replication fork stalling and restart at repressor-bound DNA *in vivo*, *EMBO J.*, 2006, **25**, 2596–2604.
70. K. Labib and B. Hodgson, Replication fork barriers: pausing for a break or stalling for time?, *EMBO Rep.*, 2007, **8**, 346–353.
71. S. Lambert, B. Froget and A. M. Carr, Arrested replication fork processing: interplay between checkpoints and recombination, *DNA Repair (Amst.)*, 2007, **6**, 1042–1061.
72. A. Calzada, B. Hodgson, M. Kanemaki, A. Bueno and K. Labib, Molecular anatomy and regulation of a stable replisome at a paused eukaryotic DNA replication fork, *Genes Dev.*, 2005, **19**, 1905–1919.
73. S. Lambert, A. Watson, D. M. Sheedy, B. Martin and A. M. Carr, Gross chromosomal rearrangements and elevated recombination at an inducible site-specific replication fork barrier, *Cell*, 2005, **121**, 689–702.

74. L. F. Liu, S. D. Desai, T. K. Li, Y. Mao, M. Sun and S. P. Sim, Mechanism of action of camptothecin, *Ann. N. Y. Acad. Sci.*, 2000, **922**, 1–10.
75. A. A. Davies, D. Huttner, Y. Daigaku, S. Chen and H. D. Ulrich, Activation of ubiquitin-dependent DNA damage bypass is mediated by replication protein A, *Mol. Cell*, 2008, **29**, 625–636.
76. C. Redon, D. R. Pilch, E. P. Rogakou, A. H. Orr, N. F. Lowndes and W. M. Bonner, Yeast histone 2A serine 129 is essential for the efficient repair of checkpoint-blind DNA damage, *EMBO Rep.*, 2003, **4**, 678–684.
77. A. M. Rodriguez-Lopez, D. A. Jackson, F. Iborra and L. S. Cox, Asymmetry of DNA replication fork progression in Werner's syndrome, *Aging Cell*, 2002, **1**, 30–39.
78. J. M. Sidorova, N. Li, A. Folch and R. J. Monnat Jr., The RecQ helicase WRN is required for normal replication fork progression after DNA damage or replication fork arrest, *Cell Cycle*, 2008, **7**, 796–807.
79. F. Cortes-Ledesma and A. Aguilera, Double-strand breaks arising by replication through a nick are repaired by cohesin-dependent sister-chromatid exchange, *EMBO Rep.*, 2006, **7**, 919–926.
80. G. K. Alderton, H. Joenje, R. Varon, A. D. Borglum, P. A. Jeggo and M. O'Driscoll, Seckel syndrome exhibits cellular features demonstrating defects in the ATR-signalling pathway, *Hum. Mol. Genet.*, 2004, **13**, 3127–3138.
81. S. G. Durkin, M. F. Arlt, N. G. Howlett and T. W. Glover, Depletion of CHK1, but not CHK2, induces chromosomal instability and breaks at common fragile sites, *Oncogene*, 2006, **25**, 4381–4388.
82. P. Nghiem, P. K. Park, Y. Kim, C. Vaziri and S. L. Schreiber, ATR inhibition selectively sensitizes G1 checkpoint-deficient cells to lethal premature chromatin condensation, *Proc. Natl. Acad. Sci. U.S.A.*, 2001, **98**, 9092–9097.
83. M. F. Arlt, S. G. Durkin, R. L. Ragland and T. W. Glover, Common fragile sites as targets for chromosome rearrangements, *DNA Repair (Amst.)*, 2006, **5**, 1126–1135.
84. A. M. Casper, P. Nghiem, M. F. Arlt and T. W. Glover, ATR regulates fragile site stability, *Cell*, 2002, **111**, 779–789.
85. R. S. Cha and N. Kleckner, ATR homolog Mec1 promotes fork progression, thus averting breaks in replication slow zones, *Science*, 2002, **297**, 602–606.
86. M. Lahiri, T. L. Gustafson, E. R. Majors and C. H. Freudenreich, Expanded CAG repeats activate the DNA damage checkpoint pathway, *Mol. Cell*, 2004, **15**, 287–293.
87. J. A. Tercero and J. F. Diffley, Regulation of DNA replication fork progression through damaged DNA by the Mec1/Rad53 checkpoint, *Nature*, 2001, **412**, 553–557.
88. C. Santocanale and J. F. Diffley, A Mec1- and Rad53-dependent checkpoint controls late-firing origins of DNA replication, *Nature*, 1998, **395**, 615–618.

89. J. A. Cobb, L. Bjergbaek, K. Shimada, C. Frei and S. M. Gasser, DNA polymerase stabilization at stalled replication forks requires Mec1 and the RecQ helicase Sgs1, *EMBO J.*, 2003, **22**, 4325–4336.
90. V. Paciotti, M. Clerici, M. Scotti, G. Lucchini and M. P. Longhese, Characterization of mec1 kinase-deficient mutants and of new hypomorphic *mec1* alleles impairing subsets of the DNA damage response pathway, *Mol. Cell. Biol.*, 2001, **21**, 3913–3925.
91. R. C. Heller and K. J. Marians, Replisome assembly and the direct restart of stalled replication forks, *Nat. Rev. Mol. Cell Biol.*, 2006, **7**, 932–943.
92. K. J. Marians, Mechanisms of replication fork restart in *Escherichia coli*, *Philos. Trans. R. Soc. Lond. Ser. B*, 2004, **359**, 71–77.
93. B. Michel, G. Grompone, M. J. Flores and V. Bidnenko, Multiple pathways process stalled replication forks, *Proc. Natl. Acad. Sci. U.S.A.*, 2004, **101**, 12783–12788.
94. W. D. Rupp and P. Howard-Flanders, Discontinuities in the DNA synthesized in an excision-defective strain of *Escherichia coli* following ultraviolet irradiation, *J. Mol. Biol.*, 1968, **31**, 291–304.
95. L. di Caprio and B. S. Cox, DNA synthesis in UV-irradiated yeast, *Mutat. Res.*, 1981, **82**, 69–85.
96. A. R. Lehmann, Postreplication repair of DNA in ultraviolet-irradiated mammalian cells, *J Mol Biol*, 1972, **66**, 319–337.
97. L. Prakash, Characterization of postreplication repair in *Saccharomyces cerevisiae* and effects of *rad6, rad18, rev3* and *rad52* mutations, *Mol. Gen. Genet.*, 1981, **184**, 471–478.
98. C. Lawrence, The RAD6 DNA repair pathway in *Saccharomyces cerevisiae*: what does it do, and how does it do it?, *Bioessays*, 1994, **16**, 253–258.
99. R. C. Heller and K. J. Marians, Replication fork reactivation downstream of a blocked nascent leading strand, *Nature*, 2006, **439**, 557–562.
100. J. H. White and K. Dixon, Gap filling and not replication fork progression is the rate-limiting step in the replication of UV-damaged simian virus 40 DNA, *Mol. Cell. Biol.*, 1984, **4**, 1286–1292.
101. B. A. Bridges, Mechanisms of radiation mutagenesis in cellular and subcellular systems, *Annu. Rev. Nucl. Sci.*, 1969, **19**, 139–178.
102. M. Radman, SOS repair hypothesis: phenomenology of an inducible DNA repair which is accompanied by mutagenesis, *Basic Life Sci.*, 1975, **5A**, 355–367.
103. A. Bagg, C. J. Kenyon and G. C. Walker, Inducibility of a gene product required for UV and chemical mutagenesis in *Escherichia coli*, *Proc. Natl. Acad. Sci. U.S.A.*, 1981, **78**, 5749–5753.
104. S. J. Elledge and G. C. Walker, Proteins required for ultraviolet light and chemical mutagenesis. Identification of the products of the *umuC* locus of *Escherichia coli*, *J. Mol. Biol.*, 1983, **164**, 175–192.
105. C. W. Lawrence and R. B. Christensen, UV-mutagenesis in radiation-sensitive strains of yeast, *Genetics*, 1976, **82**, 207–232.

106. J. F. Lemontt, Mutants of yeast defective in mutation induced by ultraviolet light, *Genetics*, 1971, **68**, 21–33.
107. R. Woodgate, M. Rajagopalan, C. Lu and H. Echols, UmuC mutagenesis protein of *Escherichia coli*: purification and interaction with UmuD and UmuD, *Proc. Natl. Acad. Sci. U.S.A.*, 1989, **86**, 7301–7305.
108. A. Morrison, R. B. Christensen, J. Alley, A. K. Beck, E. G. Bernstine, J. F. Lemontt and C. W. Lawrence, *REV3*, a *Saccharomyces cerevisiae* gene whose function is required for induced mutagenesis, is predicted to encode a non-essential DNA polymerase, *J. Bacteriol.*, 1989, **171**, 5659–5667.
109. A. R. Lehmann, Replication of damaged DNA, *Cell Cycle*, 2003, **2**, 300–302.
110. V. Pages and R. P. Fuchs, How DNA lesions are turned into mutations within cells?, *Oncogene*, 2002, **21**, 8957–8966.
111. S. Prakash, R. E. Johnson and L. Prakash, Eukaryotic translesion synthesis DNA polymerases: specificity of structure and function, *Annu. Rev. Biochem.*, 2005, **74**, 317–353.
112. P. V. Shcherbakova and I. J. Fijalkowska, Translesion synthesis DNA polymerases and control of genome stability, *Front. Biosci.*, 2006, **11**, 2496–2517.
113. H. Ohmori, E. C. Friedberg, R. P. Fuchs, M. F. Goodman, F. Hanaoka, D. Hinkle, T. A. Kunkel, C. W. Lawrence, Z. Livneh, T. Nohmi, L. Prakash, S. Prakash, T. Todo, G. C. Walker, Z. Wang and R. Woodgate, The Y-family of DNA polymerases, *Mol. Cell*, 2001, **8**, 7–8.
114. A. Alt, K. Lammens, C. Chiocchini, A. Lammens, J. C. Pieck, D. Kuch, K. P. Hopfner and T. Carell, Bypass of DNA lesions generated during anticancer treatment with cisplatin by DNA polymerase eta, *Science*, 2007, **318**, 967–970.
115. H. Ling, F. Boudsocq, R. Woodgate and W. Yang, Crystal structure of a Y-family DNA polymerase in action: a mechanism for error-prone and lesion-bypass replication, *Cell*, 2001, **107**, 91–102.
116. S. Lone, S. A. Townson, S. N. Uljon, R. E. Johnson, A. Brahma, D. T. Nair, S. Prakash, L. Prakash and A. K. Aggarwal, Human DNA polymerase kappa encircles DNA: implications for mismatch extension and lesion bypass, *Mol. Cell*, 2007, **25**, 601–614.
117. D. T. Nair, R. E. Johnson, L. Prakash, S. Prakash and A. K. Aggarwal, Rev1 employs a novel mechanism of DNA synthesis using a protein template, *Science*, 2005, **309**, 2219–2222.
118. D. T. Nair, R. E. Johnson, S. Prakash, L. Prakash and A. K. Aggarwal, Replication by human DNA polymerase-iota occurs by Hoogsteen base-pairing, *Nature*, 2004, **430**, 377–380.
119. L. F. Silvian, E. A. Toth, P. Pham, M. F. Goodman and T. Ellenberger, Crystal structure of a DinB family error-prone DNA polymerase from *Sulfolobus solfataricus*, *Nat. Struct. Biol.*, 2001, **8**, 984–989.
120. J. Trincao, R. E. Johnson, C. R. Escalante, S. Prakash, L. Prakash and A. K. Aggarwal, Structure of the catalytic core of *S. cerevisiae* DNA

polymerase eta: implications for translesion DNA synthesis, *Mol. Cell*, 2001, **8**, 417–426.

121. S. N. Uljon, R. E. Johnson, T. A. Edwards, S. Prakash, L. Prakash and A. K. Aggarwal, Crystal structure of the catalytic core of human DNA polymerase kappa, *Structure*, 2004, **12**, 1395–1404.
122. B. L. Zhou, J. D. Pata and T. A. Steitz, Crystal structure of a DinB lesion bypass DNA polymerase catalytic fragment reveals a classic polymerase catalytic domain, *Mol. Cell*, 2001, **8**, 427–437.
123. W. Yang and R. Woodgate, What a difference a decade makes: insights into translesion DNA synthesis, *Proc. Natl. Acad. Sci. U.S.A.*, 2007, **104**, 15591–15598.
124. J. P. McDonald, A. S. Levine and R. Woodgate, The *Saccharomyces cerevisiaeRAD30* gene, a homologue of *Escherichia colidinB* and *umuC*, is DNA damage inducible and functions in a novel error-free postreplication repair mechanism, *Genetics*, 1997, **147**, 1557–1568.
125. A. A. Roush, M. Suarez, E. C. Friedberg, M. Radman and W. Siede, Deletion of the *Saccharomyces cerevisiae* gene *RAD30* encoding an *Escherichia coli DinB* homolog confers UV radiation sensitivity and altered mutability, *Mol. Gen. Genet.*, 1998, **257**, 686–692.
126. R. E. Johnson, S. Prakash and L. Prakash, Efficient bypass of a thymine-thymine dimer by yeast DNA polymerase, *Pol*η, *Science*, 1999, **283**, 1001–1004.
127. A. R. Lehmann, S. Kirk-Bell, C. F. Arlett, M. C. Paterson, P. H. Lohman, E. A. de Weerd-Kastelein and D. Bootsma, Xeroderma pigmentosum cells with normal levels of excision repair have a defect in DNA synthesis after UV-irradiation, *Proc. Natl. Acad. Sci. U.S.A.*, 1975, **72**, 219–223.
128. C. Masutani, M. Araki, A. Yamada, R. Kusumoto, T. Nogimori, T. Maekawa, S. Iwai and F. Hanaoka, Xeroderma pigmentosum variant (XP-V) correcting protein from HeLa cells has a thymine dimer bypass DNA polymerase activity, *EMBO J.*, 1999, **18**, 3491–3501.
129. C. Masutani, R. Kusumoto, A. Yamada, N. Dohmae, M. Yokoi, M. Yuasa, M. Araki, S. Iwai, K. Takio and F. Hanaoka, The XPV (xeroderma pigmentosum variant) gene encodes human DNA polymerase eta, *Nature*, 1999, **399**, 700–704.
130. J. R. Nelson, C. W. Lawrence and D. C. Hinkle, Deoxycytidyl transferase activity of yeast REV1 protein, *Nature*, 1996, **382**, 729–731.
131. J. R. Nelson, P. E. Gibbs, A. M. Nowicka, D. C. Hinkle and C. W. Lawrence, Evidence for a second function for *Saccharomyces cerevisiae* Rev1p, *Mol. Microbiol.*, 2000, **37**, 549–554.
132. J. R. Nelson, C. W. Lawrence and D. C. Hinkle, Thymine-thymine dimer bypass by yeast DNA polymerase, *Science*, 1996, **272**, 1646–1649.
133. C. W. Lawrence and V. M. Maher, Mutagenesis in eukaryotes dependent on DNA polymerase ζ and Rev1p, *Philos. Trans. R. Soc. Lond. Ser. B.*, 2001, **356**, 41–46.

134. A. Stary, P. Kannouche, A. R. Lehmann and A. Sarasin, Role of DNA polymerase eta in the UV mutation spectrum in human cells, *J. Biol. Chem.*, 2003, **278**, 18767–18775.
135. M. R. Northam, P. Garg, D. M. Baitin, P. M. Burgers and P. V. Shcherbakova, A novel function of DNA polymerase ζ regulated by PCNA, *EMBO J.*, 2006, **25**, 4316–4325.
136. Y. I. Pavlov, P. V. Shcherbakova and T. A. Kunkel, *In vivo* consequences of putative active site mutations in yeast DNA polymerases alpha, epsilon, delta, and zeta, *Genetics*, 2001, **159**, 47–64.
137. J. G. Jansen, P. Langerak, A. Tsaalbi-Shtylik, P. van den Berk, H. Jacobs and N. de Wind, Strand-biased defect in C/G transversions in hypermutating immunoglobulin genes in Rev1-deficient mice, *J. Exp. Med.*, 2006, **203**, 319–323.
138. A. L. Ross and J. E. Sale, The catalytic activity of REV1 is employed during immunoglobulin gene diversification in DT40, *Mol. Immunol.*, 2006, **43**, 1587–1594.
139. X. Zeng, D. B. Winter, C. Kasmer, K. H. Kraemer, A. R. Lehmann and P. J. Gearhart, DNA polymerase eta is an A-T mutator in somatic hypermutation of immunoglobulin variable genes, *Nat. Immunol.*, 2001, **2**, 537–541.
140. T. Ogi and A. R. Lehmann, The Y-family DNA polymerase kappa (pol kappa) functions in mammalian nucleotide-excision repair, *Nat. Cell Biol.*, 2006, **8**, 640–642.
141. T. Kawamoto, K. Araki, E. Sonoda, Y. M. Yamashita, K. Harada, K. Kikuchi, C. Masutani, F. Hanaoka, K. Nozaki, N. Hashimoto and S. Takeda, Dual roles for DNA polymerase eta in homologous DNA recombination and translesion DNA synthesis, *Mol. Cell*, 2005, **20**, 793–799.
142. M. J. McIlwraith, A. Vaisman, Y. Liu, E. Fanning, R. Woodgate and S. C. West, Human DNA polymerase eta promotes DNA synthesis from strand invasion intermediates of homologous recombination, *Mol. Cell*, 2005, **20**, 783–792.
143. J. M. Sogo, M. Lopes and M. Foiani, Fork reversal and ssDNA accumulation at stalled replication forks owing to checkpoint defects, *Science*, 2002, **297**, 599–602.
144. H. Zhang and C. W. Lawrence, The error-free component of the *RAD6/RAD18* DNA damage tolerance pathway of budding yeast employs sister-strand recombination, *Proc. Natl. Acad. Sci. U.S.A.*, 2005, **102**, 15954–15959.
145. L. S. Waters and G. C. Walker, The critical mutagenic translesion DNA polymerase Rev1 is highly expressed during G(2)/M phase rather than S phase, *Proc. Natl. Acad. Sci. U.S.A.*, 2006, **103**, 8971–8976.
146. S. Sabbioneda, I. Bortolomai, M. Giannattasio, P. Plevani and M. Muzi-Falconi, Yeast Rev1 is cell cycle regulated, phosphorylated in response to DNA damage and its binding to chromosomes is dependent upon *MEC1*, *DNA Repair (Amst.)*, 2007, **6**, 121–127.

147. A. Skoneczna, J. McIntyre, M. Skoneczny, Z. Policinska and E. Sledziewska-Gojska, Polymerase eta is a short-lived, proteasomally degraded protein that is temporarily stabilized following UV irradiation in *Saccharomyces cerevisiae*, *J. Mol. Biol.*, 2007, **366**, 1074–1086.
148. P. Garg, C. M. Stith, J. Majka and P. M. Burgers, Proliferating cell nuclear antigen promotes translesion synthesis by DNA polymerase zeta, *J. Biol. Chem.*, 2005, **280**, 23446–23450.
149. L. Haracska, C. M. Kondratick, I. Unk, S. Prakash and L. Prakash, Interaction with PCNA is essential for yeast DNA polymerase eta function, *Mol. Cell*, 2001, **8**, 407–415.
150. L. S. Cox, Who binds wins: competition for PCNA rings out cell-cycle changes, *Trends Cell Biol.*, 1997, **7**, 493–498.
151. E. Warbrick, The puzzle of PCNA's many partners, *Bioessays*, 2000, **22**, 997–1006.
152. P. Kannouche, A. R. Fernandez de Henestrosa, B. Coull, A. E. Vidal, C. Gray, D. Zicha, R. Woodgate and A. R. Lehmann, Localization of DNA polymerases eta and iota to the replication machinery is tightly co-ordinated in human cells, *EMBO J.*, 2003, **22**, 1223–1233.
153. H. D. Ulrich, The *RAD6* pathway: control of DNA damage bypass and mutagenesis by ubiquitin and SUMO, *ChemBioChem*, 2005, **6**, 1735–1743.
154. O. Kerscher, R. Felberbaum and M. Hochstrasser, Modification of proteins by ubiquitin and ubiquitin-like proteins, *Annu. Rev. Cell Dev. Biol.*, 2006, **22**, 159–180.
155. L. Sun and Z. J. Chen, The novel functions of ubiquitination in signaling, *Curr. Opin. Cell Biol.*, 2004, **16**, 119–126.
156. C. M. Pickart and D. Fushman, Polyubiquitin chains: polymeric protein signals, *Curr. Opin. Chem. Biol.*, 2004, **8**, 610–616.
157. L. Hicke, H. L. Schubert and C. P. Hill, Ubiquitin-binding domains, *Nat. Rev. Mol. Cell Biol.*, 2005, **6**, 610–621.
158. S. Jentsch, J. P. McGrath and A. Varshavsky, The yeast DNA repair gene *RAD6* encodes a ubiquitin-conjugating enzyme, *Nature*, 1987, **329**, 131–134.
159. R. M. Hofmann and C. M. Pickart, Noncanonical *MMS2*-encoded ubiquitin-conjugating enzyme functions in assembly of novel polyubiquitin chains for DNA repair, *Cell*, 1999, **96**, 645–653.
160. V. Bailly, J. Lamb, P. Sung, S. Prakash and L. Prakash, Specific complex formation between yeast RAD6 and RAD18 proteins: a potential mechanism for targeting RAD6 ubiquitin-conjugating activity to DNA damage sites, *Genes Dev.*, 1994, **8**, 811–820.
161. H. D. Ulrich and S. Jentsch, Two RING finger proteins mediate cooperation between ubiquitin-conjugating enzymes in DNA repair, *EMBO J.*, 2000, **19**, 3388–3397.
162. C. Hoege, B. Pfander, G. L. Moldovan, G. Pyrowolakis and S. Jentsch, *RAD6*-dependent DNA repair is linked to modification of PCNA by ubiquitin and SUMO, *Nature*, 2002, **419**, 135–141.

163. H. Arakawa, G. L. Moldovan, H. Saribasak, N. N. Saribasak, S. Jentsch and J. M. Buerstedde, A role for PCNA ubiquitination in immunoglobulin hypermutation, *PLoS Biol.*, 2006, **4**, e366.
164. X. Bi, L. R. Barkley, D. M. Slater, S. Tateishi, M. Yamaizumi, H. Ohmori and C. Vaziri, Rad18 regulates DNA polymerase κ and is required for recovery from S-phase checkpoint-mediated arrest, *Mol. Cell. Biol.*, 2006, **26**, 3527–3540.
165. D. J. Chang, P. J. Lupardus and K. A. Cimprich, Monoubiquitination of proliferating cell nuclear antigen induced by stalled replication requires uncoupling of DNA polymerase and mini-chromosome maintenance helicase activities, *J. Biol. Chem.*, 2006, **281**, 32081–32088.
166. P. L. Kannouche, J. Wing and A. R. Lehmann, Interaction of human DNA polymerase η with monoubiquitinated PCNA: a possible mechanism for the polymerase switch in response to DNA damage, *Mol. Cell*, 2004, **14**, 491–500.
167. C. A. Leach and W. M. Michael, Ubiquitin/SUMO modification of PCNA promotes replication fork progression in *Xenopus laevis* egg extracts, *J. Cell Biol.*, 2005, **171**, 947–954.
168. L. J. Simpson, A. L. Ross, D. Szuts, C. A. Alviani, V. H. Oestergaard, K. J. Patel and J. E. Sale, RAD18-independent ubiquitination of proliferating-cell nuclear antigen in the avian cell line DT40, *EMBO Rep*, 2006, **7**, 927–932.
169. K. Watanabe, S. Tateishi, M. Kawasuji, T. Tsurimoto, H. Inoue and M. Yamaizumi, Rad18 guides polη to replication stalling sites through physical interaction and PCNA monoubiquitination, *EMBO J.*, 2004, **23**, 3886–3896.
170. R. K. Chiu, J. Brun, C. Ramaekers, J. Theys, L. Weng, P. Lambin, D. A. Gray and B. G. Wouters, Lysine 63-polyubiquitination guards against translesion synthesis-induced mutations, *PLoS Genet.*, 2006, **2**, e116.
171. Z. Li, W. Xiao, J. J. McCormick and V. M. Maher, Identification of a protein essential for a major pathway used by human cells to avoid UV-induced DNA damage, *Proc. Natl. Acad. Sci. U.S.A.*, 2002, **99**, 4459–4464.
172. A. Motegi, R. Sood, H. Moinova, S. D. Markowitz, P. P. Liu and K. Myung, Human SHPRH suppresses genomic instability through proliferating cell nuclear antigen polyubiquitination, *J. Cell Biol.*, 2006, **175**, 703–708.
173. I. Unk, I. Hajdu, K. Fatyol, B. Szakal, A. Blastyak, V. Bermudez, J. Hurwitz, L. Prakash, S. Prakash and L. Haracska, Human SHPRH is a ubiquitin ligase for Mms2-Ubc13-dependent polyubiquitylation of proliferating cell nuclear antigen, *Proc. Natl. Acad. Sci. U.S.A.*, 2006, **103**, 18107–18112.
174. R. M. Hofmann and C. M. Pickart, *In vitro* assembly and recognition of Lys-63 polyubiquitin chains, *J. Biol. Chem.*, 2001, **276**, 27936–27943.

175. A. Podlaska, J. McIntyre, A. Skoneczna and E. Sledziewska-Gojska, The link between 20S proteasome activity and post-replication DNA repair in *Saccharomyces cerevisiae*, *Mol. Microbiol.*, 2003, **49**, 1321–1332.
176. A. Blastyak, L. Pinter, I. Unk, L. Prakash, S. Prakash and L. Haracska, Yeast Rad5 protein required for postreplication repair has a DNA helicase activity specific for replication fork regression, *Mol. Cell*, 2007, **28**, 167–175.
177. S. Chen, A. A. Davies, D. Sagan and H. D. Ulrich, The RING finger ATPase Rad5p of *Saccharomyces cerevisiae* contributes to DNA double-strand break repair in a ubiquitin-independent manner, *Nucleic Acids Res.*, 2005, **33**, 5878–5886.
178. P. Stelter and H. D. Ulrich, Control of spontaneous and damage-induced mutagenesis by SUMO and ubiquitin conjugation, *Nature*, 2003, **425**, 188–191.
179. E. C. Friedberg, A. R. Lehmann and R. P. Fuchs, Trading places: how do DNA polymerases switch during translesion DNA synthesis?, *Mol. Cell*, 2005, **18**, 499–505.
180. M. Bienko, C. M. Green, N. Crosetto, F. Rudolf, G. Zapart, B. Coull, P. Kannouche, G. Wider, M. Peter, A. R. Lehmann, K. Hofmann and I. Dikic, Ubiquitin-binding domains in Y-family polymerases regulate translesion synthesis, *Science*, 2005, **310**, 1821–1824.
181. C. Guo, T. S. Tang, M. Bienko, J. L. Parker, A. B. Bielen, E. Sonoda, S. Takeda, H. D. Ulrich, I. Dikic and E. C. Friedberg, Ubiquitin-binding motifs in REV1 protein are required for its role in the tolerance of DNA damage, *Mol. Cell. Biol.*, 2006, **26**, 8892–8900.
182. B. S. Plosky, A. E. Vidal, A. R. de Henestrosa, M. P. McLenigan, J. P. McDonald, S. Mead and R. Woodgate, Controlling the subcellular localization of DNA polymerases ι and η via interactions with ubiquitin, *EMBO J.*, 2006, **25**, 2847–2855.
183. P. Garg and P. M. Burgers, Ubiquitinated proliferating cell nuclear antigen activates translesion DNA polymerases η and REV1, *Proc. Natl. Acad. Sci. U.S.A.*, 2005, **102**, 18361–18366.
184. J. L. Parker, A. B. Bielen, I. Dikic and H. D. Ulrich, Contributions of ubiquitin- and PCNA-binding domains to the activity of Polymerase η in *Saccharomyces cerevisiae*, *Nucleic Acids Res.*, 2007, **35**, 881–889.
185. J. Frampton, A. Irmisch, C. M. Green, A. Neiss, M. Trickey, H. D. Ulrich, K. Furuya, F. Z. Watts, A. M. Carr and A. R. Lehmann, Post-replication repair and PCNA modification in *Schizosaccharomyces pombe*, *Mol. Biol. Cell*, 2006, **17**, 2976–2985.
186. P. Kannouche and A. Stary, Xeroderma pigmentosum variant and error-prone DNA polymerases, *Biochimie*, 2003, **85**, 1123–1132.
187. D. A. Solomon, M. C. Cardoso and E. S. Knudsen, Dynamic targeting of the replication machinery to sites of DNA damage, *J. Cell Biol.*, 2004, **166**, 455–463.
188. S. Sarkar, A. A. Davies, H. D. Ulrich and P. J. McHugh, DNA inter-strand crosslink repair during G1 involves nucleotide excision repair and DNA polymerase ζ, *EMBO J.*, 2006, **25**, 1285–1294.

189. S. Avkin, Z. Sevilya, L. Toube, N. Geacintov, S. G. Chaney, M. Oren and Z. Livneh, p53 and p21 regulate error-prone DNA repair to yield a lower mutation load, *Mol. Cell*, 2006, **22**, 407–413.
190. G. Soria, O. Podhajcer, C. Prives and V. Gottifredi, P21Cip1/WAF1 downregulation is required for efficient PCNA ubiquitination after UV irradiation, *Oncogene*, 2006, **25**, 2829–2838.
191. M. Kai and T. S. Wang, Checkpoint activation regulates mutagenic translesion synthesis, *Genes Dev.*, 2003, **17**, 64–76.
192. S. Sabbioneda, B. K. Minesinger, M. Giannattasio, P. Plevani, M. Muzi-Falconi and S. Jinks-Robertson, The 9-1-1 checkpoint clamp physically interacts with polζ and is partially required for spontaneous polζ-dependent mutagenesis in *Saccharomyces cerevisiae*, *J. Biol. Chem.*, 2005, **280**, 38657–38665.
193. T. T. Huang, S. M. Nijman, K. D. Mirchandani, P. J. Galardy, M. A. Cohn, W. Haas, S. P. Gygi, H. L. Ploegh, R. Bernards and A. D. D'Andrea, Regulation of monoubiquitinated PCNA by DUB autocleavage, *Nat. Cell Biol.*, 2006, **8**, 339–347.
194. E. S. Johnson, Protein modification by SUMO, *Annu. Rev. Biochem.*, 2004, **73**, 355–382.
195. E. Papouli, S. Chen, A. A. Davies, D. Huttner, L. Krejci, P. Sung and H. D. Ulrich, Crosstalk between SUMO and ubiquitin on PCNA is mediated by recruitment of the helicase Srs2p, *Mol. Cell*, 2005, **19**, 123–133.
196. B. Pfander, G. L. Moldovan, M. Sacher, C. Hoege and S. Jentsch, SUMO-modified PCNA recruits Srs2 to prevent recombination during S phase, *Nature*, 2005, **436**, 428–433.
197. L. Krejci, S. Van Komen, Y. Li, J. Villemain, M. S. Reddy, H. Klein, T. Ellenberger and P. Sung, DNA helicase Srs2 disrupts the Rad51 presynaptic filament, *Nature*, 2003, **423**, 305–309.
198. X. Veaute, J. Jeusset, C. Soustelle, S. C. Kowalczykowski, E. Le Cam and F. Fabre, The Srs2 helicase prevents recombination by disrupting Rad51 nucleoprotein filaments, *Nature*, 2003, **423**, 309–312.
199. M. A. Macris and P. Sung, Multifaceted role of the *Saccharomyces cerevisiae* Srs2 helicase in homologous recombination regulation, *Biochem. Soc. Trans.*, 2005, **33**, 1447–1450.
200. T. Robert, D. Dervins, F. Fabre and S. Gangloff, Mrc1 and Srs2 are major actors in the regulation of spontaneous crossover, *EMBO J.*, 2006, **25**, 2837–2846.
201. G. L. Moldovan, B. Pfander and S. Jentsch, PCNA controls establishment of sister chromatid cohesion during S phase, *Mol. Cell*, 2006, **23**, 723–732.
202. H. Flores-Rozas and R. D. Kolodner, Links between replication, recombination and genome instability in eukaryotes, *Trends Biochem. Sci.*, 2000, **25**, 196–200.
203. E. Kraus, W. Y. Leung and J. E. Haber, Break-induced replication: a review and an example in budding yeast, *Proc. Natl. Acad. Sci. U.S.A.*, 2001, **98**, 8255–8262.

204. B. Llorente, C. E. Smith and L. S. Symington, Break-induced replication: what is it and what is it for?, *Cell Cycle*, 2008, **7**, 859–864.
205. C. Lengauer, K. W. Kinzler and B. Vogelstein, Genetic instabilities in human cancers, *Nature*, 1998, **396**, 643–649.
206. M. G. Bomar, M. T. Pai, S. R. Tzeng, S. S. Li and P. Zhou, Structure of the ubiquitin-binding zinc finger domain of human DNA Y-polymerase eta, *EMBO Rep.*, 2007, **8**, 247–251.
207. S. Doublie, S. Tabor, A. M. Long, C. C. Richardson and T. Ellenberger, Crystal structure of a bacteriophage T7 DNA replication complex at 2.2 A resolution, *Nature*, 1998, **391**, 251–258.

CHAPTER 8

Telomeres and the End Replication Problem

TRACY M. BRYAN

Children's Medical Research Institute, 214 Hawkesbury Road, Westmead NSW 2145, and University of Sydney, NSW 2006, Australia

8.1 Introduction

Telomeres are the DNA-protein structures at the ends of linear chromosomes of eukaryotic organisms and are essential for maintaining genomic stability.[1,2] Telomeres play a pivotal role in determining cellular lifespan and are intimately involved in the development of cancer. In order to understand fundamental aspects of cellular behaviour such as genome stability, cellular lifespan, and immortalisation, it is essential to understand the relationship between the structure of telomeres and their cellular function. This is a dynamic field of research, and our knowledge of telomere structure and function is accumulating rapidly. This chapter briefly reviews the mechanistic aspects of telomere replication, with an emphasis on recent data that are extending telomere biology into the structural and biochemical realm.

8.2 Telomere Involvement in Senescence and Immortalisation

Telomeres in most organisms are composed of short G-rich DNA repeats (TTAGGG in humans) complexed with proteins. The ends of linear DNA molecules cannot be replicated fully by the semi-conservative DNA replication

Molecular Themes in DNA Replication
Edited by Lynne S. Cox

Published by the Royal Society of Chemistry, www.rsc.org

machinery. At the very end of the lagging strand, the RNA primer is not replaced with DNA as there is no incoming 5′–3′ acting DNA polymerase δ to displace it. This RNA primer is then degraded, which regenerates the pre-existing 3′ single-stranded overhang. The leading strand, however, remains blunt-ended, and regeneration of the telomeric overhang by an exonuclease results in DNA loss. This is termed the 'end replication problem' (Figure 8.1), and indeed in many normal human somatic cells, telomeres do shorten with every cell division.[3] Normal somatic cells replicate only a limited

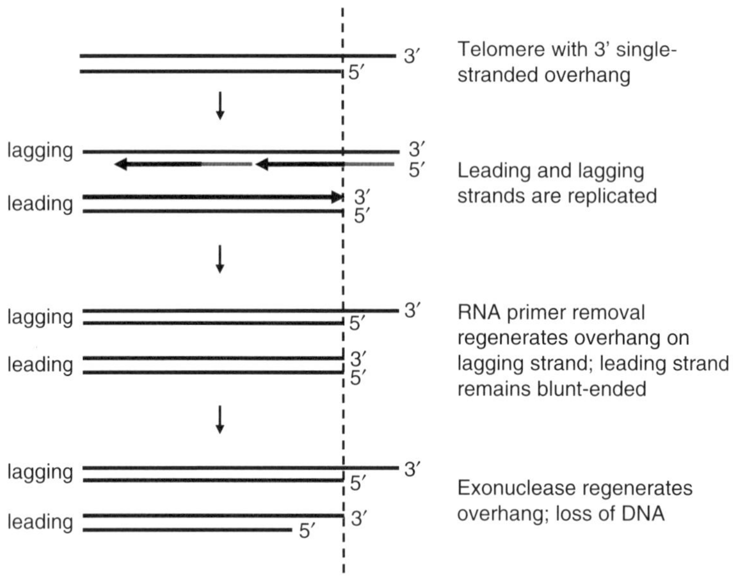

Figure 8.1 **The end replication problem.** DNA at the end of telomeres (right side) exists as a single-stranded 3′ overhang. Parental duplex (black) is unwound at the replication fork and replication proceeds continuously on the leading strand and discontinuously on the lagging strand. As one lagging strand Okazaki fragment is completed, the primase complex reassociates at the next priming site 5′ of the previous Okazaki fragment to synthesise an RNA primer (red) which is extended by DNA polymerase to generate a DNA fragment (blue). Following primer displacement by the incoming polymerase, and cleavage by Fen1 and RNaseH (see Chapter 5), the fragments are ligated. On the lagging strand at the telomere, there is no possibility of the primase complex binding DNA 5′ of the pre-existing Okazaki fragment as there is no DNA to bind to. Thus no priming occurs and no loading of DNA polymerase, and so there is no incoming polymerase to displace the RNA primer of the last Okazaki fragment. This RNA is degraded by RNases leaving a region of the lagging strand template unreplicated, which regenerates the 3′ overhang on this strand. However, the replicated leading strand remains blunt-ended. An exonuclease processes this strand to regenerate the 3′ overhang, resulting in loss of telomeric DNA.

number of times before they permanently withdraw from the cell cycle and exhibit features of cellular ageing (senescence). The limited replicative capacity of normal human cells is a major tumour suppressor mechanism; conversely, escape from the senescence barrier (immortalisation) is a critically important aspect of tumorigenesis.[4] There is solid evidence that telomere shortening is at least one component of the 'molecular clock' that leads to senescence.[5–7]

Unicellular eukaryotic organisms such as ciliated protozoa, human germline cells and immortalised human cell lines all possess mechanisms to overcome telomere shortening. This is most often achieved by the ribonucleoprotein enzyme telomerase, which adds telomeric repeats to chromosome ends by reverse transcription of an integral RNA template.[8] Telomerase is a unique and fascinating enzyme. It contains a catalytic protein subunit (known as TERT for telomerase reverse transcriptase), which bears similarities to other reverse transcriptases, but which also has large regions representing apparently novel protein folds.[9,10] The RNA subunit (TER) is also distinctive in that it contains not only the template for reverse transcription, but also many structural elements that contribute to telomerase activity in ways that are only just beginning to be understood.

Two main attributes of telomerase distinguish its mechanism of action from that of other reverse transcriptases: (i) the RNA template forms an integral part of the enzyme complex; and (ii) the enzyme has the capability to iteratively add many telomeric DNA repeats to the same molecule without dissociation, a property known as 'repeat addition processivity'.[11] This property predicts that telomerase undergoes large changes in conformation during repeat addition—a fascinating topic for future research.

An unexpected later discovery was that human and yeast cells also possess telomerase-independent alternative telomere lengthening mechanisms (in humans known as ALT for alternative lengthening of telomeres), which involve recombination between telomeres.[12,13] The molecular details of human ALT are just beginning to be elucidated, guided by the more advanced studies of yeast telomeric recombination.

Part of the motivation for the intense research activity focussing on telomerase and ALT comes from their implications for potential future cancer treatments. High levels of telomerase activity have been detected in ~85% of human tumours, whereas normal human somatic cells generally exhibit low or undetectable levels of telomerase.[14,15] This makes telomerase a highly specific target for cancer therapeutics. While the ALT mechanism is used by only 5–10% of human carcinomas, which make up the majority of human cancers, there are some types of sarcomas where the prevalence of ALT is as high as 77%.[16,17] Combined with the possibility that tumours will be under selection pressure to activate ALT when telomerase activity is suppressed, this makes it imperative that we design both ALT and telomerase inhibitors. Whilst design of ALT inhibitors awaits a more detailed understanding of the factors controlling ALT, some of the more promising telomerase inhibitors are discussed at the end of this chapter.

8.3 Structure and Stability of Telomeres

8.3.1 T-loops and the Shelterin Complex

A distinctive feature of telomeric DNA is that it terminates in a 3′ G-rich single-stranded overhang.[18] This single-stranded DNA would be predicted to be subject to nucleolytic attack and recombinational activities. Cells have therefore evolved various strategies of protecting this overhang, including folding it into higher order structures such as T-loops and G-quadruplexes.

During the formation of a T-loop, it is proposed that the 3′ G-rich overhang invades the double-stranded portion of the same telomere, base-pairing with the C-rich strand and resulting in a large duplex DNA loop at the chromosome terminus (Figure 8.2A). T-loops were first identified in the telomeric DNA of human and mouse cells by electron microscopy,[19] and have subsequently been observed in ciliates, trypanosomes and plants.[20] The proposed function of T-loops is to protect telomeres and regulate telomerase access. It is not known, however, what proportion of human telomeres form T-loops or even whether they are the predominant telomere capping structure.

Whether or not the telomere is folded into a T-loop, it is known to be a proteinaceous structure. Over the last decade or so, it has emerged that there are six specific telomere-binding proteins in humans (Figure 8.2B). Many other

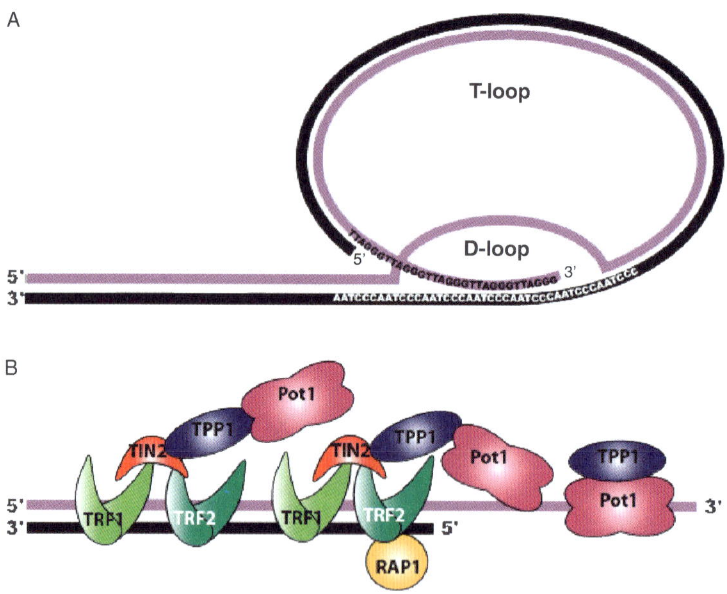

Figure 8.2 **Human telomere protection.** (**A**) Proposed DNA topology of the human T-loop. G-rich DNA strands are in purple, C-rich strands are in black. (**B**) Telomeric proteins coating a telomeric double strand-single strand boundary; see text for details.

proteins show telomere association, but unlike these six proteins, their functions are not predominantly confined to telomeres.

The first of the telomere-specific proteins to be identified were TRF1 and TRF2, which are DNA binding proteins with a high specificity for double-stranded telomeric DNA.[21–24] Rap1 is associated with the telomere *via* an interaction with TRF2; its function has remained enigmatic until recently.[25,26] Pot1 directly binds telomeric DNA and also has strong specificity for telomeric DNA, but in this case for single-stranded DNA (ssDNA).[27] TPP1 is a more recently discovered binding partner of Pot1,[28–32] and Tin2 functions to hold the complex together, tethering the Pot1/TPP1 complex to TRF1 and TRF2.[33,34] All six of these proteins have been identified in a single complex, which has become known as the 'telosome' or 'shelterin'.[33,34] It is also possible that subcomplexes of shelterin perform unique functions.

The shelterin proteins provide much of the 'capping' function of telomeres. They perform this function in a plethora of ways, and the details of the functions of the components differ between organisms. Readers are referred to several comprehensive recent reviews on the emerging complexities of shelterin component functions.[35–38] This discussion briefly highlights some of the ways in which mammalian shelterin protects telomeres.

What do telomeres need protection from? It has emerged that the cellular consequences of either shortened telomeres or deprotected telomeres are the same: a DNA damage response involving the recruitment of DNA damage proteins to telomeres and an ATM- or ATR-dependent upregulation of the p53 pathway, which then triggers either senescence or apoptosis.[39–46] Some of the ways in which the shelterin components prevent these adverse cellular consequences are detailed below.

8.3.1.1 Formation of T-loops

TRF2 has the ability to remodel a linear artificial telomere into T-loops, possibly through its ability to bind replication forks or to generate DNA supercoiling.[19,47–49] TRF1 also has DNA remodelling capabilities, being able to induce DNA bending[50] and promote pairing of telomeric tracts, an activity that is stimulated by Tin2.[51,52] It has thus been proposed that TRF2 and TRF1 may cooperate in the folding of a telomere into a T-loop *in vivo*, with the assistance of Pot1 binding to the single-stranded portion of the T-loop.[53] Since homologous recombination proteins localise to telomeres in a cell cycle phase-dependent manner, it is possible that TRF1 and TRF2 require the help of these proteins for the strand invasion that is necessary to form a T-loop.[54,55]

8.3.1.2 Protection of the 3′ Telomeric Overhang

Inhibition of either TRF2 or Pot1 can lead to a loss in the amount of telomeric ssDNA.[39,53,56] The overhang protection function of these proteins could be due to formation of a T-loop or, in the case of Pot1, direct coating of the ssDNA.

8.3.1.3 Protection from End-to-end Fusions

Inhibition of TRF2, possibly the most protective member of the shelterin complex, leads to end-to-end chromosomal fusions *via* the non-homologous end joining (NHEJ) pathway after removal of the 3′ telomeric overhang.[39,57,58] If TRF2 is completely deleted, the fusions dramatically involve every telomere in the cell.[45] Pot1 deletion results in a more modest level of chromosomal fusions.[59] Recently, Rap1 was also shown to be involved in protection against NHEJ *in vitro*, in cooperation with TRF2.[26] Interestingly, this protective function required neither the presence of a 3′ overhang nor a T-loop, demonstrating that, at least for the Rap1/TRF2 complex, protection is mediated by the proteins themselves rather than a higher-order structure of the DNA.

8.3.1.4 Protection from Homologous Recombination

The junction of a T-loop resembles a recombination intermediate (compare Figure 8.2A with Figure 8.4A). Resolution of this junction to release the T-loop would result in a shortened telomere and an extrachromosomal circle made up of telomeric DNA. Such circles have been detected at very low levels in human cell lines.[60] Telomere circle formation depends on homologous recombination proteins, supporting the proposed T-loop mediated recombination mechanism, termed 'T-loop HR'.[60] The abundance of extrachromosomal circles is enhanced upon expression of a mutant allele of TRF2[60] or deletion of Pot1,[61] suggesting that these proteins repress T-loop HR. Another form of homologous replication at telomeres is postreplicative exchange of telomeric DNA, known as telomere sister chromatid exchange (T-SCE). While present at low or undetectable levels in normal human cells,[62,63] the number of T-SCE events is greatly enhanced in the absence of Pot1[61,64] or both TRF2 and Ku.[65]

8.3.1.5 Protection from a DNA Damage Response

The discovery that DNA damage response proteins localise to normal telomeres[66] presented a paradox: how are normal telomeres protected from being processed in the same way as damaged DNA?

It is now becoming clear that shelterin components act to forestall the downstream effects of the DNA response proteins. TRF2 and Pot1 independently repress the ATM- and ATR-mediated DNA damage pathways, respectively; in the case of TRF2 this is by direct binding to ATM.[67–69] The ability of TRF2 and POT1 to inhibit a DNA damage response at the telomere is mediated by their interactions with TIN2[70] and TPP1,[71] respectively. Telomere dysfunction induced by TRF2 inhibition results in the accumulation of DNA damage response proteins at the telomeres, including 53BP1, γ-H2AX, Rad17, ATM and Mre11; the sites of accumulation were named telomere dysfunction-induced foci (TIFs).[43,44] Downregulation of Pot1 also leads to the induction of TIFs.[56,59,61,64]

8.3.1.6 Telomere Length Control

Telomere lengths are kept within a narrow range by *cis*-acting regulators of telomerase. TRF1 and TRF2 are both negative regulators of telomere lengthening.[72,73] Pot1 regulates telomere length in a telomerase-dependent manner, but the manner in which it does so seems to be complex. Pot1 has been reported as both a negative and positive regulator of telomerase,[74,75] and which role it plays seems to depend on a number of factors such as its position of binding on the telomere,[76] the structure of the telomeric DNA,[77] its binding partner TPP1[31,32] and the amount of Pot1 at the telomere.[78]

8.3.2 G-quadruplexes

An alternative structure for the single-stranded overhang of mammalian telomeres is a G-quadruplex. Guanine-rich DNA has a well-characterised tendency *in vitro* for the four guanines to hydrogen bond with each other in a planar cyclical arrangement called a G-quartet; the cavity at the core is the binding site for monovalent cations (*e.g.* Na^+, K^+) that stabilise this structure.[79,80] Multiple layers of G-quartets stack to form G-quadruplexes, in which the DNA strands assemble together in either an intramolecular (when a single strand folds upon itself) or an intermolecular (when two or more strands associate) configuration. G-quadruplexes exhibit extensive structural polymorphism; one way of classifying them is whether the DNA strand orientation is anti-parallel (Figure 8.3A), parallel (Figure 8.3B), or hybrid (Figure 8.3C).

Examination of the structural, biochemical and thermodynamic attributes of this intriguing DNA structure has raised the question of whether G-quadruplexes are physiologically relevant or merely an *in vitro* artefact. The existence of characteristically guanine-rich motifs in key regions of the eukaryotic genome, including the immunoglobulin heavy chain switch region, some promoter regions, ribosomal DNA, minisatellites and telomeres, suggests that G-quadruplexes may have important functions *in vivo*.[81,82] The large number of proteins that can bind, promote, resolve or cleave G-quadruplexes also lends support to their *in vivo* relevance.[83] Ligands that bind preferentially to G-quadruplex DNA have been shown to induce telomere-related effects in human cells.[83] Direct evidence for the *in vivo* existence of G-quadruplexes in the nuclei of ciliated protozoa was provided by their detection with an anti-G-quadruplex antibody.[84] Telomere binding proteins were involved in the formation of these G-quadruplexes and their resolution during DNA replication.[85]

The formation of an intramolecular G-quadruplex within a telomeric primer prevents its extension by human telomerase *in vitro*;[77] hence, stabilisation of this quadruplex at human chromosomal ends presents an attractive anti-cancer strategy. As a consequence, the intramolecular DNA quadruplex based on the human telomeric repeat sequence $(TTAGGG)_n$ has been the subject of extensive biophysical and biological scrutiny. There has been much controversy as to the exact conformation of this G-quadruplex under physiological conditions,

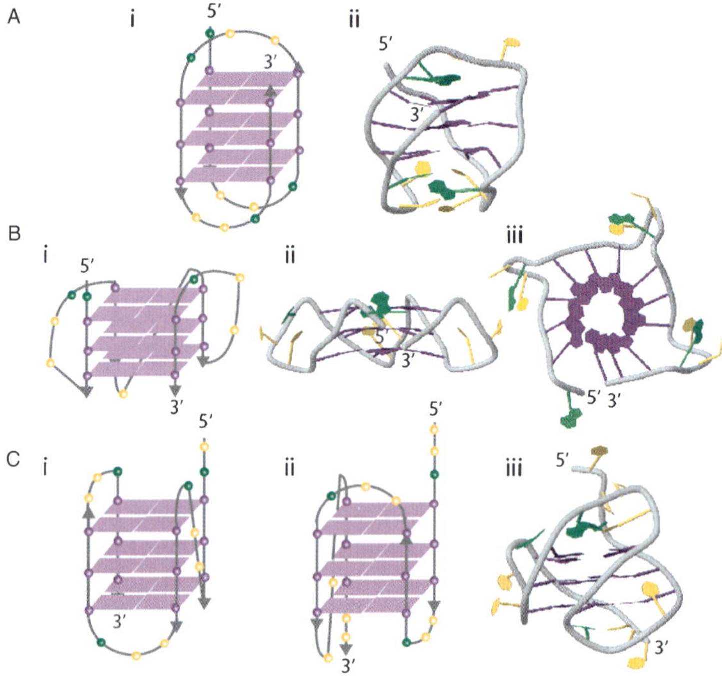

Figure 8.3 **Human telomeric intramolecular G-quadruplexes.** (**A**) Topology (i) and NMR structure (ii) of oligonucleotide AGGG(TTAGGG)$_3$ in sodium solution, demonstrating an antiparallel conformation.[86] (**B**) Topology (i) and crystal structure (ii, iii) of oligonucleotide AGGG(TTAGGG)$_3$ in potassium, showing a parallel 'propeller' structure.[87] The crystal structure is shown as a side view (ii) and a top view (iii). (**C**) Hybrid conformations in potassium solution. Hybrid 1 (i) and hybrid 2 (ii) topologies illustrate differences in loop structures.[94,95] The NMR structure of hybrid 2 is shown in (iii).[95]

and it has become increasingly apparent that the human quadruplex may be more structurally heterogeneous than that of some other species.

Both crystal and solution structures of the oligonucleotide AGGG(TTAGGG)$_3$ have been solved and reveal dramatically different topologies. The NMR solution structure of this sequence in the presence of sodium is an anti-parallel basket-type quadruplex[86] (Figure 8.3A), while the crystal structure in the presence of potassium represents a parallel propeller-type intramolecular G-quadruplex[87] (Figure 8.3B). Recently, two variations of a third conformation of human intramolecular telomeric G-quadruplex have been detected in potassium solution; these are known as 'hybrid' forms since they have some parallel and antiparallel strands.[88–91] Indeed, single molecule fluorescence resonance energy transfer (FRET) experiments had shown that two conformations with different half-lives can co-exist and interconvert in solution.[92,93] The solution structures of the two forms ('hybrid 1', Figure 8.3 Ci; 'hybrid 2', Figure 8.3Cii, iii) have recently been solved and reveal an identical

G-quadruplex core structure with differences in the connecting loops.[94,95] The equilibrium between hybrid 1 and hybrid 2 was greatly influenced by the 3′ sequence of the oligonucleotide, with GGG ends favouring formation of hybrid 1.[94,95] Since only ~5% of telomeres in human cells end in GGG,[96] this might imply that hybrid 2 predominates *in vivo*. The *in vivo* equilibrium may also be affected by temperature, ionic conditions and the presence of particular proteins.

Since potassium levels in mammalian cells are ~150 mM and generally higher than sodium levels,[97] one of the potassium structures may be the more physiologically relevant conformation. But which one? It has been argued that the parallel conformation seen in the crystal structure is not biologically relevant and may simply represent an artefact of the crowding conditions introduced by the crystalline state.[98] However, the presence of 40% polyethylene glycol induced a shift from hybrid to parallel G-quadruplexes and the authors of this study postulated that molecular crowding conditions may in fact more accurately represent the *in vivo* situation.[99]

It is possible that intramolecular G-quadruplexes predominate over intermolecular forms *in vivo*; the overhangs of human telomeres are ~150–200 nucleotides long[100,101] and it has recently been shown that such long telomeric tracts preferentially form into strings of interconnected intramolecular G-quadruplexes *in vitro*.[102] Nevertheless, the potential exists for the formation of intermolecular G-quadruplexes in some *in vivo* settings. The solution structures of some intermolecular variants of the human telomeric G-quadruplex have been solved. The single repeat oligonucleotide TTAGGG forms a parallel tetramer in the presence of potassium,[103] whereas the two repeat TAGGGTTAGGGT sequences exists as two isoforms—parallel and anti-parallel dimers that co-exist and interconvert in potassium solution.[104] A three-repeat oligonucleotide forms a structure similar to the hybrid structure reported in intramolecular G-quadruplexes.[105]

The function(s) of potential G-quadruplex structures in human cells are unknown. The different conformations may carry out distinct roles. Intermolecular G-quadruplexes could facilitate telomere–telomere associations; such interactions have been observed in the telomere-rich environment of the macronuclei of ciliated protozoa and there is evidence they are mediated by G-quadruplexes.[84,106] It has also been postulated that intermolecular parallel G-quadruplexes may be involved in the alignment of sister chromatids during meiosis.[107] The hybrid intramolecular forms of the human telomeric sequence discussed above have the potential to stack end-to-end in long arrays, which may accomplish compaction of telomeric DNA.[94,95,102]

G-quadruplexes may form an alternative to T-loops in conferring capping and protective functions to telomeres. They may sequester the telomere from inappropriate elongation by telomerase, or protect it from nuclease degradation or end-to-end fusion. Intramolecular antiparallel G-quadruplexes have indeed been shown to be resistant to telomerase elongation *in vivo*,[77,108,109] although parallel intermolecular structures are extended by telomerase.[109] There is some evidence that G-quadruplexes play a protective role; incubation of duplex DNA with human cell extract elicited a DNA damage response which was alleviated by addition of a 3′ tail capable of forming a G-quadruplex.[110] However, the possibility that the protective function was mediated by Pot1 in the extract was not ruled out.

Finally, G-quadruplexes may play an important role in inhibiting the activation of ALT (see discussion below) by sequestering the overhang in a form resistant to recombination. Several proteins that are potentially involved in the ALT mechanism are also known to unwind G-quadruplexes, including Replication Protein A (RPA)[111] and the RecQ helicases BLM[112] and WRN.[113] Unfolding of telomeric G-quadruplexes may allow access to the telomere by recombination proteins and enable initiation of the ALT mechanism of telomere elongation.

8.4 ALT Pathways of Telomere Maintenance

8.4.1 Features of ALT

The first distinguishing feature noted for the ALT pathway was the unusual and heterogeneous telomere length distribution, ranging from very short or undetectable to extremely long.[13,114] An early examination of the telomere dynamics in ALT cells revealed rapid deletions and additions of large segments of telomeric DNA occurring on a background of gradual telomere attrition.[115] Such rapid deletions might suggest the presence of extrachromosomal telomeric repeat (ECTR) DNA in ALT cells, which was indeed later detected, in linear[116,117] or circular[60,118] form.

Another phenotype that appears to be unique to ALT cells is the presence of ALT-associated PML bodies (APBs).[119] A PML body is a doughnut-shaped structure of uncertain function in the nucleus of most human cells, containing the promyelocytic leukaemia protein (PML).[120] A subset of ALT cells contain PML bodies that include telomeric DNA (APBs), which are not observed in mortal or telomerase-positive cells.[119] APBs also contain an array of proteins involved in telomere function (TRF1,[119] TRF2[119] and Rap1[121]), or DNA recombination and repair (including Rad52,[119] Rad51,[119] RPA,[119] BRCA1,[121] BLM,[122] WRN,[123] Mre11/RAD50/Nbs1,[54,121] Rad51D,[124] SMC5/6[125] and 9-1-1[126]).

The role of APBs has not yet been elucidated; they have been proposed to perform a storage function for excess products of the ALT process, or to play a direct role in the ALT telomere maintenance mechanism.[119] The extrachromosomal DNA found in APBs is linear rather than circular,[127] suggesting that it may be sequestered in APBs to protect it from recognition as damaged DNA. APBs are located both extrachromosomally and in association with intact telomeres in metaphase spreads;[126] the association of telomeres with APBs is very dynamic,[128] supporting a direct role in the ALT mechanism. Indeed, ongoing DNA replication has been detected in APBs.[126,129] In addition, APBs have been reported to be enriched in cells at the late S/G2 phases of the cell cycle,[129,130] which is when homologous recombination is known to occur,[131] although data from methionine starvation of ALT cells link APBs with the G_0/G_1 phases.[132]

8.4.2 Evidence that ALT Involves Recombination

Human ALT cells were not the first cells reported to lengthen their telomeres in the absence of telomerase activity. A seminal study in 1993[12] demonstrated that *Saccharomyces cerevisiae* lacking telomerase undergo telomere shortening and most cells die, but survivors emerge that have lengthened their telomeres by recombination. These survivors can arise by amplification of either the subtelomeric regions ('Type I survivors'[12]) or the telomere repeats themselves ('Type II survivors'[133]). The same phenomenon occurs in other yeast including *Kluyveromyces lactis*[134] and probably *Schizosaccharomyces pombe*.[135] The involvement of homologous recombination (HR) in this process was demonstrated by a genetic requirement for the HR-related genes Rad51 (Type I), Rad50 (Type II) or Rad52 (both Type I and Type II).[136–138]

The telomere length profile of human ALT cells is very reminiscent of that in Type II survivors, and HR has always been seen as a possible mechanism for ALT.[13] The presence of recombination proteins in APBs also lends indirect support to this concept. HR involves the invasion of an ssDNA end into a homologous region of duplex DNA, which it uses as a template for elongation (Figure 8.4A). It is one of the key repair pathways for double-strand breaks (DSB) in mammalian cells and also provides a pathway through which stalled replication forks can be re-established; HR is therefore intimately involved in both DNA replication and repair. The proteins involved and the mechanisms of HR are becoming increasingly understood on a structural and molecular level (see Table 8.1, Figure 8.4A and other chapters of this book; reviewed in ref. 131). This knowledge regarding HR in general is in turn increasing our understanding of the details of ALT.

The most direct evidence to date that ALT involves HR is an elegant study in which a DNA tag was inserted within a telomere in an ALT cell line. During proliferation of the cells *in vitro*, the tag increased in copy number and appeared on multiple new chromosome ends.[139] This phenomenon was specific for ALT cells. The most likely explanation for these data is that the single-stranded overhangs on telomeres in ALT cells can invade the duplex region of another telomere, as in HR, and use it as a template for elongation (Figure 8.4Bi). There are several other recent lines of evidence that ALT utilises an HR mechanism (see Sections 8.4.2.1 to 8.4.2.4).

8.4.2.1 Recombination Proteins have been Shown to Play a Direct Role in ALT

Thus far, all proteins demonstrated to play a role in ALT are also involved in HR. The first proteins shown to be required for the ALT process were the Mre11, Rad50 and Nbs1 proteins, which form the MRN complex (Table 8.1).

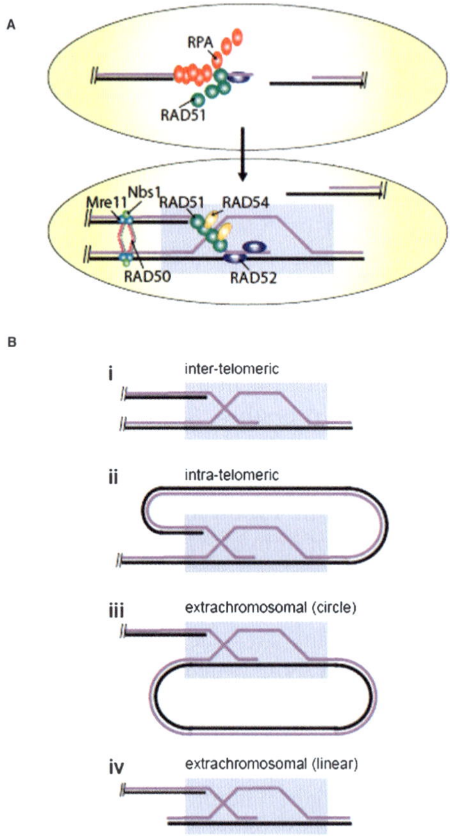

Figure 8.4 **Recombination and ALT.** (**A**) Schematic of some proteins involved in HR, in this case to carry out DSB repair. See Table 8.1 for details. (**B**) Templates that may be used for recombinational telomere lengthening in ALT: (**i**) interchromosomal telomere copying; (**ii**) intratelomeric telomeric copying using the T-loop as a template, with no resolution of the resulting Holliday junction, resulting in addition of many telomere repeats; (**iii**) copying using the extrachromasomal telomeric circles known to be present in ALT cells, similar to the 'roll and spread' used by some yeast; (**iv**) copying of the extrachromosomal linear telomeric DNA also known to be present in ALT cells. Part B reprinted by permission from Macmillan Publishers Ltd: A. A. Neumann and R. R. Reddel, Telomere maintenance and cancer—look, no telomerase, *Nature Reviews in Cancer*, **2**, 879–884, copyright (2002).[161]

Inhibition of this complex either by sequestration with overexpressed Sp100 protein, transient small interfering RNA (siRNA) inhibition or stable knockdown with short hairpin RNAs (shRNAs) leads to inhibition of ALT[125,140–142] manifested as a reduction in APBs, a decrease in telomere length and a decrease in circular ECTR DNA.

The 'structural maintenance of chromosomes' SMC5/6 complex is thought to participate in HR by recruitment of a cohesin complex that brings the DNA

Table 8.1 Some of the proteins involved in homologous recombination (HR) in mammalian cells (see Figure 8.4).[a,b]

Protein	*Function in homologous recombination*
Rad51	Forms a long nucleoprotein filament on the ssDNA end to be repaired. Promotes invasion into the duplex.
Rad51D	Rad51 paralogue. Involved in Holliday junction resolution.
Rad52	Enhances Rad51 specificity for binding ssDNA. Also promotes single-stranded annealing reaction.
RPA	Binds and protects ssDNA. RPA must be displaced from ssDNA to allow formation of the Rad51 nucleoprotein filament, which initiates homologous pairing.
Rad50 **Mre11** **Nbs1**	Form a stable complex (MRN complex). May tether the two DNA strands via dimerisation domain at end of extended Rad50 coiled-coil.
BLM	Helicase that interacts with Rad51 and localises to ssDNA. Precise role in HR unclear; may increase migration of Holliday junctions.
WRN	Helicase and endonuclease. Precise role in HR unclear; may resolve recombination intermediate structures.
SMC5/6	Required specifically for sister chromatid homologous recombination. Recruits the cohesin complex to bring DNA ends together.
BRCA1	Interaction with MRN complex might mediate resection of broken DNA ends.

[a]For references, see 133, 141, 146, 269, 270.
[b]All the proteins have been linked with ALT, either functionally (protein names in bold) or via their presence in APBs.

ends together.[143] Suppression of SMC5/6 levels reduced the number of APBs and shortened telomeres substantially in an ALT cell line, leading to senescence.[125] The ability of SMC5/6 to recruit cohesin was apparently not required for this effect, but its ability to SUMOylate other proteins was necessary, suggesting that the involvement of SMC5/6 in telomeric HR may differ mechanistically from its involvement in other HR processes.

Additional HR proteins showing evidence for a role in ALT include the RecQ helicase BLM, which promotes a helicase-dependent increase in the number and size of APBs when overexpressed in ALT cells.[122] The related RecQ helicase, WRN, on the other hand, has been postulated to suppress recombination at telomeres.[144] Mice lacking WRN develop tumours that have ALT-like telomeres and APBs. The topoisomerase TOP3, a binding partner of RecQ helicases, is thought to play a role in Holliday junction resolution during HR.[145] Downregulation of TOP3 in a human ALT cell line leads to elimination of APBs; the yeast homologue of TOP3 was also shown to be essential for the Type II survivor pathway.[146]

The involvement of the HR-associated protein RPA in ALT is likely to be complex, since a recent study suggests that its ability to bind ssDNA confers a telomere capping role specifically to ALT cells.[147] Its loss results in an increase in some ALT-specific features, indicating that it may actually be inhibiting rather than promoting HR in this setting. The involvement of Rad51 in ALT is also uncertain since its inhibition does not reduce the number of APBs;[125] this is consistent with its lack of involvement in Type II recombination in yeast,[138] which is regarded as similar to human ALT.

It should be borne in mind that most of the above proteins are involved in many DNA replication and repair processes, and it has not been directly demonstrated that it is their involvement in HR that is related to ALT. For example, RPA and the helicases BLM and WRN have been shown to unwind G-quadruplex structures; BLM and WRN actually have some specificity for this structure[111–113] (see also Chapters 3 and 5). The yeast homologue of these proteins, sgs1, also resolves G-quadruplexes[148] and is necessary for Type II recombinational survival.[149] It may be the G-quadruplex resolvase activity of these proteins that is necessary for ALT.

8.4.2.2 Direct Evidence for Postreplicative Telomere Exchanges in ALT Cells

Postreplicative exchange events involving telomeric DNA have been detected at a very high level in ALT cells, but at very low or nonexistent levels in mortal and telomerase-positive cells.[62,63,150] In ALT cells, these exchanges are specific for telomeric sequences. While referred to as telomere sister chromatid exchanges (T-SCEs), it is also possible that the exchanges involve extra chromosomal telomere repeat (ECTR) DNA or a non-sister chromatid. Sister chromatid exchange is mediated by HR,[151] so it is highly likely that T-SCE also reflects HR at telomeres.

8.4.2.3 Recombination-dependent Extrachromosomal Telomeric Circles

The production of the extrachromosomal telomeric circles that are characteristic of ALT is dependent on the HR proteins XRCC3 and Nbs1, both in situations where normal telomere capping has been perturbed[60,61] as well as in untreated ALT cells.[142] The fact that the distribution of circle sizes is similar to the distribution of T-loop sizes in ALT cells[60,118] suggests that the circles arise through the resolution of T-loop junctions by HR (T-loop HR) that has been detected at low levels in normal human cells.[60] Furthermore, yeast that can utilise HR as a back-up telomere maintenance mechanism also possess telomeric circles,[152–155] and T-loop HR dependent telomere circle formation has been shown to occur in budding yeast when telomere capping function is disturbed.[156]

8.4.2.4 Homologous Recombination is a Part of Normal Telomere Biology

HR proteins, including Mre11, Nbs1, Rad51 and Rad52, have been detected at the telomeres of non-ALT cells (including normal fibroblasts) during S/G2, but they dissociate by M phase.[54,55,66] This led to the proposal that telomeres become uncapped to enable replication during S phase and are then recapped by the HR machinery.[54,55] In support of this, deleting Rad51D or Rad54 led to telomere fusions, indicative of uncapped telomeres.[124,157] Given the resemblance of T-loop junctions to HR intermediates, it is plausible that this recapping involves formation of T-loops. Accordingly, HR proteins are necessary for strand invasion of a model telomere end into duplex telomeric DNA *in vitro*, and the ability of cell extracts to promote this peaks in S/G2.[55] These findings suggest that ALT involves a dysregulation of the HR that occurs at normal telomeres.

While determining the proteins responsible for promoting ALT is crucial for understanding the mechanism, it is becoming clear that these proteins are not specific to the ALT mechanism. Determining what leads to HR dysregulation at telomeres will have the greatest significance for cancer therapies aimed at targeting ALT. Cell fusion studies have provided evidence that ALT results from recessive gene mutations.[158] Identifying these mutations remains one of the biggest challenges in ALT research.

8.4.3 Proposed Mechanisms of Recombination-mediated Telomere Lengthening

One distinguishing feature of ALT telomeres is the extreme length of a subset of them, and we know that telomere lengthening must occur for a cell to avoid replicative senescence. The different forms of HR occurring at ALT telomeres (intertelomeric copying, T-loop HR, postreplicative telomeric exchanges) may be a result of general dysregulation of HR at telomeres, *i.e.* not functionally linked to telomere lengthening. The only HR process for which there is direct evidence of a net gain of telomere sequence is HR-mediated intertelomeric lengthening.[139]

Figure 8.4B illustrates some hypothesised models for ways in which telomere HR may use alternative templates to contribute to telomere length changes.[159–161] The T-loop itself may be used as a template, with the HR machinery mediating the migration of the Holliday junction followed by repetitive copying around the loop (Figure 8.4Bii). Extrachromosomal circular telomeric DNA may be used as a template in a 'roll and spread' mechanism similar to that used at telomeres of telomerase-deficient *K. lactis*[162,163] (Figure 8.4Biii). Alternatively, linear ECTR could invade a telomere and become copied, possibly within an APB[127] (Figure 8.4Biv).

It has also been proposed that unequal telomeric sister chromatid exchange, followed by non-random segregation of the chromosomes, could result in net telomere lengthening in ALT cells.[62,63,160,164] The above models are not mutually exclusive; they could theoretically all be happening simultaneously within ALT cells.

8.5 Telomerase: Structural and Biochemical Studies

The core catalytic components of telomerase, TER and TERT, were first identified in 1989[165] and 1997,[166] respectively. Due to their relatively large sizes, technical difficulties with the expression of the TERT protein and the fact that the two subunits form an intricate ribonucleoprotein (RNP) complex, structural characterisation of these molecules has only recently started to be achieved. For both molecules, the key to success in determining structure has been their reduction into smaller domains. Given the intense clinical interest in telomerase, such structural determination represents a significant advance in the field, providing a template for rational drug design.

8.5.1 Telomerase RNA (TER)

The RNA subunits of telomerase from different species vary in size from ~150 nucleotides to >1300 nucleotides, but remarkably seem to share common secondary structures.[167,168] The template regions of the TERs make up only a very small proportion of this total size; this was an early indication that other regions of the RNA probably play important functional roles within the RNP, a concept borne out by mutational analyses. In addition to the template, common structural elements of different TERs include a template boundary sequence, a pseudoknot and a long-range base-pairing interaction that encloses the template in a loop.

Vertebrate, ciliate and yeast TERs may also share an intriguing 3′ stem-loop region that plays a critical, but as yet incompletely defined, role in telomerase action. This region (known as stem-loop IV in ciliates, CR4-CR5 in vertebrates and CS5-CS6 in yeast) is essential for activity in all three groups,[169–173] and the *Tetrahymena thermophila* and human structures have been shown to support activity when provided in *trans*.[171,174,175] The structures of all or parts of this region from both *Tetrahymena* and human have been solved by NMR,[176–179] supporting a potential functionally analogous role (Figures 8.5 and 8.6, respectively). *Tetrahymena* stem IV contains a bulge that forms a rigid kink in the RNA that is important for activity[176,179] (Figure 8.5, cyan), as had previously been inferred from mutational studies.[170,180] This bulge forms part of the binding site for the TER-binding protein p65, and is essential for p65-mediated stimulation of TERT and TER binding.[181] Eliminating the bulge has only a modest effect on binding of TER to TERT alone,[170] indicating that the function of the rigid kink in the stem is structural rather than as a binding site for TERT. The loop that terminates stem IV forms an unusually well-defined structure (Figure 8.5, red). Nucleotides in loop IV have their Watson–Crick faces exposed to the solvent and are essential for telomerase activity;[169,170,175,176] they may interact with TERT, but this is at most a low-affinity interaction.[170,176,182,183] Interestingly, removal of loop IV results in the loss of TERT-induced base-pairing in the pseudoknot region, implicating this region as being involved in mediating an active conformation in distant parts of the RNA, but this loss is also not sufficient to account for the dramatic loss

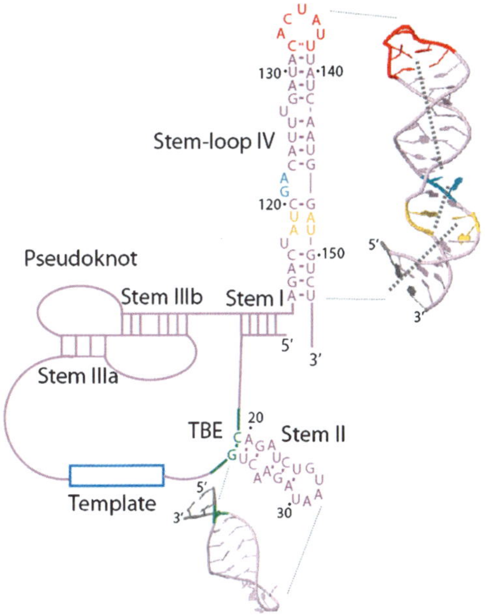

Figure 8.5 Secondary and solved tertiary structures of *Tetrahymena* telomerase RNA. The template and template boundary element (TBE) are blue and green, respectively. In stem-loop IV, the GA bulge that forms a rigid kink is cyan, the flexible proximal region of the stem is yellow, and loop residues are red. The axes of the helices are shown as dotted lines on the tertiary structure to emphasise the bend in the stem. Non-native nucleotides included during structure determination are gray. The NMR structures were rendered with Jmol (www.jmol.org) using PDB database entries 2FEY (stem-loop IV[176]) and 2FRL (stem II[323]).

of activity of loop IV mutants.[170] One striking aspect of the NMR structure of stem IV is that it demonstrated a flexible template–proximal region (Figure 8.5, yellow); the flexibility of this region influenced telomerase activity.[176] Thus a model has emerged whereby the movement of stem IV positions loop IV close to the active site during certain stages of catalysis;[176] it may interact with other parts of the RNA including the pseudoknot region or may be involved in positioning TERT. Telomerase activity is characterised by large movements between components during stages of repeat addition; the data above provide the first molecular details of some of these movements, although RNA structural determination in the presence of proteins will be also informative in this regard.

Two separate but neighbouring stem-loops of the human CR4-CR5 domain have been structurally characterised.[177,178] One of these (P6a/P6b) contains two base-paired helices that are bent in relation to each other, reminiscent of the kink in *Tetrahymena* stem IV (Figure 8.6). The other stem-loop (P6.1) is essential for telomerase activity and has a potential functional analogue in both *Saccharomyces* and *Kluyveromyces* yeasts.[171–173] Remarkably, base-pairing

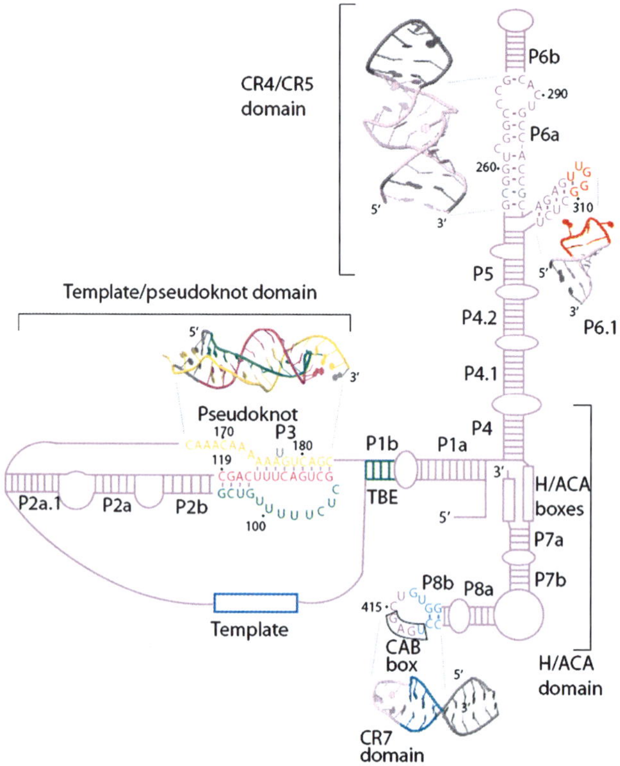

Figure 8.6 **Secondary and solved tertiary structures of human telomerase RNA.** The template and template boundary element (TBE) are blue and green, respectively. The three strands of the pseudoknot triple helix are shown in green, gold and magenta. Note that the bulged nucleotide U177 was deleted from the molecule used for structure determination. The essential loop of stem P6.1 is red. In the CR7 domain, nucleotides involved in RNA end processing are cyan. Non-native nucleotides included during structure determination are grey. The NMR structures were rendered with Jmol (www.jmol.org) using PDB database entries 1YMO (pseudoknot[198]), 1OQ0 (P6.1[177]), 1Z31 (P6a/P6b[178]) and 2QH2 (CR7[210]).

residues in this helix are absolutely conserved in 13 yeast species and 35 vertebrate species, attesting to its conserved function.[173] The loop that terminates human stem 6.1 forms a rigid structure and its nucleotides have their Watson–Crick faces exposed to the solvent, similarly to *Tetrahymena* loop IV[177] (Figure 8.6, red). These nucleotides are protected *in vivo*,[184] but this is unlikely to be due to interactions with TERT; the more proximal stem 6.1 region has instead been shown to be important for TERT binding.[172,185]

An important advance in understanding the functions of loop 6.1 was the demonstration that its nucleotides undergo a long-range base-pairing interaction with nucleotides close to the template;[186] it will be informative to determine if *Tetrahymena* loop IV nucleotides are involved in such interactions.

It will also be interesting to determine if human loop 6.1 is positioned distal to a flexible region that may control its long-range interactions.

Another conserved and important secondary structure element in telomerase RNA is a pseudoknot, an RNA structure containing two stem-loops in which the first stem's loop forms part of the second stem (Figures 8.5 and 8.6). Such a structure was proposed based on sequence conservation among telomerase RNAs of ciliated protozoa;[187–189] mutagenesis studies subsequently showed that the *Tetrahymena* pseudoknot is important for *in vitro* activity[169,182,190] and essential for formation of the telomerase RNP *in vivo*.[191] A secondary structure model of vertebrate telomerase RNAs also predicted the existence of a pseudoknot near the template,[192] which was shown to be necessary for telomerase activity.[193,194] Structural analysis of this region of hTER has led to the proposal that it undergoes a 'conformational switch' between a structure in which the P3 helix forms (forming a pseudoknot), and an alternate structure in which stem P2b forms a hairpin and a loop with extensive intraloop basepairing.[195,196] However, the functional importance of the hairpin loop was not supported by mutational analysis[197] and the P3-containing pseudoknot appears to be the dominant structure in wild-type hTER.[196]

The NMR solution structure of this pseudoknot revealed an unusual extended triple helix; its base triples were supported by mutagenesis and the stability of the structure in the mutants correlated strongly with telomerase activity[198] (Figure 8.6). The RNA used for this NMR study contained a single-nucleotide mutation that was known to stabilise the pseudoknot, facilitating structural analysis; however, molecular modelling of the wild-type hTER pseudoknot predicts a slightly different triple helix structure that would be more flexible.[199] Since the mutated hTER has greatly reduced activity,[195,196] this implies that flexibility in this region is important for telomerase activity. This is consistent with data showing that the pseudoknot is important for telomerase repeat addition processivity in both human and *Tetrahymena*.[182,185]

A pseudoknot near the template has also been found to be essential for activity in the yeast *Kluyveromyces lactis*.[200] Secondary structure models for *Saccharomyces* telomerase RNA suggest the presence of a pseudoknot in the same region,[168,201,202] although mutational analyses have not supported the functional importance of the pseudoknot and it is not present in an alternative *Saccharomyces* secondary structure model.[168,201,203] Notably, computer modelling has shown that the tertiary structure of the pseudoknot also involves triple helices in *K. lactis* as well as 30 species of ciliate;[204,205] mutational analysis in *K. lactis* supports the formation of base triples.[204] The structural and functional conservation of the pseudoknot in telomerase RNAs of different species is dramatically illustrated by the fact that the ciliate pseudoknot can substitute for those of yeast and human, and the human pseudoknot can function in place of that of yeast.[201,206]

Vertebrate telomerase RNAs possess a domain that appears to be absent from other species, the H/ACA domain at the 3′ end of the RNA (Figure 8.6). This domain is homologous to those found in the H/ACA class of small nucleolar RNAs (snoRNAs) and is necessary for hTER processing, accumulation and localisation *in vivo* but is dispensable for *in vitro* activity.[171,193,207–209] The stem-loop that terminates this region (known as the CR7 domain) contains a four-nucleotide sequence, UGAG (the CAB box, see Figure 8.6), which is required for localisation of hTER to nucleoplasmic Cajal bodies.[209] The solution structure of the CR7 domain has recently been solved, revealing a well-structured loop.[210] Extensive mutagenesis has demonstrated that nucleotides throughout the H/ACA domain, including the H and ACA boxes themselves and the CR7 loop, are necessary for *in vivo* hTER accumulation.[171,193,207,208,210] On the other hand, the signals that determine correct 3′ end processing and localisation to the Cajal body are separable; the former involves stem P8b and an unpaired configuration of the first two nucleotides in the loop[210] (Figure 8.6, cyan) and the latter localises to the CAB box. However, the two processes are linked in that incorrectly processed RNAs never localise correctly to the Cajal body.[210] The function of Cajal body accumulation of hTER is unknown, although it has been suggested to be required for eventual telomerase localisation to the telomere,[211] and there remains much to be determined about the biogenesis and trafficking of hTER throughout the cell.

8.5.2 Telomerase Reverse Transcriptase (TERT)

The catalytic protein subunit of telomerase, TERT, contains a central portion that is homologous to other reverse transcriptases, containing the same conserved motifs (1,2,A-E)[166] (Figure 8.7). The crystal structure of HIV RT[212] enables predictions to be made regarding the structure and function of homologous amino acids in TERT. Indeed, mutations within the reverse transcriptase (RT) domain of TERT have confirmed its involvement in catalytic activity,[166,213,214] nucleotide binding[215,216] and DNA binding.[217–220] While the RT domain of TERT would be predicted to interact with the template region of the RNA, this appears to be only a loose association.[221,222]

TERT also contains long N- and C-terminal extensions that are not present in other RTs, the functions of which were initially unknown. Sequence alignment of the N-terminal extension of TERT from different species combined with mutational scanning have enabled the identification of conserved motifs and functional domains.[215,223–228] It is now clear that these motifs can be grouped into two separate structural and functional domains: the Telomerase Essential N-terminal or TEN domain (which includes regions named in various studies as Region I, GQ, RID1, T2 or N) and the RNA Binding Domain or RBD (which includes Regions II, III and IV, RID2, and the T, QFP, CP and CP2 motifs). Both of these domains from *Tetrahymena* TERT can be expressed as stable and functional protein fragments in bacteria,[10,183,229] in contrast to the full-length TERT protein. As a consequence, the *Tetrahymena* TEN and RBD domains are the first regions of TERT to have had their structures solved by crystallography (Figure 8.7).

The TEN domain represents a novel fold, containing mixed α helices and β sheets.[9] Phylogenetically conserved amino acids are present in a groove

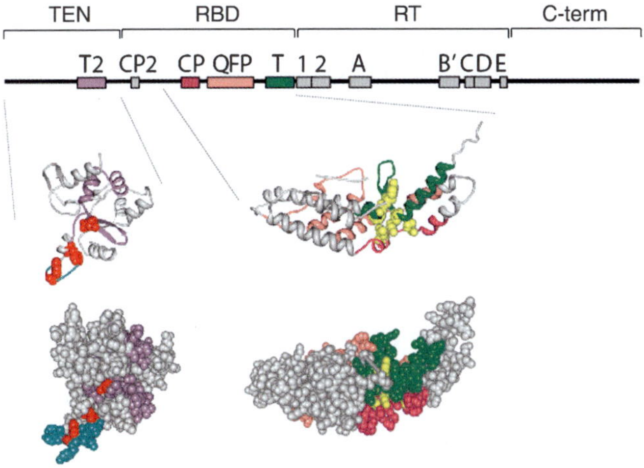

Figure 8.7 Domain arrangement and solved tertiary structures of *Tetrahymena* TERT protein. Functional domains are as follows: TERT Essential N-terminal domain (TEN), RNA Binding Domain (RBD), Reverse Transcriptase Domain (RT) and C-terminal extension (C-term). Conserved sequence motifs are labelled on the primary structure representation of the whole protein, and correspond in colour to coloured amino acids on the two tertiary structures (T2: purple; CP: magenta; QFP: salmon; T: green). On the structure of the TEN domain, amino acids involved in DNA binding are red, and a region implicated in RNA binding is cyan.[9] On the structure of the RBD domain, amino acids that are especially important for RNA binding are shown in yellow.[222,236] The crystal structures were rendered with MBT Protein Workshop using PDB database entries 2B2A (TEN[9]) and 2R4G (RBD[10]).

on the surface of the structure, and several of these (Q168, F178 and W187) were shown to be important for crosslinking to a telomeric DNA primer (Figure 8.7, red). The amino acid Q168 is universally conserved among TERTs. The crosslink to W187 was confirmed independently,[230] although mutating this amino acid does not reduce telomerase activity,[230,231] indicating that it may be close to the primer but not responsible for high-affinity binding. The whole TEN domain contributes only a small amount to the overall DNA binding ability of TERT; deleting it has only a modest effect on primer affinity,[220] even though this deletion results in a complete loss of telomerase activity.[183,229] Thus the TEN domain seems to have other functions in activity in addition to DNA binding. It has been proposed to contain a low affinity RNA interaction site[182,183,224,227,232] (Figure 8.7, cyan), but RNA interactions with the TEN domain that was used for crystallization are non-specific.[9] Potential *in vivo* functions for the TEN domain include binding to proteins that are responsible for telomerase regulation or recruitment to the telomere[228,233] or influencing the binding to the telomere of other proteins such as yeast Rap1.[234]

The RBD is essential for activity *in vitro* and *in vivo* and has been shown to constitute a high-affinity RNA binding site in TERTs of different species.[222,224,227,235] In *Tetrahymena*, the RBD binds with low nanomolar affinity to a single-stranded region at the base of stem II[183,235,236] (Figure 8.5, green), whereas the human RBD is able to bind the CR4/CR5 region of hTER.[235] Given the conservation of this region and its function, it is somewhat surprising that it binds to structurally distinct RNA regions in different species. The explanation for this may lie in the fact that non-conserved regions in a putative protein 'linker' upstream of the sequence used in crystallography studies are also essential for high affinity RNA binding;[235] this region may be responsible for species-specific recognition of different RNA structures.

Motifs within the RBD that are known to be critical for RNA binding are the T and CP motifs;[222,237] the QFP motif also has an effect on RNA binding.[227,237] The structure of the RBD reveals another novel protein fold, consisting almost entirely of α helices.[10] The T motif forms a well-defined pocket (Figure 8.7, green) and hence may bind single-stranded RNA (such as the region at the base of stem II). The CP motif forms a shallow, wide cavity adjacent to the T-pocket, of a shape consistent with binding to double-stranded RNA (Figure 8.7, magenta). The QFP motif, on the other hand, is hydrophobic and largely buried (Figure 8.7, salmon), indicating that it may be involved in folding the domain but is not likely to contact nucleic acids.

The functions of the C-terminal extension of the TERT proteins are relatively unexplored. Most of the domain is essential for *in vitro* and *in vivo* activity,[238,239] and it has been implicated in processivity and DNA binding.[239–241] The DNA binding ability of this domain is, however, a low-affinity one.[219,220] *In vivo*, a region of the C-terminal domain is involved in nuclear localisation of hTERT.[242]

8.5.3 Telomerase Ribonucleoprotein Complex

The structural studies described above have added great value to the mutational studies that preceded them, but each one characterises only an isolated domain of telomerase. While the structures of the RNA domains are generally consistent with chemical and enzymatic footprinting of the molecules,[170,176,180,243] there is evidence that interaction with proteins can modify elements of the structure.[170,184,243] In addition, none of the structural determinations has been carried out in the presence of DNA or nucleotide substrates of telomerase, the only sure way to determine their mode of interaction with the enzyme. It is therefore essential that we proceed to the determination of the structure of an entire telomerase complex. To that end, there has been substantial research effort devoted to characterising the composition of the active enzyme complex.

Despite the conservation of TERT and TER structures and functions, other proteins that constitute the telomerase complex are more diverse among species. Telomerase complexes have been purified from two ciliated protozoa, *Euplotes aediculatus* and *Tetrahymena thermophila*. The protein p43, which has homology with the La family of RNA-binding proteins, co-purifies with *Euplotes* telomerase in stoichiometric amounts.[244,245] p43 may be involved in structural stabilisation of the RNA and subsequent enhancement of telomerase activity and processivity.[246] Human La protein has also been reported to associate with telomerase.[247] In *Tetrahymena* four major proteins, p75, p65, p45 and p20, associate with telomerase.[248,249] p65 is important in the accumulation and stability of *Tetrahymena* TER.[248] Additionally, p65 drives the hierarchical assembly of the telomerase complex by inducing a conformational change in TER that permits its stable association with TERT.[181,250,251] The *Saccharomyces cerevisiae* telomerase complex contains two additional proteins, Est1p and Est3p, that are necessary for *in vivo* telomerase function but are dispensable for *in vitro* activity.[252,253] Est1p mediates telomerase recruitment to the telomere by bridging an interaction between TER and the telomeric protein cdc13p.[254,255] A human homologue of Est1p also has a telomeric phenotype, but the details of its interaction with telomerase are not yet well understood.[256,257]

The very small number of active telomerase molecules in human cells[258] has until recently precluded a direct purification approach analogous to that carried out in ciliates. The composition of the active endogenous human telomerase complex was recently identified by large-scale cultivation of an immortal human cell line followed by the development of an efficient telomerase purification scheme and analysis by mass spectrometry.[258] The core active telomerase complex consisted of only three subunits: hTERT, hTER and the protein dyskerin. Dyskerin is a putative pseudouridine synthase that had previously been found to have a second role in telomerase biochemistry.[259] While dyskerin is known to stabilise levels of hTER *in vivo*,[259] the details of its role in telomerase action, and the biochemical basis for its retention in the active complex, remain a mystery.

A different strategy of purification of overexpressed human telomerase followed by mass spectrometry revealed, in addition to the three subunits described above, several other proteins: the dyskerin-associated proteins Nop10 and Nhp2, hnRNPs U and C, several proteins from the Sm family and two NTPases.[260]

The discrepancy between these two studies may be resolved by recognising that telomerase complexes that are at different phases of biogenesis have been isolated using the two purification procedures, with the latter complex (isolated by Fu and colleagues) containing several regulatory molecules that do not remain in the core enzyme complex and are not required for core telomerase activity. The Cohen purification protocol[258] included a selection step specific for active telomerase complexes, whereas the Fu procedure[260] would isolate both active and inactive complexes. It is unlikely that the Cohen purification procedure led to dissociation of a substantial number of telomerase associated proteins, since the size of the complex remained the same between crude cell lysate and purified telomerase.[258]

In addition to proteins that form an integral part of the active telomerase enzyme complex, there have been more than 30 other proteins that are proposed to interact with human telomerase (see Table S1 in ref. 258; reviewed in ref. 261). These proteins carry out telomerase regulatory functions such as chaperone-mediated assembly of the telomerase complex (p23 and hsp90[262,263]), intracellular localisation of telomerase (14-3-3[242]) and unwinding of the telomerase DNA–RNA hybrid to achieve negative regulation of telomerase (Pif1[264,265]). Although the chaperones hsp90 and p23 have been reported to remain associated with active telomerase,[262] it appears that none of these proteins is necessary for the activity of the core telomerase complex.

8.6 Telomerase-targeted Cancer Therapeutics

Telomerase is a promising target for treatment of cancer. Maintenance of telomere length is necessary for unlimited growth of cells. A large majority of human tumours express active telomerase, whereas the corresponding normal tissues are devoid of detectable telomerase activity.[14,15] Thus, telomerase inhibitors have the potential to be specific and non-toxic drugs that are effective against many types of cancer. Several strategies for telomerase inhibition have led to the predicted telomere shortening and cell death in preclinical trials; at least one of these has reached the point of human clinical trials.

8.6.1 Oligonucleotides Targeting Telomerase RNA

The template region of the telomerase RNA subunit is accessible for reverse transcription into telomeric repeats; oligonucleotides that hybridise with this region are therefore obvious candidates for competitive inhibition of telomerase activity at telomeres. Unmodified DNA oligonucleotides are known to undergo nucleolytic degradation, and indeed were ineffective at inhibiting telomerase activity in cell extracts.[266,267] Backbone modifications known to increase oligonucleotide stability *in vivo* [peptide nucleic acid (PNA) linkage, 2′-O-methyl RNA (2′-O-MeRNA), and 2-O-methoxyethyl] yielded molecules that potently inhibited telomerase in cell-free extracts, with low nanomolar IC_{50}s.[266–270] The optimal target sequence included nine nucleotides of template and four nucleotides of non-template sequence on the 5′ side.[266,269] Phosphorothioate (PS) linkage is another backbone chemistry that is often utilised to improve stability of antisense oligonucleotides, but interestingly such oligonucleotides showed no sequence-specificity in their inhibition of telomerase activity.[266,267,270]

A combination of 2′-O-MeRNA sequence with a small number of PS linkages did yield a potent sequence-specific telomerase inhibitor,[267] and this oligonucleotide was the first to be trialled for telomerase inhibition within human cancer cell lines.[6] Telomerase inhibition led to telomere shortening and an eventual onset of apoptosis,[6] providing the first proof-of-principle that this strategy may be of clinical use. A *thio*-phosphoramidate oligonucleotide of the same sequence as the optimal 13-mer referred to above (GRN163; see Figure 8.8A), is also capable of causing telomere shortening and cell death in cultured cancer cells, with a low nanomolar cell-based IC_{50} in the presence of a lipophilic carrier.[271–273] While telomere shortening, cell death and a reduction in size of xenograft tumours occurred in the absence of such a carrier,[272,274–276] cellular uptake and potency of GRN163 was greatly enhanced in the presence of carrier.[271,272] This dependence on a transfection reagent would be likely to cause practical problems for clinical use, so a derivative of this oligonucleotide bearing a 5′ lipid moiety (GRN163L) was synthesised;[277] this derivative showed improved uptake and potency in both cells in culture and xenograft tumours.[277,278] GRN163L has been shown to be effective in causing telomere shortening, cell death and/or reduction in xenograft tumour size in preclinical trials of breast, lung and hepatocellular carcinomas.[277–282] Furthermore, in *in vitro* and *in vivo* model systems, it reduced metastasis and invasiveness and enhanced radiation sensitivity.[279–283] This oligonucleotide is the most advanced telomerase inhibitor in the clinical pipeline, currently being tested in Phase I/II clinical trials in chronic lymphocytic leukaemia, solid tumours, multiple myeloma and non small-cell lung carcinoma (www.geron.com).

While some of the above-mentioned preclinical trials found no evidence for cell death upon GRN163L treatment of a small number of telomerase-negative cell lines,[272,280,282] a full evaluation of potential adverse side-effects awaits the outcomes of these clinical trials. Most of the phenotypic effects of GRN163L in the above preclinical trials had a delayed appearance due to their requirement for gradual telomere shortening. Intriguingly, however, colony formation of lung and breast cancer cell lines was rapidly reduced.[279,280] It has recently been shown that this effect is due to inhibition of cellular adhesion and attachment, an effect that is totally independent of telomerase inhibition.[283] The anti-adhesive effect was not observed in one normal fibroblast cell strain and it was hypothesised that it may be responsible for GRN163L-mediated metastasis reduction,[283] but again, full evaluation of off-target effects of this oligonucleotide requires further study.

8.6.2 Small Molecule Inhibitors

A number of small molecules that inhibit telomerase activity *in vitro* and cause telomere shortening and cell death in cultured telomerase-positive cells have been described, including nucleoside analogues, tea catechins and other dietary phenols, and nitrostyrenes (reviewed in ref. 284). To my knowledge, none of these have reached clinical trials. The molecule 2-[(*E*)-3-naphtalen-2-yl-but-2-enoylamino]-benzoic acid (BIBR1532, see Figure 8.8B) is possibly the best-characterised of these telomerase inhibitors. BIBR1532 is a highly specific telomerase inhibitor that inhibits nucleotide binding through interaction with a distinct site on telomerase.[285,286] It is a reasonably potent inhibitor of purified telomerase (IC_{50} ~100 nM), although its IC_{50} in cell extracts is less impressive (~5 μM).[285–287] Low doses (2.5–10 μM) cause telomere shortening and senescence of cultured cells; these effects were specific to telomerase-positive cells and there was no short-term toxicity.[285,288] However, tumours with initially long telomeres showed no growth defect despite telomere shortening over a year in culture with BIBR1532, implying that this strategy will not be effective against all tumours.[289] Higher doses of BIBR1532 have cytotoxic effects that are independent of telomerase activity and involve telomere capping changes, although these effects are somewhat specific to cancer cells and hence may be useful in overcoming the time-lag that is required for telomere shortening to induce cell death.[290,291]

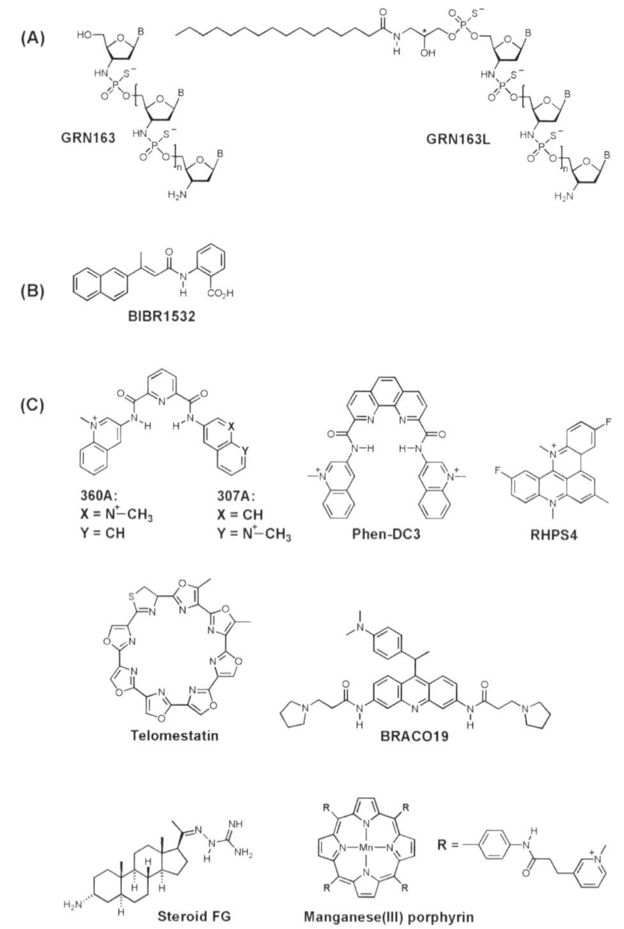

Figure 8.8 Chemical structure of some telomerase inhibitors. (**A**) *Thio*-phosphoramidate oligonucleotides complementary to human telomerase RNA, either with a conjugated 5′ lipid (GRN163L[277]) or without (GRN163[271]). Both oligonucleotides have the sequence 5′-TAGGGTTAGACAA-3′. *Stereochemistry not defined. (**B**) BIBR1532, a mixed-type non-competitive inhibitor of telomerase.[285,286] (**C**) Compounds that interact with and stabilise G-quadruplex DNA: bisquinolinium compounds 360A, 307A[298] and PhenDC3;[299] the pentacyclic acridine RHPS4;[319] the natural product telomestatin;[300] trisubstituted acridine BRACO19;[297] the steroid derivative FG;[317] and a manganese(III) porphyrin.[302]

8.6.3 G-quadruplex Stabilising Molecules

In 1991, Zahler *et al.* demonstrated for the first time that an intramolecular telomeric G-quadruplex could not be utilised by *Oxytricha* telomerase *in vitro*.[108] Based on this finding, a substantial effort has been made to identify synthetic and natural compounds that lock telomeric DNA in a G-quadruplex conformation and thus impede telomere elongation *in vivo*.

A large number of G-quadruplex-interacting ligands have been described.[284,292] Although many of these have been reported to inhibit human telomerase *in vitro*, the majority of these studies used the PCR-based TRAP assay[i] for measuring telomerase activity.[284] It has recently been unequivocally demonstrated that the TRAP assay is subject to artefacts when used to evaluate these ligands due to ligand-mediated inhibition of *Taq* polymerase extension of telomeric DNA;[293] thus, many of the reported IC_{50} values for telomerase inhibition will need to be re-evaluated. Those ligands that have been demonstrated to inhibit telomerase in a direct, non-PCR based assay include:

- 2,6-diamidoanthraquinone BSU-1051;[294]
- perylene diimide PIPER;[295]
- the porphyrin TMPyP4;[296]
- trisubstituted acridine BRACO19;[293,297]
- bisquinolinium compounds such as 360A, 307A and the PhenDC series;[293,298,299]
- the natural product telomestatin.[293,300]

Of these compounds, those with the greatest demonstrated specificity for G-quadruplex over duplex DNA (30–100-fold) are BRACO19,[297] 360A and 307A,[298] PhenDC3[299] and telomestatin[301] (Figure 8.8C). A manganese porphyrin (Figure 8.8C) has recently been shown to possess a remarkable thousand-fold specificity for G-quadruplex DNA;[302] it will be interesting to determine the biological capabilities of this compound.

Telomestatin, the natural product isolated from *Streptomyces anulatus 3533-SV4*,[300] is one of the most well-studied G-quadruplex ligands due to its ability to greatly stabilise G-quadruplexes and its high specificity for these structures. Telomestatin induces and specifically recognises the human intramolecular[303] basket-type[304] G-quadruplex conformation. Telomestatin initially appeared to be a very potent telomerase inhibitor *in vitro* with an EC_{50} value of 5 nM in the TRAP assay,[300] although this is now known to be at least an order of magnitude greater.[293] Nevertheless, at relatively low doses ($\leq 2\,\mu M$), telomestatin causes gradual telomere shortening and growth arrest or apoptosis in a large number of cancer cell lines,[301,305–309] supporting its use as a telomerase inhibitor *in vivo*.

It has recently become clear, however, that classical telomerase inhibition is only part of the telomeric mechanism of action of telomestatin and related drugs. Higher doses of telomestatin ($\geq 5\,\mu M$) lead to proliferation defects within a time frame that is too short for the effects to be explained by telomere shortening,[306,308] reminiscent of the effects of the telomerase inhibitor BIBR1532 (discussed above). The rapid telomestatin-induced phenotype is independent of the telomerase status of the cells and is likely to be due to direct uncapping of the chromosome termini in tumour cells. There are now several lines of evidence to support the uncapping mechanism; namely, treatment

[i]Telomere Repeat Amplification Protocol

with telomestatin has been shown to cause degradation of the telomeric 3′ G-overhang,[306,308,310] rapid dissociation of the telomere capping proteins TRF2 and Pot1 from telomeric termini,[308,310,311] and an increase in DNA damage signals at the telomeres, known as TIFs.[310]

A number of other G-quadruplex ligands act by directly targeting the telomere in addition to inhibiting telomere extension by telomerase; which of these effects predominates appears to depend on the dose of the drug. Treatment of human cancer cells with BRACO19 can result in almost immediate induction of cellular senescence and end-to-end fusions *in vitro*[312] and telomere destabilisation (in the form of anaphase bridge formation and atypical mitoses) in xenograft tumours *in vivo*.[313] The pentacyclic acridine RHPS4 (Figure 8.8C) causes a dramatic increase in TIFs, rapid loss of Pot1 and subsequently TRF2 from telomeres, loss of telomeric 3′-overhangs, and telomere fusions.[314–316] These effects are independent of telomerase status but dependent on the kinase ATR. The G-quadruplex-interacting steroid derivative FG (Figure 8.8C) also induces telomeric overhang shortening and anaphase bridges, effects that are abrogated by excess Pot1.[317]

It was initially envisaged that telomerase inhibition by G-quadruplex stabilisers would be a very specific cancer therapy due to the absence of active telomerase in most normal tissues. A general effect on telomere structure raises the worrying possibility of toxic effects on non-cancer cells. Nevertheless, several of the aforementioned drugs show good selectivity for cancer cell lines over normal cells, but for unknown reasons.[305,308,309,315] This may be due to a different telomere cap structure in normal versus cancer cells, or the existence of intact checkpoint pathways; these possibilities remain to be explored. This raises the exciting possibility that G-quadruplex-stabilisers will constitute a specific cancer therapy that has the capability of overcoming the time-lag required for telomere shortening to occur. Indeed, telomestatin and BRACO19 have been shown to sensitise tumour cells to chemotherapeutic agents;[305,318] when tested in xenograft mouse tumour models these drugs, and also RHPS4, were able to appreciably reduce tumour volume.[309,313,315] Ligands that cause telomere uncapping may have the additional advantage that their function will be independent of the telomere maintenance mechanism used by a given cancer (*i.e.* ALT or telomerase).[298,301,319]

Given the large number of non-telomeric G-quadruplex-forming sequences in the human genome,[81,82] an important consideration when evaluating potential telomere-targeted drugs is their specificity for particular G-quadruplex conformations. For example, the porphyrin TMPyP4 interacts with telomeric G-quadruplexes with a minimal degree of specificity over its interaction with a G-quadruplex in the promoter of the c-*Myc* oncogene.[320,321] The cellular effects of other ligands, however, are clearly mediated primarily through the telomeres; for example, overexpression of telomere proteins TRF2 and Pot1 rendered xenograft tumours resistant to the effects of RHPS4.[315] Given the prevalence of potential G-quadruplex-forming sequences in the promoters of oncogenes, and their underrepresentation in the promoters of tumour suppressor genes,[322] a lack of conformational specificity may in fact contribute to anti-cancer effects of these drugs. However, these issues remain to be thoroughly explored. Furthermore, the implications of the extension of some types of G-quadruplexes by telomerase are also unknown.[109] While G-quadruplex-stabilising molecules are showing great promise as anti-cancer drugs, their mechanisms of cellular action and the likelihood of adverse effects on healthy, proliferating cells should be further investigated prior to clinical use.

8.7 Conclusions and Future Challenges

Telomeres are critically important for cell survival, making them an intense area of research. The higher-order structure of telomeric DNA is becoming very well-defined at the *in vitro* level, and seems to involve both T-loops and G-quadruplexes. The big remaining challenge is to relate this information to the *in vivo* situation, and to integrate it with the increasing knowledge of protective telomeric proteins. In what situations do either T-loops or G-quadruplexes serve telomere capping functions? Does telomere protection in human cells always involve one of these structures or can protection be provided by linear telomeric DNA coated with proteins? What controls the disassembly of the capping structure to allow replication of telomeric DNA in normal cells during the cell cycle? And what leads to the dysregulation of this process in cells displaying the ALT phenotype?

In addition to conventional replication of the bulk of the telomeric DNA at each cell cycle, in some situations the extreme ends of telomeres are replenished to avoid the end replication problem, usually by the enzyme telomerase. Our knowledge of the biochemistry of this intriguing enzyme has increased rapidly in recent years. Areas that are as yet relatively unexplored relate to the mechanistic details of unique properties of telomerase such as repeat addition processivity, and the factors controlling interaction of telomerase with its substrate *in vivo*. The structure of the complete human telomerase enzyme complex remains a formidable but invaluable goal. Whilst much progress has already been made in developing cancer therapies targeting telomerase, undoubtedly the clinical relevance of telomerase to cancer will continue to drive rapid progress in these areas.

Acknowledgements

I am grateful to Axel Neumann and Scott Cohen for their help in preparing the figures. I thank Roger Reddel, Scott Cohen, Axel Neumann and Tony Cesare for critical reading of the manuscript.

References

1. M. J. McEachern, A. Krauskopf and E. H. Blackburn, Telomeres and their control, *Annu. Rev. Genet.*, 2000, **34**, 331.
2. L. L. Sandell and V. A. Zakian, Loss of a yeast telomere: arrest, recovery, and chromosome loss, *Cell*, 1993, **75**, 729.

3. C. B. Harley, A. B. Futcher and C. W. Greider, Telomeres shorten during ageing of human fibroblasts, *Nature*, 1990, **345**, 458.
4. R. R. Reddel, The role of senescence and immortalization in carcinogenesis, *Carcinogenesis*, 2000, **21**, 477.
5. M. S. Singer and D. E. Gottschling, *TLC1*: Template RNA component of *Saccharomyces cerevisiae* telomerase, *Science*, 1994, **266**, 404.
6. B. S. Herbert, A. E. Pitts, S. I. Baker, S. E. Hamilton, W. E. Wright, J. W. Shay and D. R. Corey, Inhibition of human telomerase in immortal human cells leads to progressive telomere shortening and cell death, *Proc. Natl. Acad. Sci. U.S.A.*, 1999, **96**, 14276.
7. X. Zhang, V. Mar, W. Zhou, L. Harrington and M. O. Robinson, Telomere shortening and apoptosis in telomerase-inhibited human tumor cells, *Genes Dev.*, 1999, **13**, 2388.
8. C. W. Greider and E. H. Blackburn, Identification of a specific telomere terminal transferase activity in *Tetrahymena* extracts, *Cell*, 1985, **43**, 405.
9. S. A. Jacobs, E. R. Podell and T. R. Cech, Crystal structure of the essential N-terminal domain of telomerase reverse transcriptase, *Nat. Struct. Mol. Biol.*, 2006, **13**, 218.
10. S. Rouda and E. Skordalakes, Structure of the RNA-binding domain of telomerase: implications for RNA recognition and binding, *Structure*, 2007, **15**, 1403.
11. C. W. Greider, Telomerase is processive, *Mol. Cell. Biol.*, 1991, **11**, 4572.
12. V. Lundblad and E. H. Blackburn, An alternative pathway for yeast telomere maintenance rescues *est1*$^{-}$ senescence, *Cell*, 1993, **73**, 347.
13. T. M. Bryan, A. Englezou, J. Gupta, S. Bacchetti and R. R. Reddel, Telomere elongation in immortal human cells without detectable telomerase activity, *EMBO J.*, 1995, **14**, 4240.
14. J. W. Shay and S. Bacchetti, A survey of telomerase activity in human cancer, *Eur. J. Cancer*, 1997, **33**, 787.
15. N. W. Kim, M. A. Piatyszek, K. R. Prowse, C. B. Harley, M. D. West, P. L. Ho, G. M. Coviello, W. E. Wright, S. L. Weinrich and J. W. Shay, Specific association of human telomerase activity with immortal cells and cancer, *Science*, 1994, **266**, 2011.
16. A. Costa, M. G. Daidone, L. Daprai, R. Villa, S. Cantu, S. Pilotti, L. Mariani, A. Gronchi, J. D. Henson, R. R. Reddel and N. Zaffaroni, Telomere maintenance mechanisms in liposarcomas: association with histologic subtypes and disease progression, *Cancer Res.*, 2006, **66**, 8918.
17. J. D. Henson, J. A. Hannay, S. W. McCarthy, J. A. Royds, T. R. Yeager, R. A. Robinson, S. B. Wharton, D. A. Jellinek, S. M. Arbuckle, J. Yoo, B. G. Robinson, D. L. Learoyd, P. D. Stalley, S. F. Bonar, D. Yu, R. E. Pollock and R. R. Reddel, A robust assay for alternative lengthening of telomeres (ALT) in tumors demonstrates the significance of ALT in sarcomas and astrocytomas, *Clin. Cancer Res.*, 2005, **11**, 217.
18. E. H. Blackburn, Switching and signaling at the telomere, *Cell*, 2001 **106**, 661.

19. J. D. Griffith, L. Comeau, S. Rosenfield, R. M. Stansel, A. Bianchi, H. Moss and T. de Lange, Mammalian telomeres end in a large duplex loop, *Cell*, 1999, **97**, 503.
20. T. de Lange, T-loops and the origin of telomeres, *Nat. Rev. Mol. Cell Biol.*, 2004, **5**, 323.
21. Z. Zhong, L. Shiue, S. Kaplan and T. de Lange, A mammalian factor that binds telomeric TTAGGG repeats *in vitro*, *Mol. Cell. Biol.*, 1992, **12**, 4834.
22. L. Chong, B. van Steensel, D. Broccoli, H. Erdjument-Bromage, J. Hanish, P. Tempst and T. de Lange, A human telomeric protein, *Science*, 1995, **270**, 1663.
23. T. Bilaud, C. Brun, K. Ancelin, C. E. Koering, T. Laroche and E. Gilson, Telomeric localization of TRF2, a novel human telobox protein, *Nat. Genet.*, 1997, **17**, 236.
24. D. Broccoli, A. Smogorzewska, L. Chong and T. de Lange, Human telomeres contain two distinct Myb-related proteins, TRF1 and TRF2, *Nat. Genet.*, 1997, **17**, 231.
25. B. Li, S. Oestreich and T. de Lange, Identification of human Rap1: implications for telomere evolution, *Cell*, 2000, **101**, 471.
26. N. S. Bae and P. Baumann, A RAP1/TRF2 complex inhibits nonhomologous end-joining at human telomeric DNA ends, *Mol. Cell*, 2007, **26**, 323.
27. P. Baumann and T. R. Cech, Pot1, the putative telomere end-binding protein in fission yeast and humans, *Science*, 2001, **292**, 1171.
28. B. R. Houghtaling, L. Cuttonaro, W. Chang and S. Smith, A dynamic molecular link between the telomere length regulator TRF1 and the chromosome end protector TRF2, *Curr. Biol.*, 2004, **14**, 1621.
29. D. Liu, A. Safari, M. S. O'Connor, D. W. Chan, A. Laegeler, J. Qin and Z. Songyang, PTOP interacts with POT1 and regulates its localization to telomeres, *Nat. Cell Biol.*, 2004, **6**, 673.
30. J. Z. Ye, D. Hockemeyer, A. N. Krutchinsky, D. Loayza, S. M. Hooper, B. T. Chait and T. de Lange, POT1-interacting protein PIP1: a telomere length regulator that recruits POT1 to the TIN2/TRF1 complex, *Genes Dev.*, 2004, **18**, 1649.
31. F. Wang, E. R. Podell, A. J. Zaug, Y. Yang, P. Baciu, T. R. Cech and M. Lei, The POT1-TPP1 telomere complex is a telomerase processivity factor, *Nature*, 2007, **445**, 506.
32. H. Xin, D. Liu, M. Wan, A. Safari, H. Kim, W. Sun, M. S. O'Connor and Z. Songyang, TPP1 is a homologue of ciliate TEBP-β and interacts with POT1 to recruit telomerase, *Nature*, 2007, **445**, 559.
33. D. Liu, M. S. O'Connor, J. Qin and Z. Songyang, Telosome, a mammalian telomere associated complex formed by multiple telomeric proteins, *J. Biol. Chem.*, 2004, **279**, 51338.
34. J. Z. Ye, J. R. Donigian, M. Van Overbeek, D. Loayza, Y. Luo, A. N. Krutchinsky, B. T. Chait and T. de Lange, TIN2 binds TRF1 and TRF2

simultaneously and stabilizes the TRF2 complex on telomeres, *J. Biol. Chem.*, 2004, **279**, 47264.
35. T. de Lange, Shelterin: the protein complex that shapes and safeguards human telomeres, *Genes Dev.*, 2005, **19**, 2100.
36. P. Baumann, Are mouse telomeres going to pot?, *Cell*, 2006, **126**, 33.
37. C. M. Price, Stirring the POT1: surprises in telomere protection, *Nat. Struct. Mol. Biol.*, 2006, **13**, 673.
38. R. E. Verdun and J. Karlseder, Replication and protection of telomeres, *Nature*, 2007, **447**, 924.
39. B. van Steensel, A. Smogorzewska and T. de Lange, TRF2 protects human telomeres from end-to-end fusions, *Cell*, 1998, **92**, 401.
40. L. Chin, S. E. Artandi, Q. Shen, A. Tam, S. L. Lee, G. J. Gottlieb, C. W. Greider and R. A. DePinho, p53 deficiency rescues the adverse effects of telomere loss and cooperates with telomere dysfunction to accelerate carcinogenesis, *Cell*, 1999, **97**, 527.
41. J. Karlseder, D. Broccoli, Y. Dai, S. Hardy and T. de Lange, p53- and ATM-dependent apoptosis induced by telomeres lacking TRF2, *Science*, 1999, **283**, 1321.
42. A. Smogorzewska and T. de Lange, Different telomere damage signaling pathways in human and mouse cells, *EMBO J.*, 2002, **21**, 4338.
43. H. Takai, A. Smogorzewska and T. de Lange, DNA damage foci at dysfunctional telomeres, *Curr. Biol.*, 2003, **13**, 1549.
44. F. d'Adda di Fagagna, P. M. Reaper, L. Clay-Farrace, H. Fiegler, P. Carr, T. Von Zglinicki, G. Saretzki, N. P. Carter and S. P. Jackson, A DNA damage checkpoint response in telomere-initiated senescence, *Nature*, 2003, **426**, 194.
45. G. B. Celli and T. de Lange, DNA processing is not required for ATM-mediated telomere damage response after TRF2 deletion, *Nat. Cell Biol.*, 2005, **7**, 712.
46. T. von Zglinicki, G. Saretzki, J. Ladhoff, F. Fagagna and S. P. Jackson, Human cell senescence as a DNA damage response, *Mech. Ageing Dev.*, 2005, **126**, 111.
47. R. M. Stansel, T. de Lange and J. D. Griffith, T-loop assembly *in vitro* involves binding of TRF2 near the 3′ telomeric overhang, *EMBO J.*, 2001, **20**, 5532.
48. N. Fouche, A. J. Cesare, S. Willcox, S. Ozgur, S. A. Compton and J. D. Griffith, The basic domain of TRF2 directs binding to DNA junctions irrespective of the presence of TTAGGG repeats, *J. Biol. Chem.*, 2006, **281**, 37486.
49. S. Amiard, M. Doudeau, S. Pinte, A. Poulet, C. Lenain, C. Faivre-Moskalenko, D. Angelov, N. Hug, A. Vindigni, P. Bouvet, J. Paoletti, E. Gilson and M. J. Giraud-Panis, A topological mechanism for TRF2-enhanced strand invasion, *Nat. Struct. Mol. Biol.*, 2007, **14**, 147.
50. A. Bianchi, S. Smith, L. Chong, P. Elias and T. de Lange, TRF1 is a dimer and bends telomeric DNA, *EMBO J.*, 1997, **16**, 1785.

51. J. Griffith, A. Bianchi and T. de Lange, TRF1 promotes parallel pairing of telomeric tracts *in vitro*, *J. Mol. Biol.*, 1998, **278**, 79.
52. S. H. Kim, S. Han, Y. H. You, D. J. Chen and J. Campisi, The human telomere-associated protein TIN2 stimulates interactions between telomeric DNA tracts *in vitro*, *EMBO Rep.*, 2003, **4**, 685.
53. Q. Yang, Y. L. Zheng and C. C. Harris, POT1 and TRF2 cooperate to maintain telomeric integrity, *Mol. Cell. Biol.*, 2005, **25**, 1070.
54. X. D. Zhu, B. Kuster, M. Mann, J. H. Petrini and T. de Lange, Cell-cycle-regulated association of RAD50/MRE11/NBS1 with TRF2 and human telomeres, *Nat. Genet.*, 2000, **25**, 347.
55. R. E. Verdun and J. Karlseder, The DNA damage machinery and homologous recombination pathway act consecutively to protect human telomeres, *Cell*, 2006, **127**, 709.
56. D. Hockemeyer, A. J. Sfeir, J. W. Shay, W. E. Wright and T. de Lange, POT1 protects telomeres from a transient DNA damage response and determines how human chromosomes end, *EMBO J.*, 2005, **24**, 2667.
57. A. Smogorzewska, J. Karlseder, H. Holtgreve-Grez, A. Jauch and T. de Lange, DNA ligase IV-dependent NHEJ of deprotected mammalian telomeres in G1 and G2, *Curr. Biol.*, 2002, **12**, 1635.
58. X. D. Zhu, L. Niedernhofer, B. Kuster, M. Mann, J. H. Hoeijmakers and T. de Lange, ERCC1/XPF removes the 3′ overhang from uncapped telomeres and represses formation of telomeric DNA-containing double minute chromosomes, *Mol. Cell*, 2003, **12**, 1489.
59. D. Hockemeyer, J. P. Daniels, H. Takai and T. de Lange, Recent expansion of the telomeric complex in rodents: two distinct POT1 proteins protect mouse telomeres, *Cell*, 2006, **126**, 63.
60. R. C. Wang, A. Smogorzewska and T. de Lange, Homologous recombination generates T-loop-sized deletions at human telomeres, *Cell*, 2004, **119**, 355.
61. L. Wu, A. S. Multani, H. He, W. Cosme-Blanco, Y. Deng, J. M. Deng, O. Bachilo, S. Pathak, H. Tahara, S. M. Bailey, Y. Deng, R. R. Behringer and S. Chang, Pot1 deficiency initiates DNA damage checkpoint activation and aberrant homologous recombination at telomeres, *Cell*, 2006, **126**, 49.
62. S. M. Bailey, M. A. Brenneman and E. H. Goodwin, Frequent recombination in telomeric DNA may extend the proliferative life of telomerase-negative cells, *Nucleic Acids Res.*, 2004, **32**, 3743.
63. J. A. Londono-Vallejo, H. Der-Sarkissian, L. Cazes, S. Bacchetti and R. R. Reddel, Alternative lengthening of telomeres is characterized by high rates of telomeric exchange, *Cancer Res.*, 2004, **64**, 2324.
64. H. He, A. S. Multani, W. Cosme-Blanco, H. Tahara, J. Ma, S. Pathak, Y. Deng and S. Chang, POT1b protects telomeres from end-to-end chromosomal fusions and aberrant homologous recombination, *EMBO J.*, 2006, **25**, 5180.

65. G. B. Celli, E. L. Denchi and T. de Lange, Ku70 stimulates fusion of dysfunctional telomeres yet protects chromosome ends from homologous recombination, *Nat. Cell Biol.*, 2006, **8**, 885.
66. R. E. Verdun, L. Crabbe, C. Haggblom and J. Karlseder, Functional human telomeres are recognized as DNA damage in G2 of the cell cycle, *Mol. Cell*, 2005, **20**, 551.
67. J. Karlseder, K. Hoke, O. K. Mirzoeva, C. Bakkenist, M. B. Kastan, J. H. Petrini and T. de Lange, The telomeric protein TRF2 binds the ATM kinase and can inhibit the ATM-dependent DNA damage response, *PLoS Biol.*, 2004, **2**, E240.
68. E. L. Denchi and T. de Lange, Protection of telomeres through independent control of ATM and ATR by TRF2 and POT1, *Nature*, 2007, **448**, 1068.
69. X. Guo, Y. Deng, Y. Lin, W. Cosme-Blanco, S. Chan, H. He, G. Yuan, E. J. Brown and S. Chang, Dysfunctional telomeres activate an ATM-ATR-dependent DNA damage response to suppress tumorigenesis, *EMBO J.*, 2007, **26**, 4709.
70. S. H. Kim, C. Beausejour, A. R. Davalos, P. Kaminker, S. J. Heo and J. Campisi, TIN2 mediates functions of TRF2 at human telomeres, *J. Biol. Chem.*, 2004, **279**, 43799.
71. D. Hockemeyer, W. Palm, T. Else, J. P. Daniels, K. K. Takai, J. Z. Ye, C. E. Keegan, T. de Lange and G. D. Hammer, Telomere protection by mammalian Pot1 requires interaction with Tpp1, *Nat. Struct. Mol. Biol.*, 2007, **14**, 754.
72. A. Smogorzewska, B. van Steensel, A. Bianchi, S. Oelmann, M. R. Schaefer, G. Schnapp and T. de Lange, Control of human telomere length by TRF1 and TRF2, *Mol. Cell. Biol.*, 2000, **20**, 1659.
73. B. van Steensel and T. de Lange, Control of telomere length by the human telomeric protein TRF1, *Nature*, 1997, **385**, 740.
74. D. Loayza and T. de Lange, POT1 as a terminal transducer of TRF1 telomere length control, *Nature*, 2003, **424**, 1013.
75. L. M. Colgin, K. Baran, P. Baumann, T. R. Cech and R. R. Reddel, Human POT1 facilitates telomere elongation by telomerase, *Curr. Biol.*, 2003, **13**, 942.
76. M. Lei, A. J. Zaug, E. R. Podell and T. R. Cech, Switching human telomerase on and off with hPOT1 protein *in vitro*, *J. Biol. Chem.*, 2005, **280**, 20449.
77. A. J. Zaug, E. R. Podell and T. R. Cech, Human POT1 disrupts telomeric G-quadruplexes allowing telomerase extension *in vitro*, *Proc. Natl. Acad. Sci. U.S.A.*, 2005, **102**, 10864.
78. J. T. Bunch, N. S. Bae, J. Leonardi and P. Baumann, Distinct requirements for Pot1 in limiting telomere length and maintaining chromosome stability, *Mol. Cell. Biol.*, 2005, **25**, 5567.
79. M. Gellert, M. N. Lipsett and D. R. Davies, Helix formation by guanylic acid, *Proc. Natl. Acad. Sci. U.S.A.*, 1962, **48**, 2013.

80. J. R. Williamson, M. K. Raghuraman and T. R. Cech, Monovalent cation-induced structure of telomeric DNA: the G-quartet model, *Cell*, 1989, **59**, 871.
81. J. L. Huppert and S. Balasubramanian, Prevalence of quadruplexes in the human genome, *Nucleic Acids Res.*, 2005, **33**, 2908.
82. A. K. Todd, M. Johnston and S. Neidle, Highly prevalent putative quadruplex sequence motifs in human DNA, *Nucleic Acids Res.*, 2005 **33**, 2901.
83. L. Oganesian and T. M. Bryan, Physiological relevance of telomeric G-quadruplex formation: a potential drug target, *Bioessays*, 2007 **29**, 155.
84. C. Schaffitzel, I. Berger, J. Postberg, J. Hanes, H. J. Lipps and A. Pluckthun, *In vitro* generated antibodies specific for telomeric guanine-quadruplex DNA react with *Stylonychia lemnae* macronuclei, *Proc. Natl. Acad. Sci. U.S.A.*, 2001, **98**, 8572.
85. K. Paeschke, T. Simonsson, J. Postberg, D. Rhodes and H. J. Lipps, Telomere end-binding proteins control the formation of G-quadruplex DNA structures *in vivo*, *Nat. Struct. Mol. Biol.*, 2005, **12**, 847.
86. Y. Wang and D. J. Patel, Solution structure of the human telomeric repeat d[AG3(T2AG3)3] G-tetraplex, *Structure*, 1993, **1**, 263.
87. G. N. Parkinson, M. P. Lee and S. Neidle, Crystal structure of parallel quadruplexes from human telomeric DNA, *Nature*, 2002, **417**, 876.
88. Y. Xu, Y. Noguchi and H. Sugiyama, The new models of the human telomere d[AGGG(TTAGGG)3] in K(+) solution, *Bioorg. Med. Chem.*, 2006, **14**, 5584.
89. A. Ambrus, D. Chen, J. Dai, T. Bialis, R. A. Jones and D. Yang, Human telomeric sequence forms a hybrid-type intramolecular G-quadruplex structure with mixed parallel/antiparallel strands in potassium solution, *Nucleic Acids Res.*, 2006, **34**, 2723.
90. K. N. Luu, A. T. Phan, V. Kuryavyi, L. Lacroix and D. J. Patel, Structure of the human telomere in K(+) solution: an intramolecular (3+1) g-quadruplex scaffold, *J. Am. Chem. Soc.*, 2006, **128**, 9963.
91. A. T. Phan, K. N. Luu and D. J. Patel, Different loop arrangements of intramolecular human telomeric (3+1) G-quadruplexes in K+ solution, *Nucleic Acids Res.*, 2006, **34**, 5715.
92. L. Ying, J. J. Green, H. Li, D. Klenerman and S. Balasubramanian, Studies on the structure and dynamics of the human telomeric G quadruplex by single-molecule fluorescence resonance energy transfer, *Proc. Natl. Acad. Sci. U.S.A.*, 2003, **100**, 14629.
93. J. Y. Lee, B. Okumus, D. S. Kim and T. Ha, Extreme conformational diversity in human telomeric DNA, *Proc. Natl. Acad. Sci. U.S.A.*, 2005, **102**, 18938.
94. A. T. Phan, V. Kuryavyi, K. N. Luu and D. J. Patel, Structure of two intramolecular G-quadruplexes formed by natural human telomere sequences in K+ solution, *Nucleic Acids Res.*, 2007, **35**, 6517.

95. J. Dai, M. Carver, C. Punchihewa, R. A. Jones and D. Yang, Structure of the Hybrid-2 type intramolecular human telomeric G-quadruplex in K+ solution: insights into structure polymorphism of the human telomeric sequence, *Nucleic Acids Res.*, 2007, **35**, 4927.
96. A. J. Sfeir, W. Chai, J. W. Shay and W. E. Wright, Telomere-end processing the terminal nucleotides of human chromosomes, *Mol. Cell*, 2005, **18**, 131.
97. S. N. Orlov and P. Hamet, Intracellular monovalent ions as second messengers, *J. Membr. Biol.*, 2006, **210**, 161.
98. J. Li, J. J. Correia, L. Wang, J. O. Trent and J. B. Chaires, Not so crystal clear: the structure of the human telomere G-quadruplex in solution differs from that present in a crystal, *Nucleic Acids Res.*, 2005, **33**, 4649.
99. Y. Xue, Z. Y. Kan, Q. Wang, Y. Yao, J. Liu, Y. H. Hao and Z. Tan, Human telomeric DNA forms parallel-stranded intramolecular G-quadruplex in K(+) solution under molecular crowding condition, *J. Am. Chem. Soc.*, 2007, **129**, 11185.
100. W. E. Wright, V. M. Tesmer, K. E. Huffman, S. D. Levene and J. W. Shay, Normal human chromosomes have long G-rich telomeric overhangs at one end, *Genes Dev.*, 1997, **11**, 2801.
101. V. L. Makarov, Y. Hirose and J. P. Langmore, Long G tails at both ends of human chromosomes suggest a C strand degradation mechanism for telomere shortening, *Cell*, 1997, **88**, 657.
102. H. Q. Yu, D. Miyoshi and N. Sugimoto, Properties of long human telomeric DNAs under cell-mimicking conditions, *Nucleic Acids Symp. Ser.(Oxf.)*, 2006, **50**, 207.
103. Y. Wang and D. J. Patel, Guanine residues in d(T2AG3) and d(T2G4) form parallel-stranded potassium cation stabilized G-quadruplexes with anti glycosidic torsion angles in solution, *Biochemistry*, 1992 **31**, 8112.
104. A. T. Phan and D. J. Patel, Two-repeat human telomeric d(TAGGGT-TAGGGT) sequence forms interconverting parallel and antiparallel G-quadruplexes in solution: distinct topologies, thermodynamic properties and folding/unfolding kinetics, *J. Am. Chem. Soc.*, 2003, **125**, 15021.
105. N. Zhang, A. T. Phan and D. J. Patel, (3+1) Assembly of three human telomeric repeats into an asymmetric dimeric G-quadruplex, *J. Am. Chem. Soc.*, 2005, **127**, 17277.
106. H. J. Lipps, *In vitro* aggregation of the gene-sized DNA molecules of the ciliate *Stylonychia mytilus*, *Proc. Natl. Acad. Sci. U.S.A.*, 1980, **77**, 4104.
107. D. Sen and W. Gilbert, Formation of parallel four-stranded complexes by guanine-rich motifs in DNA and its implications for meiosis, *Nature*, 1988, **334**, 364.
108. A. M. Zahler, J. R. Williamson, T. R. Cech and D. M. Prescott, Inhibition of telomerase by G-quartet DNA structures, *Nature*, 1991, **350**, 718.
109. L. Oganesian, I. K. Moon, T. M. Bryan and M. B. Jarstfer, Extension of G-quadruplex DNA by ciliate telomerase, *EMBO J.*, 2006, **25**, 1148.

110. Y. C. Tsai, H. Qi and L. F. Liu, Protection of DNA ends by telomeric 3′ G-tail sequences, *J. Biol. Chem.*, 2007, **282**, 18786.
111. T. R. Salas, I. Petruseva, O. Lavrik, A. Bourdoncle, J. L. Mergny, A. Favre and C. Saintome, Human replication protein A unfolds telomeric G-quadruplexes, *Nucleic Acids Res.*, 2006, **34**, 4857.
112. H. Sun, J. K. Karow, I. D. Hickson and N. Maizels, The Bloom's syndrome helicase unwinds G4 DNA, *J. Biol. Chem.*, 1998, **273**, 27587.
113. P. Mohaghegh, J. K. Karow, R. M. Brosh Jr., V. A. Bohr and I. D. Hickson, The Bloom's and Werner's syndrome proteins are DNA structure-specific helicases, *Nucleic Acids Res.*, 2001, **29**, 2843.
114. K. Perrem, L. M. Colgin, A. A. Neumann, T. R. Yeager and R. R. Reddel, Coexistence of alternative lengthening of telomeres and telomerase in hTERT-transfected GM847 cells, *Mol. Cell. Biol.*, 2001, **21**, 3862.
115. J. P. Murnane, L. Sabatier, B. A. Marder and W. F. Morgan, Telomere dynamics in an immortal human cell line, *EMBO J.*, 1994, **13**, 4953.
116. H. Ogino, K. Nakabayashi, M. Suzuki, E. I. Takahashi, M. Fujii, T. Suzuki and D. Ayusawa, Release of telomeric DNA from chromosomes in immortal human cells lacking telomerase activity, *Biochem. Biophys. Res. Commun.*, 1998, **248**, 223.
117. Y. Tokutake, T. Matsumoto, T. Watanabe, S. Maeda, H. Tahara, S. Sakamoto, H. Niida, M. Sugimoto, T. Ide and Y. Furuichi, Extra-chromosome telomere repeat DNA in telomerase-negative immortalized cell lines, *Biochem. Biophys. Res. Commun.*, 1998, **247**, 765.
118. A. J. Cesare and J. D. Griffith, Telomeric DNA in ALT cells is characterized by free telomeric circles and heterogeneous T-loops, *Mol. Cell. Biol.*, 2004, **24**, 9948.
119. T. R. Yeager, A. A. Neumann, A. Englezou, L. I. Huschtscha, J. R. Noble and R. R. Reddel, Telomerase-negative immortalized human cells contain a novel type of promyelocytic leukemia (PML) body, *Cancer Res.*, 1999, **59**, 4175.
120. P. Salomoni and P. P. Pandolfi, The role of PML in tumor suppression, *Cell*, 2002, **108**, 165.
121. G. Wu, X. Jiang, W. H. Lee and P. L. Chen, Assembly of functional ALT-associated promyelocytic leukemia bodies requires Nijmegen breakage syndrome 1, *Cancer Res.*, 2003, **63**, 2589.
122. D. J. Stavropoulos, P. S. Bradshaw, X. Li, I. Pasic, K. Truong, M. Ikura, M. Ungrin and M. S. Meyn, The Bloom syndrome helicase BLM interacts with TRF2 in ALT cells and promotes telomeric DNA synthesis, *Hum. Mol. Genet.*, 2002, **11**, 3135.
123. F. B. Johnson, R. A. Marciniak, M. McVey, S. A. Stewart, W. C. Hahn and L. Guarente, The *Saccharomyces cerevisiae* WRN homolog Sgs1p participates in telomere maintenance in cells lacking telomerase, *EMBO J.*, 2001, **20**, 905.
124. M. Tarsounas, P. Munoz, A. Claas, P. G. Smiraldo, D. L. Pittman, M. A. Blasco and S. C. West, Telomere maintenance requires the RAD51D recombination/repair protein, *Cell*, 2004, **117**, 337.

125. P. R. Potts and H. Yu, The SMC5/6 complex maintains telomere length in ALT cancer cells through SUMOylation of telomere-binding proteins, *Nat. Struct. Mol. Biol.*, 2007, **14**, 581.
126. A. Nabetani, O. Yokoyama and F. Ishikawa, Localization of hRad9, hHus1, hRad1 and hRad17 and caffeine-sensitive DNA replication at ALT (alternative lengthening of telomeres)-associated promyelocytic leukemia body, *J. Biol. Chem.*, 2004, **279**, 25849.
127. C. L. Fasching, A. A. Neumann, A. Muntoni, T. R. Yeager and R. R. Reddel, DNA damage induces alternative lengthening of telomeres (ALT) associated promyelocytic leukemia bodies that preferentially associate with linear telomeric DNA, *Cancer Res.*, 2007, **67**, 7072.
128. C. Molenaar, K. Wiesmeijer, N. P. Verwoerd, S. Khazen, R. Eils, H. J. Tanke and R. W. Dirks, Visualizing telomere dynamics in living mammalian cells using PNA probes, *EMBO J.*, 2003, **22**, 6631.
129. G. Wu, W. H. Lee and P. L. Chen, NBS1 and TRF1 colocalize at promyelocytic leukemia bodies during late S/G2 phases in immortalized telomerase-negative cells. Implication of NBS1 in alternative lengthening of telomeres, *J. Biol. Chem.*, 2000, **275**, 30618.
130. J. V. Grobelny, A. K. Godwin and D. Broccoli, ALT-associated PML bodies are present in viable cells and are enriched in cells in the G_2/M phase of the cell cycle, *J. Cell Sci.*, 2000, **113**, 4577.
131. S. C. West, Molecular views of recombination proteins and their control, *Nat. Rev. Mol. Cell Biol.*, 2003, **4**, 435.
132. W. Q. Jiang, Z. H. Zhong, J. D. Henson and R. R. Reddel, Identification of candidate alternative lengthening of telomeres genes by methionine restriction and RNA interference, *Oncogene*, 2007, **26**, 4635.
133. S. C. Teng and V. A. Zakian, Telomere-telomere recombination is an efficient bypass pathway for telomere maintenance in *Saccharomyces cerevisiae*, *Mol. Cell. Biol.*, 1999, **19**, 8083.
134. M. J. McEachern and E. H. Blackburn, Cap-prevented recombination between terminal telomeric repeat arrays (telomere CPR) maintains telomeres in *Kluyveromyces lactis* lacking telomerase, *Genes Dev.*, 1996, **10**, 1822.
135. T. M. Nakamura, J. P. Cooper and T. R. Cech, Two modes of survival of fission yeast without telomerase, *Science*, 1998, **282**, 493.
136. S. Le, J. K. Moore, J. E. Haber and C. W. Greider, *RAD50* and *RAD51* define two pathways that collaborate to maintain telomeres in the absence of telomerase, *Genetics*, 1999, **152**, 143.
137. S. C. Teng, J. Chang, B. McCowan and V. A. Zakian, Telomerase-independent lengthening of yeast telomeres occurs by an abrupt Rad50p-dependent, Rif-inhibited recombinational process, *Mol. Cell*, 2000, **6**, 947.
138. Q. Chen, A. Ijpma and C. W. Greider, Two survivor pathways that allow growth in the absence of telomerase are generated by distinct telomere recombination events, *Mol. Cell. Biol.*, 2001, **21**, 1819.
139. M. A. Dunham, A. A. Neumann, C. L. Fasching and R. R. Reddel, Telomere maintenance by recombination in human cells, *Nat. Genet.*, 2000, **26**, 447.

140. W. Q. Jiang, Z. H. Zhong, J. D. Henson, A. A. Neumann, A. C. Chang and R. R. Reddel, Suppression of alternative lengthening of telomeres by Sp100-mediated sequestration of MRE11/RAD50/NBS1 complex., *Mol. Cell. Biol.*, 2005, **25**, 2708.
141. Z. H. Zhong, W. Q. Jiang, A. J. Cesare, A. A. Neumann, R. Wadhwa and R. R. Reddel, Disruption of telomere maintenance by depletion of the MRE11/RAD50/NBS1 complex in cells that use alternative lengthening of telomeres, *J. Biol. Chem.*, 2007, **282**, 29314.
142. S. A. Compton, J. H. Choi, A. J. Cesare, S. Ozgur and J. D. Griffith, Xrcc3 and Nbs1 are required for the production of extrachromosomal telomeric circles in human alternative lengthening of telomere cells, *Cancer Res.*, 2007, **67**, 1513.
143. P. R. Potts, M. H. Porteus and H. Yu, Human SMC5/6 complex promotes sister chromatid homologous recombination by recruiting the SMC1/3 cohesin complex to double-strand breaks, *EMBO J.*, 2006, **25**, 3377.
144. P. R. Laud, A. S. Multani, S. M. Bailey, L. Wu, J. Ma, C. Kingsley, M. Lebel, S. Pathak, R. A. DePinho and S. Chang, Elevated telomere-telomere recombination in WRN-deficient, telomere dysfunctional cells promotes escape from senescence and engagement of the ALT pathway, *Genes Dev.*, 2005, **19**, 2560.
145. F. G. Harmon, R. J. DiGate and S. C. Kowalczykowski, RecQ helicase and topoisomerase III comprise a novel DNA strand passage function: a conserved mechanism for control of DNA recombination, *Mol. Cell*, 1999, **3**, 611.
146. H. J. Tsai, W. H. Huang, T. K. Li, Y. L. Tsai, K. J. Wu, S. F. Tseng and S. C. Teng, Involvement of topoisomerase III in telomere-telomere recombination, *J. Biol. Chem.*, 2006, **281**, 13717.
147. A. Grudic, A. Jul-Larsen, S. J. Haring, M. S. Wold, P. E. Lonning, R. Bjerkvig and S. O. Boe, Replication protein A prevents accumulation of single-stranded telomeric DNA in cells that use alternative lengthening of telomeres, *Nucleic Acids Res.*, 2007, **35**, 7267.
148. H. Sun, R. J. Bennett and N. Maizels, The *Saccharomyces cerevisiae* Sgs1 helicase efficiently unwinds G-G paired DNAs, *Nucleic Acids Res.*, 1999, **27**, 1978.
149. H. Cohen and D. A. Sinclair, Recombination-mediated lengthening of terminal telomeric repeats requires the Sgs1 DNA helicase, *Proc. Natl. Acad. Sci. U.S.A.*, 2001, **98**, 3174.
150. O. E. Bechter, Y. Zou, W. Walker, W. E. Wright and J. W. Shay, Telomeric recombination in mismatch repair deficient human colon cancer cells after telomerase inhibition, *Cancer Res.*, 2004, **64**, 3444.
151. D. M. Wilson III and L. H. Thompson, Molecular mechanisms of sister-chromatid exchange, *Mutat. Res.*, 2007, **616**, 11.
152. L. Tomaska, J. Nosek, A. M. Makhov, A. Pastorakova and J. D. Griffith, Extragenomic double-stranded DNA circles in yeast with linear mitochondrial genomes: potential involvement in telomere maintenance, *Nucleic Acids Res.*, 2000, **28**, 4479.

153. C. Y. Lin, H. H. Chang, K. J. Wu, S. F. Tseng, C. C. Lin, C. P. Lin and S. C. Teng, Extrachromosomal telomeric circles contribute to Rad52-, Rad50- and polymerase delta-mediated telomere-telomere recombination in *Saccharomyces cerevisiae*, *Eukaryot. Cell*, 2005, **4**, 327.
154. M. Larrivee and R. J. Wellinger, Telomerase- and capping-independent yeast survivors with alternate telomere states, *Nat. Cell Biol.*, 2006, **8**, 741.
155. C. Groff-Vindman, A. J. Cesare, S. Natarajan, J. D. Griffith and M. J. McEachern, Recombination at long mutant telomeres produces tiny single- and double-stranded telomeric circles, *Mol. Cell. Biol.*, 2005, **25**, 4406.
156. A. J. Cesare, C. Groff-Vindman, S. A. Compton, M. J. McEachern and J. D. Griffith, T-loops and homologous recombination dependent t-circles in a *Kluyveromyces lactis* telomere mutant strain, *Mol. Cell. Biol.*, 2008, **128**, 546.
157. I. Jaco, P. Munoz, F. Goytisolo, J. Wesoly, S. Bailey, G. Taccioli and M. A. Blasco, Role of mammalian Rad54 in telomere length maintenance, *Mol. Cell. Biol.*, 2003, **23**, 5572.
158. K. Perrem, T. M. Bryan, A. Englezou, T. Hackl, E. L. Moy and R. R. Reddel, Repression of an alternative mechanism for lengthening of telomeres in somatic cell hybrids, *Oncogene*, 1999, **18**, 3383.
159. J. D. Henson, A. A. Neumann, T. R. Yeager and R. R. Reddel, Alternative lengthening of telomeres in mammalian cells, *Oncogene*, 2002, **21**, 598.
160. A. Muntoni and R. R. Reddel, The first molecular details of ALT in human tumor cells, *Hum. Mol. Genet.*, 2005, **14**(Suppl 2), R191.
161. A. A. Neumann and R. R. Reddel, Telomere maintenance and cancer -- look, no telomerase, *Nat. Rev. Cancer*, 2002, **2**, 879.
162. S. Natarajan and M. J. McEachern, Recombinational telomere elongation promoted by DNA circles, *Mol. Cell. Biol.*, 2002, **22**, 4512.
163. Z. Topcu, K. Nickles, C. Davis and M. J. McEachern, Abrupt disruption of capping and a single source for recombinationally elongated telomeres in *Kluyveromyces lactis*, *Proc. Natl. Acad. Sci. U.S.A.*, 2005, **102**, 3348.
164. K. B. Blagoev and E. H. Goodwin, Telomere exchange and asymmetric segregation of chromosomes can account for the unlimited proliferative potential of ALT cell populations, *DNA Repair (Amst.)*, 2008, **7**, 199.
165. C. W. Greider and E. H. Blackburn, A telomeric sequence in the RNA of *Tetrahymena* telomerase required for telomere repeat synthesis, *Nature*, 1989, **337**, 331.
166. J. Lingner, T. R. Hughes, A. Shevchenko, M. Mann, V. Lundblad and T. R. Cech, Reverse transcriptase motifs in the catalytic subunit of telomerase, *Science*, 1997, **276**, 561.
167. J. L. Chen and C. W. Greider, An emerging consensus for telomerase RNA structure, *Proc. Natl. Acad. Sci. U.S.A.*, 2004, **101**, 14683.
168. J. Lin, H. Ly, A. Hussain, M. Abraham, S. Pearl, Y. Tzfati, T. G. Parslow and E. H. Blackburn, A universal telomerase RNA core structure includes

structured motifs required for binding the telomerase reverse transcriptase protein, *Proc. Natl. Acad. Sci. U.S.A.*, 2004, **101**, 14713.

169. C. Autexier and C. W. Greider, Mutational analysis of the *Tetrahymena* telomerase RNA: identification of residues affecting telomerase activity *in vitro*, *Nucleic Acids Res.*, 1998, **26**, 787.
170. J. M. Sperger and T. R. Cech, A stem-loop of *Tetrahymena* telomerase RNA distant from the template potentiates RNA folding and telomerase activity, *Biochemistry*, 2001, **40**, 7005.
171. J. R. Mitchell and K. Collins, Human telomerase activation requires two independent interactions between telomerase RNA and telomerase reverse transcriptase, *Mol. Cell*, 2000, **6**, 361.
172. J. L. Chen, K. K. Opperman and C. W. Greider, A critical stem-loop structure in the CR4-CR5 domain of mammalian telomerase RNA, *Nucleic Acids Res.*, 2002, **30**, 592.
173. Y. Brown, M. Abraham, S. Pearl, M. M. Kabaha, E. Elboher and Y. Tzfati, A critical three-way junction is conserved in budding yeast and vertebrate telomerase RNAs, *Nucleic Acids Res.*, 2007, **35**, 6280.
174. V. M. Tesmer, L. P. Ford, S. E. Holt, B. C. Frank, X. Yi, D. L. Aisner, M. Ouellette, J. W. Shay and W. E. Wright, Two inactive fragments of the integral RNA cooperate to assemble active telomerase with the human protein catalytic subunit (hTERT) *in vitro*, *Mol. Cell. Biol.*, 1999, **19**, 6207.
175. D. X. Mason, E. Goneska and C. W. Greider, Stem-loop IV of *Tetrahymena* telomerase RNA stimulates processivity in *trans*, *Mol. Cell. Biol.*, 2003, **23**, 5606.
176. Y. Chen, J. Fender, J. D. Legassie, M. B. Jarstfer, T. M. Bryan and G. Varani, Structure of stem-loop IV of *Tetrahymen*a telomerase RNA, *EMBO J.*, 2006, **25**, 3156.
177. T. Leeper, N. Leulliot and G. Varani, The solution structure of an essential stem-loop of human telomerase RNA, *Nucleic Acids Res.*, 2003, **31**, 2614.
178. T. C. Leeper and G. Varani, The structure of an enzyme-activating fragment of human telomerase RNA, *RNA*, 2005, **11**, 394.
179. R. J. Richards, H. Wu, L. Trantirek, C. M. O'Connor, K. Collins and J. Feigon, Structural study of elements of *Tetrahymena* telomerase RNA stem-loop IV domain important for function, *RNA*, 2006, **12**, 1475.
180. A. Bhattacharyya and E. H. Blackburn, Architecture of telomerase RNA, *EMBO J.*, 1994, **13**, 5721.
181. R. Prathapam, K. L. Witkin, C. M. O'Connor and K. Collins, A telomerase holoenzyme protein enhances telomerase RNA assembly with telomerase reverse transcriptase, *Nat. Struct. Mol. Biol.*, 2005, **12**, 252.
182. C. K. Lai, M. C. Miller and K. Collins, Roles for RNA in telomerase nucleotide and repeat addition processivity, *Mol. Cell*, 2003, **11**, 1673.
183. C. M. O'Connor, C. K. Lai and K. Collins, Two purified domains of telomerase reverse transcriptase reconstitute sequence-specific interactions with RNA, *J. Biol. Chem.*, 2005, **280**, 17533.

184. M. Antal, E. Boros, F. Solymosy and T. Kiss, Analysis of the structure of human telomerase RNA *in vivo*, *Nucleic Acids Res.*, 2002, **30**, 912.
185. T. J. Moriarty, D. T. Marie-Egyptienne and C. Autexier, Functional organization of repeat addition processivity and DNA synthesis determinants in the human telomerase multimer, *Mol. Cell. Biol.*, 2004, **24**, 3720.
186. C. T. Ueda and R. W. Roberts, Analysis of a long-range interaction between conserved domains of human telomerase RNA, *RNA*, 2004, **10**, 139.
187. E. ten Dam, A. van Belkum and K. Pleij, A conserved pseudoknot in telomerase RNA, *Nucleic Acids Res.*, 1991, **19**, 6951.
188. D. P. Romero and E. H. Blackburn, A conserved secondary structure for telomerase RNA, *Cell*, 1991, **67**, 343.
189. J. Lingner, L. L. Hendrick and T. R. Cech, Telomerase RNAs of different ciliates have a common secondary structure and a permuted template, *Genes Dev.*, 1994, **8**, 1984.
190. J. D. Licht and K. Collins, Telomerase RNA function in recombinant *Tetrahymena* telomerase, *Genes Dev.*, 1999, **13**, 1116.
191. D. Gilley and E. H. Blackburn, The telomerase RNA pseudoknot is critical for the stable assembly of a catalytically active ribonucleoprotein, *Proc. Natl. Acad. Sci. U.S.A.*, 1999, **96**, 6621.
192. J. L. Chen, M. A. Blasco and C. W. Greider, Secondary structure of vertebrate telomerase RNA, *Cell*, 2000, **100**, 503.
193. L. Martin-Rivera and M. A. Blasco, Identification of functional domains and dominant negative mutations in vertebrate telomerase RNA using an *in vivo* reconstitution system, *J. Biol. Chem.*, 2001, **276**, 5856.
194. H. Ly, E. H. Blackburn and T. G. Parslow, Comprehensive structure-function analysis of the core domain of human telomerase RNA, *Mol. Cell. Biol.*, 2003, **23**, 6849.
195. L. R. Comolli, I. Smirnov, L. Xu, E. H. Blackburn and T. L. James, A molecular switch underlies a human telomerase disease, *Proc. Natl. Acad. Sci. U.S.A.*, 2002, **99**, 16998.
196. C. A. Theimer, L. D. Finger, L. Trantirek and J. Feigon, Mutations linked to dyskeratosis congenita cause changes in the structural equilibrium in telomerase RNA, *Proc. Natl. Acad. Sci. U.S.A.*, 2003 **100**, 449.
197. J. L. Chen and C. W. Greider, Functional analysis of the pseudoknot structure in human telomerase RNA, *Proc. Natl. Acad. Sci. U.S.A.*, 2005, **102**, 8080.
198. C. A. Theimer, C. A. Blois and J. Feigon, Structure of the human telomerase RNA pseudoknot reveals conserved tertiary interactions essential for function, *Mol. Cell*, 2005, **17**, 671.
199. Y. G. Yingling and B. A. Shapiro, The prediction of the wild-type telomerase RNA pseudoknot structure and the pivotal role of the bulge in its formation, *J. Mol. Graph. Model.*, 2006, **25**, 261.

200. Y. Tzfati, Z. Knight, J. Roy and E. H. Blackburn, A novel pseudoknot element is essential for the action of a yeast telomerase, *Genes Dev.*, 2003, **17**, 1779.
201. A. S. Chappell and V. Lundblad, Structural elements required for association of the *Saccharomyces cerevisiae* telomerase RNA with the Est2 reverse transcriptase, *Mol. Cell. Biol.*, 2004, **24**, 7720.
202. A. T. Dandjinou, N. Levesque, S. Larose, J. F. Lucier, S. A. Elela and R. J. Wellinger, A phylogenetically based secondary structure for the yeast telomerase RNA, *Curr. Biol.*, 2004, **14**, 1148.
203. D. C. Zappulla and T. R. Cech, Yeast telomerase RNA: a flexible scaffold for protein subunits, *Proc. Natl. Acad. Sci. U.S.A.*, 2004, **101**, 10024.
204. K. Shefer, Y. Brown, V. Gorkovoy, T. Nussbaum, N. B. Ulyanov and Y. Tzfati, A triple helix within a pseudoknot is a conserved and essential element of telomerase RNA, *Mol. Cell. Biol.*, 2007, **27**, 2130.
205. N. B. Ulyanov, K. Shefer, T. L. James and Y. Tzfati, Pseudoknot structures with conserved base triples in telomerase RNAs of ciliates, *Nucleic Acids Res.*, 2007, **35**, 6150.
206. D. T. Marie-Egyptienne, M. A. Cerone, J. A. Londono-Vallejo and C. Autexier, A human-*Tetrahymena* pseudoknot chimeric telomerase RNA reconstitutes a nonprocessive enzyme *in vitro* that is defective in telomere elongation, *Nucleic Acids Res.*, 2005, **33**, 5446.
207. J. R. Mitchell, J. Cheng and K. Collins, A box H/ACA small nucleolar RNA-like domain at the human telomerase RNA 3′ end, *Mol. Cell. Biol.*, 1999, **19**, 567.
208. D. Fu and K. Collins, Distinct biogenesis pathways for human telomerase RNA and H/ACA small nucleolar RNAs, *Mol. Cell*, 2003, **11**, 1361.
209. B. E. Jady, E. Bertrand and T. Kiss, Human telomerase RNA and box H/ACA scaRNAs share a common Cajal body-specific localization signal, *J. Cell Biol.*, 2004, **164**, 647.
210. C. A. Theimer, B. E. Jady, N. Chim, P. Richard, K. E. Breece, T. Kiss and J. Feigon, Structural and functional characterization of human telomerase RNA processing and cajal body localization signals, *Mol. Cell*, 2007, **27**, 869.
211. G. Cristofari, E. Adolf, P. Reichenbach, K. Sikora, R. M. Terns, M. P. Terns and J. Lingner, Human telomerase RNA accumulation in Cajal bodies facilitates telomerase recruitment to telomeres and telomere elongation, *Mol. Cell*, 2007, **27**, 882.
212. A. Jacobo-Molina, J. Ding, R. G. Nanni, A. D. Clark Jr, X. Lu, C. Tantillo, R. L. Williams, G. Kamer, A. L. Ferris, P. Clark, A. Hizi, S. H. Hughes and E. Arnold, Crystal structure of human immunodeficiency virus type 1 reverse transcriptase complexed with double-stranded DNA at 3.0 A resolution shows bent DNA, *Proc. Natl. Acad. Sci. U.S.A.*, 1993, **90**, 6320.
213. C. M. Counter, M. Meyerson, E. N. Eaton and R. A. Weinberg, The catalytic subunit of yeast telomerase, *Proc. Natl. Acad. Sci. U.S.A.*, 1997, **94**, 9202.

214. L. Harrington, W. Zhou, T. McPhail, R. Oulton, D. S. Yeung, V. Mar, M. B. Bass and M. O. Robinson, Human telomerase contains evolutionarily conserved catalytic and structural subunits, *Genes Dev.*, 1997, **11**, 3109.
215. M. C. Miller, J. K. Liu and K. Collins, Template definition by *Tetrahymena* telomerase reverse transcriptase, *EMBO J.*, 2000, **19**, 4412.
216. D. Bosoy and N. F. Lue, Functional analysis of conserved residues in the putative 'finger' domain of telomerase reverse transcriptase, *J. Biol. Chem.*, 2001, **276**, 46305.
217. T. M. Bryan, K. J. Goodrich and T. R. Cech, A mutant of *Tetrahymena* telomerase reverse transcriptase with increased processivity, *J. Biol. Chem.*, 2000, **275**, 24199.
218. N. F. Lue, Y. C. Lin and I. S. Mian, A conserved telomerase motif within the catalytic domain of telomerase reverse transcriptase is specifically required for repeat addition processivity, *Mol. Cell. Biol.*, 2003, **23**, 8440.
219. H. D. Wyatt, D. A. Lobb and T. L. Beattie, Characterization of physical and functional anchor site interactions in human telomerase, *Mol. Cell. Biol.*, 2007, **27**, 3226.
220. S. N. Finger and T. M. Bryan, Multiple DNA-binding sites in *Tetrahymena* telomerase, *Nucleic Acids Res.*, 2008, **36**, 1260.
221. K. Collins and L. Gandhi, The reverse transcriptase component of the *Tetrahymena* telomerase ribonucleoprotein complex, *Proc. Natl. Acad. Sci. U.S.A.*, 1998, **95**, 8485.
222. T. M. Bryan, K. J. Goodrich and T. R. Cech, Telomerase RNA bound by protein motifs specific to telomerase reverse transcriptase, *Mol. Cell*, 2000, **6**, 493.
223. T. M. Nakamura, G. B. Morin, K. B. Chapman, S. L. Weinrich, W. H. Andrews, J. Lingner, C. B. Harley and T. R. Cech, Telomerase catalytic subunit homologs from fission yeast and human, *Science*, 1997, **277**, 955.
224. K. L. Friedman and T. R. Cech, Essential functions of amino-terminal domains in the yeast telomerase catalytic subunit revealed by selection for viable mutants, *Genes Dev.*, 1999, **13**, 2863.
225. J. Xia, Y. Peng, I. S. Mian and N. F. Lue, Identification of functionally important domains in the N-terminal region of telomerase reverse transcriptase, *Mol. Cell. Biol.*, 2000, **20**, 5196.
226. B. N. Armbruster, S. S. Banik, C. Guo, A. C. Smith and C. M. Counter, N-terminal domains of the human telomerase catalytic subunit required for enzyme activity *in vivo*, *Mol. Cell. Biol.*, 2001, **21**, 7775.
227. T. J. Moriarty, S. Huard, S. Dupuis and C. Autexier, Functional multimerization of human telomerase requires an RNA interaction domain in the N terminus of the catalytic subunit, *Mol. Cell. Biol.*, 2002, **22**, 1253.
228. K. L. Friedman, J. J. Heit, D. M. Long and T. R. Cech, N-terminal domain of yeast telomerase reverse transcriptase: recruitment of Est3p to the telomerase complex, *Mol. Biol. Cell*, 2003, **14**, 1.

229. S. A. Jacobs, E. R. Podell, D. S. Wuttke and T. R. Cech, Soluble domains of telomerase reverse transcriptase identified by high-throughput screening, *Protein Sci.*, 2005, **14**, 2051.
230. E. Romi, N. Baran, M. Gantman, M. Shmoish, B. Min, K. Collins and H. Manor, High-resolution physical and functional mapping of the template adjacent DNA binding site in catalytically active telomerase, *Proc. Natl. Acad. Sci. U.S.A.*, 2007, **104**, 8791.
231. S. A. Jacobs, E. R. Podell and T. R. Cech, Addendum: Crystal structure of the essential N-terminal domain of telomerase reverse transcriptase, *Nat. Struct. Mol. Biol.*, 2007, **14**, 984.
232. T. L. Beattie, W. Zhou, M. O. Robinson and L. Harrington, Functional multimerization of the human telomerase reverse transcriptase, *Mol. Cell. Biol.*, 2001, **21**, 6151.
233. B. N. Armbruster, K. T. Etheridge, D. Broccoli and C. M. Counter, Putative telomere-recruiting domain in the catalytic subunit of human telomerase, *Mol. Cell. Biol.*, 2003, **23**, 3237.
234. H. Ji, C. J. Adkins, B. R. Cartwright and K. L. Friedman, Yeast Est2p affects telomere length by influencing association of Rap1p with telomeric chromatin, *Mol. Cell. Biol.*, 2008, **28**, 2380.
235. C. K. Lai, J. R. Mitchell and K. Collins, RNA binding domain of telomerase reverse transcriptase, *Mol. Cell. Biol.*, 2001, **21**, 990.
236. C. K. Lai, M. C. Miller and K. Collins, Template boundary definition in *Tetrahymena* telomerase, *Genes Dev.*, 2002, **16**, 415.
237. D. Bosoy, Y. Peng, I. S. Mian and N. F. Lue, Conserved N-terminal motifs of telomerase reverse transcriptase required for ribonucleoprotein assembly *in vivo*, *J. Biol. Chem.*, 2003, **278**, 3882.
238. S. S. Banik, C. Guo, A. C. Smith, S. S. Margolis, D. A. Richardson, C. A. Tirado and C. M. Counter, C-terminal regions of the human telomerase catalytic subunit essential for *in vivo* enzyme activity, *Mol. Cell. Biol.*, 2002, **22**, 6234.
239. S. Huard, T. J. Moriarty and C. Autexier, The C terminus of the human telomerase reverse transcriptase is a determinant of enzyme processivity, *Nucleic Acids Res.*, 2003, **31**, 4059.
240. S. Hossain, S. M. Singh and N. F. Lue, Functional analysis of the C-terminal extension of telomerase reverse transcriptase: a putative 'thumb' domain, *J. Biol. Chem.*, 2002, **277**, 36174.
241. T. J. Moriarty, R. J. Ward, M. A. Taboski and C. Autexier, An anchor site-type defect in human telomerase that disrupts telomere length maintenance and cellular immortalization, *Mol. Biol. Cell*, 2005, **16**, 3152.
242. H. Seimiya, H. Sawada, Y. Muramatsu, M. Shimizu, K. Ohko, K. Yamane and T. Tsuruo, Involvement of 14-3-3 proteins in nuclear localization of telomerase, *EMBO J.*, 2000, **19**, 2652.
243. A. J. Zaug and T. R. Cech, Analysis of the structure of *Tetrahymena* nuclear RNAs *in vivo*: telomerase RNA, the self-splicing rRNA intron and U2 snRNA, *RNA*, 1995, **1**, 363.

244. J. Lingner and T. R. Cech, Purification of telomerase from *Euplotes aediculatus*: Requirement of a primer 3′ overhang, *Proc. Natl. Acad. Sci. U.S.A.*, 1996, **93**, 10712.
245. S. Aigner, J. Lingner, K. J. Goodrich, C. A. Grosshans, A. Shevchenko, M. Mann and T. R. Cech, *Euplotes* telomerase contains an La motif protein produced by apparent translational frameshifting, *EMBO J.*, 2000, **19**, 6230.
246. S. Aigner and T. R. Cech, The *Euplotes* telomerase subunit p43 stimulates enzymatic activity and processivity *in vitro*, *RNA*, 2004, **10**, 1108.
247. L. P. Ford, J. W. Shay and W. E. Wright, The La antigen associates with the human telomerase ribonucleoprotein and influences telomere length *in vivo*, *RNA*, 2001, **7**, 1068.
248. K. L. Witkin and K. Collins, Holoenzyme proteins required for the physiological assembly and activity of telomerase, *Genes Dev.*, 2004, **18**, 1107.
249. K. L. Witkin, R. Prathapam and K. Collins, Positive and negative regulation of *Tetrahymena* telomerase holoenzyme, *Mol. Cell. Biol.*, 2007, **27**, 2074.
250. C. M. O'Connor and K. Collins, A novel RNA binding domain in *Tetrahymena* telomerase p65 initiates hierarchical assembly of telomerase holoenzyme, *Mol. Cell. Biol.*, 2006, **26**, 2029.
251. M. D. Stone, M. Mihalusova, C. M. O'Connor, R. Prathapam, K. Collins and X. Zhuang, Stepwise protein-mediated RNA folding directs assembly of telomerase ribonucleoprotein, *Nature*, 2007, **446**, 458.
252. T. S. Lendvay, D. K. Morris, J. Sah, B. Balasubramanian and V. Lundblad, Senescence mutants of *Saccharomyces cerevisiae* with a defect in telomere replication identify three additional EST genes, *Genetics*, 1996, **144**, 1399.
253. J. Lingner, T. R. Cech, T. R. Hughes and V. Lundblad, Three ever shorter telomere (*EST*) genes are dispensable for *in vitro* yeast telomerase activity, *Proc. Natl. Acad. Sci. U.S.A.*, 1997, **94**, 11190.
254. S. K. Evans and V. Lundblad, Est1 and Cdc13 as comediators of telomerase access, *Science*, 1999, **286**, 117.
255. E. Pennock, K. Buckley and V. Lundblad, Cdc13 delivers separate complexes to the telomere for end protection and replication, *Cell*, 2001, **104**, 387.
256. P. Reichenbach, M. Hoss, C. M. Azzalin, M. Nabholz, P. Bucher and J. Lingner, A human homolog of yeast est1 associates with telomerase and uncaps chromosome ends when overexpressed, *Curr. Biol.*, 2003, **13**, 568.
257. B. E. Snow, N. Erdmann, J. Cruickshank, H. Goldman, R. M. Gill, M. O. Robinson and L. Harrington, Functional conservation of the telomerase protein est1p in humans, *Curr. Biol.*, 2003, **13**, 698.
258. S. B. Cohen, M. E. Graham, G. O. Lovrecz, N. Bache, P. J. Robinson and R. R. Reddel, Protein composition of catalytically active human telomerase from immortal cells, *Science*, 2007, **315**, 1850.
259. J. R. Mitchell, E. Wood and K. Collins, A telomerase component is defective in the human disease dyskeratosis congenita, *Nature*, 1999, **402**, 551.

260. D. Fu and K. Collins, Purification of human telomerase complexes identifies factors involved in telomerase biogenesis and telomere length regulation, *Mol. Cell*, 2007, **28**, 773.
261. C. Autexier and N. F. Lue, The structure and function of telomerase reverse transcriptase, *Annu. Rev. Biochem.*, 2006, **75**, 493.
262. S. E. Holt, D. L. Aisner, J. Baur, V. M. Tesmer, M. Dy, M. Ouellette, J. B. Trager, G. B. Morin, D. O. Toft, J. W. Shay, W. E. Wright and M. A. White, Functional requirement of p23 and Hsp90 in telomerase complexes, *Genes Dev.*, 1999, **13**, 817.
263. H. L. Forsythe, J. L. Jarvis, J. W. Turner, L. W. Elmore and S. E. Holt, Stable association of hsp90 and p23, but not hsp70, with active human telomerase, *J. Biol. Chem.*, 2001, **276**, 15571.
264. J. B. Boule, L. R. Vega and V. A. Zakian, The yeast Pif1p helicase removes telomerase from telomeric DNA, *Nature*, 2005, **438**, 57.
265. M. K. Mateyak and V. A. Zakian, Human PIF helicase is cell cycle regulated and associates with telomerase, *Cell Cycle*, 2006, **5**, 2796.
266. J. C. Norton, M. A. Piatyszek, W. E. Wright, J. W. Shay and D. R. Corey, Inhibition of human telomerase activity by peptide nucleic acids, *Nat. Biotechnol.*, 1996, **14**, 615.
267. A. E. Pitts and D. R. Corey, Inhibition of human telomerase by 2'-*O*-methyl-RNA, *Proc. Natl. Acad. Sci. U.S.A.*, 1998, **95**, 11549.
268. A. N. Elayadi, A. Demieville, E. V. Wancewicz, B. P. Monia and D. R. Corey, Inhibition of telomerase by 2'-O-(2-methoxyethyl) RNA oligomers: effect of length, phosphorothioate substitution and time inside cells, *Nucleic Acids Res.*, 2001, **29**, 1683.
269. S. E. Hamilton, A. E. Pitts, R. R. Katipally, X. Jia, J. P. Rutter, B. A. Davies, J. W. Shay, W. E. Wright and D. R. Corey, Identification of determinants for inhibitor binding within the RNA active site of human telomerase using PNA scanning, *Biochemistry*, 1997, **36**, 11873.
270. E. Matthes and Ch. Lehmann, Telomerase protein rather than its RNA is the target of phosphorothioate-modified oligonucleotides, *Nucleic Acids Res.*, 1999, **27**, 1152.
271. B. S. Herbert, K. Pongracz, J. W. Shay and S. M. Gryaznov, Oligonucleotide N3′-->P5′ phosphoramidates as efficient telomerase inhibitors, *Oncogene*, 2002, **21**, 638.
272. A. Asai, Y. Oshima, Y. Yamamoto, T. A. Uochi, H. Kusaka, S. Akinaga, Y. Yamashita, K. Pongracz, R. Pruzan, E. Wunder, M. Piatyszek, S. Li, A. C. Chin, C. B. Harley and S. Gryaznov, A novel telomerase template antagonist (GRN163) as a potential anticancer agent, *Cancer Res.*, 2003, **63**, 3931.
273. B. K. Canales, Y. Li, M. G. Thompson, J. M. Gleason, Z. Chen, B. Malaeb, D. R. Corey, B. S. Herbert, J. W. Shay and K. S. Koeneman, Small molecule, oligonucleotide-based telomerase template inhibition in combination with cytolytic therapy in an in vitro androgen-independent prostate cancer model, *Urol. Oncol.*, 2006, **24**, 141.

274. M. Akiyama, T. Hideshima, M. A. Shammas, T. Hayashi, M. Hamasaki, Y. T. Tai, P. Richardson, S. Gryaznov, N. C. Munshi and K. C. Anderson, Effects of oligonucleotide N3′→P5′ thio-phosphoramidate (GRN163) targeting telomerase RNA in human multiple myeloma cells, *Cancer Res.*, 2003, **63**, 6187.
275. T. Ozawa, S. M. Gryaznov, L. J. Hu, K. Pongracz, R. A. Santos, A. W. Bollen, K. R. Lamborn and D. F. Deen, Antitumor effects of specific telomerase inhibitor GRN163 in human glioblastoma xenografts, *Neuro-oncology*, 2004, **6**, 218.
276. E. S. Wang, K. Wu, A. C. Chin, S. Chen-Kiang, K. Pongracz, S. Gryaznov and M. A. Moore, Telomerase inhibition with an oligonucleotide telomerase template antagonist: *in vitro* and *in vivo* studies in multiple myeloma and lymphoma, *Blood*, 2003, **103**, 258.
277. B. S. Herbert, G. C. Gellert, A. Hochreiter, K. Pongracz, W. E. Wright, D. Zielinska, A. C. Chin, C. B. Harley, J. W. Shay and S. M. Gryaznov, Lipid modification of GRN163, an N3′ → P5′ thio-phosphoramidate oligonucleotide, enhances the potency of telomerase inhibition, *Oncogene*, 2005, **24**, 5262.
278. M. W. Djojosubroto, A. C. Chin, N. Go, S. Schaetzlein, M. P. Manns, S. Gryaznov, C. B. Harley and K. L. Rudolph, Telomerase antagonists GRN163 and GRN163L inhibit tumor growth and increase chemosensitivity of human hepatoma, *Hepatology*, 2005, **42**, 1127.
279. Z. G. Dikmen, G. C. Gellert, S. Jackson, S. Gryaznov, R. Tressler, P. Dogan, W. E. Wright and J. W. Shay, *In vivo* inhibition of lung cancer by GRN163L: a novel human telomerase inhibitor, *Cancer Res.*, 2005, **65**, 7866.
280. G. C. Gellert, Z. G. Dikmen, W. E. Wright, S. Gryaznov and J. W. Shay, Effects of a novel telomerase inhibitor, GRN163L, in human breast cancer, *Breast Cancer Res. Treat.*, 2006, **96**, 73.
281. J. Gomez-Millan, E. M. Goldblatt, S. M. Gryaznov, M. S. Mendonca and B. S. Herbert, Specific telomere dysfunction induced by GRN163L increases radiation sensitivity in breast cancer cells, *Int. J. Radiat. Oncol. Biol. Phys.*, 2007, **67**, 897.
282. A. E. Hochreiter, H. Xiao, E. M. Goldblatt, S. M. Gryaznov, K. D. Miller, S. Badve, G. W. Sledge and B. S. Herbert, Telomerase template antagonist GRN163L disrupts telomere maintenance, tumor growth and metastasis of breast cancer, *Clin. Cancer Res.*, 2006, **12**, 3184.
283. S. R. Jackson, C. H. Zhu, V. Paulson, L. Watkins, Z. G. Dikmen, S. M. Gryaznov, W. E. Wright and J. W. Shay, Antiadhesive effects of GRN163L--an oligonucleotide N3′->P5′ thio-phosphoramidate targeting telomerase, *Cancer Res.*, 2007, **67**, 1121.
284. A. De Cian, L. Lacroix, C. Douarre, N. Temime-Smaali, C. Trentesaux, J. F. Riou and J. L. Mergny, Targeting telomeres and telomerase, *Biochimie*, 2008, **90**, 131.
285. K. Damm, U. Hemmann, P. Garin-Chesa, N. Hauel, I. Kauffmann, H. Priepke, C. Niestroj, C. Daiber, B. Enenkel, B. Guilliard, I. Lauritsch,

E. Muller, E. Pascolo, G. Sauter, M. Pantic, U. M. Martens, C. Wenz, J. Lingner, N. Kraut, W. J. Rettig and A. Schnapp, A highly selective telomerase inhibitor limiting human cancer cell proliferation, *EMBO J.*, 2001, **20**, 6958.

286. E. Pascolo, C. Wenz, J. Lingner, N. Hauel, H. Priepke, I. Kauffmann, P. Garin-Chesa, W. J. Rettig, K. Damm and A. Schnapp, Mechanism of human telomerase inhibition by BIBR1532, a synthetic, non-nucleosidic drug candidate, *J. Biol. Chem.*, 2002, **277**, 15566.
287. D. K. Barma, A. Elayadi, J. R. Falck and D. R. Corey, Inhibition of telomerase by BIBR 1532 and related analogues, *Bioorg. Med. Chem. Lett.*, 2003, **13**, 1333.
288. R. J. Ward and C. Autexier, Pharmacological telomerase inhibition can sensitize drug-resistant and drug-sensitive cells to chemotherapeutic treatment, *Mol. Pharmacol.*, 2005, **68**, 779.
289. S. Mueller, U. Hartmann, F. Mayer, S. Balabanov, J. T. Hartmann, T. H. Brummendorf and C. Bokemeyer, Targeting telomerase activity by BIBR1532 as a therapeutic approach in germ cell tumors, *Invest. New Drugs*, 2007, **25**, 519.
290. H. El-Daly, M. Kull, S. Zimmermann, M. Pantic, C. F. Waller and U. M. Martens, Selective cytotoxicity and telomere damage in leukemia cells using the telomerase inhibitor BIBR1532, *Blood*, 2005, **105**, 1742.
291. A. Roth, J. Durig, H. Himmelreich, S. Bug, R. Siebert, U. Duhrsen, P. M. Lansdorp and G. M. Baerlocher, Short telomeres and high telomerase activity in T-cell prolymphocytic leukemia, *Leukemia*, 2007, **21**, 2456.
292. D. Monchaud and M. P. Teulade-Fichou, A hitchhiker's guide to G-quadruplex ligands, *Org. Biomol. Chem.*, 2008, **6**, 627.
293. A. De Cian, G. Cristofari, P. Reichenbach, E. De Lemos, D. Monchaud, M. P. Teulade-Fichou, K. Shin-Ya, L. Lacroix, J. Lingner and J. L. Mergny, Reevaluation of telomerase inhibition by quadruplex ligands and their mechanisms of action, *Proc. Natl. Acad. Sci. U.S.A.*, 2007, **104**, 17347.
294. D. Sun, B. Thompson, B. E. Cathers, M. Salazar, S. M. Kerwin, J. O. Trent, T. C. Jenkins, S. Neidle and L. H. Hurley, Inhibition of human telomerase by a G-quadruplex-interactive compound, *J. Med. Chem.*, 1997, **40**, 2113.
295. O. Y. Fedoroff, M. Salazar, H. Y. Han, V. V. Chemeris, S. M. Kerwin and L. H. Hurley, NMR-based model of a telomerase-inhibiting compound bound to G-quadruplex DNA, *Biochemistry*, 1998, **37**, 12367.
296. R. T. Wheelhouse, D. K. Sun, H. Y. Han, F. X. Han and L. H. Hurley, Cationic porphyrins as telomerase inhibitors: the interaction of tetra-(N-methyl-4-pyridyl)porphine with quadruplex DNA, *J. Am. Chem. Soc.*, 1998, **120**, 3261.
297. M. Read, R. J. Harrison, B. Romagnoli, F. A. Tanious, S. H. Gowan, A. P. Reszka, W. D. Wilson, L. R. Kelland and S. Neidle, Structure-based design of selective and potent G quadruplex-mediated telomerase inhibitors, *Proc. Natl. Acad. Sci. U.S.A.*, 2001, **98**, 4844.

298. G. Pennarun, C. Granotier, L. R. Gauthier, D. Gomez, F. Hoffschir, E. Mandine, J. F. Riou, J. L. Mergny, P. Mailliet and F. D. Boussin, Apoptosis related to telomere instability and cell cycle alterations in human glioma cells treated by new highly selective G-quadruplex ligands, *Oncogene*, 2005, **24**, 2917.
299. A. De Cian, E. Delemos, J. L. Mergny, M. P. Teulade-Fichou and D. Monchaud, Highly efficient G-quadruplex recognition by bisquinolinium compounds, *J. Am. Chem. Soc.*, 2007, **129**, 1856.
300. K. Shin-Ya, K. Wierzba, K. Matsuo, T. Ohtani, Y. Yamada, K. Furihata, Y. Hayakawa and H. Seto, Telomestatin, a novel telomerase inhibitor from *Streptomyces anulatus*, *J. Am. Chem. Soc.*, 2001, **123**, 1262.
301. M. Y. Kim, M. Gleason-Guzman, E. Izbicka, D. Nishioka and L. H. Hurley, The different biological effects of telomestatin and TMPyP4 can be attributed to their selectivity for interaction with intramolecular or intermolecular G-quadruplex structures, *Cancer Res.*, 2003, **63**, 3247.
302. I. M. Dixon, F. Lopez, A. M. Tejera, J. P. Esteve, M. A. Blasco, G. Pratviel and B. Meunier, A G-quadruplex ligand with 10000-fold selectivity over duplex DNA, *J. Am. Chem. Soc.*, 2007, **129**, 1502.
303. M. Y. Kim, H. Vankayalapati, K. Shin-Ya, K. Wierzba and L. H. Hurley, Telomestatin, a potent telomerase inhibitor that interacts quite specifically with the human telomeric intramolecular G-quadruplex, *J. Am. Chem. Soc.*, 2002, **124**, 2098.
304. E. M. Rezler, J. Seenisamy, S. Bashyam, M. Y. Kim, E. White, W. D. Wilson and L. H. Hurley, Telomestatin and diseleno sapphyrin bind selectively to two different forms of the human telomeric G-quadruplex structure, *J. Am. Chem. Soc.*, 2005, **127**, 9439.
305. T. Tauchi, K. Shin-Ya, G. Sashida, M. Sumi, A. Nakajima, T. Shimamoto, J. H. Ohyashiki and K. Ohyashiki, Activity of a novel G-quadruplex-interactive telomerase inhibitor, telomestatin (SOT-095), against human leukemia cells: involvement of ATM-dependent DNA damage response pathways, *Oncogene*, 2003, **22**, 5338.
306. D. Gomez, R. Paterski, T. Lemarteleur, K. Shin-Ya, J. L. Mergny and J. F. Riou, Interaction of telomestatin with the telomeric single-strand overhang, *J. Biol. Chem.*, 2004, **279**, 41487.
307. M. A. Shammas, R. J. Reis, C. Li, H. Koley, L. H. Hurley, K. C. Anderson and N. C. Munshi, Telomerase inhibition and cell growth arrest after telomestatin treatment in multiple myeloma, *Clin. Cancer Res.*, 2004, **10**, 770.
308. H. Tahara, K. Shin-Ya, H. Seimiya, H. Yamada, T. Tsuruo and T. Ide, G-Quadruplex stabilization by telomestatin induces TRF2 protein dissociation from telomeres and anaphase bridge formation accompanied by loss of the 3′ telomeric overhang in cancer cells, *Oncogene*, 2006 **25**, 1955.
309. T. Tauchi, K. Shin-Ya, G. Sashida, M. Sumi, S. Okabe, J. H. Ohyashiki and K. Ohyashiki, Telomerase inhibition with a novel G-quadruplex-

interactive agent, telomestatin: *in vitro* and *in vivo* studies in acute leukemia, *Oncogene*, 2006, **25**, 5719.

310. D. Gomez, T. Wenner, B. Brassart, C. Douarre, M. F. O'Donohue, V. El Khoury, K. Shin-Ya, H. Morjani, C. Trentesaux and J. F. Riou, Telomestatin induced telomere uncapping is modulated by POT1 through G-overhang extension in HT1080 human tumor cells, *J. Biol. Chem.*, 2006, **281**, 38721.
311. D. Gomez, M. F. O'Donohue, T. Wenner, C. Douarre, J. Macadre, P. Koebel, M. J. Giraud-Panis, H. Kaplan, A. Kolkes, K. Shin-Ya and J. F. Riou, The G-quadruplex ligand telomestatin inhibits POT1 binding to telomeric sequences *in vitro* and induces GFP-POT1 dissociation from telomeres in human cells, *Cancer Res.*, 2006, **66**, 6908.
312. C. M. Incles, C. M. Schultes, H. Kempski, H. Koehler, L. R. Kelland and S. Neidle, A G-quadruplex telomere targeting agent produces p16-associated senescence and chromosomal fusions in human prostate cancer cells, *Mol. Cancer Ther.*, 2004, **3**, 1201.
313. A. M. Burger, F. Dai, C. M. Schultes, A. P. Reszka, M. J. Moore, J. A. Double and S. Neidle, The G-quadruplex-interactive molecule BRACO-19 inhibits tumor growth, consistent with telomere targeting and interference with telomerase function, *Cancer Res.*, 2005, **65**, 1489.
314. C. Leonetti, S. Amodei, C. D'Angelo, A. Rizzo, B. Benassi, A. Antonelli, R. Elli, M. Stevens, M. D'Incalci, G. Zupi and A. Biroccio, Biological activity of the G-quadruplex ligand RHPS4 (3,11-difluoro-6,8,13-trimethyl-8H-quino[4,3,2-kl]acridinium methosulfate) is associated with telomere capping alteration, *Mol. Pharmacol.*, 2004, **66**, 1138.
315. E. Salvati, C. Leonetti, A. Rizzo, M. Scarsella, M. Mottolese, R. Galati, I. Sperduti, M. F. Stevens, M. D'Incalci, M. Blasco, G. Chiorino, S. Bauwens, B. Horard, E. Gilson, A. Stoppacciaro, G. Zupi and A. Biroccio, Telomere damage induced by the G-quadruplex ligand RHPS4 has an antitumor effect, *J. Clin. Invest.*, 2007, **117**, 3236.
316. P. Phatak, J. C. Cookson, F. Dai, V. Smith, R. B. Gartenhaus, M. F. Stevens and A. M. Burger, Telomere uncapping by the G-quadruplex ligand RHPS4 inhibits clonogenic tumour cell growth *in vitro* and *in vivo* consistent with a cancer stem cell targeting mechanism, *Br. J. Cancer*, 2007, **96**, 1223.
317. B. Brassart, D. Gomez, A. De Cian, R. Paterski, A. Montagnac, K. H. Qui, N. Temime-Smaali, C. Trentesaux, J. L. Mergny, F. Gueritte and J. F. Riou, A new steroid derivative stabilizes G-quadruplexes and induces telomere uncapping in human tumor cells, *Mol. Pharmacol.*, 2007, **72**, 631.
318. S. M. Gowan, J. R. Harrison, L. Patterson, M. Valenti, M. A. Read, S. Neidle and L. R. Kelland, A G-quadruplex-interactive potent small-molecule inhibitor of telomerase exhibiting *in vitro* and *in vivo* antitumor activity, *Mol. Pharmacol.*, 2002, **61**, 1154.

319. S. M. Gowan, R. Heald, M. F. Stevens and L. R. Kelland, Potent inhibition of telomerase by small-molecule pentacyclic acridines capable of interacting with G-quadruplexes, *Mol. Pharmacol.*, 2001, **60**, 981.
320. T. Lemarteleur, D. Gomez, R. Paterski, E. Mandine, P. Mailliet and J. F. Riou, Stabilization of the c-myc gene promoter quadruplex by specific ligands' inhibitors of telomerase, *Biochem. Biophys. Res. Commun.*, 2004, **323**, 802.
321. K. Halder and S. Chowdhury, Quadruplex-coupled kinetics distinguishes ligand binding between G4 DNA motifs, *Biochemistry*, 2007, **46**, 14762.
322. J. Eddy and N. Maizels, Gene function correlates with potential for G4 DNA formation in the human genome, *Nucleic Acids Res.*, 2006, **34**, 3887.
323. R. J. Richards, C. A. Theimer, L. D. Finger and J. Feigon, Structure of the *Tetrahymena thermophila* telomerase RNA helix II template boundary element, *Nucleic Acids Res.*, 2006, **34**, 816.

CHAPTER 9

Keeping Replicated Chromatids Together Until Mitosis

CHRISTIAN H. HAERING

European Molecular Biology Laboratory (EMBL), Cell Biology and Biophysics Unit, Meyerhofstrasse 1, 69117 Heidelberg, Germany

9.1 Introduction

Following replication of the genome, the two identical sister chromatids must be segregated into different daughter cells. This ensures that daughter cells receive a complete set of chromosomes during every cell division and thereby guarantees faithful propagation of genetic information from one generation to the next. Failures in chromosome segregation lead to the generation of aneuploid cells bearing too few or too many chromosomes, a situation frequently found in malignant cancer cells and thought to contribute to their development.[1] It is therefore not surprising that cells possess a sophisticated and precisely regulated chromosome segregation machinery, which depends to a large extent on establishing and maintaining a tight connection between sister chromatid pairs until anaphase. This chapter assesses the current understanding of establishment of this sister chromatid cohesion during S phase and its destruction at anaphase to trigger chromosome segregation.

9.2 The Cohesin Complex

DNA replication leaves the newly generated sister chromatids intertwined, and it might seem conceivable that sisters are simply held together by their concatenation. Segregation at anaphase could then be triggered by a sudden

Molecular Themes in DNA Replication
Edited by Lynne S. Cox

Published by the Royal Society of Chemistry, www.rsc.org

activation of the decatenating enzyme topoisomerase II.[2] However, the bulk of circular minichromosomes isolated from yeast cells arrested before anaphase are not concatenated.[3,4] This suggests that sister chromatids must be held together by a different mechanism, probably by linker proteins. Genetic screens for yeast mutants that separate sister chromatids prematurely identified four candidates for such linker proteins, namely Scc1/Mcd1, Scc3, Smc1 and Smc3[5–7] (Table 9.1). As one would expect if they were holding sister chromatids together, the four proteins localize to chromosomes until metaphase and dissociate from chromosomes just before sister chromatids separate at the onset of anaphase. The fact that they can be co-immunoprecipitated and their binding to chromosomes depends on each other suggested they are subunits of one protein complex. This was consequently named the cohesin complex.[6]

A fifth subunit of the cohesin complex, Pds5, was identified independently by a genetic screen for yeast mutants that separate sister chromatids precociously and by a homology search for proteins involved in chromosome segregation in other fungi.[8,9] Even though Pds5 is recruited to and released from chromosomes together with cohesin, it associates with the other four cohesin subunits sub-stoichiometrically; its association is sensitive to salt treatment, suggesting lower affinity binding.[10] Unlike the four core cohesin subunits, Pds5 is not essential for viability in all organisms, including fission yeast[11] and an ancient eukaryote called *Encephalitozoon cuniculi.* Pds5 is predicted to be largely composed of HEAT[i] repeats, which are protein motifs thought to act as scaffolds for protein–protein interactions.[12]

Cohesin complexes have been isolated from a variety of organisms, and genes encoding their core subunits are found in every eukaryotic genome that has been sequenced to date. Some organisms even possess multiple isoforms of the same subunit.[10,13–16] Mutation of cohesin genes in budding and fission yeasts, antibody depletion of cohesin proteins from frog egg extract, or knock-down of cohesin subunits in fruit fly or mammalian cultured cells by RNA interference all lead to precocious separation of sister chromatids (reviewed in ref. 17), suggesting a universality of action.

9.2.1 Structure of the Cohesin Complex

Analysis of the architecture of the yeast cohesin complex brought the first hints for how it could connect sister chromatids. Cohesin's Smc1 and Smc3 subunits belong to the Structural Maintenance of Chromosomes (SMC) protein family (Table 9.1), whose members are found in all kingdoms of life. SMCs are characterized by two long alpha-helical regions that fold around a central 'hinge' domain into a 35 nm long intra-molecular anti-parallel coiled coil[18–20] (Figure 9.1A). This brings together the N- and C-terminal domains to form a globular ATPase 'head' domain of the ATP Binding Cassette (ABC) family. Smc1 and Smc3 dimerize *via* their hinge domains, giving rise to V-shaped molecules. Scc1 binds to Smc3's head *via* its N-terminal domain and to Smc1's

---

[i] Huntington elongation A subunit TOR

Table 9.1 Proteins involved in cohesion.

Name	Function	Interactions	Regulation	*S. cerevisiae* MW (kDa)	*S. cerevisiae* aa	*S. cerevisiae* molecules per cell[118]
Scc1/Mcd1/ Rad21	α-Kleisin subunit of the cohesin complex	Smc3 via its N- and Smc1 via its C-terminus	Cleaved by separase	63.2	566	1040
Smc1	Coiled-coil subunit of the cohesin complex	Scc1 C-terminus		141.2	1225	5710
Smc3	Coiled-coil subunit of the cohesin complex	Scc1 N-terminus		141.3	1230	2660
Scc3 SA1/2	Subunit of the cohesin complex	Scc1		133.0	1150	4090
Pds5 Pds5A/B	HEAT repeat containing sub-stoichiometric subunit of the cohesin complex			147.0	1277	7720
Scc2 Mis4 Nipped-B NIPLB	HEAT repeat containing subunit of the cohesin loading factor	Scc4		171.1	1493	3310
Scc4	Subunit of the cohesin loading factor	Scc2		72.1	624	907
Chl1/Ctf1	DEAD box DNA helicase	Eco1, PCNA		98.8	861	98
Eco1/Ctf7 Eso1 ESCO1/2	Acetyltransferase essential for the establishment of cohesion	Chl1, PCNA via conserved PIP box	Human E(S)CO1 phosphorylated in mitosis	31.8	281	
Cdc5 PLK	Polo-like kinase, phosphorylates Scc1 and SA1/2	Scc1 SA1/2		81.0	705	1480
Esp1 Separase	Caspase-like cysteine protease, cleaves Scc1	securin	Inhibited by its binding to securin, inhibited by CDK phosphorylation	187.4	1630	
Pds1 Securin	Inhibits separase	separase	Ubiquitinated and destroyed at anaphase (via N-terminal D-box), phosphorylated by CDK	41.8	373	

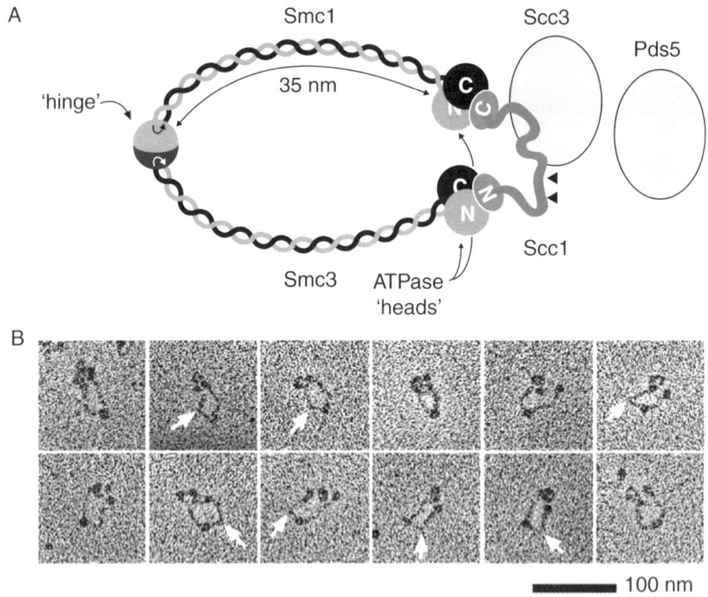

Figure 9.1 **The cohesin complex.** (**A**) Cohesin's Smc1 and Smc3 subunits dimerize via their hinge domains at one end of a 35 nm long anti-parallel intramolecular coiled coil. Scc1 connects their ATPase head domains at the other end of the coiled coil, thereby creating a large ring structure. It also recruits Scc3. Triangles indicate two sites in the central domain of Scc1 that are cleaved by the separase protease at the onset of anaphase. How Pds5 associates with the complex is not known. (**B**) Electron micrographs of vertebrate cohesin complexes, reproduced from ref. 22 with permission, © Anderson *et al.*, 2002. Originally published in *J. Cell Biol.*, 2002, **156**, 419.

head *via* its C-terminal winged-helix domain, thereby connecting the two head domains situated at the apices of the V to form a large tripartite ring structure.[19,21] (Figure 9.1A). Such rings can also be observed in electron micrographs of vertebrate cohesin complexes[22] (Figure 9.1B). Scc3 binds to the complex *via* its association with Scc1's central domain.[19]

9.2.2 Models for Sister Chromatid Cohesion *via* Cohesin

The gigantic ring-like structure of cohesin led to the proposal that it might hold sister chromatids together by topologically entrapping them inside its ring[19] (Figure 9.2A). Variations of this model envisage that the two sister DNAs are entrapped within separate cohesin rings, which then dimerize by protein–protein interactions or intertwining of the two cohesin rings (Figure 9.2B). Alternatively, cohesin might bind the two sister DNAs non-topologically, *e.g. via* positively charged surface patches at the head or hinge domains.[23]

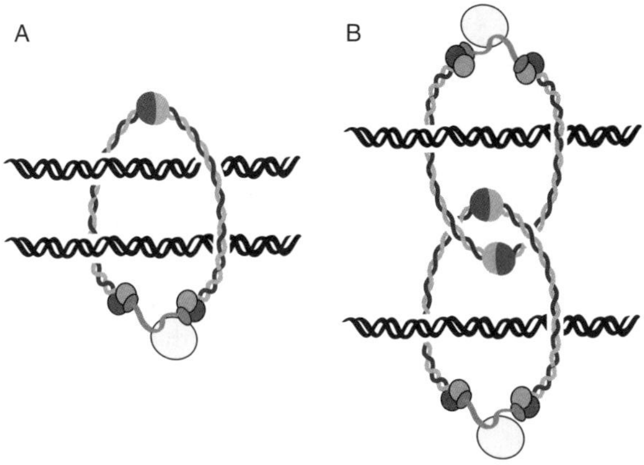

Figure 9.2 **Models for sister chromatids entrapment within cohesin rings.** (**A**) According to the original cohesin ring model, two sister chromatid DNA fibres are held together by their entrapment within the same cohesin ring. (**B**) A variation of this model suggests the entrapment of each sister chromatid within separate rings that are somehow connected, possibly by intertwining of the two cohesin rings.

Several lines of evidence support the topological mode of cohesin-DNA association suggested by the ring entrapment model. Firstly, opening of the ring by proteolytic cleavage of Scc1 (see Section 9.5) or by cleavage of engineered sites within Smc3's coiled coil releases cohesin from DNA and destroys cohesion.[21] Secondly, linearization of circular minichromosome DNA bound by cohesin *in vitro* releases the DNA from cohesin, suggesting that DNA might slide through the cohesin ring.[24] A similar sliding mechanism might be used *in vivo*, where it has been proposed that the RNA polymerase machinery might push cohesin to the 3′ end of actively transcribed genes.[25] Thirdly, covalent connection of the three interfaces between the three cohesin ring subunits to form a continuously closed ring generates cohesion between two circular minichromosomes that is resistant to protein denaturation.[26] This result is inconsistent with a direct binding of cohesin to chromosomal DNA (as denaturation would have destroyed any binding surface), but is easy to explain if sister chromatids were topologically entrapped by the cohesin ring.

9.3 Loading Cohesin onto Chromosomes

If the entrapment model is correct, one of the key questions is how chromosomal DNA gets into the cohesin ring. In principle, cohesin ring subunits could assemble *de novo* around chromosomes. However, the fact that most cohesin can be isolated from soluble cell extracts as intact complexes[7,21,27] suggests that the ring exists even when not bound to chromosomes, in which case at least one of the ring subunit interactions has to temporarily disengage to let chromosomal DNA pass into the ring.

To test which of the three interfaces functions as an entry gate for DNA, Gruber *et al.* closed the Smc3–Scc1 or Scc1–Smc1 interfaces by connecting the C-terminus of one subunit with the N-terminus of the other using a flexible peptide linker.[28] Surprisingly, yeast cells expressing either fusion showed no significant cohesion defects. Closure of the Smc1–Smc3 interface, however, by insertion of Frb and FKBP12 domains that dimerize with nanomolar affinity upon binding to the small molecule rapamycin, greatly diminished the amount of cohesin loaded onto chromosomes.[28] Strikingly, rapamycin addition led to loss of cohesion when added before but not after cells had loaded cohesin and established cohesion. These data strongly suggest that the Smc1–Smc3 hinge is the entry gate for DNA into the cohesin ring.

9.3.1 Role of the Smc1/Smc3 ATPases in Cohesin Loading

The opening of such a gate would have to be tightly controlled to prevent escape of entrapped DNA once cohesion has been established. The finding that mutations in the Smc1 or Smc3 ATPase head domains that prevent ATP hydrolysis also block loading of cohesin onto DNA *in vivo*[29,30] suggests that it might be this activity that controls ring opening. Supporting evidence for an interplay between SMC hinge and ATPase head domains comes from the study of a bacterial SMC homodimer. Mutation of positively charged surface residues in the hinge domain abolish the association of *Bacillus subtilis* SMCs with DNA (as measured by an *in vitro* gel retardation assay) and reduce the (very modest) stimulation of ATPase activity by double-stranded DNA.[31] It is, however, puzzling how ATP hydrolysis by the SMC head domains, which are separated from the SMC hinge domains by 35 nm long coiled coils, could break the latter's high affinity contact and how this reaction might be regulated during loading onto chromosomes. It is therefore necessary to have a detailed look at the SMC ATPase.

9.3.2 Structural Analysis of SMC ATPase Domains

Like all nucleotide-binding domains of the ABC ATPase family, the SMC head domains contain bipartite motifs for ATP binding and hydrolysis. The nucleotide is bound by contacts between its α- and β-phosphates to residues of the Walker A motif (also called P-loop) in one SMC head domain. Crystal structures of the SMC-like Rad50 head domain solved in the presence of the slowly hydrolysable ATP analogue AMP-PNP and Mg^{2+}, or in the presence of ATP but absence of Mg^{2+}, first revealed that the ATP binding pocket is completed by residues of the so-called LSGG signature motif from a second head domain, which contacts the ATP γ-phosphate.[32] Two SMC head domains hence dimerize by sandwiching two molecules of ATP within a central groove. This arrangement was confirmed by homodimeric structures of an archaeal SMC head domain[33] and cohesin's Smc1 head domain[34] (Figure 9.3). Binding of cohesin's Scc1 C-terminal winged helix domain to the Smc1 head domain

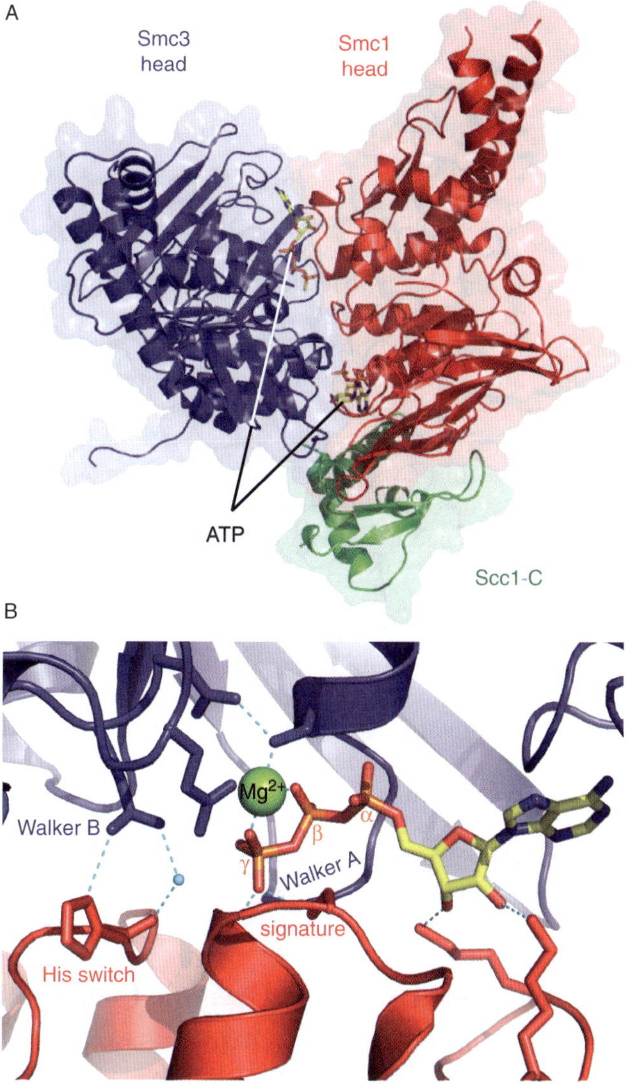

Figure 9.3 **SMC ATPase head structure**. (**A**) Cartoon model of the yeast cohesin Smc1/Smc3 head domain dimer bound by the C-terminal winged helix domain of Scc1 generated with PyMOL (http://pymol.org/).[117] The second Smc1 head domain in the actual crystal structure (PDB accession 1W1W) is replaced by a model of the Smc3 head.[118] Two ATP molecules are sandwiched between the two head domains. (**B**) Close-up view of the bipartite ATPase site of the archaeal SMC head homodimer (PDB accession 1XEX). The γ-phosphate moiety of Mg^{2+}-ATP bound to the Walker A motif and side-chains of other residues of one head domain (blue) is bound by hydrogen bonds (cyan) to the signature motif of the opposite head domain (red). The attacking water molecule (cyan sphere) is positioned for in-line attack on the γ-phosphate by hydrogen bonds to the catalytic glutamate residue of Walker B (blue, here mutated to glutamine) and His switch (red) motifs.

enhances ATP binding to the head domain, possibly by stabilizing the Walker A motif.[35]

9.3.3 Conformational Changes May Activate the SMC ATPase upon DNA Binding

Within SMC head domains, mutagenesis of a glutamate residue in the Walker B motif traps the nucleotide binding domains in the dimer state,[33,36,37] and also greatly reduces ATP hydrolysis by bacterial SMCs[38] or cohesin's Smc1/Smc3 dimer.[35] In the structurally similar ABC ATPase-transporter HisP, this glutamate is important for ATP hydrolysis by polarizing a water molecule for nucleophilic in-line attack on the γ-phosphate moiety[39] (see Figure 9.3B). However, the water molecule positioned collinear with the scissile bond in the SMC-like Rad50 is not contacted by the catalytic glutamate residue.[32] It has therefore been suggested that a conformational change might be required to position the glutamate side-chain for catalytic activation. What might trigger such a conformational change to activate the ATPase activity is not known, but it could be the binding of DNA to the SMC head domains, because the presence of DNA stimulates the (very low) ATPase activity displayed by prokaryotic SMC dimers *in vitro*.[33,40] Another suggestion is that a flexible surface loop that contacts the α-phosphate in the ATP-bound archaeal SMC head structure could be the activator of the ATPase upon DNA binding to the head domain.[33] Consistently, mutation of a conserved arginine residue at the centre of this loop significantly reduces the DNA-dependent stimulation of its ATPase activity. However, the effect on ATPase activity by single- or double-stranded DNA was relatively modest in every case (5–15 fold increase of the hydrolysis rate[33,40,41]) and no DNA-dependent stimulation of ATPase activity could be observed for the Smc1/Smc3 dimer of yeast cohesin.[35]

9.3.4 How Might ATP Hydrolysis Open the Cohesin Ring?

ATP hydrolysis is thought to drive the two SMC heads apart. In the structurally related ABC transporters, the cycle of nucleotide binding domain (NBD) dimerization and dissociation drives a conformational flip in the attached transmembrane domains for cargo transport (reviewed in ref. 42). ATP-dependent dimerization is thought to provide the power stroke for opening the transmembrane domain on the outside for substrate binding (the high energy state), while ATP hydrolysis and NBD dissociation resets the transmembrane domain to release bound substrate to the inside (the ground state). Whether similar conformational changes are driven by SMC ATPase head domains is unknown. It is possible that ATP binding or hydrolysis induces a force (or twist) along the coiled coils to induce SMC hinge opening, but the facts that the coiled coils are non-continuous[21,22] and no large conformational changes in the SMC heads themselves can be observed when the structures of ATP-free and ATP-bound archaeal SMC head domains are compared,[33] make this scenario unlikely.

An alternative is that SMC head dimerization may generate a binding platform for associated proteins, or possibly their own hinge domain.[28] Atomic force microscopy of the cohesin-related condensin complex (see Section 9.6) shows a large number of molecules whose hinge domains appear folded back onto their head domains.[43] Sequential ATP hydrolysis first at one active site may open one of the two interfaces in the doughnut-shaped hinge dimer, while ATP hydrolysis at the second active site may open the other and re-close the first hinge interface to promote a directional transport of DNA through the hinge.[28] The advantage of such a system would be that the energy required to break the nanomolar affinity hinge–hinge interaction would be significantly reduced and that the hinge domains would disengage completely only for a very short time. It has to be noted, however, that the central channel in the SMC hinge doughnut is not large enough to accommodate a double-stranded DNA helix when both interfaces are connected.[19] An alternative model proposes that the SMC hinge first makes contact with DNA, which then triggers ATP hydrolysis at the head domains to allow passage of DNA through the dissociated heads. Further opening (partial dissociation) of the SMC hinge domain may then make a permanent contact with DNA.[31]

9.3.5 The Scc2/Scc4 Cohesin Loading Factor

The differences between the two current models emphasize how little is still known about the mechanics of SMC proteins. A huge step forward would be the availability of a controlled loading reaction with defined, purified components. One of the difficulties so far has been that cohesin complexes are quite inert when it comes to binding DNA *in vitro*. Cohesin association with DNA could only be detected in gelshift assays upon addition of an excess of protein over DNA (50 : 1 or more[44]), resulting in the formation of large cohesin–DNA aggregates. Crucially, the DNA binding activity was not ATP-dependent. Additional factors that regulate cohesin binding to chromosomes are presumably missing in these reactions. One of these factors is most probably a heterodimeric protein complex composed of Scc2 and Scc4 proteins (Table 9.1). Scc2 was identified in one of the initial yeast screens for mutants defective in cohesion[6] and was found to form a complex with Scc4.[45] The same complex was recently identified in fission yeast[46] and metazoa.[47,48] Mutation or depletion by RNA interference of Scc2 or Scc4 drastically reduces the amounts of cohesin associated with chromosomes and causes cohesion defects.[45,47,48] How Scc2/Scc4 loads cohesin onto chromosomes is not known. If the ring embracement model is true, one possibility is that Scc2/Scc4 regulates opening of the ring for entry of chromosomal DNA.

9.4 Establishment of Sister Chromatid Cohesion

Even though cohesin can associate with chromosomes at any stage of the cell cycle, its binding to chromosomes alone is not sufficient to generate cohesion. In yeast cells, Scc1 re-accumulates and readily binds to chromosomes before the initiation of DNA replication (at late G1 phase) following its degradation at anaphase of the previous cell cycle.[6] When expression of Scc1 is delayed until after DNA replication, Scc1 can still bind to chromosomes but cannot establish sister chromatid cohesion.[49] This is also the case when sister chromatids were held together by pre-existing cohesion.[34] In mammalian cells, cohesin released by the prophase pathway can re-associate with chromosomes as early as at the end of the same mitosis (during telophase). These findings suggest that cohesin's property to establish sister chromatid cohesion is limited to the time of DNA replication.

9.4.1 Proteins Involved in Cohesion Establishment

Several proteins play a role in the establishment of sister chromatid cohesion. Yeast strains expressing a temperature sensitive mutation in the Eco1/Ctf7 protein (Table 9.1) have cohesion defects, despite apparently normal association of cohesin complexes with chromosomes, when the mutant protein is inactivated before or during S phase, but not when it is inactivated after cells have completed DNA replication.[7,50] Eco1 possesses an acetyltransferase domain and can acetylate itself and cohesin subunits *in vitro*,[51] but its *in vivo* targets and how this activity could function for cohesion establishment are not known. Genetic and biochemical data suggest that Eco1 could be recruited to the replication fork by its direct interaction with the DNA polymerase sliding clamp PCNA and/or a variant of PCNA's RF-C (Replication Factor-C) loading complex containing Ctf18, Ctf8, and Dcc1 proteins[50,52–56] (Figure 9.4). Eco1 homologues are also essential for sister chromatid cohesion in the fruit fly,[57] cultured human cells[58] and fission yeast cells.[59] The surprising finding that Eco1 becomes no longer essential in fission yeast cells lacking Pds5 suggests that Eco1 could counteract a negative function of Pds5 on cohesion formation.[60]

9.4.2 Replication-dependent Establishment of Cohesion

How could sister chromatid cohesion be established during DNA replication? In light of the ring model, the simplest solution would be the passage of the replication fork through cohesin rings that have encircled chromosomes, leaving the two sister chromatids entrapped inside cohesin rings at its wake (Figure 9.5A). This simple model would provide an elegant way to ensure that two sisters, and not just any chromatids, end up entrapped within the same cohesin rings, but it raises questions as to whether the cohesin ring might be large enough to accommodate a replisome (or replication factory), and why

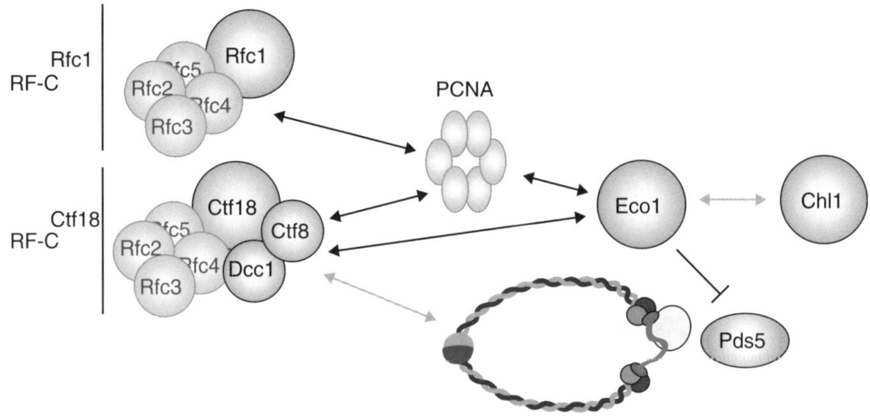

Figure 9.4 **Proteins involved in cohesion establishment**. The acetyltransferase Eco1 is central for the generation of cohesion, potentially by counteracting an inhibitory function of the non-stoichiometric cohesin subunit Pds5 on cohesion establishment. Eco1 genetically and physically interacts with the DNA polymerase sliding clamp PCNA and a Ctf18/Ctf8/Dcc1-containing variant of the RF-C, which like the canonical RF-C^{Rfc1} can load PCNA onto DNA (see also Chapters 2 and 3). Mutants of Ctf18, Ctf8 and Dcc1 have cohesion defects and are synthetically lethal with cohesin mutants. Likewise, mutants in the helicase Chl1 have cohesion defects and are synthetically lethal with Eco1.

establishment of cohesion would require specialized proteins like Eco1 or RF-C^{Ctf18}. One suggestion is that a DNA polymerase switch might occur at cohesin sites, that requires RF-C^{Ctf18}.[52,61] Alternatively, cohesin might be unloaded in front of the approaching replication fork and be transferred to the two sister DNAs as they emerge from the replication machinery (Figure 9.5B). A different idea is that cohesin rings are by default bound to one of the two sister DNAs and that the passage of the replication fork somehow connects two cohesin rings bound to opposite sisters. Cohesion defects were also observed in mutants of other factors that play an important role during DNA replication, including the DNA polymerase α-associated protein Ctf4 and the helicase Chl1.[52,55,61,62] How these proteins might assist in the establishment of cohesion is not known.

9.4.3 DNA Double-strand Break-dependent Establishment of Cohesion

Replication fork-dependent establishment of cohesion, whilst physiologically important, is not the only mechanism by which sister chromatid cohesion is set up. Cohesin is recruited to sites of DNA double strand breaks (DSBs), lesions that can occur both at damaged and collapsed replication forks, and also

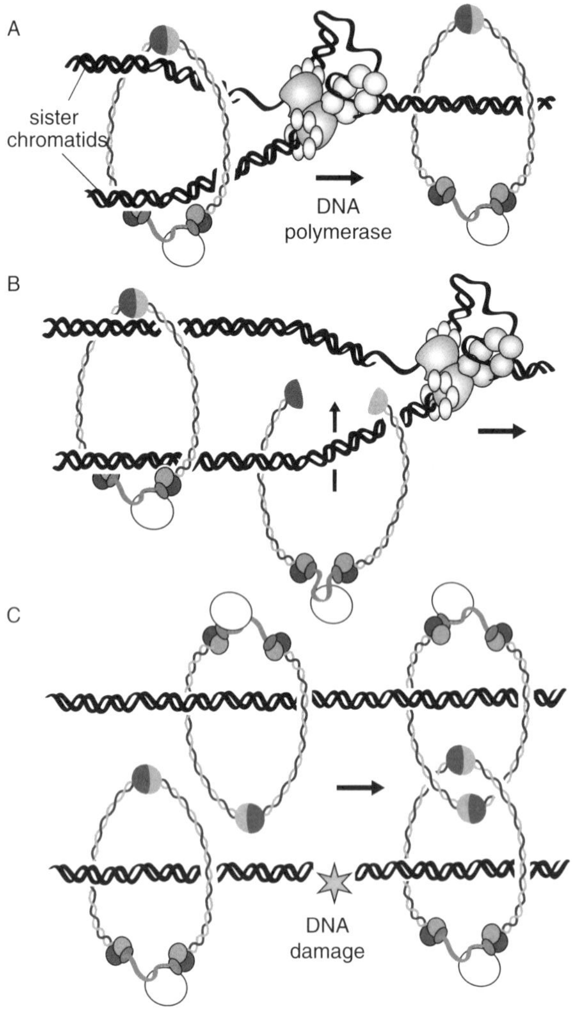

Figure 9.5 **Replication-dependent and -independent cohesion establishment**. (**A**) In the simplest scenario, replication through pre-assembled cohesin rings would automatically ensure that two sister chromatids are entrapped within the same cohesin ring. Associated proteins like Eco1 might be required to promote passage of the replication fork through the ring and RF-C^{Ctf18} might be needed to (re-)assemble PCNA at cohesin-bound sites. (**B**) A second scenario sees the (re-)loading of cohesin rings around both sister chromatids as they emerge from the replication machinery. The immediate proximity of the two sister strands would ensure that only the correct chromatids are entrapped within the same ring. (**C**) An alternative way to generate cohesion, possibly in the absence of DNA replication upon DNA damage, could be the direct or indirect (not shown) association of two cohesin rings that each have entrapped one sister chromatid. Note that double-strand breaks (DSBs) probably occur at high frequency throughout replication as forks encounter lesions or upon fork stalling and collapse.

following genotoxic insult. Cytological studies in mammalian cells show that cohesin accumulates at nuclear sites irradiated with a laser to induce DNA damage.[63] Use of a site-specific endonuclease to create a DSB in yeast cells arrested in a mitotic state induces the enrichment of cohesin in a large domain (several 10 kb) on both sides of the break site.[64,65] Cohesin enrichment at DSBs depends on the loading factor Scc2, phosphorylation of a variant of histone H2A (γH2AX), and the DSB repair endonuclease Mre11.[64,65] Repair of DSBs is severely impaired when cohesin loading is prevented after DSB formation in an *scc2* mutant.[65] Strikingly, cohesin loaded onto chromosomes after induction of DSBs is able to build *de novo* sister chromatid cohesion not only at the break site[65,66] but also on other chromosomes.[67,68] The damage-induced generation of cohesion moreover does not require recombination-dependent DNA replication, but requires the acetyltransferase activity of Eco1. It was suggested that Eco1 would normally be only active during DNA replication, and that post-replicative DNA damage might re-activate Eco1 *via* the DNA damage checkpoint Mec1/ATR kinase pathway, which then could generate new cohesive bridges between sister chromatids.[67,68]

The replication-independent formation of sister chromatid cohesion undermines the suggestion that replication fork passage through cohesin rings is needed to ensure that both sister chromatids are entrapped inside rings (Figure 9.5A). There must be an alternative mechanism either to get two sister chromatids inside rings that involves ring opening and re-closing, or a mechanism that connects two cohesin rings that each bind one sister chromatid (Figure 9.5C). One suggestion is that additional factors like Pds5 might generate such a link.

9.5 Destruction of Cohesion Triggers Chromosome Segregation

To ensure that sister chromatids are segregated into different daughter cells, all sister kinetochores need to attach to microtubules of the mitotic spindle with opposite orientations (bi-orientation or amphitelic attachment) before sister chromatid cohesion is lost and segregation towards the cell poles is initiated. Cohesion counteracts the pulling microtubule forces only if sisters are bi-oriented (Figure 9.6A, left) and thereby generates tension at the kinetochores. Microtubule attachments that have come under tension are stable, while those that have not are resolved again (Figure 9.6A, right). Unattached kinetochores initiate a signalling cascade known as the spindle assembly checkpoint that ultimately inhibits a large ubiquitin ligase complex called anaphase-promoting complex or cyclosome (APC/C). Once the last kinetochore pair has stably bi-oriented on the mitotic spindle, APC/C becomes active and polyubiquinates (amongst others) its target protein securin (Table 9.1) and thereby marks it for degradation by the 26S proteasome. Degradation of securin liberates a protein called separase, which triggers destruction of sister chromatid cohesion and initiates the pole-ward movement of sister chromatids at anaphase (Figure 9.6B).

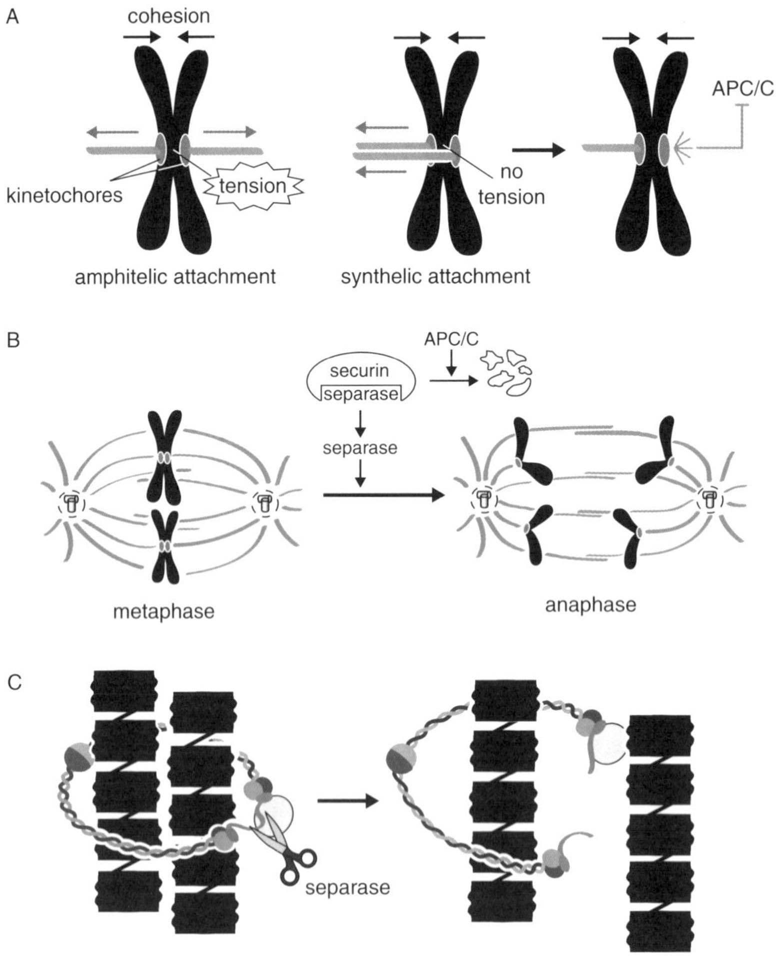

Figure 9.6 **Separase cleavage of Scc1 triggers chromosome segregation.** (**A**) Cohesion counteracts the pulling forces of spindle microtubules of bi-oriented sister chromatids (left), which generates tension at their kinetochores. Attachment of sister kinetochores to microtubules emanating from the same pole (right) cannot generate tension and such attachments are destabilized. The spindle assembly checkpoint recognizes unattached kinetochores and prevents the activation of the APC/C. (**B**) Once all kinetochore pairs have bi-oriented on the mitotic spindle at metaphase, ubiquitination by the APC/C triggers degradation of securin. Securin degradation liberates separase, which resolves sister chromatid cohesion and thereby initiates the pole-ward movement of chromatids at anaphase. (**C**) The embrace model proposes that cohesin holds sister chromatids together by entrapping them within its ring structure. The ring should be large enough to encompass two 10 nm chromatin fibres (*e.g.* DNA bounds to nucleosomes, indicated as cylinders). Opening of the ring by separase cleavage of Scc1 would trigger the release of the entrapped sister DNAs.

9.5.1 Separase Cleaves Scc1

How does the activation of separase destroy cohesion? The first hint to this question came from the finding that extracts prepared from yeast strains that express large amounts of separase solubilized cohesin from a chromatin-containing pellet. Surprisingly, Scc1 released from chromosomes in such a way had been cut into two halves.[69] This explained the *in vivo* observation in yeast that Scc1 is an unstable protein that requires *de novo* synthesis at late G1 stage of the next cell cycle, prior to the next round of DNA replication.[6] Preventing Scc1 cleavage by mutation of its cleavage sites inhibited cohesin release from chromosomes and sister separation upon separase activation not only in yeast but also in human cells.[70] It turned out that separase itself is the site-specific thiol protease that cleaves Scc1 at one of two target sites[71] (Figure 9.6C). Crucially, cleavage by a foreign (tobacco etch virus, TEV) protease at an engineered target site within Scc1 is sufficient to release cohesin from chromosomes and to trigger sister splitting in the absence of separase activation.[71]

9.5.2 Regulation of Separase Activity

Cleavage of Scc1 by separase is irreversible, and therefore activation of the protease has to be tightly controlled. The discoveries that non-degradable securin blocks the onset of anaphase,[72] and that separase release following securin degradation by the proteasome is essential for triggering sister separation,[73] first suggested that securin is the major regulator of separase activity. Strikingly, yeast as well as mice lacking securin are viable[74,75] and human cells lacking securin still seem to separate their sister chromatids in a cell-cycle regulated manner, but show high rates of aneuploidy during the divisions immediately following loss of both securin alleles.[1] The frequent loss of chromosomes is, surprisingly, not due to a premature loss of cohesion but, quite the contrary, due to a failure to activate separase. Securin must therefore have an inhibitory as well as an activating effect on separase. Studies of the yeast proteins showed that while securin binding to separase prevents access of substrates to its C-terminal active site, it also blocks interaction of separase's N- and C-terminal domains, which is thought to be required to fully turn on its protease activity.[76] Securin might hence be considered an inhibitory chaperone for separase. A second, securin-independent level of separase regulation has been discovered in human cells. Separase stays inactive even after securin release in extracts with high cyclin-dependent kinase (CDK) activity. This is due to phosphorylation of a single serine residue within separase by cyclin-CDK and the formation of an inhibitory complex between separase and cyclin-CDK.[77,78] Thus human separase is only fully functional after the metaphase to anaphase transition when CDK activity drops abruptly on degradation of cyclins (see Figure 1.7).

Not only separase activity, but also the susceptibility of its target Scc1 to cleavage, is under regulatory control. Yeast Scc1 is phosphorylated by polo-like kinase (PLK) at several serine residues *in vitro*, including one within its separase target sequence. Mutation of this serine residue to alanine prevents sister separation in the absence of securin, while PLK phosphorylation renders Scc1 a better substrate for separase *in vitro*.[79]

9.5.3 A Separase-independent Pathway to Remove Cohesin from Vertebrate Chromosomes

It was initially puzzling that the bulk of cohesin dissociates from vertebrate chromosomes at the onset of mitosis, *i.e.* at a time when separase has not yet been activated. This prophase pathway of cohesin removal depends on phosphorylation of cohesin's Scc3/SA subunit by PLK[80–82] and a recently identified protein named Wapl.[83] However, residual amounts of cohesin remain at centromeres and distinct sites along chromosome arms, and are sufficient and necessary to hold sister chromatids together until anaphase. A protein called shugoshin is the key player in protecting centromeric cohesin from prophase release,[84] potentially by recruiting PP2A phosphatase to counteract Scc3/SA phosphorylation (Table 9.1). Like in yeast, separase is finally required to trigger anaphase also in mammalian cells by resolving the residual cohesion.[85,86]

9.6 Maintenance of Chromatid Stability during Segregation: Condensin

Another important aspect for chromosome segregation is to prevent chromatids from getting tangled up during their movement towards the cell poles, and to keep them compact enough not to get into the way of, and be trapped by, the dividing cell during cytokinesis. This requires a five-subunit SMC protein complex very similar to cohesin. Its Smc2 and Smc4 subunits (Table 9.2) were discovered as major protein components of mitotic chromosomes isolated from chicken cell lines[87] or assembled *in vitro* by incubation of sperm chromatin in frog egg mitotic extract.[88] Antibody depletion of its SMC or non-SMC subunits from mitotic extracts prevented the formation of compact, rod-shaped chromatin.[89] This suggested that the complex is essential for mitotic chromosome condensation and was hence named condensin.

Fission or budding yeast mutants of condensin subunits similarly show less compacted chromosomes.[90–92] Surprisingly, chromosomes appear to compact to a more or less normal extent when condensin subunits are mutated or depleted by RNA interference or promoter repression in fruit fly, chicken, roundworm or human cells.[93–100] Yet in all cases, sister chromatids fail to segregate, their centromeres are overstretched under tension, and chromosomes lose their structural integrity when exposed to hypotonic conditions. The consensus view is that condensin is required to provide structural integrity to mitotic chromosomes and keep them in a compacted state, but it seems unlikely that condensin is required for the initial chromosome compaction activity.

Table 9.2 The condensin complex

Name	Homologue in condensin II	Function	Interactions	*S. cerevisiae* MW (kDa)	aa	Molecules per cell [118]
Brn1 Cnd2 Barren CAP-H	CAP-H2	γ-Kleisin subunit of the condensin complex	Smc2 via its N- and Smc4 via its C-terminus	83.0	728	704
Smc4 Cut3 CAP-C		Coiled-coil subunit of the condensin complex	Brn1 C-terminus	162.1	1418	573
Smc2 Cut14 CAP-E		Coiled-coil subunit of the condensin complex	Brn1 N-terminus	133.9	1170	3290
Ycg1 Cnd3 CAP-G	CAP-G2	HEAT repeat containing subunit of the condensin complex	Brn1 C-terminus	119.5	1051	1920
Ycs4 Cnd1 CAP-D2	CAP-D3	HEAT repeat containing subunit of the condensin complex	Brn1 N-terminus	132.9	1176	2540

9.6.1 Association of Condensin with Chromosomes

How could condensins shape a chromosome? Three non-SMC subunits, which come as two sets of different isoforms in vertebrates (condensin I and II; for review see ref. 101), bind to the head domains of Smc2/Smc4 heterodimers in electron micrographs.[22] Sequence homology of the Barren/CAP-H non-SMC subunit with cohesin's Scc1 kleisin subunit first suggested that it might connect the Smc2/Smc4 head domains.[102] Experiments with recombinant condensin complexes showed that Barren indeed binds to the Smc2 and Smc4 ATPase head domains *via* its N- and C-terminus, respectively.[103] The other two non-SMC subunits are predicted to contain HEAT repeats similar to cohesin's Pds5 subunit and are recruited to the complex *via* their binding to Barren. Electron spectroscopic imaging suggests that they form a globular structure at the Smc2/Smc4 head domains that can wrap DNA around them *in vitro*.[104] At high protein : DNA ratios, condensin or Smc2/Smc4 dimers can introduce positive supercoils or knots into circular DNA in the presence of topoisomerases *in vitro*.[41,105–107] Whether DNA wrapping, supercoiling or knotting activities play a role in structuring a mitotic chromosome, or might just be consequences of condensin's association with DNA *in vitro*, is not known.

Another possibility is that condensin might provide integrity to a chromosome by holding onto, and thereby bringing together, two distant chromosomal sites, which would loop the region between the two sites. Support that condensin could loop chromosomes comes from single molecule experiments. When a linear DNA fragment is stretched between a glass slide and a magnetic bead, and condensin immunopurified from mitotic extracts and ATP are added, the magnetic bead is pulled rapidly towards the glass slide due to shortening of the DNA.[108]

It has been suggested that several condensin complexes might associate to form higher-order scaffolds along the central axis of a chromosome with loops of chromosomal DNA emerging from them[109] (Figure 9.7A). This is consistent with the staining of condensin along the central axis of mitotic chromosomes.[110,111] Alternatively, condensin might function as monomeric complexes that crosslink two chromosomal regions in a less organized manner[112] (Figure 9.7B). Given that condensin presumably forms a large ring structure like cohesin, it is tempting to speculate that it might do so by trapping chromosomal regions within its ring. In either case, it appears that condensin's binding to chromosomes is a dynamic process. Photobleaching experiments with fluorescently labelled condensin I show that it rapidly turns over on mitotic chromosomes in mammalian or fruit fly cells.[100,113] Cohesin, on the other hand, is stably bound to chromosomes after DNA replication.[114]

9.7 Outlook

Starting from their discovery only ten years ago, cohesin and condensin have quickly moved to the centre stage of chromosome biology. The fact that SMC

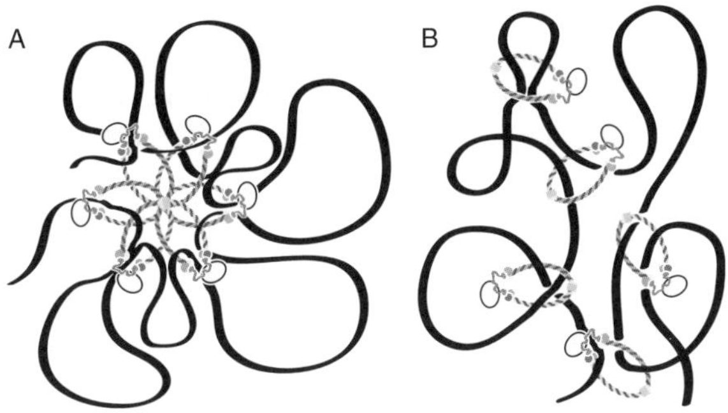

Figure 9.7 **Models for mitotic chromosome organization by condensin.** (**A**) Condensin could assemble into multimeric assemblies along the centre of the chromatid axis to stabilize radial loops of chromatin. (**B**) Alternatively, individual condensin molecules could clamp together different regions of a chromatid to provide structural stability along the chromatid.

protein complexes can be found in nearly every living cell on the planet underlines their importance as chromosomal organizers during cell division and beyond. Accumulating evidence now suggests that cohesin is not only vital for segregating chromosomes during mitosis, but in addition plays a crucial role in regulating gene transcription during interphase (reviewed in ref. 115). Mutations in Smc1, Scc2/NIPLB and Eco1/ESCO2 have been linked to Cornelia de Lange and Roberts syndromes (reviewed in ref. 116). The causes for these developmental disorders are most probably changes in gene expression rather than defects in chromosome segregation.

To understand how SMC complexes can fulfil their various functions in chromosome metabolism, it will be essential to find out how they work at the molecular level. The discovery that cohesin forms a large ring structure raises the possibility that it could entrap sister chromatids within its embrace and thereby work as a protein-DNA 'concatenase'. A large body of biochemical data suggests that this is indeed the case. It is quite likely that other SMC complexes like condensin might interact with chromosomal DNA in a similar topological fashion. Inevitably, this notion raises crucial questions as to how DNA enters and exits these rings, how such an entrapment could be used to structure chromosomes and control transcription, whether SMC complexes function as separate entities or multimeric assemblies, and what the consequences of ATP binding and hydrolysis by the SMC head domains are.

An important step forward to answer these questions will be the development of a functional *in vitro* system with defined components that recapitulates the *in vivo* properties of cohesin and condensin. In addition, it will be necessary to obtain more detailed pictures of these complexes bound to chromosomes at the light or electron microscopical level, which might give important clues as to how cohesin and condensin work. Finally, the development of tools to manipulate the function of cohesin and condensin in living cells and follow the consequences by real time imaging will be required. Chemical biology might be the key to the generation of such tools. For example, specific small molecule SMC ATPase inhibitors or modified peptides to disrupt specific protein–protein interactions may very well stand behind the breakthrough in deciphering the molecular machinery behind SMC complex functions.

References

1. P. V. Jallepalli, I. C. Waizenegger, F. Bunz, S. Langer, M. R. Speicher, J. M. Peters, K. W. Kinzler, B. Vogelstein and C. Lengauer, Securin is required for chromosomal stability in human cells, *Cell*, 2001, **105**, 445–457.
2. A. W. Murray and J. W. Szostak, Chromosome segregation in mitosis and meiosis, *Annu. Rev. Cell. Biol.*, 1985, **1**, 289–315.
3. S. DiNardo, K. Voelkel and R. Sternglanz, DNA topoisomerase II mutant of *Saccharomyces cerevisiae*: topoisomerase II is required for segregation of daughter molecules at the termination of DNA replication, *Proc. Natl. Acad. Sci. U.S.A.*, 1984, **81**, 2616–2620.
4. D. E. Koshland and L. H. Hartwell, The structure of sister minichromosome DNA before anaphase in *Saccharomyces cerevisiae*, *Science*, 1987, **238**, 1713–1716.
5. V. Guacci, D. E. Koshland and A. Strunnikov, A direct link between sister chromatid cohesion and chromosome condensation revealed through the analysis of MCD1 in *S. cerevisiae*, *Cell*, 1997, **91**, 47–57.
6. C. Michaelis, R. Ciosk and K. Nasmyth, Cohesins: chromosomal proteins that prevent premature separation of sister chromatids, *Cell*, 1997, **91**, 35–45.
7. A. Tóth, R. Ciosk, F. Uhlmann, M. Galova, A. Schleiffer and K. Nasmyth, Yeast cohesin complex requires a conserved protein, Eco1p(Ctf7), to establish cohesion between sister chromatids during DNA replication, *Genes Dev.*, 1999, **13**, 320–333.
8. T. Hartman, K. Stead, D. E. Koshland and V. Guacci, Pds5p is an essential chromosomal protein required for both sister chromatid cohesion and condensation in *Saccharomyces cerevisiae*, *J Cell Biol*, 2000, **151**, 613–626.
9. S. Panizza, T. Tanaka, A. Hochwagen, F. Eisenhaber and K. Nasmyth, Pds5 cooperates with cohesin in maintaining sister chromatid cohesion, *Curr. Biol.*, 2000, **10**, 1557–1564.
10. I. Sumara, E. Vorlaufer, C. Gieffers, B. H. Peters and J. M. Peters, Characterization of vertebrate cohesin complexes and their regulation in prophase, *J. Cell Biol.*, 2000, **151**, 749–762.
11. S. W. Wang, R. L. Read and C. J. Norbury, Fission yeast Pds5 is required for accurate chromosome segregation and for survival after DNA damage or metaphase arrest, *J. Cell Sci.*, 2002, **115**, 587–598.
12. A. F. Neuwald and T. Hirano, HEAT repeats associated with condensins, cohesins, and other complexes involved in chromosome-related functions, *Genome Res.*, 2000, **10**, 1445–1452.

13. F. Klein, P. Mahr, M. Gálová, S. B. Buonomo, C. Michaelis, K. Nairz and K. Nasmyth, A central role for cohesins in sister chromatid cohesion, formation of axial elements, and recombination during yeast meiosis, *Cell*, 1999, **98**, 91–103.
14. A. Losada, T. Yokochi, R. Kobayashi and T. Hirano, Identification and characterization of SA/Scc3p subunits in the *Xenopus* and human cohesin complexes, *J. Cell Biol.*, 2000, **150**, 405–416.
15. A. Losada, T. Yokochi and T. Hirano, Functional contribution of Pds5 to cohesin-mediated cohesion in human cells and *Xenopus* egg extracts, *J. Cell Sci.*, 2005, **118**, 2133–2141.
16. E. Revenkova, M. Eijpe, C. Heyting, B. Gross and R. Jessberger, Novel meiosis-specific isoform of mammalian SMC1, *Mol. Cell. Biol.*, 2001, **21**, 6984–6998.
17. K. Nasmyth and C. H. Haering, The structure and function of SMC and kleisin complexes, *Annu. Rev. Biochem.*, 2005, **74**, 595–648.
18. T. E. Melby, C. N. Ciampaglio, G. Briscoe and H. P. Erickson, The symmetrical structure of structural maintenance of chromosomes (SMC) and MukB proteins: long, antiparallel coiled coils, folded at a flexible hinge, *J. Cell Biol.*, 1998, **142**, 1595–1604.
19. C. H. Haering, J. Löwe, A. Hochwagen and K. Nasmyth, Molecular architecture of SMC proteins and the yeast cohesin complex, *Mol. Cell*, 2002, **9**, 773–788.
20. M. Hirano and T. Hirano, Hinge-mediated dimerization of SMC protein is essential for its dynamic interaction with DNA, *EMBO J.*, 2002, **21**, 5733–5744.
21. S. Gruber, C. H. Haering and K. Nasmyth, Chromosomal cohesin forms a ring, *Cell*, 2003, **112**, 765–777.
22. D. E. Anderson, A. Losada, H. P. Erickson and T. Hirano, Condensin and cohesin display different arm conformations with characteristic hinge angles, *J. Cell Biol.*, 2002, **156**, 419–424.
23. M. Milutinovich and D. E. Koshland, Molecular biology. SMC complexes—wrapped up in controversy, *Science*, 2003, **300**, 1101–1102.
24. D. Ivanov and K. Nasmyth, A topological interaction between cohesin rings and a circular minichromosome, *Cell*, 2005, **122**, 849–860.
25. A. Lengronne, Y. Katou, S. Mori, S. Yokobayashi, G. P. Kelly, T. Itoh, Y. Watanabe, K. Shirahige and F. Uhlmann, Cohesin relocation from sites of chromosomal loading to places of convergent transcription, *Nature*, 2004, **430**, 573–578.
26. C. H. Haering, A.-M. Farcas, P. Arumugam, J. Metson and K. Nasmyth, The cohesin ring concatenates individual sister DNA molecules, *Nature*, 2008, **454**, 297–301.
27. A. Losada, M. Hirano and T. Hirano, Identification of *Xenopus* SMC protein complexes required for sister chromatid cohesion, *Genes Dev.*, 1998, **12**, 1986–1997.
28. S. Gruber, P. Arumugam, Y. Katou, D. Kuglitsch, W. Helmhart, K. Shirahige and K. Nasmyth, Evidence that loading of cohesin

onto chromosomes involves opening of its SMC hinge, *Cell*, 2006, **127**, 523–537.

29. P. Arumugam, S. Gruber, K. Tanaka, C. H. Haering, K. Mechtler and K. Nasmyth, ATP hydrolysis is required for cohesin's association with chromosomes, *Curr. Biol.*, 2003, **13**, 1941–1953.
30. S. Weitzer, C. Lehane and F. Uhlmann, A model for ATP hydrolysis-dependent binding of cohesin to DNA, *Curr. Biol.*, 2003, **13**, 1930–1940.
31. M. Hirano and T. Hirano, Opening closed arms: long-distance activation of SMC ATPase by hinge-DNA interactions, *Mol. Cell*, 2006, **21**, 175–186.
32. K. P. Hopfner, A. Karcher, D. S. Shin, L. Craig, L. M. Arthur, J. P. Carney and J. A. Tainer, Structural biology of Rad50 ATPase: ATP-driven conformational control in DNA double-strand break repair and the ABC-ATPase superfamily, *Cell*, 2000, **101**, 789–800.
33. A. Lammens, A. Schele and K. P. Hopfner, Structural biochemistry of ATP-driven dimerization and DNA-stimulated activation of SMC ATPases, *Curr. Biol.*, 2004, **14**, 1778–1782.
34. C. H. Haering, D. Schoffnegger, T. Nishino, W. Helmhart, K. Nasmyth and J. Löwe, Structure and stability of cohesin's Smc1–kleisin interaction, *Mol. Cell*, 2004, **15**, 951–964.
35. P. Arumugam, T. Nishino, C. H. Haering, S. Gruber and K. Nasmyth, Cohesin's ATPase activity is stimulated by the C-terminal winged-helix domain of its kleisin subunit, *Curr. Biol.*, 2006, **16**, 1998–2008.
36. J. E. Moody, L. Millen, D. Binns, J. F. Hunt and P. J. Thomas, Cooperative, ATP-dependent association of the nucleotide binding cassettes during the catalytic cycle of ATP-binding cassette transporters, *J. Biol. Chem.*, 2002, **277**, 21111–21114.
37. P. C. Smith, N. Karpowich, L. Millen, J. E. Moody, J. Rosen, P. J. Thomas and J. F. Hunt, ATP binding to the motor domain from an ABC transporter drives formation of a nucleotide sandwich dimer, *Mol. Cell*, 2002, **10**, 139–149.
38. M. Hirano and T. Hirano, Positive and negative regulation of SMC–DNA interactions by ATP and accessory proteins, *EMBO J.*, 2004, **23**, 2664–2673.
39. L. W. Hung, I. X. Wang, K. Nikaido, P. Q. Liu, G. F. Ames and S. H. Kim, Crystal structure of the ATP-binding subunit of an ABC transporter, *Nature*, 1998, **396**, 703–707.
40. M. Hirano, D. E. Anderson, H. P. Erickson and T. Hirano, Bimodal activation of SMC ATPase by intra- and inter-molecular interactions, *EMBO J.*, 2001, **20**, 3238–3250.
41. K. Kimura and T. Hirano, ATP-dependent positive supercoiling of DNA by 13S condensin: a biochemical implication for chromosome condensation, *Cell*, 1997, **90**, 625–634.
42. K. Hollenstein, R. J. Dawson and K. P. Locher, Structure and mechanism of ABC transporter proteins, *Curr. Opin. Struct. Biol.*, 2007, **17**, 412–418.

43. S. H. Yoshimura, K. Hizume, A. Murakami, T. Sutani, K. Takeyasu and M. Yanagida, Condensin architecture and interaction with DNA: regulatory non-SMC subunits bind to the head of SMC heterodimer, *Curr. Biol.*, 2002, **12**, 508–513.
44. A. Losada and T. Hirano, Intermolecular DNA interactions stimulated by the cohesin complex *in vitro*: implications for sister chromatid cohesion, *Curr. Biol.*, 2001, **11**, 268–272.
45. R. Ciosk, M. Shirayama, A. Shevchenko, T. Tanaka, A. Toth and K. Nasmyth, Cohesin's binding to chromosomes depends on a separate complex consisting of Scc2 and Scc4 proteins, *Mol. Cell*, 2000, **5**, 243–254.
46. P. Bernard, J. Drogat, J. F. Maure, S. Dheur, S. Vaur, S. Genier and J. P. Javerzat, A screen for cohesion mutants uncovers Ssl3, the fission yeast counterpart of the cohesin loading factor Scc4, *Curr. Biol.*, 2006, **16**, 875–881.
47. E. Watrin, A. Schleiffer, K. Tanaka, F. Eisenhaber, K. Nasmyth and J. M. Peters, Human Scc4 is required for cohesin binding to chromatin, sister-chromatid cohesion, and mitotic progression, *Curr. Biol.*, 2006, **16**, 863–874.
48. V. C. Seitan, P. Banks, S. Laval, N. A. Majid, D. Dorsett, A. Rana, J. Smith, A. Bateman, S. Krpic, A. Hostert, R. A. Rollins, H. Erdjument-Bromage, P. Tempst, C. Y. Benard, S. Hekimi, S. F. Newbury and T. Strachan, Metazoan Scc4 homologs link sister chromatid cohesion to cell and axon migration guidance, *PLoS Biol.*, 2006, **4**, e242.
49. F. Uhlmann and K. Nasmyth, Cohesion between sister chromatids must be established during DNA replication, *Curr. Biol.*, 1998, **8**, 1095–1101.
50. R. V. Skibbens, L. B. Corson, D. E. Koshland and P. Hieter, Ctf7p is essential for sister chromatid cohesion and links mitotic chromosome structure to the DNA replication machinery, *Genes Dev.*, 1999, **13**, 307–319.
51. D. Ivanov, A. Schleiffer, F. Eisenhaber, K. Mechtler, C. H. Haering and K. Nasmyth, Eco1 is a novel acetyltransferase that can acetylate proteins involved in cohesion, *Curr. Biol.*, 2002, **12**, 323–328.
52. M. L. Mayer, S. P. Gygi, R. Aebersold and P. Hieter, Identification of RFC(Ctf18p, Ctf8p, Dcc1p): an alternative RFC complex required for sister chromatid cohesion in *S. cerevisiae*, *Mol. Cell*, 2001, **7**, 959–970.
53. V. P. Bermudez, Y. Maniwa, I. Tappin, K. Ozato, K. Yokomori and J. Hurwitz, The alternative Ctf18-Dcc1-Ctf8-replication factor C complex required for sister chromatid cohesion loads proliferating cell nuclear antigen onto DNA, *Proc. Natl. Acad. Sci. U.S.A.*, 2003, **100**, 10237–10242.
54. M. A. Kenna and R. V. Skibbens, Mechanical link between cohesion establishment and DNA replication: Ctf7p/Eco1p, a cohesion establishment factor, associates with three different replication factor C complexes, *Mol. Cell. Biol.*, 2003, **23**, 2999–3007.
55. M. Petronczki, B. Chwalla, M. F. Siomos, S. Yokobayashi, W. Helmhart, A. M. Deutschbauer, R. W. Davis, Y. Watanabe and K. Nasmyth, Sister-chromatid cohesion mediated by the alternative RF-CCtf18/Dcc1/Ctf8, the helicase Chl1 and the polymerase-alpha-associated protein Ctf4 is

essential for chromatid disjunction during meiosis II, *J. Cell Sci.*, 2004, **117**, 3547–3559.

56. G. L. Moldovan, B. Pfander and S. Jentsch, PCNA controls establishment of sister chromatid cohesion during S phase, *Mol. Cell*, 2006, **23**, 723–732.
57. B. C. Williams, C. M. Garrett-Engele, Z. Li, E. V. Williams, E. D. Rosenman and M. L. Goldberg, Two putative acetyltransferases, san and deco, are required for establishing sister chromatid cohesion in Drosophila, *Curr. Biol.*, 2003, **13**, 2025–2036.
58. F. Hou and H. Zou, Two human orthologues of Eco1/Ctf7 acetyltransferases are both required for proper sister-chromatid cohesion, *Mol. Biol. Cell*, 2005, **16**, 3908–3918.
59. K. Tanaka, T. Yonekawa, Y. Kawasaki, M. Kai, K. Furuya, M. Iwasaki, H. Murakami, M. Yanagida and H. Okayama, Fission yeast Eso1p is required for establishing sister chromatid cohesion during S phase, *Mol. Cell. Biol.*, 2000, **20**, 3459–3469.
60. K. Tanaka, Z. Hao, M. Kai and H. Okayama, Establishment and maintenance of sister chromatid cohesion in fission yeast by a unique mechanism, *EMBO J.*, 2001, **20**, 5779–5790.
61. J. S. Hanna, E. S. Kroll, V. Lundblad and F. A. Spencer, *Saccharomyces cerevisiae* CTF18 and CTF4 are required for sister chromatid cohesion, *Mol. Cell. Biol.*, 2001, **21**, 3144–3158.
62. R. V. Skibbens, Chl1p, a DNA helicase-like protein in budding yeast, functions in sister-chromatid cohesion, *Genetics*, 2004, **166**, 33–42.
63. J. S. Kim, T. B. Krasieva, V. LaMorte, A. M. Taylor and K. Yokomori, Specific recruitment of human cohesin to laser-induced DNA damage, *J. Biol. Chem.*, 2002, **277**, 45149–45153.
64. E. Unal, A. Arbel-Eden, U. Sattler, R. Shroff, M. Lichten, J. E. Haber and D. E. Koshland, DNA damage response pathway uses histone modification to assemble a double-strand break-specific cohesin domain, *Mol. Cell*, 2004, **16**, 991–1002.
65. L. Ström, H. B. Lindroos, K. Shirahige and C. Sjögren, Postreplicative recruitment of cohesin to double-strand breaks is required for DNA repair, *Mol. Cell*, 2004, **16**, 1003–1015.
66. L. Ström and C. Sjögren, DNA damage-induced cohesion, *Cell Cycle*, 2005, **4**, 536–539.
67. L. Ström, C. Karlsson, H. B. Lindroos, S. Wedahl, Y. Katou, K. Shirahige and C. Sjögren, Postreplicative formation of cohesion is required for repair and induced by a single DNA break, *Science*, 2007, **317**, 242–245.
68. E. Unal, J. M. Heidinger-Pauli and D. E. Koshland, DNA double-strand breaks trigger genome-wide sister-chromatid cohesion through Eco1 (Ctf7), *Science*, 2007, **317**, 245–248.
69. F. Uhlmann, F. Lottspeich and K. Nasmyth, Sister-chromatid separation at anaphase onset is promoted by cleavage of the cohesin subunit Scc1, *Nature*, 1999, **400**, 37–42.

70. S. Hauf, I. C. Waizenegger and J. M. Peters, Cohesin cleavage by separase required for anaphase and cytokinesis in human cells, *Science*, 2001, **293**, 1320–1323.
71. F. Uhlmann, D. Wernic, M. A. Poupart, E. V. Koonin and K. Nasmyth, Cleavage of cohesin by the CD clan protease separin triggers anaphase in yeast, *Cell*, 2000, **103**, 375–386.
72. O. Cohen-Fix, J. M. Peters, M. W. Kirschner and D. Koshland, Anaphase initiation in *Saccharomyces cerevisiae* is controlled by the APC-dependent degradation of the anaphase inhibitor Pds1p, *Genes Dev.*, 1996, **10**, 3081–3093.
73. R. Ciosk, W. Zachariae, C. Michaelis, A. Shevchenko, M. Mann and K. Nasmyth, An ESP1/PDS1 complex regulates loss of sister chromatid cohesion at the metaphase to anaphase transition in yeast, *Cell*, 1998, **93**, 1067–1076.
74. A. Yamamoto, V. Guacci and D. Koshland, Pds1p is required for faithful execution of anaphase in the yeast, *Saccharomyces cerevisiae*, *J. Cell Biol.*, 1996, **133**, 85–97.
75. J. Mei, X. Huang and P. Zhang, Securin is not required for cellular viability, but is required for normal growth of mouse embryonic fibroblasts, *Curr. Biol.*, 2001, **11**, 1197–1201.
76. N. C. Hornig, P. P. Knowles, N. Q. McDonald and F. Uhlmann, The dual mechanism of separase regulation by securin, *Curr. Biol.*, 2002, **12**, 973–982.
77. O. Stemmann, H. Zou, S. A. Gerber, S. P. Gygi and M. W. Kirschner, Dual inhibition of sister chromatid separation at metaphase, *Cell*, 2001, **107**, 715–726.
78. I. H. Gorr, D. Boos and O. Stemmann, Mutual inhibition of separase and Cdk1 by two-step complex formation, *Mol. Cell*, 2005, **19**, 135–141.
79. G. Alexandru, F. Uhlmann, K. Mechtler, M. A. Poupart and K. Nasmyth, Phosphorylation of the cohesin subunit Scc1 by Polo/Cdc5 kinase regulates sister chromatid separation in yeast, *Cell*, 2001, **105**, 459–472.
80. A. Losada, M. Hirano and T. Hirano, Cohesin release is required for sister chromatid resolution, but not for condensin-mediated compaction, at the onset of mitosis, *Genes Dev.*, 2002, **16**, 3004–3016.
81. I. Sumara, E. Vorlaufer, P. T. Stukenberg, O. Kelm, N. Redemann, E. A. Nigg and J. M. Peters, The dissociation of cohesin from chromosomes in prophase is regulated by Polo-like kinase, *Mol. Cell*, 2002, **9**, 515–525.
82. S. Hauf, E. Roitinger, B. Koch, C. M. Dittrich, K. Mechtler and J. M. Peters, Dissociation of cohesin from chromosome arms and loss of arm cohesion during early mitosis depends on phosphorylation of SA2, *PLoS Biol.*, 2005, **3**, e69.
83. S. Kueng, B. Hegemann, B. H. Peters, J. J. Lipp, A. Schleiffer, K. Mechtler and J. M. Peters, Wapl controls the dynamic association of cohesin with chromatin, *Cell*, 2006, **127**, 955–967.
84. B. E. McGuinness, T. Hirota, N. R. Kudo, J. M. Peters and K. Nasmyth, Shugoshin prevents dissociation of cohesin from centromeres during mitosis in vertebrate cells, *PLoS Biol.*, 2005, **3**, e86.

85. K. G. Wirth, G. Wutz, N. R. Kudo, C. Desdouets, A. Zetterberg, S. Taghybeeglu, J. Seznec, G. M. Ducos, R. Ricci, N. Firnberg, J. M. Peters and K. Nasmyth, Separase: a universal trigger for sister chromatid disjunction but not chromosome cycle progression, *J. Cell Biol.*, 2006, **172**, 847–860.
86. I. C. Waizenegger, S. Hauf, A. Meinke and J. M. Peters, Two distinct pathways remove mammalian cohesin from chromosome arms in prophase and from centromeres in anaphase, *Cell*, 2000, **103**, 399–410.
87. N. Saitoh, I. G. Goldberg, E. R. Wood and W. C. Earnshaw, ScII: an abundant chromosome scaffold protein is a member of a family of putative ATPases with an unusual predicted tertiary structure, *J. Cell Biol.*, 1994, **127**, 303–318.
88. T. Hirano and T. J. Mitchison, A heterodimeric coiled-coil protein required for mitotic chromosome condensation *in vitro*, *Cell*, 1994, **79**, 449–458.
89. T. Hirano, R. Kobayashi and M. Hirano, Condensins, chromosome condensation protein complexes containing XCAP-C, XCAP-E and a *Xenopus* homolog of the *Drosophila* Barren protein, *Cell*, 1997, **89**, 511–521.
90. Y. Saka, T. Sutani, Y. Yamashita, S. Saitoh, M. Takeuchi, Y. Nakaseko and M. Yanagida, Fission yeast cut3 and cut14, members of a ubiquitous protein family, are required for chromosome condensation and segregation in mitosis, *EMBO J.*, 1994, **13**, 4938–4952.
91. A. V. Strunnikov, E. Hogan and D. E. Koshland, SMC2, a *Saccharomyces cerevisiae* gene essential for chromosome segregation and condensation, defines a subgroup within the SMC family, *Genes Dev.*, 1995, **9**, 587–599.
92. B. D. Lavoie, K. M. Tuffo, S. Oh, D. E. Koshland and C. Holm, Mitotic chromosome condensation requires Brn1p, the yeast homologue of Barren, *Mol. Biol. Cell*, 2000, **11**, 1293–1304.
93. M. A. Bhat, A. V. Philp, D. M. Glover and H. J. Bellen, Chromatid segregation at anaphase requires the barren product, a novel chromosome-associated protein that interacts with Topoisomerase II, *Cell*, 1996, **87**, 1103–1114.
94. S. Steffensen, P. A. Coelho, N. Cobbe, S. Vass, M. Costa, B. Hassan, S. N. Prokopenko, H. Bellen, M. M. Heck and C. E. Sunkel, A role for *Drosophila* SMC4 in the resolution of sister chromatids in mitosis, *Curr. Biol.*, 2001, **11**, 295–307.
95. M. P. Somma, B. Fasulo, G. Siriaco and G. Cenci, Chromosome condensation defects in barren RNA-interfered *Drosophila* cells, *Genetics*, 2003, **165**, 1607–1611.
96. R. A. Oliveira, P. A. Coelho and C. E. Sunkel, The condensin I subunit Barren/CAP-H is essential for the structural integrity of centromeric heterochromatin during mitosis, *Mol. Cell. Biol.*, 2005, **25**, 8971–8984.

97. D. F. Hudson, P. Vagnarelli, R. Gassmann and W. C. Earnshaw, Condensin is required for nonhistone protein assembly and structural integrity of vertebrate mitotic chromosomes, *Dev. Cell*, 2003, **5**, 323–336.
98. K. A. Hagstrom, V. F. Holmes, N. R. Cozzarelli and B. J. Meyer, C. elegans condensin promotes mitotic chromosome architecture, centromere organization, and sister chromatid segregation during mitosis and meiosis, *Genes Dev.*, 2002, **16**, 729–742.
99. T. Hirota, D. Gerlich, B. Koch, J. Ellenberg and J. M. Peters, Distinct functions of condensin I and II in mitotic chromosome assembly, *J. Cell Sci.*, 2004, **117**, 6435–6445.
100. D. Gerlich, T. Hirota, B. Koch, J. M. Peters and J. Ellenberg, Condensin I stabilizes chromosomes mechanically through a dynamic interaction in live cells, *Curr. Biol.*, 2006, **16**, 333–344.
101. T. Hirano, Chromosome shaping by two condensins, *Cell Cycle*, 2004, **3**, 26–28.
102. A. Schleiffer, S. Kaitna, S. Maurer-Stroh, M. Glotzer, K. Nasmyth and F. Eisenhaber, Kleisins: a superfamily of bacterial and eukaryotic SMC protein partners, *Mol. Cell*, 2003, **11**, 571–575.
103. I. Onn, N. Aono, M. Hirano and T. Hirano, Reconstitution and subunit geometry of human condensin complexes, *EMBO J.*, 2007, **26**, 1024–1034.
104. D. P. Bazett-Jones, K. Kimura and T. Hirano, Efficient supercoiling of DNA by a single condensin complex as revealed by electron spectroscopic imaging, *Mol. Cell*, 2002, **9**, 1183–1190.
105. K. Kimura, V. V. Rybenkov, N. J. Crisona, T. Hirano and N. R. Cozzarelli, 13S condensin actively reconfigures DNA by introducing global positive writhe: implications for chromosome condensation, *Cell*, 1999, **98**, 239–248.
106. J. E. Stray, N. J. Crisona, B. P. Belotserkovskii, J. E. Lindsley and N. R. Cozzarelli, The *Saccharomyces cerevisiae* Smc2/4 condensin compacts DNA into (+) chiral structures without net supercoiling, *J. Biol. Chem.*, 2005, **280**, 34723–34734.
107. J. E. Stray and J. E. Lindsley, Biochemical analysis of the yeast condensin Smc2/4 complex: an ATPase that promotes knotting of circular DNA, *J. Biol. Chem.*, 2003, **278**, 26238–26248.
108. T. R. Strick, T. Kawaguchi and T. Hirano, Real-time detection of single-molecule DNA compaction by condensin I, *Curr. Biol.*, 2004, **14**, 874–880.
109. T. Hirano, SMC proteins and chromosome mechanics: from bacteria to humans, *Philos. Trans. R. Soc. Lond. Ser. Bi*, 2005, **360**, 507–514.
110. K. Maeshima and U. K. Laemmli, A two-step scaffolding model for mitotic chromosome assembly, *Dev. Cell*, 2003, **4**, 467–480.
111. T. Ono, Y. Fang, D. L. Spector and T. Hirano, Spatial and temporal regulation of Condensins I and II in mitotic chromosome assembly in human cells, *Mol. Biol. Cell*, 2004, **15**, 3296–3308.
112. C. H. Haering and K. Nasmyth, Building and breaking bridges between sister chromatids, *Bioessays*, 2003, **25**, 1178–1191.

113. R. A. Oliveira, S. Heidmann and C. E. Sunkel, Condensin I binds chromatin early in prophase and displays a highly dynamic association with *Drosophila* mitotic chromosomes, *Chromosoma*, 2007, **116**, 259–274.
114. D. Gerlich, B. Koch, F. Dupeux, J. M. Peters and J. Ellenberg, Live-cell imaging reveals a stable cohesin-chromatin interaction after but not before DNA replication, *Curr. Biol.*, 2006, **16**, 1571–1578.
115. D. Dorsett, Roles of the sister chromatid cohesion apparatus in gene expression, development, and human syndromes, *Chromosoma*, 2007, **116**, 1–13.
116. J. Liu and I. D. Krantz, Cohesin and human disease, *Annu. Rev. Genomics Hum. Genet.*, 2008, **9**, 303–320.
117. W. L. DeLano, http://pymol.org/, 2002.
118. N. Eswar, M. A. Marti-Renom, B. Webb, M. S. Madhusudhan, D. Eramian, M. Shen, U. Pieper and A. Sali, in *Current Protocols in Bioinformatics*, John Wiley & Sons, Inc., 2006, Unit 5.6.

CHAPTER 10

Replication of Chromatin

ANJA GROTH[a] AND GENEVIÈVE ALMOUZNI[b]

[a] Biotech Research and Innovation Centre (BRIC), University of Copenhagen, Ole Maaløes Vej 5, DK-2200, Copenhagen, Denmark; [b] Laboratory of Nuclear Dynamics and Genome Plasticity, UMR 218 CNRS/Institut Curie, 26 rue d'Ulm, 75248 Paris, Cedex 5, France

10.1 Introduction

The inheritance and maintenance of both DNA sequence and its organization into chromatin are central for eukaryotic life. Both the replication of DNA and chromatin organization represent major challenges, partly in relation to the accessibility of DNA in the context of chromatin and partly the complexity of chromatin organization. To meet the challenge of maintenance, cells have evolved efficient nucleosome assembly pathways and chromatin maturation mechanisms that reproduce chromatin organization in the wake of DNA replication.

In this chapter, we focus on how the basic framework of chromatin, nucleosomal organization, is duplicated during replication. This entails three critical steps:

- disruption of parental nucleosomes ahead of the fork;
- transfer and recycling of parental histones onto daughter strands;
- *de novo* deposition of histones.

The proper coordination of all these events with replication fork progression and replication of the underlying DNA sequence is key to ensure reproduction of a physiological nucleosome density on new DNA. Moreover, this represents

Molecular Themes in DNA Replication
Edited by Lynne S. Cox

Published by the Royal Society of Chemistry, www.rsc.org

a critical period during which both genetic and epigenetic stability can be compromised.

The basic building block of chromatin is the nucleosome core particle. This unit contains an octamer of histone proteins, that is an $(H3\text{-}H4)_2$ tetramer flanked on either side by two H2A-H2B dimers, around which 147 base pairs of double-stranded DNA are wrapped in 1.65 left-handed superhelical turns (see Figure 1.4). These units are repeated regularly with intervening linker DNA and incorporation of additional chromatin proteins, including the linker histone H1 and other non-histone proteins. This structure restricts access to the replication machinery. Chromatin can thus potentially represent a barrier both to initiation and elongation of the replication fork. Here, we focus mainly on the dynamics of histone and nucleosomes during replication elongation in order to highlight how duplication of DNA sequence and nucleosomal organization is coordinated. Chromatin changes (in particular histone acetylation) are implicated in initiation control, but mechanistically the link between these changes and the exact impact on the replication machinery is not well understood. This is covered briefly in the section dealing with chromatin disruption.

Two fundamentally distinct processes affect chromatin structure during DNA replication (Figure 10.1). The first is the 'transient disruption' of pre-existing nucleosomes located ahead of replication forks and their transfer onto nascent DNA, a reaction known as parental histone segregation and here called 'recycling'. The second is the deposition of newly synthesized histones through a pathway known as replication-dependent *de novo* nucleosome assembly.

Both parental histone recycling and *de novo* assembly affect the whole genome during each passage through S-phase. Therefore, these two processes potentially have a widespread and profound impact on the ability of proliferating cells to propagate or modify epigenetic states that depend upon specific chromatin structures. In this chapter, we start with the part concerning *de novo* histone deposition, for which many advances over the years have helped to gain a mechanistic understanding. We then move to the fate of parental histones—an area that is still in progress.

10.2 *De novo* Histone Deposition

10.2.1 Provision of Histones

Doubling the amount of DNA in the context of chromatin also requires doubling the number of nucleosomes. This requires a large supply of new histones and an effective delivery system to secure efficient provision at the replication fork. However, too large a supply of new histones is not desirable because excess undeposited/free histones can jeopardize genome stability.[1–3] Mechanisms have therefore evolved to ensure highly efficient production of histones in a manner that is tightly linked to ongoing DNA synthesis. These multilayered regulatory mechanisms operate both at the transcriptional and post-transcriptional level, with different strategies exploited in mammals and budding yeast.[4,5]

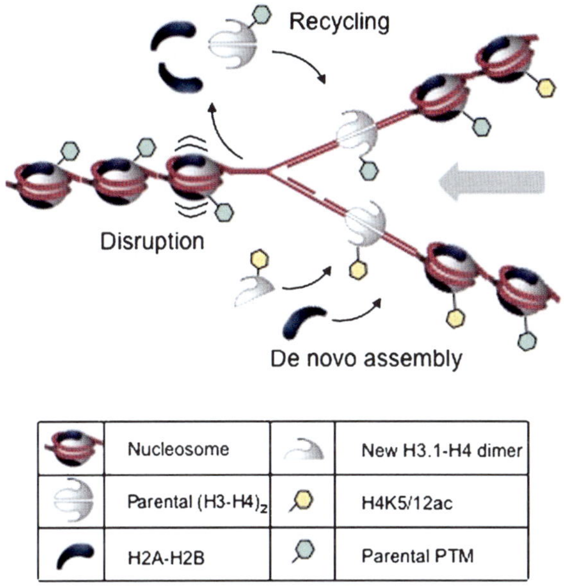

Figure 10.1 **Current model of histone dynamics at the replication fork.** Progression of the replication fork in chromatin entails disruption of parental nucleosomes immediately ahead of the fork. Parental histones released in this process are transferred efficiently to daughter strands in a random fashion. Provision of new histones and *de novo* assembly operate in parallel to ensure restoration of nucleosomal density. Recycling may involve H3-H4 dimers, but as new and old H3-H4 dimers in general do not mix, the separation of the parental tetramer must be transient. (Figure modified from ref. 59.)

In mammalian cells a set of replicative histones, so-called canonical histones, are encoded by the replication-dependent histone genes. These genes are characterized by:

(i) being clustered together;
(ii) being free of introns; and
(iii) encoding mRNAs that are not polyadenylated but carry a conserved stem-loop in their 3′ end.

Levels of these replication-dependent histone mRNAs increase up to 35-fold as cells progress from G1 into S phase due to transcriptional activation, increased efficiency of pre-mRNA processing, and mRNA stabilization.[5] One of the key co-activators of core histone genes, NPAT, is phosphorylated and activated by the master regulator of S phase entry, cyclin E-cdk2.[6–8] Transcription takes place in distinct structures called histone locus bodies, which are enriched in factors dedicated to histone pre-mRNA processing. The stem-loop binding protein (SLBP) is required for proper processing and stabilization of the mature histone mRNA.[9] Cell cycle regulation of SLBP expression and stability[10,11] plays a significant role in restricting high histone expression to S phase. In addition, histone gene expression is subject to additional levels of control including checkpoint signaling,[12–16] ensuring that production is rapidly shut down if DNA replication is perturbed and histone demand reduced. In addition, normal polyadenylated mRNAs are expressed more uniformly throughout the cell cycle and in quiescent cells from a minority of histone genes encoding replacement variants (*i.e.* H3.3, CENPA, H2AZ, macroH2A and H1°).

In yeast, histone mRNAs are polyadenylated, but sequences in their 3′ region participate to regulate their stability.[17] Histone chaperones play a role in regulating histone expression in yeast, suggesting a link between protein levels and gene expression. Hir1, Hir2 (Histone regulatory) and Asf1 (Anti-silencing factor) are required for proper periodic transcription of histone genes.[18,19] This may be a conserved feature of these H3-H4 chaperones, as HIRA (Hir homologue) can repress histone expression in human cells[20] and *Drosophila* Asf1 is found at histone gene puffs on polytene chromosomes.[21] Yeast also regulate histone provision at the level of protein stability. Whereas nucleosomal histones are generally highly stable, excess histones that cannot be incorporated or protected by a chaperone are degraded *via* a pathway that depends on the checkpoint kinase Rad53.[1] Although buffering of excess histones can be observed in mammals,[22] whether degradation is involved is an open question.

By matching histone supply to demand, these multifaceted control mechanisms ensure that at a steady state, the pool of non-nucleosomal core histones is kept minimal in somatic cells (approximately 1–2% of total histone protein),[23,24] even though a huge histone supply is necessary in S phase.

10.2.2 Histone Acetylation and Methylation in Regulating Histone Supply

Newly synthesized H3 and H4 are transiently acetylated at specific lysine residues within their amino-terminal tails.[24–26] These acetylations are significant for replication-coupled histone dynamics, as mutations of multiple lysine residues that compromise the acetylation of both H3 and H4 lead to a loss of cell viability associated with severe defects in chromatin structure during passage through S phase.[27] The acetylation pattern on H3 varies between different species and even between the human histone H3 variants.[24–26] In contrast, acetylation of newly synthesized histone H4 at lysine 5 and 12 is highly conserved among species.[25] Mass spectrometry analysis of soluble HeLa histone H3.1-H4 and H3.3-H4 dimers showed an almost exclusive enrichment of H4 diacetylated at K5/12.[24] In yeast, a two-subunit enzyme known as HAT1-RbAp46 is the prime candidate for acetylating newly synthesized H4 at K5/K12.[28] Although HAT1 is dispensable for replication-coupled nucleosome assembly, it is required for surviving DNA damage during replication in chicken DT40 cells.[29]

It is possible that these acetylations either help to distinguish new H3-H4 from old and/or ensure flexibility of nascent chromatin to allow time for repair and maturation processes. Intriguingly, in yeast an additional acetylation is imposed at lysine 56 on virtually all newly synthesized histone H3.[30–32] This acetylation is catalyzed by the unconventional KAT (lysine acetyl transferase), Rtt109 and requires Asf1.[32–37] Mutations that abolish H3K56 cause hypersensitivity to genotoxic agents that interfere with DNA replication in a manner that is epistatic with mutation of Asf1.[32,33] The exact function of this acetyl mark remains enigmatic, but it could generate a favourable environment to repair DNA lesions and/or deal with stalled replication forks. Several recent reports implicate H3K56ac directly in the process of nucleosome assembly and disassembly during replication, repair and transcription;[38–40] thus incomplete packaging of new DNA into chromatin may be responsible for the acute DNA-damage sensitivity of yeast lacking this post-translational modification. Given that only very little H3K56ac is present in proliferating human cells and that no Rtt109 homologue is present,[41,42] it is not clear whether a similar mechanism operates in vertebrates.

In addition to acetylation, methylation of histone H3 on K9 can also be detected on non-nucleosomal histones.[24] Thus, in addition to considering how the provision of new histones contributes to *de novo* histone deposition, their post-translational modification must also be considered. A key issue concerns the mechanisms at work following deposition of these new histones to ensure the maintenance of particular 'parental' histone marks in a chromatin domain.

10.3 The Assembly Line for *de novo* Histone Deposition

Early *in vivo* studies of histone dynamics using metabolic labelling and density gradient centrifugation showed that new H3-H4 is deposited almost exclusively at newly replicated DNA.[43,44] In contrast, a significant fraction of new H2A-H2B is also deposited onto 'old' non-replicating chromatin, as there is significant exchange of H2A-H2B outside replication, in particular associated with transcription.[45,46]

These *in vivo* approaches provided important knowledge about the contribution of new and old histones to nucleosome formation, but mechanistic insights awaited the development of *in vitro* systems to recapitulate chromatin assembly. Following the first *in vitro* system using *Xenopus* egg extract to assemble chromatin,[47] replication/DNA synthesis-coupled systems were established with the SV40-based cell free replication system[48] and *Xenopus* egg extract[49] which demonstrated coupling between histone deposition and DNA synthesis. This quickly led to the identification of CAF-1 (Chromatin Assembly Factor 1)[50] and later the isolation of Asf1,[51] two key histone chaperones in *de novo* assembly.[52]

10.3.1 CAF-1

CAF-1 is an evolutionarily conserved three-subunit protein with the unique ability to preferentially deposit newly synthesized H3-H4 onto replicating

DNA.[53–55] Consistent with this, human CAF-1 is found in a specific predeposition complex containing the major S-phase histone H3.1 and H4.[56] Notably, CAF-1 was not detected with the replacement histone variant H3.3, which is incorporated into chromatin independently of DNA replication.

CAF-1 can be targeted to sites of DNA synthesis *via* a direct interaction with proliferating cell nuclear antigen (PCNA),[54,57,58] a ring-shaped homotrimeric protein that encircles DNA and serves as a polymerase processivity factor and recruitment platform in DNA and chromatin maturation[59] (see Chapters 3, 6 and 7). The interaction of CAF-1 with PCNA can be modulated by Cdc7-Dbf4,[60] a protein kinase that is essential for DNA replication. Phosphorylation by Cdc7-Dbf4 enhances CAF-1 binding to PCNA by disrupting the ability to dimerize in its largest subunit.[60] This could be a mechanism to fine-tune histone deposition with ongoing DNA replication and/or other PCNA-dependent maturation processes. Indeed, *de novo* deposition of replicative histone H3.1 at sites of damage *in vivo* was shown to be a CAF-1-dependent event.[61]

10.3.2 Asf1 Maintains Histones as Dimers

A notable feature of newly synthesized H3-H4 is that, prior to their deposition onto DNA, they are found as dimers. First evidence for this came from the composition of non-nucleosomal complexes containing epitope-tagged H3.1 (replication-dependent variant), H3.3 (replacement variant) or H4.[26,56] This challenged the standing paradigm that H3-H4 away from DNA is constitutively in a tetrameric form and showed that the basic building blocks for *de novo* nucleosome assembly are H3-H4 dimers (Figure 10.1).

Interaction with histone chaperones may to a large extent explain why tetramer formation does not necessarily occur prior to deposition (Figure 10.2). Structural studies of Asf1 bound to H3-H4 indeed showed a physical interference with tetramer formation.[62–65] On one side of Asf1, a hydrophobic concave surface binds the C-terminal alpha-helices of H3 and another hydrophobic pocket forms a secondary interaction with the C-terminal beta-strand of H4.[62,65] Within the nucleosome, this part of H3 forms the H3 : H3 dimerization interface whereas the H4 C-terminus binds H2A.[66] The use of Asf1 as a histone escort thus precludes tetramerization of H3-H4 dimers while they are shuttled through the assembly line (Figure 10.2). Asf1 cannot alone promote DNA synthesis-dependent histone deposition, but it may dock onto its downstream chaperone and deliver its histone cargo to the next chaperone. Asf1 can synergize with CAF-1 in repair and replication-coupled nucleosome assembly *in vitro*.[51,67] The delivery could be effected *via* the p60 subunit of CAF-1 which binds Asf1 on the opposite side to histone H3-H4.[68,69] A similar interaction mode is probably involved for Asf1 and HIRA (Figure 10.2), which participates in replication-independent deposition of H3.3-H4.[56,70] Regardless of the assembly pathway, two dimers of H3-H4 have to be delivered onto DNA so that H2A-H2B can be added in order to form a complete nucleosomal core particle.

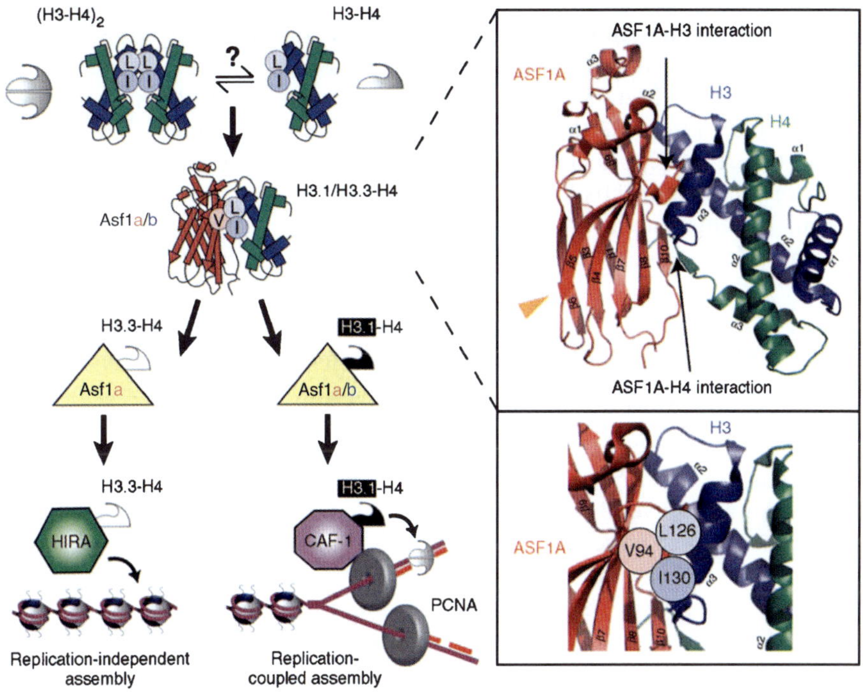

Figure 10.2 **Asf1: structural insight into the assembly line.** Structural analysis of human Asf1a and yeast Asf1 show that H3-H4 are bound in a manner that precludes tetramer formation. The HIRA binding site is indicated by an orange triangle on the structure. Asf1 loaded with histone H3.1-H4 synergizes with CAF-1 in replication-coupled (RC) *de novo* nucleosome assembly. Similarly, Asf1 loaded with the H3.3-H4 aids HIRA mediated replication-independent (RI) chromatin assembly. How and when Asf1 first associates with its histone cargo remains unknown. (Figure from ref. 52)

10.4 The Fate of Parental Histones

10.4.1 Parental Nucleosome Disruption

Nucleosome disruption leads to transient release of parental histones that carry information in the form of post-translational modifications (PTMs) and the variant subtype. Loss of such information could cause changes in genome function such as inappropriate gene expression or chromosome segregation defects. It is thus important to understand better to what extent the disruption is controlled and how parental histones are handled during replication.

The consensus from early studies is that nucleosomes are transiently disrupted immediately ahead of the replication fork and that histones are transferred with high efficiency in a random manner onto the two daughter strands,[59,71] thereby avoiding loss of parental histones. Electron microscopy analysis of replicating SV40 minichromosomes indicate that about 1–2 nucleosomes are destabilized ahead of the replication fork,[72,73] yet it remains open whether destabilization reflects solely H1 loss or stepwise disassembly of the core octamer. Metabolic labelling studies indicate that parental octamers are disrupted into H2A-H2B dimers and $(H3\text{-}H4)_2$ tetramers.[43,44,74] These histones must be kept in close proximity to the replication machinery, as high levels of competitor DNA were necessary to interfere with recycling onto nascent daughter stands.[75,76] To what extent these general principles apply genome-wide is certainly an issue that needs to be investigated.

10.4.2 Helicases and Histone Chaperones in Nucleosome Disassembly and Histone Recycling

How the force of the moving replication fork contributes to evict parental histones and whether other factors are critical in this process are still matters of debate. An attractive idea is that nucleosome disruption is coordinated with DNA unwinding in close proximity to the fork. Indeed, the replicative helicase of the SV40 virus, Large T antigen, can unwind DNA in a nucleosomal template in the presence of SSB (single-strand binding protein) from *E. coli*.[77] This, along with its capacity to bind histone H3, first suggested a role for replicative helicases in nucleosome disruption.[77] While this is possible, it does not seem to be obligatory, as replication of SV40 chromatin in human cell extracts can occur on templates with crosslinked histone octamers, although at a reduced speed.[78] Obviously, given the viral nature of this helicase, the question that arises is how the cellular replicative helicases deal with chromatin. The Mcm2-7 (mini-chromosome maintenance) complex is widely regarded as the replicative helicase in eukaryotes[79–81] (see Chapter 3). This conserved ATP-driven hexameric complex has now been linked to two histone chaperones, FACT (Facilitates Chromatin Transcription) and Asf1.[3,82,83] This provides mechanistic support to the hypothesis that chaperones may facilitate nucleosome disruption by acting as histone acceptors. They would also aid histone transfer onto daughter strands and thereby secure recycling of parental histones.

FACT is an evolutionarily conserved H2A-H2B chaperone complex that facilitates progression of the RNA polymerase during transcription by mediating H2A-H2B transfer.[84] Evidence from both *in vitro* studies in *Xenopus* and genetic studies in *Saccharomyces cerevisiae* support a similar role for FACT in replication.[85] FACT can be isolated in complex with Mcm2-7 proteins from yeast and human cells.[82,83] In human cells, this complex reflects a direct interaction between Spt16, a FACT subunit and Mcm4. This interaction with the replicative helicase places FACT in a key position to peel off H2A-H2B from parental nucleosomes and potentially aid re-deposition. This then leaves a $(H3\text{-}H4)_2$ tetramer as the last obstacle to unwinding.

Asf1 is ideally placed to help to relieve this hindrance in vertebrates. Human Asf1 (a and b) form a complex with Mcm2-7 proteins and histone H3-H4.[3] Within this complex, dimeric H3-H4 bridges the interaction between Asf1 and

MCMs, as a histone-binding mutant of Asf1 does not interact with MCMs. This is in agreement with structural data since Mcm2 binds histone H3 with high affinity in a region that is accessible in the Asf1-(H3-H4) complex.[62,65,86] How Asf1 would be recruited to the replication fork is not clear, but an attractive hypothesis is that Asf1 initially binds the C-terminus of H4 that becomes available upon removal of H2A-H2B. In the face of Mcm2-7-mediated ATP-dependent pulling on DNA, Asf1 may then split the parental $(H3\text{-}H4)_2$ tetramer, leading to formation of an intermediate Asf1-(H3-H4)-Mcm2-7 complex before dimers are transferred onto nascent DNA. The prediction would be that the dimers stay in close vicinity during transfer such that the 'old' tetramer can be reformed (see above).

However, many questions remain and other chaperones could well operate together with Asf1 and FACT in disassembly and recycling. It is noteworthy that human cells depleted for Asf1 replicate slowly and show impaired unwinding of DNA at replication sites.[3] This phenotype could reflect greater difficulties in removing the barrier to the progressing replication fork posed by parental nucleosomes, although assembly defects may also contribute. In chicken DT40 cells, Asf1 is required for DNA replication independently of its role in CAF-1-mediated *de novo* assembly.[87,88] Moreover, ectopic expression of histone H3.1-H4 interferes with DNA replication in human cells in a fashion that phenocopies Asf1 depletion.[3] Given that Asf1 binds excess new histones,[22] elongation defects could arise if insufficient Asf1 is available for disassembly.

10.4.3 Histone Acetylation and Nucleosome Remodelling in Replication Initiation and Fork Progression

By analogy to transcription, ATP-dependent chromatin remodelling enzymes and histone modifying enzymes, *i.e.* lysine acetyltransferases (KATs), could aid access of the replisome to DNA. HBO1 (histone acetyltransferase binding to ORC1) is one of the primary enzymes acetylating H4K5, H4K8 and H4K12 *in vivo*. This KAT is found in a complex with Mcm2-7 proteins and is required for S phase progression.[89] However, whether HBO1 operates on parental chromatin ahead or behind of the fork during elongation remains open. HBO1 also binds ORC1 (Origin Recognition Complex) and is involved in initiation control by facilitating Mcm2-7 loading onto chromatin during licensing.[90,91] An attractive possibility is that histone acetylation enhances chromatin flexibility, which in turn facilitates Mcm2-7 loading and hence origin firing. However, possible non-histone protein targets of acetyltransferases like HBO1 should also be considered. Indeed, the importance of Mcm3 acetylation has yet to be evaluated,[92] although it is driven by a distinct enzyme.

There are many correlations between early replication timing and the transcriptional activity that brings about hyperacetylation. A key issue is whether acetylation of nucleosomal histones around origins participates directly in setting the time of origin firing. Indeed, artificial targeting of histone acetyltransferases to chromatin around origins can advance replication timing in both yeast and mammals.[93,94] Whether targeted histone acetylation participates to establish replication timing in G1 and how this may integrate with transcription-coupled acetylation is still open. There is evidence that ATP-dependent nucleosome remodelling is needed during replication, but exact functions remain to be defined. In mammals the ISWI (imitation switch) complex, SNF2–ACF[i] is required for replication through, and may aid disruption of, heterochromatin domains.[95] A closely related complex, WSTF-SNF2 (Williams syndrome transcription factor), is recruited to PCNA and operates in chromatin maturation.[96] In yeast the Swi/Snf type complex, INO80[ii] also plays a role in timely progression through S phase that is shared with the ISWI complex, Isw2.[97,98] INO80 appears to be particularly important in cells challenged with replication stress, possibly to facilitate restart of stalled forks.[98,99] An unresolved issue regarding most of these remodelers is how they are targeted to the replication fork and whether they operate on parental nucleosomes ahead of, or nascent chromatin behind, the fork.

10.4.4 Parental Histone Transfer

Parental histones carry post-translational modifications that contribute to the control of gene expression and could possibly serve as a blueprint from which epigenetic information is copied onto newly synthesized histones. Histone variants incorporated at specific chromatin loci (*e.g.* CenpA at centromeres) provide additional information that, at least in some cases, needs to be maintained on both new copies of DNA. The exact mechanism by which parental histones are reassembled onto nascent DNA can thus have a major impact on the ability of cells to stably propagate histone-based information through DNA replication.

Histone segregation during chromatin replication has been studied by metabolic labelling and electron microscopy of crosslinked replicating SV40 minichromosomes.[59,71] The consensus from these studies is that parental histones are transferred behind the replication fork randomly onto either the leading or the lagging strand. This transfer occurs almost as soon as enough DNA has emerged from the replisome to allow the formation of nucleosomes.[73] Recycling of parental $(H3\text{-}H4)_2$ tetramers is highly efficient.[45] Global analysis showed that, on average, the $(H3\text{-}H4)_2$ tetramer is segregated as a stable entity and newly synthesized H3-H4 is generally not found within the same nucleosome as parental H3-H4[74,100,101] (Figure 10.2).

For Asf1 to function in handling parental H3-H4[3] as dimers, a transfer mechanism should favour re-association of dimers of parental origin. However, the available data represent an average and thus do not exclude the possibility that parental H3-H4 dimers could associate with newly synthesized H3-H4 dimers in specific regions of the genome. In principle, the presence of both parental and newly synthesized H3-H4 molecules in the same nucleosome could

[i] Sucrose nonfermenting–ATP-utilizing chromatin assembly and modifying factor
[ii] Inositol-requiring protein 80

facilitate the duplication of pre-existing post-translational modifications onto the new histones.[56] However, parental tetramers were found intermingled with new tetramers in metabolic labelling studies[45] and they distribute completely randomly to the two daughter strands under conditions where new deposition is prevented by cycloheximide treatment.[73] This pattern of random segregation is important when considering inheritance of histone-based information, as it could allow spreading of marks from old to new tetramers and at the same time be a source of stochastic variation in gene expression.

One insight into understanding this issue is that Asf1 can be found in complex with histones presenting parental marks when replication is blocked.[3] It will be interesting to explore how Asf1 handles information from new and old histones. Obviously, a similar issue exists for histone variants and their inheritance. However, one should also bear in mind that transmission of information may not be strictly coupled to replication, as highly specialized variants such as centromeric histone CenpA most probably take a different path.[102]

10.5 Concluding Remarks

We have discussed here how the passage of the replication fork deals with chromatin at the nucleosomal level to handle both parental and new histones. Sophisticated mechanisms are at work involving histone chaperones to escort histones in their choreography and ensure an interface with the replication machinery. While we have now identified several key players on the scene, discovering how their function can be regulated to adapt to various contexts will be a future challenge. Finally, this information should also pave the way for an understanding of how the epigenetic landscape may be stably maintained even in the face of dramatic changes in chromatin structure.

Acknowledgements

We apologize to colleagues who could not be cited due to manuscript length limitations. Support to AG is from the Lundbeck Foundation, the Danish Research Council and the Danish Cancer Society and to GA from the CNRS & Curie Institute along with the ANR and Network of Excellence Epigenome, La Ligue contre le Cancer and the Canceropole d'Ile de France.

References

1. A. Gunjan and A. Verreault, A Rad53 kinase-dependent surveillance mechanism that regulates histone protein levels in *S. cerevisiae*, *Cell*, 2003, **115**, 537–549.
2. D. Meeks-Wagner and L. H. Hartwell, Normal stoichiometry of histone dimer sets is necessary for high fidelity of mitotic chromosome transmission, *Cell*, 1986, **44**, 43–52.

3. A. Groth, A. Corpet, A. J. Cook, D. Roche, J. Bartek, J. Lukas and G. Almouzni, Regulation of replication fork progression through histone supply and demand, *Science*, 2007, **318**, 1928–1931.
4. A. Gunjan, J. Paik and A. Verreault, Regulation of histone synthesis and nucleosome assembly, *Biochimie*, 2005, **87**, 625–635.
5. W. F. Marzluff and R. J. Duronio, Histone mRNA expression: multiple levels of cell cycle regulation and important developmental consequences, *Curr. Opin. Cell Biol.*, 2002, **14**, 692–699.
6. T. Ma, B. A. Van Tine, Y. Wei, M. D. Garrett, D. Nelson, P. D. Adams, J. Wang, J. Qin, L. T. Chow and J. W. Harper, Cell cycle-regulated phosphorylation of p220(NPAT) by cyclin E/Cdk2 in Cajal bodies promotes histone gene transcription, *Genes Dev.*, 2000, **14**, 2298–2313.
7. J. Zhao, B. Dynlacht, T. Imai, T. Hori and E. Harlow, Expression of NPAT, a novel substrate of cyclin E-CDK2, promotes S-phase entry, *Genes Dev.*, 1998, **12**, 456–461.
8. J. Zhao, B. K. Kennedy, B. D. Lawrence, D. A. Barbie, A. G. Matera, J. A. Fletcher and E. Harlow, NPAT links cyclin E-Cdk2 to the regulation of replication-dependent histone gene transcription, *Genes Dev.*, 2000, **14**, 2283–2297.
9. W. F. Marzluff, Metazoan replication-dependent histone mRNAs: a distinct set of RNA polymerase II transcripts, *Curr. Opin. Cell Biol.*, 2005, **17**, 274–280.
10. M. L. Whitfield, L. X. Zheng, A. Baldwin, T. Ohta, M. M. Hurt and W. F. Marzluff, Stem-loop binding protein, the protein that binds the 3′ end of histone mRNA, is cell cycle regulated by both translational and posttranslational mechanisms, *Mol. Cell. Biol.*, 2000, **20**, 4188–4198.
11. L. Zheng, Z. Dominski, X. C. Yang, P. Elms, C. S. Raska, C. H. Borchers and W. F. Marzluff, Phosphorylation of stem-loop binding protein (SLBP) on two threonines triggers degradation of SLBP, the sole cell cycle-regulated factor required for regulation of histone mRNA processing at the end of S phase, *Mol. Cell. Biol.*, 2003, **23**, 1590–1601.
12. H. Kaygun and W. F. Marzluff, Regulated degradation of replication-dependent histone mRNAs requires both ATR and Upf1, *Nat. Struct. Mol. Biol.*, 2005, **12**, 794–800.
13. C. Su, G. Gao, S. Schneider, C. Helt, C. Weiss, M. A. O'Reilly, D. Bohmann and J. Zhao, DNA damage induces downregulation of histone gene expression through the G(1) checkpoint pathway, *EMBO J.*, 2004, **23**, 1133–1143.
14. N. Heintz, H. L. Sive and R. G. Roeder, Regulation of human histone gene expression: kinetics of accumulation and changes in the rate of synthesis and in the half-lives of individual histone mRNAs during the HeLa cell cycle, *Mol. Cell. Biol.*, 1983, **3**, 539–550.
15. M. F. Lavin, J. Houldsworth, S. Kumar, J. L. Stein and G. S. Stein, Coupling of histone mRNA levels to radioresistant DNA synthesis in ataxia-telangiectasia cells, *Mol. Cell. Biochem.*, 1987, **73**, 45–54.

16. D. B. Sittman, R. A. Graves and W. F. Marzluff, Histone mRNA concentrations are regulated at the level of transcription and mRNA degradation, *Proc. Natl. Acad. Sci. U.S.A.*, 1983, **80**, 1849–1853.
17. M. A. Osley, The regulation of histone synthesis in the cell cycle, *Annu. Rev. Biochem.*, 1991, **60**, 827–861.
18. A. Sutton, J. Bucaria, M. A. Osley and R. Sternglanz, Yeast ASF1 protein is required for cell cycle regulation of histone gene transcription, *Genetics*, 2001, **158**, 587–596.
19. M. S. Spector, A. Raff, H. DeSilva, K. Lee and M. A. Osley, Hir1p and Hir2p function as transcriptional corepressors to regulate histone gene transcription in the *Saccharomyces cerevisiae* cell cycle, *Mol. Cell. Biol.*, 1997, **17**, 545–552.
20. D. M. Nelson, X. Ye, C. Hall, H. Santos, T. Ma, G. D. Kao, T. J. Yen, J. W. Harper and P. D. Adams, Coupling of DNA synthesis and histone synthesis in S phase independent of cyclin/cdk2 activity, *Mol. Cell. Biol.*, 2002, **22**, 7459–7472.
21. Y. M. Moshkin, J. A. Armstrong, R. K. Maeda, J. W. Tamkun, P. Verrijzer, J. A. Kennison and F. Karch, Histone chaperone ASF1 cooperates with the Brahma chromatin-remodelling machinery, *Genes Dev.*, 2002, **16**, 2621–2626.
22. A. Groth, D. Ray-Gallet, J. P. Quivy, J. Lukas, J. Bartek and G. Almouzni, Human Asf1 regulates the flow of S phase histones during replicational stress, *Mol. Cell*, 2005, **17**, 301–311.
23. W. M. Bonner, R. S. Wu, H. T. Panusz and C. Muneses, Kinetics of accumulation and depletion of soluble newly synthesized histone in the reciprocal regulation of histone and DNA synthesis, *Biochemistry*, 1988, **27**, 6542–6550.
24. A. Loyola, T. Bonaldi, D. Roche, A. Imhof and G. Almouzni, PTMs on H3 variants before chromatin assembly potentiate their final epigenetic state, *Mol. Cell*, 2006, **24**, 309–316.
25. R. E. Sobel, R. G. Cook, C. A. Perry, A. T. Annunziato and C. D. Allis, Conservation of deposition-related acetylation sites in newly synthesized histones H3 and H4, *Proc. Natl. Acad. Sci. U.S.A.*, 1995, **92**, 1237–1241.
26. L. J. Benson, Y. Gu, T. Yakovleva, K. Tong, C. Barrows, C. L. Strack, R. G. Cook, C. A. Mizzen and A. T. Annunziato, Modifications of H3 and H4 during chromatin replication, nucleosome assembly and histone exchange, *J. Biol. Chem.*, 2006, **281**, 9287–9296.
27. X. J. Ma, J. Wu, B. A. Altheim, M. C. Schultz and M. Grunstein, Deposition-related sites K5/K12 in histone H4 are not required for nucleosome deposition in yeast, *Proc. Natl. Acad. Sci. U.S.A.*, 1998, **95**, 6693–6698.
28. M. R. Parthun, Hat1: the emerging cellular roles of a type B histone acetyltransferase, *Oncogene*, 2007, **26**, 5319–5328.
29. H. K. Barman, Y. Takami, T. Ono, H. Nishijima, F. Sanematsu, K. Shibahara and T. Nakayama, Histone acetyltransferase 1 is dispensable for replication-coupled chromatin assembly but contributes to recover

DNA damages created following replication blockage in vertebrate cells, *Biochem. Biophys. Res. Commun.*, 2006, **345**, 1547–1557.

30. A. Rufiange, P. E. Jacques, W. Bhat, F. Robert and A. Nourani, Genome-wide replication-independent histone H3 exchange occurs predominantly at promoters and implicates H3 K56 acetylation and Asf1, *Mol. Cell*, 2007, **27**, 393–405.
31. H. Masumoto, D. Hawke, R. Kobayashi and A. Verreault, A role for cell-cycle-regulated histone H3 lysine 56 acetylation in the DNA damage response, *Nature*, 2005, **436**, 294–298.
32. I. Celic, H. Masumoto, W. P. Griffith, P. Meluh, R. J. Cotter, J. D. Boeke and A. Verreault, The sirtuins hst3 and Hst4p preserve genome integrity by controlling histone h3 lysine 56 deacetylation, *Curr. Biol.*, 2006, **16**, 1280–1289.
33. J. Recht, T. Tsubota, J. C. Tanny, R. L. Diaz, J. M. Berger, X. Zhang, B. A. Garcia, J. Shabanowitz, A. L. Burlingame, D. F. Hunt, P. D. Kaufman and C. D. Allis, Histone chaperone Asf1 is required for histone H3 lysine 56 acetylation, a modification associated with S phase in mitosis and meiosis, *Proc. Natl. Acad. Sci. U.S.A.*, 2006, **103**, 6988–6993.
34. J. Schneider, P. Bajwa, F. C. Johnson, S. R. Bhaumik and A. Shilatifard, Rtt109 is required for proper H3K56 acetylation: a chromatin mark associated with the elongating RNA polymerase II, *J. Biol. Chem.*, 2006, **281**, 37270–37274.
35. R. Driscoll, A. Hudson and S. P. Jackson, Yeast Rtt109 promotes genome stability by acetylating histone H3 on lysine 56, Science (New York, N.Y, 2007, **315**, 649–652.
36. J. Han, H. Zhou, B. Horazdovsky, K. Zhang, R. M. Xu and Z. Zhang, Rtt109 acetylates histone H3 lysine 56 and functions in DNA replication, *Science*, 2007, **315**, 653–655.
37. S. R. Collins, K. M. Miller, N. L. Maas, A. Roguev, J. Fillingham, C. S. Chu, M. Schuldiner, M. Gebbia, J. Recht, M. Shales, H. Ding, H. Xu, J. Han, K. Ingvarsdottir, B. Cheng, B. Andrews, C. Boone, S. L. Berger, P. Hieter, Z. Zhang, G. W. Brown, C. J. Ingles, A. Emili, C. D. Allis, D. P. Toczyski, J. S. Weissman, J. F. Greenblatt and N. J. Krogan, Functional dissection of protein complexes involved in yeast chromosome biology using a genetic interaction map, *Nature*, 2007, **446**, 806–810.
38. S. K. Williams, D. Truong and J. K. Tyler, Acetylation in the globular core of histone H3 on lysine-56 promotes chromatin disassembly during transcriptional activation, *Proc. Natl. Acad. Sci. U.S.A.*, 2008, **105**, 9000–9005.
39. C. C. Chen, J. J. Carson, J. Feser, B. Tamburini, S. Zabaronick, J. Linger and J. K. Tyler, Acetylated lysine 56 on histone H3 drives chromatin assembly after repair and signals for the completion of repair, *Cell*, 2008, **134**, 231–243.
40. Q. Li, H. Zhou, H. Wurtele, B. Davies, B. Horazdovsky, A. Verreault and Z. Zhang, Acetylation of histone H3 lysine 56 regulates replication-coupled nucleosome assembly, *Cell*, 2008, **134**, 244–255.

41. B. A. Garcia, S. B. Hake, R. L. Diaz, M. Kauer, S. A. Morris, J. Recht, J. Shabanowitz, N. Mishra, B. D. Strahl, C. D. Allis and D. F. Hunt, Organismal differences in post-translational modifications in histones H3 and H4, *J. Biol. Chem.*, 2007, **282**, 7641–7655.
42. W. Rocha and A. Verreault, Clothing up DNA for all seasons: Histone chaperones and nucleosome assembly pathways, *FEBS letters*, 2008.
43. V. Jackson and R. Chalkley, A reevaluation of new histone deposition on replicating chromatin, *J. Biol. Chem.*, 1981, **256**, 5095–5103.
44. V. Jackson and R. Chalkley, A new method for the isolation of replicative chromatin: selective deposition of histone on both new and old DNA, *Cell*, 1981, **23**, 121–134.
45. V. Jackson, *In vivo* studies on the dynamics of histone-DNA interaction: evidence for nucleosome dissolution during replication and transcription and a low level of dissolution independent of both, *Biochemistry*, 1990, **29**, 719–731.
46. H. Kimura and P. R. Cook, Kinetics of core histones in living human cells: little exchange of H3 and H4 and some rapid exchange of H2B, *J. Cell Biol.*, 2001, **153**, 1341–1353.
47. R. A. Laskey, B. M. Honda, A. D. Mills, N. R. Morris, A. H. Wyllie, J. E. Mertz, E. M. De Roberts and J. B. Gurdon, Chromatin assembly and transcription in eggs and oocytes of *Xenopus laevis*, *Cold Spring Harb. Symp. Quant. Biol.*, 1978, **42**, 171–178.
48. B. Stillman, Chromatin assembly during SV40 DNA replication *in vitro*, *Cell*, 1986, **45**, 555–565.
49. G. Almouzni and M. Mechali, Assembly of spaced chromatin promoted by DNA synthesis in extracts from *Xenopus* eggs, *EMBO J.*, 1988, **7**, 665–672.
50. S. Smith and B. Stillman, Purification and characterization of CAF-I, a human cell factor required for chromatin assembly during DNA replication *in vitro*, *Cell*, 1989, **58**, 15–25.
51. J. K. Tyler, C. R. Adams, S. R. Chen, R. Kobayashi, R. T. Kamakaka and J. T. Kadonaga, The RCAF complex mediates chromatin assembly during DNA replication and repair, *Nature*, 1999, **402**, 555–560.
52. L. De Koning, A. Corpet, J. E. Haber and G. Almouzni, Histone chaperones: an escort network regulating histone traffic, *Nat. Struct. Mol. Biol.*, 2007, **14**, 997–1007.
53. A. Verreault, P. D. Kaufman, R. Kobayashi and B. Stillman, Nucleosome assembly by a complex of CAF-1 and acetylated histones H3/H4, *Cell*, 1996, **87**, 95–104.
54. K. Shibahara and B. Stillman, Replication-dependent marking of DNA by PCNA facilitates CAF-1-coupled inheritance of chromatin, *Cell*, 1999, **96**, 575–585.
55. P. D. Kaufman, R. Kobayashi, N. Kessler and B. Stillman, The p150 and p60 subunits of chromatin assembly factor I: a molecular link between newly synthesized histones and DNA replication, *Cell*, 1995, **81**, 1105–1114.

56. H. Tagami, D. Ray-Gallet, G. Almouzni and Y. Nakatani, Histone H3.1 and H3.3 complexes mediate nucleosome assembly pathways dependent or independent of DNA synthesis, *Cell*, 2004, **116**, 51–61.
57. P. H. Gaillard, E. M. Martini, P. D. Kaufman, B. Stillman, E. Moustacchi and G. Almouzni, Chromatin assembly coupled to DNA repair: a new role for chromatin assembly factor I, *Cell*, 1996, **86**, 887–896.
58. J. G. Moggs, P. Grandi, J. P. Quivy, Z. O. Jonsson, U. Hubscher, P. B. Becker and G. Almouzni, A CAF-1-PCNA-mediated chromatin assembly pathway triggered by sensing DNA damage, *Mol. Cell. Biol.*, 2000, **20**, 1206–1218.
59. A. Groth, W. Rocha, A. Verreault and G. Almouzni, Chromatin challenges during DNA replication and repair, *Cell*, 2007, **128**, 721–733.
60. A. Gerard, S. Koundrioukoff, V. Ramillon, J. C. Sergere, N. Mailand, J. P. Quivy and G. Almouzni, The replication kinase Cdc7-Dbf4 promotes the interaction of the p150 subunit of chromatin assembly factor 1 with proliferating cell nuclear antigen, *EMBO Rep.*, 2006, **7**, 817–823.
61. S. E. Polo, D. Roche and G. Almouzni, New histone incorporation marks sites of UV repair in human cells, *Cell*, 2006, **127**, 481–493.
62. R. Natsume, M. Eitoku, Y. Akai, N. Sano, M. Horikoshi and T. Senda, Structure and function of the histone chaperone CIA/ASF1 complexed with histones H3 and H4, *Nature*, 2007, **446**, 338–341.
63. F. Mousson, A. Lautrette, J. Y. Thuret, M. Agez, R. Courbeyrette, B. Amigues, E. Becker, J. M. Neumann, R. Guerois, C. Mann and F. Ochsenbein, Structural basis for the interaction of Asf1 with histone H3 and its functional implications, *Proc. Natl. Acad. Sci. U.S.A.*, 2005, **102**, 5975–5980.
64. C. M. English, N. K. Maluf, B. Tripet, M. E. Churchill and J. K. Tyler, ASF1 Binds to a heterodimer of histones H3 and H4: a two-step mechanism for the assembly of the H3-H4 heterotetramer on DNA, *Biochemistry*, 2005, **44**, 13673–13682.
65. C. M. English, M. W. Adkins, J. J. Carson, M. E. Churchill and J. K. Tyler, Structural basis for the histone chaperone activity of asf1, *Cell*, 2006, **127**, 495–508.
66. K. Luger, A. W. Mader, R. K. Richmond, D. F. Sargent and T. J. Richmond, Crystal structure of the nucleosome core particle at 2.8 Å resolution, *Nature*, 1997, **389**, 251–260.
67. J. A. Mello, H. H. Sillje, D. M. Roche, D. B. Kirschner, E. A. Nigg and G. Almouzni, Human Asf1 and CAF-1 interact and synergize in a repair-coupled nucleosome assembly pathway, *EMBO Rep.*, 2002, **3**, 329–334.
68. Y. Tang, M. V. Poustovoitov, K. Zhao, M. Garfinkel, A. Canutescu, R. Dunbrack, P. D. Adams and R. Marmorstein, Structure of a human ASF1a-HIRA complex and insights into specificity of histone chaperone complex assembly, *Nat. Struct. Mol. Biol.*, 2006, **13**, 921–929.
69. A. D. Malay, T. Umehara, K. Matsubara, B. Padmanabhan and S. Yokoyama, Crystal structures of fission yeast histone chaperone Asf1

complexed with the Hip1 B domain or the Cac2 C-terminus, *J. Biol. Chem.*, 2008, **283**, 14022–14031.
70. D. Ray-Gallet, J. P. Quivy, C. Scamps, E. M. Martini, M. Lipinski and G. Almouzni, HIRA is critical for a nucleosome assembly pathway independent of DNA synthesis, *Mol. Cell*, 2002, **9**, 1091–1100.
71. A. T. Annunziato, Split decision: what happens to nucleosomes during DNA replication?, *J. Biol. Chem.*, 2005, **280**, 12065–12068.
72. R. Gasser, T. Koller and J. M. Sogo, The stability of nucleosomes at the replication fork, *J. Mol. Biol.*, 1996, **258**, 224–239.
73. J. M. Sogo, H. Stahl, T. Koller and R. Knippers, Structure of replicating simian virus 40 minichromosomes. The replication fork, core histone segregation and terminal structures, *J. Mol. Biol.*, 1986, **189**, 189–204.
74. V. Jackson, Deposition of newly synthesized histones: new histones H2A and H2B do not deposit in the same nucleosome with new histones H3 and H4, *Biochemistry*, 1987, **26**, 2315–2325.
75. C. Gruss, J. Wu, T. Koller and J. M. Sogo, Disruption of the nucleosomes at the replication fork, *EMBO J.*, 1993, **12**, 4533–4545.
76. S. K. Randall and T. J. Kelly, The fate of parental nucleosomes during SV40 DNA replication, *J. Biol. Chem.*, 1992, **267**, 14259–14265.
77. U. Ramsperger and H. Stahl, Unwinding of chromatin by the SV40 large T antigen DNA helicase, *EMBO J.*, 1995, **14**, 3215–3225.
78. B. Vestner, T. Waldmann and C. Gruss, Histone octamer dissociation is not required for *in vitro* replication of simian virus 40 minichromosomes, *J. Biol. Chem.*, 2000, **275**, 8190–8195.
79. K. Labib, J. A. Tercero and J. F. Diffley, Uninterrupted MCM2-7 function required for DNA replication fork progression, *Science*, 2000, **288**, 1643–1647.
80. T. S. Takahashi, D. B. Wigley and J. C. Walter, Pumps, paradoxes and ploughshares: mechanism of the MCM2-7 DNA helicase, *Trends Biochem. Sci.*, 2005, **30**, 437–444.
81. M. L. Bochman and A. Schwacha, The Mcm2-7 complex has *in vitro* helicase activity, *Mol. Cell*, 2008, **31**, 287–293.
82. B. C. Tan, C. T. Chien, S. Hirose and S. C. Lee, Functional cooperation between FACT and MCM helicase facilitates initiation of chromatin DNA replication, *EMBO J.*, 2006, **25**, 3975–3985.
83. A. Gambus, R. C. Jones, A. Sanchez-Diaz, M. Kanemaki, F. van Deursen, R. D. Edmondson and K. Labib, GINS maintains association of Cdc45 with MCM in replisome progression complexes at eukaryotic DNA replication forks, *Nat. Cell Biol.*, 2006, **8**, 358–366.
84. R. Belotserkovskaya and D. Reinberg, Facts about FACT and transcript elongation through chromatin, *Curr. Opin. Genet. Dev.*, 2004, **14**, 139–146.
85. T. Formosa, Changing the DNA landscape: putting a SPN on chromatin, *Curr. Top. Microbiol. Immunol.*, 2003, **274**, 171–201.
86. Y. Ishimi, Y. Komamura, Z. You and H. Kimura, Biochemical function of mouse minichromosome maintenance 2 protein, *J. Biol. Chem.*, 1998, **273**, 8369–8375.

87. F. Sanematsu, Y. Takami, H. K. Barman, T. Fukagawa, T. Ono, K. Shibahara and T. Nakayama, Asf1 is required for viability and chromatin assembly during DNA replication in vertebrate cells, *J. Biol. Chem.*, 2006, **281**, 13817–13827.
88. Y. Takami, T. Ono, T. Fukagawa, K. Shibahara and T. Nakayama, Essential role of chromatin assembly factor-1-mediated rapid nucleosome assembly for DNA replication and cell division in vertebrate cells, *Mol. Biol. Cell*, 2007, **18**, 129–141.
89. Y. Doyon, C. Cayrou, M. Ullah, A. J. Landry, V. Cote, W. Selleck, W. S. Lane, S. Tan, X. J. Yang and J. Cote, ING tumor suppressor proteins are critical regulators of chromatin acetylation required for genome expression and perpetuation, *Mol. Cell*, 2006, **21**, 51–64.
90. M. Iizuka, T. Matsui, H. Takisawa and M. M. Smith, Regulation of replication licensing by acetyltransferase Hbo1, *Mol. Cell. Biol.*, 2006, **26**, 1098–1108.
91. M. Iizuka and B. Stillman, Histone acetyltransferase HBO1 interacts with the ORC1 subunit of the human initiator protein, *J. Biol. Chem.*, 1999, **274**, 23027–23034.
92. Y. Takei, M. Swietlik, A. Tanoue, G. Tsujimoto, T. Kouzarides and R. Laskey, MCM3AP, a novel acetyltransferase that acetylates replication protein MCM3, *EMBO Rep.*, 2001, **2**, 119–123.
93. M. Vogelauer, L. Rubbi, I. Lucas, B. J. Brewer and M. Grunstein, Histone acetylation regulates the time of replication origin firing, *Mol. Cell*, 2002, **10**, 1223–1233.
94. A. Goren, A. Tabib, M. Hecht and H. Cedar, DNA replication timing of the human β-globin domain is controlled by histone modification at the origin, *Genes Dev.*, 2008, **22**, 1319–1324.
95. N. Collins, R. A. Poot, I. Kukimoto, C. Garcia-Jimenez, G. Dellaire and P. D. Varga-Weisz, An ACF1-ISWI chromatin-remodeling complex is required for DNA replication through heterochromatin, *Nat. Genet.*, 2002, **32**, 627–632.
96. R. A. Poot, L. Bozhenok, D. L. van den Berg, S. Steffensen, F. Ferreira, M. Grimaldi, N. Gilbert, J. Ferreira and P. D. Varga-Weisz, The Williams syndrome transcription factor interacts with PCNA to target chromatin remodelling by ISWI to replication foci, *Nat. Cell Biol.*, 2004, **6**, 1236–1244.
97. J. A. Vincent, T. J. Kwong and T. Tsukiyama, ATP-dependent chromatin remodeling shapes the DNA replication landscape, *Nat. Struct. Mol. Biol.*, 2008, **15**, 477–484.
98. M. Papamichos-Chronakis and C. L. Peterson, The Ino80 chromatin-remodeling enzyme regulates replisome function and stability, *Nat. Struct. Mol. Biol.*, 2008, **15**, 338–345.
99. K. Shimada, Y. Oma, T. Schleker, K. Kugou, K. Ohta, M. Harata and S. M. Gasser, Ino80 chromatin remodeling complex promotes recovery of stalled replication forks, *Curr. Biol.*, 2008, **18**, 566–575.

100. V. Jackson, Deposition of newly synthesized histones: hybrid nucleosomes are not tandemly arranged on daughter DNA strands, *Biochemistry*, 1988, **27**, 2109–2120.
101. C. P. Prior, C. R. Cantor, E. M. Johnson and V. G. Allfrey, Incorporation of exogenous pyrene-labeled histone into Physarum chromatin: a system for studying changes in nucleosomes assembled *in vivo*, *Cell*, 1980, **20**, 597–608.
102. M. Durand-Dubief and K. Ekwall, Heterochromatin tells CENP-A where to go, *Bioessays*, 2008, **30**, 526–529.

CHAPTER 11

Mitochondrial DNA Replication

TAKEHIRO YASUKAWA[a] AND JOANNA POULTON[b]

[a] The Wolfson Institute for Biomedical Research, University College London, Gower Street, London, WC1E 6BT, UK; [b] Nuffield Department of Obstetrics and Gynaecology, University of Oxford, The Women's Centre, John Radcliffe Hospital, Headington, Oxford, OX3 9DU, UK

11.1 Introduction

Mitochondria contain their own genome, the mitochondrial DNA (mtDNA). In humans and other mammals, it is a small closed-circular multicopy DNA located inside the inner of the two mitochondrial membranes. The 16 569 bp human mitochondrial genome[1,2] is maternally inherited and encodes components of the respiratory chain. Mitochondrial DNA maintenance is clearly central to many cellular activities: mtDNA loss abolishes oxidative phosphorylation in cultured cells, and mtDNA mutation and depletion are associated with severe human diseases, infertility and ageing.

Despite the importance of DNA replication as a biological process essential to maintain genetic information, the mechanism of mammalian mtDNA replication is still a matter of intense debate. For over 25 years, the strand-asynchronous model[3] was generally accepted as the sole mechanism of mammalian mtDNA replication, but this has recently been revisited. In this chapter, we discuss both the long-standing model and a recently proposed model of mtDNA replication, as well as related studies that have advanced our understanding of mtDNA organisation and copy number regulation. We also discuss the controversies surrounding a possible relationship between mtDNA replication and the cell cycle. Finally, we describe how defects in mtDNA replication can result in human disease.

Molecular Themes in DNA Replication
Edited by Lynne S. Cox

Published by the Royal Society of Chemistry, www.rsc.org

11.2 The Mammalian Mitochondrial Genome

11.2.1 Template Copy Number, Structure and Organisation

Mammalian mtDNA is a circular molecule of approximately 16 Kb. Unlike nuclear DNA, in which there are usually two copies of every gene per diploid cell, the copy number of mtDNA is very variable between and within different nucleated cell types, ranging from 1000 to 250 000 copies in sperm[4] and oocytes, respectively.[5,6] Nevertheless, there are typically between 1000 and 10 000 copies of mtDNA per cell.

Mammalian mtDNA encodes 13 subunits for integral components of the respiratory chain and ATP synthase, and 22 transfer RNAs and two ribosomal RNAs for translation of the subunits in mitochondria (Figure 11.1). MtDNA is highly compact with very little non-coding DNA, and in some cases, the reading frames of neighbouring genes overlap each other (for instance NADH-ubiquinone oxidoreductase subunits ND4 and ND4L genes, and ATP synthase subunits ATPase 6 and ATPase 8). Although mtDNA is very small in size

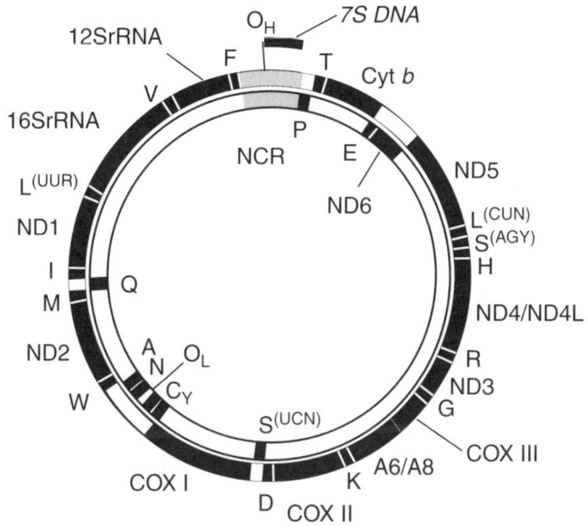

Figure 11.1 **Genetic map of mammalian mitochondrial DNA**. The two strands of mtDNA are termed heavy strand (H strand) and light strand (L strand), according to the difference in their specific gravity (dependent on GC content). Here the H strand is drawn outside and the L strand inside. Gene-encoding regions are shown in black. Proteins encoded are: ND, subunits of NADH-ubiquinone oxidoreductase (complex I); cyt *b*, subunit of ubiquinone-cytochrome c oxidoreductase (complex III); COX; subunits of cytochrome c oxidase (complex IV); A, subunits of ATP synthase (complex V). Mitochondrial tRNA genes are shown by their cognate amino acid (single letter notation). Mitochondrial rRNA genes are indicated as 12SrRNA and 16SrRNA. The position of the major non-coding region (NCR) is shown in grey and 7S DNA region is shown above as a black bar. O_H and O_L are the origins, or replication initiation sites of H and L strands, respectively under the strand-asynchronous mtDNA replication model.

compared with nuclear DNA, it provides important subunits of the respiratory complexes in the oxidative phosphorylation system. mtDNA is thus indispensable for proper function of a cell and an individual organism.

11.2.2 Is mtDNA Naked?

Mammalian mitochondria do not contain nucleosome components such as histone proteins, so for many years mammalian mtDNA was generally thought to be 'naked'. Recent studies, however, have challenged this view and it may be that mtDNA is coated entirely by mitochondrial transcription factor A (TFAM), a member of high mobility group (HMG)-box proteins.[7]

Abf2p is the abundant *Saccharomyces cerevisiae* homologue of TFAM that binds every 15 bp of mtDNA.[8] Unlike TFAM, Abf2p is not required for transcription initiation[8] but is an architectural factor: depletion of Abf2p results in a loss of mtDNA,[9] which is rescued by a bacterial histone-like protein HU.[10] This rescue is achieved by TFAM as well, implying that TFAM may share common properties with Abf2p and HU.[11]

While some groups estimated that the stoichiometry between mtDNA and TFAM is only 15–50 molecules per mtDNA,[12–14] other estimates suggested that TFAM is comparable to Abf2p, being nearer to 1000 molecules per mtDNA in HeLa cells,[15] human placenta[15] and mouse kidney mitochondria.[16] Both immunostaining and immunoprecipitation have shown that the majority of TFAM molecules are associated with mtDNA and *vice versa*.[17–19] Furthermore, a 50% reduction in TFAM levels in heterozygous TFAM +/– mice decreased mtDNA copy number by ~35%, suggesting that mtDNA is unstable if TFAM does not bind mtDNA fully.[20] Detailed analysis of TFAM function both in transgenic mice[16] and cultured cells[21] indicated that TFAM's ability to maintain mtDNA copy number is distinct from its function as a transcription activator. These studies support the idea that TFAM and Abf2 have similar relationships in wrapping and stabilising mammalian and yeast mtDNA, respectively. Furthermore, an *in vitro* study showed that TFAM is present as a homodimer that binds to DNA and is capable of forming a compact DNA-TFAM complex, suggesting that TFAM plays an important role in packaging mtDNA.[22] The frequency of TFAM binding to DNA *in vitro*[22] is in a close agreement with the data obtained from an *in vivo* experiment of co-sedimentation of mtDNA and TFAM.[15]

These studies from different groups strongly suggest that TFAM is not only a transcription factor but also functions for mtDNA wrapping and packaging. Thus, it is likely that mtDNA does not exist as 'naked' but is fully coated by TFAM which probably serves as a 'mitochondrial histone'.

11.2.3 mtDNA Forms the Nucleoid

It is now generally accepted that molecules of mammalian mtDNA are packaged into large protein-DNA complexes, called the mitochondrial nucleoids.

Nucleoids have also been identified in fungi, plants and protozoa. Early electron microscopy studies of mammalian mitochondria identified a structure with features similar to bacterial nucleoids,[23,24] suggesting that mtDNA might be packaged into a structure that was associated with the mitochondrial membrane.[25] These nucleoid structures gained credence from DAPI staining,[26] then confocal microscopy, showing that the number of ethidium bromide-stained punctate spots inside mitochondria is much less than the mtDNA copy number in human cultured cell lines; this suggested that several mtDNAs must be packaged in each unit.[27]

Quantitative studies using human cultured cells showed cell-type specific differences in both nucleoid number and DNA content: fibroblasts were observed to have ~750 nucleoid foci each containing ~2.4 mtDNA genomes,[28] HeLa and 143B cells contain ~500 nucleoid foci with 5.7~7.5 mtDNA,[28] and carcinoma-derived cells showed ~475 nucleoid foci with 6–10 mtDNA.[29] Furthermore, time-lapse microscopy of nucleoid dynamics showed both fission and fusion[19,29] and suggested that mtDNA molecules within a nucleoid do not replicate simultaneously, but independently of each other.[29] Along with these elegant studies, identification of nucleoid component proteins and characterisation of their function will allow us to understand complex features of nucleoid dynamics, then unravel how mtDNA copy number and nucleoid number are maintained, and how mtDNA (in nucleoids) is transmitted equally to daughter cells. These issues potentially have important implications for understanding of segregation of heteroplasmic mtDNA mutations in human mitochondrial diseases (see Section 11.5).

Based on studies using microscopy and biochemical purification, a number of proteins have been proposed as candidate components of the nucleoid. These include TFAM,[17,19,30–33] TWINKLE (mitochondrial DNA helicase, see Section 11.5.2.2),[27,31,32] mitochondrial single-stranded DNA binding protein (mtSSB)[19,30–32,34] and mitochondrial DNA polymerase gamma (POLG, comprising catalytic and accessory subunits: POLG 1 and 2, see Section 11.5.2.1).[19,31,32] Mitochondrial RNA polymerase,[31,32] mitochondrial transcription factors B1[31] and B2,[31,32] mitochondrial topoisomerase I[31] and Lon protease[31] were also found in some nucleoid preparations. Very recently, flap endonuclease 1 (FEN1) was shown to be present in the nucleoid.[35]

In addition to the above proteins implicated in mtDNA replication, expression and maintenance, several more without an obvious role in mtDNA metabolism were also detected (for detail, see references cited above). Further investigation is necessary to determine whether such proteins are really nucleoid factors or not. A good example of functional study is that ATAD3, identified in some nucleoid preparations[32,33] was proposed to bind to the short triple-stranded D-loop region of mtDNA and play a role in holding multiple mtDNA molecules together.[33] However, another study suggested that this protein does not contact mtDNA directly.[31]

11.3 Models of mtDNA Replication

11.3.1 Strand-asynchronous Model

In the early 1970s, a model was proposed for the mechanism of mammalian mtDNA replication, based initially on electron microscopic analysis of the replicating molecules. This model was called the strand-asynchronous, asymmetrical model of mtDNA replication.[36,37] According to this model, mtDNA replication initiates with H strand synthesis from a single origin called O_H in the major non-coding region (NCR) of mtDNA, and the H strand is synthesised continuously without synchronous synthesis of the L strand (clockwise in Figure 11.2A). Synthesis of the L strand initiates from a specific site called O_L, which is located at approximately two-thirds of the genome away from O_H. In this model, it is not until O_L is exposed on the displaced parental H strand that continuous L strand synthesis starts (anti-clockwise in Figure 11.2A). This replication model also proposed alternative origins of L-strand synthesis.[38] The characteristic feature of this model is that all the replicating mtDNA molecules would have partially single-stranded regions and that no short lagging strand Okazaki fragments would be synthesised. A short H-strand, called 7S DNA or D-loop DNA, forming a triple-stranded structure with mtDNA, was found in the NCR of many mtDNA molecules,[39,40] its 5′ DNA end mapping to O_H. It is widely held that 7S DNA represents the aborted replication intermediates of H-strand synthesis,[3] but its exact role in replication remains unclear.

Despite other earlier studies which suggested that coupled leading and lagging strand DNA synthesis might be a feature of mammalian mitochondria,[41–43] the strand-asynchronous model was until recently considered by many to be the sole mechanism of mammalian mtDNA replication. This view was recently challenged by Holt *et al.*[44]; their work and the subsequent studies by Holt, Jacobs and co-workers proposed a different model.[44–48]

Thus, mammalian mtDNA replication is currently a controversial topic among scientists and clinicians in the fields of mitochondrial biology and mitochondrial diseases.[49,50] A critical difference is that mtDNA replication intermediates are essentially duplex in the recent model, unlike the strand-asynchronous model. Since there are reviews describing the strand-asynchronous model (for example, refs. 36,51,52), we focus here on the recent model of strand-coupled replication of mammalian mtDNA.

11.3.2 Coupled Leading and Lagging Strand DNA Synthesis and the RITOLS Model

Neutral/neutral two-dimensional agarose gel electrophoresis (2D-NAGE)[53,54] has been successfully used to study DNA replication of various DNA templates including nuclear DNA, viral DNA and plasmid DNA, as well as mtDNA of yeast and sea urchins.[55–57] 2D-NAGE separates DNA not only by its molecular

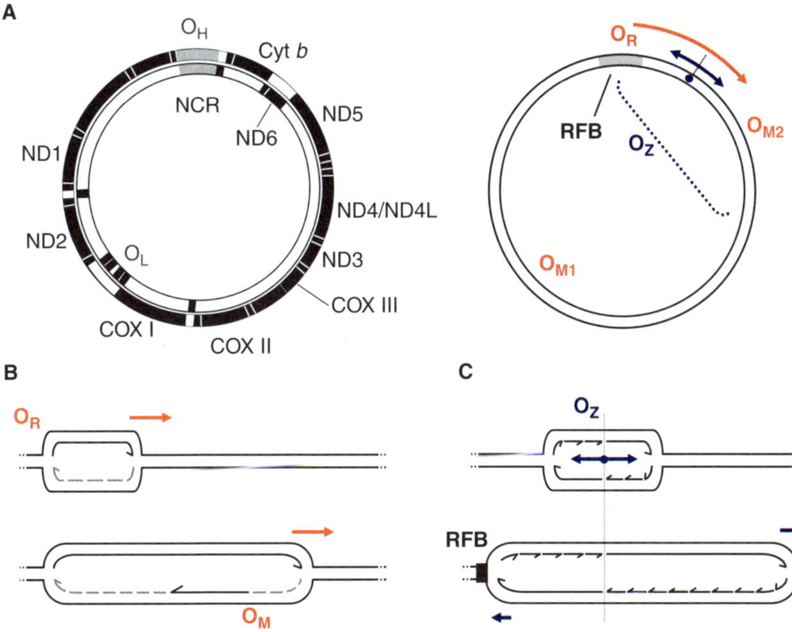

Figure 11.2 **Proposed model of mammalian mtDNA replication with duplex strand replication intermediates.** (**A**) Circular mitochondrial genome in the same orientation as represented in Figure 11.1, showing a simplified genetic map with replication origins O_H and O_L (green) as proposed for the strand-asynchronous mtDNA replication model (left panel), and the origin positions proposed for RITOLS and coupled leading and lagging strand DNA synthesis models (right panel): replication origin O_R (for RITOLS replication) within the non-coding region[48] (NCR, shaded grey); the zone of initiation (O_Z) for bidirectional replication (coupled leading and lagging strand DNA synthesis model[46]); and the presumptive replication fork barrier (RFB) proposed at O_H (or the region near $O_{H,}$ see left panel) which may serve as the replication termination point of the replication fork.[46] (**B**) In RITOLS replication, the leading strand is DNA and the lagging strand is laid down initially as RNA (grey dashed line); the replication proceeds in effect unidirectionally (red arrows).[48] The proposed predominant maturation start site (O_{M1}) is located approximately at O_L (see also Figure 11.1), and a proposed additional maturation start site (O_{M2}) in the ND5 locus[48] is indicated. Loss of the L strand RNA fragments from RITOLS mode would result in the type of replication intermediates proposed under the strand-asynchronous model. (**C**) Coupled leading and lagging strand DNA synthesis within a broad origin zone (O_Z). The blue dot here and in (**A**, right panel) indicates an example of a replication origin within this zone, and the blue arrows indicate the bidirectionality of this replication mode. Replication forks terminate at the replication fork barrier (RFB) located near O_H/O_R. (B and C show a linearized view of mtDNA).

size but also structure, enabling analysis of DNA replication intermediates with complex structure. When DNA is replicated *via* coupled leading and lagging strand synthesis (as for nuclear DNA), characteristic gel electrophoresis patterns such as Y arcs (indicating a replication fork passing through a fragment) and initiation (bubble) arcs (showing that initiation has occurred within the fragment of interest) are observed (for details of the method, see for example, ref. 58,59).

Using 2D-NAGE, Holt *et al.*[44] showed unambiguously that a substantial fraction of mammalian mtDNA replication intermediates have properties of coupled leading and lagging strand synthesis. The rest of the replication intermediates exhibited S1 nuclease sensitivity, indicating that they contained single-stranded regions, initially attributed to strand-asynchronous replication.[44]

This work revealed that two modes of mtDNA replication could operate in mammalian cells. By preventing degradation of the mtDNA replication intermediates during preparation, it was possible to characterise mtDNA replication intermediates as duplex, but containing extensive regions of DNA-RNA hybrid due to incorporation of ribonucleotides in the nascent L strand.[45] The replication intermediates with partially single-stranded region that had been observed previously were proposed to originate from a DNA–RNA hybrid region of the replication intermediates that had lost ribonucleotides.[45] Using a series of 2D-NAGE analyses, initiation arcs from various DNA fragments that cover several kilobases of mtDNA were detected, indicating the existence of multiple origins of bidirectional strand-coupled replication dispersed in a broad zone including the genes for cytochrome *b* and NADH dehydrogenase subunits 5 and 6.[46] A subsequent study proposed that such initiations could occur from an even broader region.[60] Furthermore, it was shown that avian mtDNA also has a similar initiation zone.[60]

Further study unravelled the properties of the ribonucleotide-incorporating nascent L strand and also characterised the two classes of replication intermediates in more detail. These differ in ribonucleotide content and initiation sites.[48] A certain fraction of the intermediates was resistant to ribonuclease H (RNase H) and nuclease S1 treatment. This indicates that they are composed of DNA, and that they are double-stranded, strongly supporting the earlier finding that these intermediates are the products of coupled leading and lagging strand DNA synthesis.[44] Initiation of this replication mode occurs across a broad region of mtDNA[46,48,60] (O_Z in Figure 11.2A, C). In contrast, the remaining significant fraction of the replication intermediates was sensitive to RNase H,[48] suggesting that they incorporate ribonucleotides. Initiation of this ribonucleotide-rich mode of replication was found to be confined to the NCR region, and in the case of the mouse, in a region which spans less than 100 bp from O_H in the D-loop region of the NCR (Figure 11.2A, B).[48] It was also held that an additional origin site for such replication exists close to the cytochrome *b* gene within the NCR. The latter mode of replication proceeds in effect unidirectionally from the NCR.[48]

A series of 2D-NAGE analyses showed that extensive ribonucleotide incorporation occurs in only one of the nascent strands, but this can extend throughout virtually the entire genome.[48] The initiation arc chiefly containing the 'ribonucleotide-rich' replication intermediates was then gel purified from 2D-NAGE and the properties of the nascent strand in the arc analysed, indicating that the nascent L strand (lagging strand) here is initially composed of continuous RNA laid down in tracts of at least 200–600 nucleotides. This replication mode was designated as Ribonucleotide Incorporation ThroughOut the Lagging Strand (RITOLS) replication.[48] Furthermore, based on the 2D-NAGE analysis, the RNA fragments appeared to be replaced with a DNA strand at a later stage of replication during 'maturation steps'. Details of the mechanism of this maturation are not clear, but in murine mtDNA, a defined, prominent maturation start site (O_{M1}) was found to be approximately at O_L and a probable maturation start site (O_{M2}) was identified in the ND5 gene (Figure 11.2A, B).[48] The authors proposed that the physiological role of the RNA may be to protect and stabilise the displaced parental H strand until the nascent DNA L strand is synthesised at the maturation step.[48]

An outstanding issue of the RITOLS replication model is the mechanism that gives rise to the RNA fragments. Two models were proposed.[48] One possibility is that an unknown mitochondrial primase-like factor exists that can synthesise the 200–600 nucleotide long RNA (primer) with no DNA synthesis occurring at the 3′ end of the RNA fragments. Primase activity was fractionated from mammalian mitochondria;[61,62] however, whether it is capable of generating long RNA constantly *in vivo* remains unclear. The second possibility is that preformed transcripts might be hybridised to the displaced parental H strand in the 3′–5′ direction, like threading a bootlace in parallel with the leading DNA strand (nascent H strand) synthesis ('bootlace model'[48]). An earlier electron microscopic study observed RNA complexed with mtDNA;[63] the bootlace model appears to be consistent with this observation.[48]

In summary, the new model of mammalian mtDNA replication comprises two modes of replication:

- conventional coupled leading and lagging strand DNA synthesis mode; and
- the novel replication mode, RITOLS replication.

While the reason that mammalian mitochondria retain the two replication modes is not understood, they may have different physiological roles. Transient depletion of mtDNA can be achieved by exposure of cultured cells to ethidium bromide[64] or dideoxycytidine.[65] When these drugs are removed, cells amplify mtDNA and recover the copy number back to normal. During the amplification, the majority of mtDNA replication intermediates appear to be resistant to RNase H, indicating that they arise through strand-coupled DNA synthesis.[44,47] A major bidirectional origin during the amplification was found in a confined region in the cytochrome *b* end of the NCR, designated as 'cluster II'[47] or OriB. The exact positions of the replication origin and the bidirectionality were proposed using two independent techniques, 2D-NAGE and ligation-mediated PCR. OriB is located close to the cytochrome *b* gene within the NCR, mapping to almost the same regions as the additional origin (O_R) of the RITOLS mode.[48] This finding implies that mitochondria could modulate the replication mechanism in response to various physiological conditions of cells and different cellular energy demands.

A recent study demonstrated that mitochondrial RNA polymerase (POLRMT) is capable of providing an RNA primer of 25–75 nucleotides in length.[66] This work showed that POLG, mtSSB and POLRMT could function together to synthesise an RNA primer, and then DNA using the RNA primer. It is thus possible that POLRMT is the provider of the RNA primer for the coupled leading and lagging strand DNA synthesis observed in mitochondria of cultured cells and tissues.[44–48,60,67]

It has been proposed that the RITOLS replication intermediates represent the true mtDNA replication and that the partially single-stranded replication intermediates on which the strand-asynchronous model was based is generated as a result of L-strand ribonucleotide loss from RITOLS replication intermediates.[45,48] This idea is supported by the finding that high-quality mtDNA preparations contain essentially only duplex replication intermediates, but partially single-stranded DNA molecules can be generated by treatment with RNase H.[45] In addition, cruder mitochondrial preparations gave such partially single-stranded DNA molecules.[44,45]

11.4 Regulation of mtDNA Replication

Given its critical importance in human disease and ageing (see Section 11.5), surprisingly little is known about the regulation of mtDNA replication. It is widely held that mtDNA copy number is the main way that mitochondrial gene expression is regulated in skeletal muscle,[68] and it is clear that this may be limiting in disease. Increases in copy number are an essential adaptation for the increased demands of exercise training,[68] but the triggers of these cellular changes are controversial.[69]

One possible trigger is mitochondrial dysfunction since in some tissues, such as skeletal muscle, mtDNA proliferates in response to mutations causing defects in translation of mtDNA-encoded genes, but not in response to missense mutations in protein-encoding genes. While the severity of these types of mutation may be comparable at the level of the clinical and biochemical phenotype, the mtDNA copy number response is thus very different. Alternatively, production of reactive oxygen species, rather than impaired respiratory chain function, may act as a trigger for compensatory increases in mtDNA copy number.[69] Understanding the regulators of mtDNA replication might reveal potential therapies for diseases where mtDNA replication is aberrant.

Clearly, if mtDNA copy number is to be maintained in dividing cells, there should be a link between mtDNA copy number and cell division. This implies a link between mtDNA replication or turnover, and phase of the cell cycle. However even this is controversial. Technical issues may underlie some of the controversy. When early data were generated, nothing was known about the link between *de novo* dNTP synthesis and S phase. While dNTP synthesis *via* the salvage pathway is continuous, *de novo* synthesis in S phase results in fluctuations of dNTP levels in different cellular compartments by as much as 4–20 fold.[70] Bogenhagen and Clayton[71] concluded that mtDNA replicated throughout the cell cycle independently of S phase. However Attardi and colleagues[72,73] found a five-fold higher incorporation of tritiated methylthymidine during late S and G2 phase than during G1 phase in synchronised A9 and HeLa cells. Their superior physiological synchronisation method probably explains the impressive magnitude of this difference. Results of more recent studies largely depend on the same technological limitations. Hence, by pulsing with BrdU, Taanman and co-workers[74] detected rapid mtDNA synthesis corresponding to S phase in cells defective for dGTP synthesis due to dGK (deoxyguanosine kinase) deficiency. However, they did not detect significantly more BrdU incorporation during S than G phase in dGK proficient control cells. Presumably, the cellular defect decreased non-S phase mtDNA synthesis. Another report suggested that mtDNA copy number increases chiefly during the pre-S phase.[75] We conclude that mtDNA replication may vary at different stages of the cell cycle but that this is not yet fully established experimentally.

11.5 Diseases of mtDNA Replication

Mitochondrial cytopathies comprise a heterogeneous group of disorders caused by mutations in mitochondrial[76,77] or nuclear genes involved in mtDNA maintenance.[27] In normal individuals, all of the mtDNA is effectively identical (homoplasmy), whereas in mtDNA diseases, mutant mtDNA often co-exists with normal (wild-type) mtDNA (heteroplasmy). Such heteroplasmy, with the presence of both normal and mutant mtDNA within a single individual, permits a wide variation in the tissue distribution of mutant mtDNAs and underlies a dose-dependent progression of disease in many of the disorders of mtDNA maintenance.[76] Defects in mtDNA replication can be inherited genetically[78] or acquired, with environmental causes largely resulting from drug therapy for retroviruses such as human immunodeficiency virus (HIV).

11.5.1 Drug-induced Inhibition of mtDNA Replication

The classical drugs for treating HIV infections are nucleoside reverse transcriptase inhibitors (NRTIs) that compete with endogenous dNTPs for the viral polymerases. They coincidentally inhibit POLG in mtDNA synthesis by causing chain termination, leading to a reversible time- and dose-dependent decrease in the intracellular levels of mtDNA.[79] However, the NRTIs may have additional effects. For instance zidovudine inhibits the mitochondrial adenine nucleotide translocator, binds to adenylate kinase and may also be converted into stavudine triphosphate *in vivo*. The most important organs adversely affected by these drugs are the liver, skeletal muscle, peripheral nerves and subcutaneous adipose tissue.[80] Persistently raised blood lactate (indicative of poor mitochondrial function in respiration), mtDNA depletion and occasional cases of mitochondrial encephalomyopathy have been observed in babies exposed to zidovudine in the perinatal period.[81] This is of major medical importance, as mothers with HIV are routinely treated with such drugs prior to giving birth, in order to reduce maternal–foetal transmission of the virus. In the longer term, it is possible that glucose tolerance may also be impaired in these treated children.[82]

Many drugs, including ethidium bromide and the potent anti-cancer drug doxorubicin, intercalate into the nuclear DNA of cancer cells. Some of these are also known to cause mtDNA depletion.[83] Doxorubicin is widely used, but its maximum dose is limited by its tendency to cause a late onset cardiomyopathy, accompanied by mtDNA depletion. We have recently demonstrated that doxorubicin rapidly intercalates into mtDNA of living cells,[83] causing aggregation of mtDNA nucleoids (N. Ashley and J. Poulton, unpublished). This impairs mtDNA replication and is likely to affect mtDNA segregation into daughter cells, possibly explaining the late-onset cardiomyopathy.

11.5.2 Single Gene Disorders of mtDNA Maintenance Reveal Critical Replication Factors

DNA replication requires the coordinated action of multiple proteins. This is further complicated by an interplay with cell cycle regulatory proteins if, indeed, it does turn out that mitochondrial replication is coordinated with the cell-cycle. To date, only a few essential proteins of human mitochondrial DNA replication are known, including the DNA polymerase γ, POLG, single stranded binding protein mtSSB (reviewed in ref. 84), and the Twinkle helicase,[27,85] which are all encoded by nuclear genes. Remarkably, these proteins can be used to reconstitute mitochondrial replisome activity *in vitro*[86] (with synthesis of $\sim$16 kb DNA), supporting their role in mtDNA synthesis *in vivo*. Very recently, addition of POLRMT has been shown to support simultaneous leading and lagging strand synthesis in the *in vitro* system used.[66] In addition, RNase H1 is also probably involved in mtDNA replication, because RNase H1 knockout mice exhibit mtDNA depletion *in utero*.[87] Furthermore, DNA ligase III[88,89] and type IB topoisomerase (topoisomerase I)[90,91] have been demonstrated in mitochondria. FEN1 and DNA2 were very recently proposed to function in RNA primer removal in human mitochondria,[92] presumably similarly to their role in Okazaki fragment processing in nuclear DNA replication (see Chapter 5). By analogy to nuclear DNA replication, it is thought that other factors such as mitochondrial replication initiation factors remain to be identified.

Single gene defects in mtDNA synthesis are summarised in Table 11.1. The majority of severe recessive phenotypes present clinically in childhood, primarily affecting liver and brain, while the milder conditions present in adult life as either dominant or recessive disorders and affect muscle, brain and heart.

Table 11.1 Single gene defects in mtDNA synthesis.

Common (gene) name and protein function	*Major syndromes*	*Features*	*Typical age of onset*	*Type of mtDNA maintenance defect*[a]	*Main affected tissue*
POLG 1 (*POLG1*) DNA polymerase γ large subunit[93,136]	AdPEO and ArPEO	Chronic progressive external ophthalmoplegia (CPEO) Movement disorder Parkinson's Deafness Cataracts Hypogonadism Psychiatric disorder Dysphagia Dysphonia Peripheral neuropathy	Adult	Deletions	Muscle
	SANDO	PEO Peripheral neuropathy Dysarthria Movement disorder	Adult	(Deletions)	Nerves Muscle Brain
	MIRAS	Epilepsy Movement disorder Encephalopathy	Adult	(Deletions)	Brain Muscle
	Alpers	Epilepsy Liver dysfunction Movement disorder Encephalopathy Lactic acidosis	Child	Depletion	Brain Liver (muscle)
POLG2 (*POLG2*) DNA polymerase γ small subunit[147]	AdPEO	See above	Adult	Deletions	Muscle

Table 11.1 (*continued*).

Common (gene) name and protein function	*Major syndromes*	*Features*	*Typical age of onset*	*Type of mtDNA maintenance defect*[a]	*Main affected tissue*
TWINKLE (*PEO1*) Helicase[27,137,138]	AdPEO	See above	Adult	Deletions	Muscle Heart
	Hepato-encephalopathy	Hypotonia Movement disorder Sensory neuropathy Hearing deficit Ophthalmoplegia Intractable epilepsy Liver disease	Child	Depletion	Liver
	IOSCA	Progressive degeneration of cerebellum, brain stem and spinal cord Sensory axonal neuropathy	Child	Depletion	Brain
ANT1 (*SCL25A4*)[139]	AdPEO	See above	Adult	Deletions	Muscle
TP (*ECGF1*) Thymidine phosphorylase[140]	MNGIE	Peripheral neuropathy PEO Gastrointestinal problems Thinness Lactic acidosis	Adult	Deletions (Depletion)	Nerves Smooth and skeletal muscle
TK2 (*TK2*) Thymidine kinase 2[141,142]	Myopathy	Respiratory failure Limb weakness Poor feeding Lactic acidosis	Child	Depletion	Skeletal muscle
SUCLA2 (*SUCLA2*) Succinyl-coA synthetase[119]	Encephalopathy	Hypotonia Myopathy Movement disorder Epilepsy Hyperhidrosis	Child	Depletion	Muscle Liver Brain

		Hearing impairment Lactic acidosis			
SUCLG1 (*SUCLG1*) Succinyl-coA synthetase[118]	Encephalopathy	Myopathy Death in first week Lactic acidosis	Child	Depletion	Muscle Liver Brain
Dgk (*DGUOK*) Deoxyguanosine kinase[143]	Hepato-encephalopathy	Liver disease Cerebral atrophy Microcephaly Nystagmus Hypotonia Lactic acidosis	Child	Depletion	Liver Brain
MPV17 (*MPV17*) Unknown function[94,144,145]	Hepatocerebral mtDNA depletion syndrome	Hypoglycaemia Liver disease	Child	Depletion	Brain Liver
	Navajo neurohepatopathy	Sensory motor neuropathy Corneal scaring Growth retardation Lactic acidosis			Peripheral nerve
P53RNR (*RRM2B*) P53-inducible ribo-nucleotide reductase[114,146]	Myopathy, renal tubulopathy, seizures	Hypotonia Tubulopathy Seizures Respiratory distress Diarrhoea Lactic acidosis	Child	Depletion	Muscle Kidney Brain

[a]In all conditions where it has been investigated, there is an increased incidence of point mutations compared to controls.

Interestingly, the configuration of symptoms is rather different from disorders caused by single heteroplasmic mtDNA mutations. For instance, diabetes is infrequent and peripheral neuropathy common.[93] Table 11.1 lists the main clinical features of these diseases and shows that the products of the genes involved can, for the most part, be divided into proteins of the replisome and proteins involved in dNTP synthesis. In addition, a protein of unknown function (MPV17)[94] is implicated in regulation of antioxidant levels.[95] We have also listed OPA1, a protein involved in mitochondrial dynamics that is implicated in autosomal dominant optic atrophy (ADOA), because the phenotype of patients with ADOA occasionally overlaps with the dominant phenotypes of defects in mtDNA maintenance. However, as little is known about the effects of either MPV17 or OPA1 on mtDNA maintenance, no further discussion is given here.

Defects in the genes listed in Table 11.1 may cause heteroplasmic multiple mtDNA deletions, mtDNA point mutations, mtDNA depletion, or a combination of these, with a diversity of phenotypes attributable to mutations in the same gene.[96] The most widely held hypothesis explaining why defects in so many different proteins (Table 11.1) have similar effects on mtDNA is that all are likely to cause replisome stalling.[97,98] Replication stalling has been documented by detailed analysis of the replication intermediates arising from cell lines carrying mutations in either POLG or TWINKLE.[99] This is likely to occur *in vivo* as several of these mutations have been identified in patients with autosomal dominant progressive external ophthalmoplegia (adPEO).[148]

11.5.2.1 Defects in Mitochondrial DNA Polymerase γ POLG

POLG1 is the nuclear gene encoding the 195 kDa catalytic (alpha) subunit of the mitochondrial DNA polymerase γ, located on chromosome 15q25. Three polymerase domains (exons 17–22) are separated from 3′ exonuclease domains (exons 2–8) by a linker region (residues 418–755) that interacts with the beta subunit (POLG2). Mutations in POLG1 can cause anything from relatively mild autosomal dominant and recessive progressive exophthalmogia (AdPEO and ArPEO), Parkinsonism, mitochondrial recessive ataxia syndrome (MIRAS),[100] hereditary motor sensory neuropathy, to its commonest severe autosomal recessive phenotype, Alpers syndrome (with explosive onset of epilepsy in infancy, accompanied by developmental regression and liver dysfunction[101]). Given the premature ageing observed in mice carrying a POLG mutation (the so-called mutator mice),[102,103] it is entirely possible that many of the morbidities associated with human ageing are related to defects in mtDNA maintenance. While this might be a direct effect of mtDNA mutations on specific tissues, it might equally be an indirect effect on nuclear DNA repair systems.

On the whole, the severe phenotypic defects present with mtDNA depletion (low copy number) but not mtDNA deletions; the milder ones (probably including ageing) with multiple variable deletions but little evidence of depletion, and members of both groups may have multiple point mutations. Patients with mutations only in the POLG2-interacting linker region (residues 418–755) of POLG1 have relatively mild phenotypes, may present in late childhood or adult life, and may not have mtDNA depletion.[104] While many authors have been puzzled because similar genotypes may be associated with variable recessive phenotypes,[96] severe catalytic mutations of POLG tend to cause severe phenotypes and profound mtDNA depletion. Alpers patients with early onset usually have mtDNA depletion in liver, and sometimes in both liver and fibroblasts.[104] When mtDNA depletion is present, patients generally have mutations in a catalytic domain or a nonsense mutation in at least one POLG1 allele. Patients manifesting severe mtDNA depletion in fibroblast cultures as well as liver have particularly short life expectancy.[104] To our knowledge, these patients all had mutations in a catalytic domain in both POLG1 alleles, in either the polymerase or exonuclease domain, or both.[104] One patient with a severe cellular and clinical phenotype was a compound heterozygote, with POLG1 mutations in the polymerase and exonuclease domain in trans,[104] suggesting that POLG1 requires both polymerase and 3′–5′ exonuclease activity in the same molecule. This is consistent with current functional models for eukaryotic DNA polymerases, which alternate between polymerisation and editing modes, as determined by competition between these two active sites for the 3′ end of the DNA (see Chapter 4).

Nevertheless, it is unclear why some patients with defects in the polymerase domain of POLG1 may have a mild phenotype with onset much later and multiple mtDNA deletions, rather than the mtDNA depletion of Alpers syndrome. Deletions might arise during either replication or repair of mtDNA, both of which are likely to depend on POLG1. Mutations in either or both of the catalytic domains of POLG may cause replication pausing. Dissociation of the POLG polymerase may follow replication stalling or failure of the exonuclease activity to excise a misincorporated base. If this is a frequent occurrence, replication may simply be aborted, resulting in mtDNA depletion. However, consequential strand separation of the terminal portion of the newly synthesised strand[105] creates an end which might give rise either to replication slippage, or might itself be recombinogenic. Hence, such ends could underlie the propensity for mtDNA deletions.[106] Alternatively, deletions could result from defective repair of mtDNA damage,[107] because these deletions appear to be indistinguishable from those that are associated with neurodegeneration.[108] Multiple deletions, however they are formed, then accumulate to detectable levels by clonal proliferation of mitochondria in post-mitotic tissues (hence mtDNA replication here is not linked to cell cycle). Thus, the most severe defects of mtDNA maintenance result in depletion, the milder ones in deletions; younger patients with mtDNA depletion might develop deletions if they survived long enough.

11.5.2.2 TWINKLE Helicase

During the process of DNA replication, the duplex template needs to be unwound for the replicative helicase to polymerise against a template strand.

In mitochondria, the TWINKLE 5′–3′ DNA helicase may carry out this function acting as a hexamer, like other replicative helicases (see also Chapter 3). TWINKLE has been shown to bind to single stranded DNA *via* its N-terminus, and mutations within the N-terminal domain result in low helicase processivity and a slowing in replisome progression.[109] Increasing TWINKLE expression leads to an increase in mtDNA copy number in both muscle and heart,[110] while conversely, overexpression of dominant negative TWINKLE mutants in cultured cells leads to accumulation of replication intermediates,[97] indicative of high rates of mtDNA replication fork stalling.[111] Such findings are highly relevant for human disease, where TWINKLE defects are observed in progressive external ophthalmoplegia (PEO)[110] and recessively inherited infantile-onset spinocerebellar ataxia.

11.5.2.3 Factors that Regulate Nucleotide Pools in Mitochondria

mtDNA replisome stalling has been inferred in cell cultures from patients with defects in nucleotide synthesis because the error rate of wild-type POLG is increased by cellular nucleotide imbalances,[112,113] and consequent misincorporation is believed to induce stalling and slippage of DNA polymerases.[106] Ashley *et al.*[98] demonstrated that intramitochondrial nucleotide imbalance underlies impaired mtDNA synthesis in thymidine kinase 2 (TK2) and deoxyguanosine kinase (dGK) mutant cell lines derived from patients with mitochondrial depletion syndrome (MDS). It is highly likely that intramitochondrial nucleotide imbalance is also an important feature of the other disorders of nucleotide synthesis. These include mtDNA depletion caused by mutations in genes such as the p53-dependent subunit of ribonucleotide reductase (RRM2B, encoding p53R2)[114] which contributes to *de novo* nucleotide synthesis.[115] (Note that the tumour suppressor p53 has been detected in mitochondria,[116] though its role has predominantly been associated with mitochondrial-driven apoptosis.[117]) Further factors that regulate nucleotide pools in mitochondria include components of the TCA cycle enzyme succinyl-CoA synthetase, SUCLG1[118] and SUCLA2.[119] Succinyl-CoA synthetase is complexed with mitochondrial nucleoside diphosphate kinase (NDPK), so mtDNA depletion presumably arises because of reduced NDPK activity. mtDNA packaging into nucleoids may be impaired in both types of disorder (nucleotide imbalance and enzyme defects, *e.g.* POLG1).[104,120–122]

11.5.3 Defects of mtDNA Maintenance in Polygenic and Multifactorial Disease

It is now clear that mitochondria are involved in some multifactorial and polygenic diseases. For instance, maternally inherited deafness results from interaction of the 1555 mtDNA mutation with an environmental factor, namely exposure to aminoglycoside antibiotics.[123] Poulton *et al.*[127] have demonstrated an association between a common mtDNA variant in man (the 16189 variant, found in 8% of Caucasians in the UK), and four multifactorial phenotypes,[124–126] including diabetes.[127] Despite initial opposition to this view,[128] a meta-analysis has now confirmed that this 16189 variant is associated with type 2 diabetes[129] (odds ratio in East Asians is 1.335, 95% confidence intervals for this odds ratio being 1.18–1.51, $p = 0.000003$). As this mtDNA 16189 variant maps precisely to the novel bidirectional replication origin,[47] OriB (see Section 11.3.2), it has been renamed the OriB variant. It is highly likely that the OriB variant affects mtDNA replication, but this has not yet been characterised in detail. If it is indeed shown to impact on mtDNA replication, then defects in mtDNA maintenance may be substantially commoner than previously thought, with wide clinical consequence.

11.6 Treatment for Defects of mtDNA Maintenance

There are currently no curative treatments for defects in mtDNA maintenance, medical intervention being limited to genetic management and supportive therapy of complications such as epilepsy, diabetes and deafness. Prospects for specific treatments for disorders of the intramitochondrial nucleotide pool[130] are potentially better than for genetic disorders of the replisome. Excess nucleotides can potentially be removed by dialysis,[131] deficient nucleotides supplemented,[122] and enzymes replaced by direct administration[132] or by organ transplant.[133] Understanding the mitochondrial nucleotide balance[98] and the biology of the nucleoid could give rise to new therapeutic strategies.

Mitochondrial DNA replication may also provide a useful target for drug inhibition and some agents that specifically inhibit mitochondrial TK2 or dGK are in development. For instance, 2′-O-decanoyl-BVaraU, a derivative of araT, adversely affects mitochondrial function by inhibiting TK2.[134] However, the utility of such potential therapies in treating human hyperproliferative disease is limited by the Warburg effect, where cancer cells switch from oxidative phosphorylation to glycolytic metabolism.[135]

Acknowledgements

TY is supported by a BBSRC David Phillips Fellowship. JP acknowledges her funding bodies: the Medical Research Council, the Wellcome Trust and the Angus Memorial Mitochondrial Fund. JP takes clinical and diagnostic referrals for the Oxford centre in the Rare Mitochondrial Disorders Service for Adults and Children (NCG). Further information may be found on www.obs-gyn.ox.ac.uk/research/Poulton.

References

1. S. Anderson, A. T. Bankier, B. G. Barrell, M. H. de Bruijn, A. R. Coulson, J. Drouin, I. C. Eperon, D. P. Nierlich, B. A. Roe, F. Sanger, P. H. Schreier, A. J. Smith, R. Staden and I. G. Young, Sequence and

organization of the human mitochondrial genome, *Nature*, 1981, **290**, 457–465.

2. R. M. Andrews, I. Kubacka, P. F. Chinnery, R. N. Lightowlers, D. M. Turnbull and N. Howell, Reanalysis and revision of the Cambridge reference sequence for human mitochondrial DNA, *Nature Genet.*, 1999, **23**, 147.
3. D. A. Clayton, Nuclear gene products that function in mitochondrial DNA replication, *Philos. Trans. R. Soc. Lond. Ser. B*, 1987, **317**, 473–482.
4. C. Diez-Sanchez, E. Ruiz-Pesini, A. C. Lapena, J. Montoya, A. Perez-Martos, J. A. Enriquez and M. J. Lopez-Perez, Mitochondrial DNA content of human spermatozoa, *Biol. Reprod.*, 2003, **68**, 180–185.
5. P. May-Panloup, M. F. Chretien, C. Jacques, C. Vasseur, Y. Malthiery and P. Reynier, Low oocyte mitochondrial DNA content in ovarian insufficiency, *Hum. Reprod.*, 2005, **20**, 593–597.
6. T. A. Santos, S. El Shourbagy and J. C. St John, Mitochondrial content reflects oocyte variability and fertilization outcome, *Fertil. Steril.*, 2006, **85**, 584–591.
7. R. P. Fisher and D. A. Clayton, Purification and characterization of human mitochondrial transcription factor 1, *Mol. Cell. Biol.*, 1988, **8**, 3496–3509.
8. J. F. Diffley and B. Stillman, DNA binding properties of an HMG1-related protein from yeast mitochondria, *J. Biol. Chem.*, 1992, **267**, 3368–3374.
9. J. F. Diffley and B. Stillman, A close relative of the nuclear, chromosomal high-mobility group protein HMG1 in yeast mitochondria, *Proc. Natl. Acad. Sci. U.S.A.*, 1991, **88**, 7864–7868.
10. T. L. Megraw and C. B. Chae, Functional complementarity between the HMG1-like yeast mitochondrial histone HM and the bacterial histone-like protein HU, *J. Biol. Chem.*, 1993, **268**, 12758–12763.
11. M. A. Parisi, B. Xu and D. A. Clayton, A human mitochondrial transcriptional activator can functionally replace a yeast mitochondrial HMG-box protein both *in vivo* and *in vitro*, *Mol. Cell. Biol.*, 1993, **13**, 1951–1961.
12. R. P. Fisher, T. Lisowsky, G. A. Breen and D. A. Clayton, A rapid, efficient method for purifying DNA-binding proteins. Denaturation–renaturation chromatography of human and yeast mitochondrial extracts, *J. Biol. Chem.*, 1991, **266**, 9153–9160.
13. K. Maniura-Weber, S. Goffart, H. L. Garstka, J. Montoya and R. J. Wiesner, Transient overexpression of mitochondrial transcription factor A (TFAM) is sufficient to stimulate mitochondrial DNA transcription, but not sufficient to increase mtDNA copy number in cultured cells, *Nucleic Acids Res.*, 2004, **32**, 6015–6027.
14. J. Cotney, Z. Wang and G. S. Shadel, Relative abundance of the human mitochondrial transcription system and distinct roles for h-mtTFB1 and h-mtTFB2 in mitochondrial biogenesis and gene expression, *Nucleic Acids Res.*, 2007, **35**, 4042–4054.

15. C. Takamatsu, S. Umeda, T. Ohsato, T. Ohno, Y. Abe, A. Fukuoh, H. Shinagawa, N. Hamasaki and D. Kang, Regulation of mitochondrial D-loops by transcription factor A and single-stranded DNA-binding protein, *EMBO Rep.*, 2002, **3**, 451–456.
16. M. I. Ekstrand, M. Falkenberg, A. Rantanen, C. B. Park, M. Gaspari, K. Hultenby, P. Rustin, C. M. Gustafsson and N. G. Larsson, Mitochondrial transcription factor A regulates mtDNA copy number in mammals, *Hum. Mol. Genet.*, 2004, **13**, 935–944.
17. T. I. Alam, T. Kanki, T. Muta, K. Ukaji, Y. Abe, H. Nakayama, K. Takio, N. Hamasaki and D. Kang, Human mitochondrial DNA is packaged with TFAM, *Nucleic Acids Res.*, 2003, **31**, 1640–1645.
18. K. Ohgaki, T. Kanki, A. Fukuoh, H. Kurisaki, Y. Aoki, M. Ikeuchi, S. H. Kim, N. Hamasaki and D. Kang, The C-terminal tail of mitochondrial transcription factor a markedly strengthens its general binding to DNA, *J. Biochem. (Tokyo)*, 2007, **141**, 201–211.
19. N. Garrido, L. Griparic, E. Jokitalo, J. Wartiovaara, A. M. van der Bliek and J. N. Spelbrink, Composition and dynamics of human mitochondrial nucleoids, *Mol. Biol. Cell*, 2003, **14**, 1583–1596.
20. N. G. Larsson, J. Wang, H. Wilhelmsson, A. Oldfors, P. Rustin, M. Lewandoski, G. S. Barsh and D. A. Clayton, Mitochondrial transcription factor A is necessary for mtDNA maintenance and embryogenesis in mice, *Nature Genet.*, 1998, **18**, 231–236.
21. T. Kanki, K. Ohgaki, M. Gaspari, C. M. Gustafsson, A. Fukuoh, N. Sasaki, N. Hamasaki and D. Kang, Architectural role of mitochondrial transcription factor A in maintenance of human mitochondrial DNA, *Mol. Cell. Biol.*, 2004, **24**, 9823–9834.
22. B. A. Kaufman, N. Durisic, J. M. Mativetsky, S. Costantino, M. A. Hancock, P. Grutter and E. A. Shoubridge, The mitochondrial transcription factor TFAM coordinates the assembly of multiple DNA molecules into nucleoid-like structures, *Mol. Biol. Cell*, 2007, **18**, 3225–3236.
23. M. M. Nass, The circularity of mitochondrial DNA, *Proc. Natl. Acad. Sci. U.S.A.*, 1966, **56**, 1215–1222.
24. M. M. Nass and S. Nass, Intramitochondrial fibers with DNA characteristics. I. Fixation and electron staining reactions., *J. Cell Biol.*, 1963, **19**, 593–611.
25. M. Albring, J. Griffith and G. Attardi, Association of a protein structure of probable membrane derivation with HeLa cell mitochondrial DNA near its origin of replication, *Proc. Natl. Acad. Sci. U.S.A.*, 1977, **74**, 1348–1352.
26. M. Satoh and T. Kuroiwa, Organization of multiple nucleoids and DNA molecules in mitochondria of a human cell, *Exp. Cell Res.*, 1991, **196**, 137–140.
27. J. N. Spelbrink, F. Y. Li, V. Tiranti, K. Nikali, Q. P. Yuan, M. Tariq, S. Wanrooij, N. Garrido, G. Comi, L. Morandi, L. Santoro, A. Toscano, G. M. Fabrizi, H. Somer, R. Croxen, D. Beeson, J. Poulton, A. Suomalainen,

H. T. Jacobs, M. Zeviani and C. Larsson, Human mitochondrial DNA deletions associated with mutations in the gene encoding Twinkle, a phage T7 gene 4-like protein localized in mitochondria, *Nature Genet.*, 2001, **28**, 223–231.

28. F. Legros, F. Malka, P. Frachon, A. Lombes and M. Rojo, Organization and dynamics of human mitochondrial DNA, *J. Cell Sci.*, 2004, **117**, 2653–2662.
29. F. J. Iborra, H. Kimura and P. R. Cook, The functional organization of mitochondrial genomes in human cells, *BMC Biol.*, 2004, **2**, 9.
30. D. F. Bogenhagen, Y. Wang, E. L. Shen and R. Kobayashi, Protein components of mitochondrial DNA nucleoids in higher eukaryotes, *Mol. Cell Proteomics*, 2003, **2**, 1205–1216.
31. D. F. Bogenhagen, D. Rousseau and S. Burke, The layered structure of human mitochondrial DNA nucleoids, *J. Biol. Chem.*, 2008, **283**, 3665–3675.
32. Y. Wang and D. F. Bogenhagen, Human mitochondrial DNA nucleoids are linked to protein folding machinery and metabolic enzymes at the mitochondrial inner membrane, *J. Biol. Chem.*, 2006, **281**, 25791–25802.
33. J. He, C. C. Mao, A. Reyes, H. Sembongi, M. Di Re, C. Granycome, A. B. Clippingdale, I. M. Fearnley, M. Harbour, A. J. Robinson, S. Reichelt, J. N. Spelbrink, J. E. Walker and I. J. Holt, The AAA+ protein ATAD3 has displacement loop binding properties and is involved in mitochondrial nucleoid organization, *J. Cell Biol.*, 2007, **176**, 141–146.
34. G. C. Van Tuyle and P. A. Pavco, The rat liver mitochondrial DNA-protein complex: displaced single strands of replicative intermediates are protein coated, *J. Cell Biol.*, 1985, **100**, 251–257.
35. P. Liu, L. Qian, J. S. Sung, N. C. de Souza-Pinto, L. Zheng, D. F. Bogenhagen, V. A. Bohr, D. M. Wilson 3rd, B. Shen and B. Demple, Removal of oxidative DNA damage *via* FEN1-dependent long-patch base excision repair in human cell mitochondria, *Mol. Cell. Biol.*, 2008, **28**, 4975–4987.
36. D. A. Clayton, Replication of animal mitochondrial DNA, *Cell*, 1982, **28**, 693–705.
37. D. L. Robberson, H. Kasamatsu and J. Vinograd, Replication of mitochondrial DNA. Circular replicative intermediates in mouse L cells, *Proc. Natl. Acad. Sci. U.S.A.*, 1972, **69**, 737–741.
38. T. A. Brown, C. Cecconi, A. N. Tkachuk, C. Bustamante and D. A. Clayton, Replication of mitochondrial DNA occurs by strand displacement with alternative light-strand origins, not *via* a strand-coupled mechanism, *Genes Dev.*, 2005, **19**, 2466–2476.
39. A. Arnberg, E. F. van Bruggen and P. Borst, The presence of DNA molecules with a displacement loop in standard mitochondrial DNA preparations, *Biochim. Biophys. Acta*, 1971, **246**, 353–357.
40. H. Kasamatsu, D. L. Robberson and J. Vinograd, A novel closed-circular mitochondrial DNA with properties of a replicating intermediate, *Proc. Natl. Acad. Sci. U.S.A.*, 1971, **68**, 2252–2257.

41. R. H. Kirschner, D. R. Wolstenholme and N. J. Gross, Replicating molecules of circular mitochondrial DNA, *Proc. Natl. Acad. Sci. U.S.A.*, 1968, **60**, 1466–1472.
42. K. Koike and D. R. Wolstenholme, Evidence for discontinuous replication of circular mitochondrial DNA molecules from Novikoff rat ascites hepatoma cells, *J. Cell Biol.*, 1974, **61**, 14–25.
43. D. R. Wolstenholme, K. Koike and P. Cochran-Fouts, Single strand-containing replicating molecules of circular mitochondrial DNA, *J. Cell Biol.*, 1973, **56**, 230–245.
44. I. J. Holt, H. E. Lorimer and H. T. Jacobs, Coupled leading- and lagging-strand synthesis of mammalian mitochondrial DNA, *Cell*, 2000, **100**, 515–524.
45. M. Y. Yang, M. Bowmaker, A. Reyes, L. Vergani, P. Angeli, E. Gringeri, H. T. Jacobs and I. J. Holt, Biased incorporation of ribonucleotides on the mitochondrial L-strand accounts for apparent strand-asymmetric DNA replication, *Cell*, 2002, **111**, 495–505.
46. M. Bowmaker, M. Y. Yang, T. Yasukawa, A. Reyes, H. T. Jacobs, J. A. Huberman and I. J. Holt, Mammalian mitochondrial DNA replicates bidirectionally from an initiation zone, *J. Biol. Chem.*, 2003, **278**, 50961–50969.
47. T. Yasukawa, M. Y. Yang, H. T. Jacobs and I. J. Holt, A bidirectional origin of replication maps to the major noncoding region of human mitochondrial DNA, *Mol. Cell*, 2005, **18**, 651–662.
48. T. Yasukawa, A. Reyes, T. J. Cluett, M. Y. Yang, M. Bowmaker, H. T. Jacobs and I. J. Holt, Replication of vertebrate mitochondrial DNA entails transient ribonucleotide incorporation throughout the lagging strand, *EMBO J.*, 2006, **25**, 5358–5371.
49. D. F. Bogenhagen and D. A. Clayton, The mitochondrial DNA replication bubble has not burst, *Trends Biochem. Sci.*, 2003, **28**, 357–360.
50. I. J. Holt and H. T. Jacobs, Response: the mitochondrial DNA replication bubble has not burst, *Trends Biochem. Sci.*, 2003, **28**, 355–356.
51. D. A. Clayton, Mitochondrial DNA replication: what we know, *IUBMB Life*, 2003, **55**, 213–217.
52. G. S. Shadel and D. A. Clayton, Mitochondrial DNA maintenance in vertebrates, *Annu. Rev. Biochem.*, 1997, **66**, 409–435.
53. B. J. Brewer and W. L. Fangman, The localization of replication origins on ARS plasmids in S. cerevisiae, *Cell*, 1987, **51**, 463–471.
54. B. J. Brewer and W. L. Fangman, A replication fork barrier at the 3′ end of yeast ribosomal RNA genes, *Cell*, 1988, **55**, 637–643.
55. Z. Han and C. Stachow, Analysis of *Schizosaccharomyces pombe* mitochondrial DNA replication by two dimensional gel electrophoresis, *Chromosoma*, 1994, **103**, 162–170.
56. D. Lockshon, S. G. Zweifel, L. L. Freeman-Cook, H. E. Lorimer, B. J. Brewer and W. L. Fangman, A role for recombination junctions in the segregation of mitochondrial DNA in yeast, *Cell*, 1995, **81**, 947–955.

57. A. G. Mayhook, A. M. Rinaldi and H. T. Jacobs, Replication origins and pause sites in sea urchin mitochondrial DNA, *Proc. Biol. Sci.*, 1992, **248**, 85–94.
58. K. L. Friedman and B. J. Brewer, Analysis of replication intermediates by two-dimensional agarose gel electrophoresis, *Methods Enzymol.*, 1995, **262**, 613–627.
59. A. Reyes, T. Yasukawa and I. J. Holt, Analysis of replicating mitochondrial DNA by two-dimensional agarose gel electrophoresis, *Methods Mol. Biol.*, 2007, **372**, 219–232.
60. A. Reyes, M. Y. Yang, M. Bowmaker and I. J. Holt, Bidirectional replication initiates at sites throughout the mitochondrial genome of birds, *J. Biol. Chem.*, 2005, **280**, 3242–3250.
61. B. J. Ledwith, S. Manam and G. C. Van Tuyle, Characterization of a DNA primase from rat liver mitochondria, *J. Biol. Chem.*, 1986, **261**, 6571–6577.
62. T. W. Wong and D. A. Clayton, Isolation and characterization of a DNA primase from human mitochondria, *J. Biol. Chem.*, 1985, **260**, 11530–11535.
63. D. Carre and G. Attardi, Biochemical and electron microscopic characterization of DNA-RNA complexes from HeLa cell mitochondria, *Biochemistry*, 1978, **17**, 3263–3273.
64. A. Wiseman and G. Attardi, Reversible ten-fold reduction in mitochondria DNA content of human cells treated with ethidium bromide, *Mol. Gen. Genet.*, 1978, **167**, 51–63.
65. M. V. Simpson, C. D. Chin, S. A. Keilbaugh, T. S. Lin and W. H. Prusoff, Studies on the inhibition of mitochondrial DNA replication by 3′-azido-3′-deoxythymidine and other dideoxynucleoside analogs which inhibit HIV-1 replication, *Biochem. Pharmacol.*, 1989, **38**, 1033–1036.
66. S. Wanrooij, J. M. Fuste, G. Farge, Y. Shi, C. M. Gustafsson and M. Falkenberg, Human mitochondrial RNA polymerase primes lagging-strand DNA synthesis *in vitro*, *Proc. Natl. Acad. Sci. U.S.A*, 2008, **105**, 11122–11127.
67. O. A. Kajander, P. J. Karhunen, I. J. Holt and H. T. Jacobs, Prominent mitochondrial DNA recombination intermediates in human heart muscle, *EMBO Rep*, 2001, **2**, 1007–1012.
68. R. S. Williams, Mitochondrial gene expression in mammalian striated muscle. Evidence that variation in gene dosage is the major regulatory event, *J. Biol. Chem.*, 1986, **261**, 12390–12394.
69. R. Moreno-Loshuertos, R. Acin-Perez, P. Fernandez-Silva, N. Movilla, A. Perez-Martos, S. Rodriguez de Cordoba, M. E. Gallardo and J. A. Enriquez, Differences in reactive oxygen species production explain the phenotypes associated with common mouse mitochondrial DNA variants, *Nat. Genet.*, 2006, **38**, 1261–1268.
70. P. C. Bradshaw and D. C. Samuels, A computational model of mitochondrial deoxynucleotide metabolism and DNA replication, *Am. J. Physiol. Cell Physiol.*, 2005, **288**, C989–1002.

71. D. Bogenhagen and D. A. Clayton, Mouse L cell mitochondrial DNA molecules are selected randomly for replication throughout the cell cycle, *Cell*, 1977, **11**, 719–727.
72. L. Pica-Mattoccia and G. Attardi, Expression of the mitochondrial genome in HeLa cells. IX. Replication of mitochondrial DNA in relationship to cell cycle in HeLa cells, *J. Mol. Biol.*, 1972, **64**, 465–484.
73. J. W. Posakony, J. M. England and G. Attardi, Mitochondrial growth and division during the cell cycle in HeLa cells, *J. Cell Biol.*, 1977, **74**, 468–491.
74. J. Magnusson, M. Orth, P. Lestienne and J. W. Taanman, Replication of mitochondrial DNA occurs throughout the mitochondria of cultured human cells, *Exp. Cell Res.*, 2003, **289**, 133–142.
75. M. Trinei, I. Berniakovich, P. G. Pelicci and M. Giorgio, Mitochondrial DNA copy number is regulated by cellular proliferation: a role for Ras and p66(Shc), *Biochim. Biophys. Acta*, 2006, **1757**, 624–630.
76. J. Poulton, S. O'Rahilly, K. J. Morten and A. Clark, Mitochondrial DNA, diabetes and pancreatic pathology in Kearns–Sayre syndrome, *Diabetologia*, 1995, **38**, 868–871.
77. R. N. Lightowlers, P. F. Chinnery, D. M. Turnbull and N. Howell, Mammalian mitochondrial genetics: heredity, heteroplasmy and disease, *Trends Genet.*, 1997, **13**, 450–455.
78. W. C. Copeland, Inherited mitochondrial diseases of DNA replication, *Annu. Rev. Med.*, 2008, **59**, 131–146.
79. E. Arnaudo, M. Dalakas, S. Shanske, C. T. Moraes, S. DiMauro and E. A. Schon, Depletion of muscle mitochondrial DNA in AIDS patients with zidovudine-induced myopathy, *Lancet*, 1991, **337**, 508–510.
80. U. A. Walker, Update on mitochondrial toxicity: where are we now?, *J. HIV Ther.*, 2003, **8**, 32–35.
81. S. Blanche, M. Tardieu, P. Rustin, A. Slama, B. Barret, G. Firtion, N. Ciraru-Vigneron, C. Lacroix, C. Rouzioux, L. Mandelbrot, I. Desguerre, A. Rotig, M. J. Mayaux and J. F. Delfraissy, Persistent mitochondrial dysfunction and perinatal exposure to antiretroviral nucleoside analogues, *Lancet*, 1999, **354**, 1084–1089.
82. K. Morten, P. Field, N. Ashley, K. A. Williams, D. Harris, M. Hartley, A. Clark and J. Poulton, Fetal and neonatal exposure to AZT and low-protein diet affects glucose homeostasis: a model with implications for AIDS prevention, *Am. J. Physiol. Endocrinol. Metab.*, 2005, **289**, E1115–1118.
83. N. Ashley and J. Poulton, Mitochondrial DNA is a direct target of anti-cancer anthracycline drugs, *Biochem. Biophys. Res. Commun.*, 2009, **378**, 450.
84. L. S. Kaguni, DNA polymerase gamma, the mitochondrial replicase, *Annu. Rev. Biochem.*, 2004, **73**, 293–320.
85. J. A. Korhonen, M. Gaspari and M. Falkenberg, TWINKLE Has $5 \rightarrow 3'$ DNA helicase activity and is specifically stimulated by mitochondrial single-stranded DNA-binding protein, *J. Biol. Chem.*, 2003, **278**, 48627–48632.

86. J. A. Korhonen, X. H. Pham, M. Pellegrini and M. Falkenberg, Reconstitution of a minimal mtDNA replisome *in vitro*, *EMBO J.*, 2004, **23**, 2423–2429.
87. S. M. Cerritelli, E. G. Frolova, C. Feng, A. Grinberg, P. E. Love and R. J. Crouch, Failure to produce mitochondrial DNA results in embryonic lethality in Rnaseh1 null mice, *Mol. Cell*, 2003, **11**, 807–815.
88. U. Lakshmipathy and C. Campbell, The human DNA ligase III gene encodes nuclear and mitochondrial proteins, *Mol. Cell. Biol.*, 1999, **19**, 3869–3876.
89. U. Lakshmipathy and C. Campbell, Antisense-mediated decrease in DNA ligase III expression results in reduced mitochondrial DNA integrity, *Nucleic Acids Res.*, 2001, **29**, 668–676.
90. H. Zhang, J. M. Barcelo, B. Lee, G. Kohlhagen, D. B. Zimonjic, N. C. Popescu and Y. Pommier, Human mitochondrial topoisomerase I, *Proc. Natl. Acad. Sci. U.S.A.*, 2001, **98**, 10608–10613.
91. H. Zhang, L. H. Meng and Y. Pommier, Mitochondrial topoisomerases and alternative splicing of the human TOP1mt gene, *Biochimie*, 2007, **89**, 474–481.
92. L. Zheng, M. Zhou, Z. Guo, H. Lu, L. Qian, H. Dai, J. Qiu, E. Yakubovskaya, D. F. Bogenhagen, B. Demple and B. Shen, Human DNA2 is a mitochondrial nuclease/helicase for efficient processing of DNA replication and repair intermediates, *Mol. Cell*, 2008, **32**, 325–336.
93. R. Horvath, G. Hudson, G. Ferrari, N. Futterer, S. Ahola, E. Lamantea, H. Prokisch, H. Lochmuller, R. McFarland, V. Ramesh, T. Klopstock, P. Freisinger, F. Salvi, J. A. Mayr, R. Santer, M. Tesarova, J. Zeman, B. Udd, R. W. Taylor, D. Turnbull, M. Hanna, D. Fialho, A. Suomalainen, M. Zeviani and P. F. Chinnery, Phenotypic spectrum associated with mutations of the mitochondrial polymerase gamma gene, *Brain*, 2006, **129**, 1674–1684.
94. A. Spinazzola, C. Viscomi, E. Fernandez-Vizarra, F. Carrara, P. D'Adamo, S. Calvo, R. M. Marsano, C. Donnini, H. Weiher, P. Strisciuglio, R. Parini, E. Sarzi, A. Chan, S. DiMauro, A. Rotig, P. Gasparini, I. Ferrero, V. K. Mootha, V. Tiranti and M. Zeviani, MPV17 encodes an inner mitochondrial membrane protein and is mutated in infantile hepatic mitochondrial DNA depletion, *Nat. Genet.*, 2006, **38**, 570–575.
95. R. Iida, T. Yasuda, E. Tsubota, H. Takatsuka, T. Matsuki and K. Kishi, Human Mpv17-like protein is localized in peroxisomes and regulates expression of antioxidant enzymes, *Biochem. Biophys. Res. Commun.*, 2006, **344**, 948–954.
96. P. F. Chinnery and M. Zeviani, 155th ENMC workshop: polymerase gamma and disorders of mitochondrial DNA synthesis, 21-23 September 2007, Naarden, The Netherlands, *Neuromuscul. Disord.*, 2008, **18**, 259–267.
97. S. Wanrooij, S. Goffart, J. L. Pohjoismaki, T. Yasukawa and J. N. Spelbrink, Expression of catalytic mutants of the mtDNA helicase Twinkle and polymerase POLG causes distinct replication stalling phenotypes, *Nucleic Acids Res.*, 2007, **35**, 3238–3251.

98. N. Ashley, S. Adams, A. Slama, M. Zeviani, A. Suomalainen, A. L. Andreu, R. K. Naviaux and J. Poulton, Defects in maintenance of mitochondrial DNA are associated with intramitochondrial nucleotide imbalances, *Hum. Mol. Genet.*, 2007, **16**, 1400–1411.
99. S. Wanrooij, P. Luoma, G. van Goethem, C. van Broeckhoven, A. Suomalainen and J. N. Spelbrink, Twinkle and POLG defects enhance age-dependent accumulation of mutations in the control region of mtDNA, *Nucleic Acids Res.*, 2004, **32**, 3053–3064.
100. A. H. Hakonen, S. Heiskanen, V. Juvonen, I. Lappalainen, P. T. Luoma, M. Rantamaki, G. V. Goethem, A. Lofgren, P. Hackman, A. Paetau, S. Kaakkola, K. Majamaa, T. Varilo, B. Udd, H. Kaariainen, L. A. Bindoff and A. Suomalainen, Mitochondrial DNA polymerase W748S mutation: a common cause of autosomal recessive ataxia with ancient European origin, *Am. J. Hum. Genet.*, 2005, **77**, 430–441.
101. R. K. Naviaux and K. V. Nguyen, POLG mutations associated with Alpers' syndrome and mitochondrial DNA depletion, *Ann. Neurol.*, 2004, **55**, 706–712.
102. A. Trifunovic, A. Wredenberg, M. Falkenberg, J. Spelbrink, A. Rovio, C. Bruder, M. Bohlooly-Y, S. Gidlof, A. Oldfors, R. Wibom, J. Tornell, H. Jacobs and N. Larsson, Premature ageing in mice expressing defective mitochondrial DNA polymerase, *Nature*, 2004, **429**, 417–423.
103. M. Vermulst, J. Wanagat, G. C. Kujoth, J. H. Bielas, P. S. Rabinovitch, T. A. Prolla and L. A. Loeb, DNA deletions and clonal mutations drive premature aging in mitochondrial mutator mice, *Nat. Genet.*, 2008, **40**, 392–394.
104. N. Ashley, A. O'Rourke, C. Smith, S. Adams, V. Gowda, M. Zeviani, G. K. Brown, C. Fratter and J. Poulton, Depletion of mitochondrial DNA in fibroblast cultures from patients with POLG1 mutations is a consequence of catalytic mutations, *Hum. Mol. Genet.*, 2008, **17**, 2496–2506.
105. M. V. Ponamarev, M. J. Longley, D. Nguyen, T. A. Kunkel and W. C. Copeland, Active site mutation in DNA polymerase gamma associated with progressive external ophthalmoplegia causes error-prone DNA synthesis, *J. Biol. Chem.*, 2002, **277**, 15225–15228.
106. E. Viguera, D. Canceill and S. D. Ehrlich, Replication slippage involves DNA polymerase pausing and dissociation, *EMBO J.*, 2001, **20**, 2587–2595.
107. K. J. Krishnan, A. K. Reeve, D. C. Samuels, P. F. Chinnery, J. K. Blackwood, R. W. Taylor, S. Wanrooij, J. N. Spelbrink, R. N. Lightowlers and D. M. Turnbull, What causes mitochondrial DNA deletions in human cells?, *Nat. Genet.*, 2008, **40**, 275–279.
108. N. Howell, J. L. Elson, P. F. Chinnery and D. M. Turnbull, mtDNA mutations and common neurodegenerative disorders, *Trends Genet.*, 2005, **21**, 583–586.
109. G. Farge, T. Holmlund, J. Khvorostova, R. Rofougaran, A. Hofer and M. Falkenberg, The N-terminal domain of TWINKLE contributes to single-stranded DNA binding and DNA helicase activities, *Nucleic Acids Res.*, 2008, **36**, 393–403.

110. H. Tyynismaa, H. Sembongi, M. Bokori-Brown, C. Granycome, N. Ashley, J. Poulton, A. Jalanko, J. N. Spelbrink, I. J. Holt and A. Suomalainen, Twinkle helicase is essential for mtDNA maintenance and regulates mtDNA copy number, *Hum. Mol. Genet.*, 2004, **13**, 3219–3227.
111. S. Goffart, H. M. Cooper, H. Tyynismaa, S. Wanrooij, A. Suomalainen and J. N. Spelbrink, Twinkle mutations associated with autosomal dominant progressive external ophthalmoplegia lead to impaired helicase function and *in vivo* mtDNA replication stalling, *Hum. Mol. Genet.*, 2008, **18**, 328–340.
112. C. M. Wernette, M. C. Conway and L. S. Kaguni, Mitochondrial DNA polymerase from *Drosophila melanogaster* embryos: kinetics, processivity, and fidelity of DNA polymerization, *Biochemistry*, 1988, **27**, 6046–6054.
113. T. A. Kunkel and K. Bebenek, Recent studies of the fidelity of DNA synthesis, *Biochim. Biophys. Acta*, 1988, **951**, 1–15.
114. A. Bourdon, L. Minai, V. Serre, J. P. Jais, E. Sarzi, S. Aubert, D. Chretien, P. de Lonlay, V. Paquis-Flucklinger, H. Arakawa, Y. Nakamura, A. Munnich and A. Rotig, Mutation of RRM2B, encoding p53-controlled ribonucleotide reductase (p53R2), causes severe mitochondrial DNA depletion, *Nat. Genet.*, 2007, **39**, 776–780.
115. G. Pontarin, P. Ferraro, P. Hakansson, L. Thelander, P. Reichard and V. Bianchi, p53R2-dependent ribonucleotide reduction provides deoxyribonucleotides in quiescent human fibroblasts in the absence of induced DNA damage, *J. Biol. Chem.*, 2007, **282**, 16820–16828.
116. G. Achanta, R. Sasaki, L. Feng, J. S. Carew, W. Lu, H. Pelicano, M. J. Keating and P. Huang, Novel role of p53 in maintaining mitochondrial genetic stability through interaction with DNA Pol gamma, *EMBO J.*, 2005, **24**, 3482–3492.
117. P. F. Li, R. Dietz and R. von Harsdorf, p53 regulates mitochondrial membrane potential through reactive oxygen species and induces cytochrome c-independent apoptosis blocked by Bcl-2, *EMBO J.*, 1999, **18**, 6027–6036.
118. E. Ostergaard, E. Christensen, E. Kristensen, B. Mogensen, M. Duno, E. A. Shoubridge and F. Wibrand, Deficiency of the alpha subunit of succinate-coenzyme A ligase causes fatal infantile lactic acidosis with mitochondrial DNA depletion, *Am. J. Hum. Genet.*, 2007, **81**, 383–387.
119. O. Elpeleg, C. Miller, E. Hershkovitz, M. Bitner-Glindzicz, G. Bondi-Rubinstein, S. Rahman, A. Pagnamenta, S. Eshhar and A. Saada, Deficiency of the ADP-forming succinyl-CoA synthase activity is associated with encephalomyopathy and mitochondrial DNA depletion, *Am. J. Hum. Genet.*, 2005, **76**, 1081–1086.
120. N. Ashley, D. Harris and J. Poulton, Detection of mitochondrial DNA depletion in living human cells using PicoGreen staining, *Exp. Cell Res.*, 2005, **303**, 432–446.
121. K. J. Morten, N. Ashley, F. Wijburg, N. Hadzic, J. Parr, S. Jayawant, S. Adams, L. Bindoff, H. D. Bakker, G. Mieli-Vergani, M. Zeviani and

J. Poulton, Liver mtDNA content increases during development: a comparison of methods and the importance of age- and tissue-specific controls for the diagnosis of mtDNA depletion, *Mitochondrion*, 2007, **7**, 386–395.

122. J. W. Taanman, J. R. Muddle and A. C. Muntau, Mitochondrial DNA depletion can be prevented by dGMP and dAMP supplementation in a resting culture of deoxyguanosine kinase-deficient fibroblasts, *Hum. Mol. Genet.*, 2003, **12**, 1839–1845.
123. X. Estivill, N. Govea, E. Barcelo, C. Badenas, E. Romero, L. Moral, R. Scozzri, L. D'Urbano, M. Zeviani and A. Torroni, Familial progressive sensorineural deafness is mainly due to the mtDNA A1555G mutation and is enhanced by treatment of aminoglycosides, *Am. J. Hum. Genet.*, 1998, **62**, 27–35.
124. K. Casteels, K. Ong, D. Phillips, H. Bendall and M. Pembrey, Mitochondrial 16189 variant, thinness at birth, and type-2 diabetes. ALSPAC study team. Avon Longitudinal Study of Pregnancy and Childhood, *Lancet*, 1999, **353**, 1499–1500.
125. S. S. Khogali, B. M. Mayosi, J. M. Beattie, W. J. McKenna, H. Watkins and J. Poulton, A common mitochondrial DNA variant associated with susceptibility to dilated cardiomyopathy in two different populations, *Lancet*, 2001, **357**, 1265–1267.
126. K. J. Livesey, V. L. Wimhurst, K. Carter, M. Worwood, E. Cadet, J. Rochette, A. G. Roberts, J. J. Pointon, A. T. Merryweather-Clarke, M. L. Bassett, A. M. Jouanolle, A. Mosser, V. David, J. Poulton and K. J. Robson, The 16189 variant of mitochondrial DNA occurs more frequently in C282Y homozygotes with haemochromatosis than those without iron loading, *J. Med. Genet.*, 2004, **41**, 6–10.
127. J. Poulton, J. Luan, V. Macaulay, S. Hennings, J. Mitchell and N. J. Wareham, Type 2 diabetes is associated with a common mitochondrial variant: evidence from a population-based case-control study, *Hum. Mol. Genet.*, 2002, **11**, 1581–1583.
128. P. F. Chinnery, H. R. Elliott, S. Patel, C. Lambert, S. M. Keers, S. E. Durham, M. I. McCarthy, G. A. Hitman, A. T. Hattersley and M. Walker, Role of the mitochondrial DNA 16184-16193 poly-C tract in type 2 diabetes, *Lancet*, 2005, **366**, 1650–1651.
129. K. S. Park, J. C. Chan, L. M. Chuang, S. Suzuki, E. Araki, K. Nanjo, L. Ji, M. Ng, M. Nishi, H. Furuta, T. Shirotani, B. Y. Ahn, S. S. Chung, H. K. Min, S. W. Lee, J. H. Kim, Y. M. Cho and H. K. Lee, A mitochondrial DNA variant at position 16189 is associated with type 2 diabetes mellitus in Asians, *Diabetologia*, 2008, **51**, 602–608.
130. G. Pontarin, P. Ferraro, M. L. Valentino, M. Hirano, P. Reichard and V. Bianchi, Mitochondrial DNA depletion and thymidine phosphate pool dynamics in a cellular model of mitochondrial neurogastrointestinal encephalomyopathy, *J. Biol. Chem.*, 2006, **281**, 22720–22728.
131. A. Spinazzola, R. Marti, I. Nishino, A. L. Andreu, A. Naini, S. Tadesse, I. Pela, E. Zammarchi, M. A. Donati, J. A. Oliver and M. Hirano, Altered

thymidine metabolism due to defects of thymidine phosphorylase, *J. Biol. Chem.*, 2002, **277**, 4128–4133.
132. B. E. Bax, M. D. Bain, L. D. Fairbanks, A. D. Webster, P. W. Ind, M. S. Hershfield and R. A. Chalmers, A 9-yr evaluation of carrier erythrocyte encapsulated adenosine deaminase (ADA) therapy in a patient with adult-type ADA deficiency, *Eur. J. Haematol.*, 2007, **79**, 338–348.
133. L. Salviati, S. Sacconi, M. Mancuso, D. Otaegui, P. Camano, A. Marina, S. Rabinowitz, R. Shiffman, K. Thompson, C. M. Wilson, A. Feigenbaum, A. B. Naini, M. Hirano, E. Bonilla, S. DiMauro and T. H. Vu, Mitochondrial DNA depletion and dGK gene mutations, *Ann. Neurol.*, 2002, **52**, 311–317.
134. N. Ciliberti, S. Manfredini, A. Angusti, E. Durini, N. Solaroli, S. Vertuani, L. Buzzoni, M. C. Bonache, E. Ben-Shalom, A. Karlsson, A. Saada and J. Balzarini, Novel selective human mitochondrial kinase inhibitors: design, synthesis and enzymatic activity, *Bioorg. Med. Chem.*, 2007, **15**, 3065–3081.
135. O. H. E. Warburg, *The Metabolism of Tumours,* Constable, London, 1930.
136. G. Van Goethem, B. Dermaut, A. Lofgren, J. J. Martin and C. Van Broeckhoven, Mutation of POLG is associated with progressive external ophthalmoplegia characterized by mtDNA deletions, *Nature Genet.*, 2001, **28**, 211–212.
137. A. H. Hakonen, P. Isohanni, A. Paetau, R. Herva, A. Suomalainen and T. Lonnqvist, Recessive Twinkle mutations in early onset encephalopathy with mtDNA depletion, *Brain*, 2007, **130**, 3032–3040.
138. K. Nikali, A. Suomalainen, J. Saharinen, M. Kuokkanen, J. N. Spelbrink, T. Lonnqvist and L. Peltonen, Infantile onset spinocerebellar ataxia is caused by recessive mutations in mitochondrial proteins Twinkle and Twinky, *Hum. Mol. Genet.*, 2005, **14**, 2981–2990.
139. J. Kaukonen, J. K. Juselius, V. Tiranti, A. Kyttala, M. Zeviani, G. P. Comi, S. Keranen, L. Peltonen and A. Suomalainen, Role of adenine nucleotide translocator 1 in mtDNA maintenance, *Science*, 2000, **289**, 782–785.
140. I. Nishino, A. Spinazzola and M. Hirano, Thymidine phosphorylase gene mutations in MNGIE, a human mitochondrial disorder, *Science*, 1999, **283**, 689–692.
141. A. Saada, E. Ben-Shalom, R. Zyslin, C. Miller, H. Mandel and O. Elpeleg, Mitochondrial deoxyribonucleoside triphosphate pools in thymidine kinase 2 deficiency, *Biochem. Biophys. Res. Commun.*, 2003, **310**, 963–966.
142. A. Saada, A. Shaag, H. Mandel, Y. Nevo, S. Eriksson and O. Elpeleg, Mutant mitochondrial thymidine kinase in mitochondrial DNA depletion myopathy, *Nature Genet.*, 2001, **29**, 342–344.
143. H. Mandel, R. Szargel, V. Labay, O. Elpeleg, A. Saada, A. Shalata, Y. Anbinder, D. Berkowitz, C. Hartman, M. Barak, S. Eriksson and N. Cohen, The deoxyguanosine kinase gene is mutated in individuals with

depleted hepatocerebral mitochondrial DNA, *Nature Genet.*, 2001, **29**, 337–341.

144. A. Spinazzola, V. Massa, M. Hirano and M. Zeviani, Lack of founder effect for an identical mtDNA depletion syndrome (MDS)-associated MPV17 mutation shared by Navajos and Italians, *Neuromuscul. Disord.*, 2008, **18**, 315–318.
145. L. J. Wong, N. Brunetti-Pierri, Q. Zhang, N. Yazigi, K. E. Bove, B. B. Dahms, M. A. Puchowicz, I. Gonzalez-Gomez, E. S. Schmitt, C. K. Truong, C. L. Hoppel, P. C. Chou, J. Wang, E. E. Baldwin, D. Adams, N. Leslie, R. G. Boles, D. S. Kerr and W. J. Craigen, Mutations in the MPV17 gene are responsible for rapidly progressive liver failure in infancy, *Hepatology*, 2007, **46**, 1218–1227.
146. B. Bornstein, E. Area, K. M. Flanigan, J. Ganesh, P. Jayakar, K. J. Swoboda, J. Coku, A. Naini, S. Shanske, K. Tanji, M. Hirano and S. DiMauro, Mitochondrial DNA depletion syndrome due to mutations in the RRM2B gene, *Neuromuscul. Disord.*, 2008, **18**, 453–459.
147. M. J. Longley, S. Clark, C. Yu Wai Man, G. Hudson, S. E. Durham, R. W. Taylor, S. Nightingale, D. M. Turnbull, W. C. Capeland and P. F. Chinnery, *Am. J. Hum. Genet.*, 2006, **78**(6), 1026.
148. S. Goffart, H. M. Cooper, H. Tyynismaa, S. Wanrooij, A. Suomalainen, J. N. Spelbrink, Twinkle mutations associated with autosomal dominant progressive external ophthalmoplegia lead to impaired helicase function and *in vivo* mtDNA replication stalling, *Hum. Mol. Genet.*, 2009, **18**(2), 328–340.

CHAPTER 12

DNA Replication in the Archaea: a Paradigm for Eukaryotic Replication

STEPHEN D. BELL

Sir William Dunn School of Pathology, University of Oxford, South Parks Road, Oxford, OX1 3RE, UK

12.1 Introduction

Although morphologically resembling bacteria, it has become apparent that the archaea possess a DNA replication machinery that is orthologous to that found in eukaryotes. In general, the archaeal machinery is a simplified version of that found in present day eukaryotes and, as such, can serve as a valuable model for the core conserved mechanisms that underpin the process of genome duplication.[1,2] The overall process of DNA replication can be subdivided into several key stages:

- definition of the start site or origin of replication;
- recruitment of the replicative helicase and origin unwinding;
- establishment of the replication fork (including replisome assembly);
- ensuring progression of the replication fork;
- termination of replication.

With the exception of the process of termination, much has been elucidated about the key molecular players and events in these stages of archaeal DNA replication. This chapter summarizes recent progress.

Molecular Themes in DNA Replication
Edited by Lynne S. Cox

Published by the Royal Society of Chemistry, www.rsc.org

12.2 Archaeal Origins of Replication

Like bacteria, archaea lack nuclei and, like most bacteria, have simple circular chromosomes with polycistronic gene clusters. Bacterial chromosomes are replicated from a single origin of replication (*oriC*); this contrasts with the situation in eukaryotes where multiple origins are used to replicate the linear eukaryotic chromosomes.

The first work on archaeal origins of replication was performed in the laboratory of Patrick Forterre and revealed, through a combination of *in vivo* labelling, two-dimensional (2D) gel studies and bioinformatic analyses, that the single circular chromosome of *Pyrococcus* was replicated from a single replication origin, termed *oriC*.[3,4] In *Pyrococcus, oriC* was located immediately upstream of the gene for the single *Pyrococcus* Orc1/Cdc6 homologue.

In general, archaeal species possess at least one gene that is homologous to the eukaryotic replication initiation factors, Orc1 and Cdc6.[5] It has been known for some time that eukaryotic Orc1 and Cdc6 were related to one another in sequence—presumably indicative of them having evolved from a common ancestor. Thus, it appears that archaea may possess just such an ancestral molecule. The juxtaposition of the origin of replication with the gene for the candidate initiator protein is in fact reminiscent of the situation in many bacteria, where the gene for the DnaA initiator protein is located adjacent to the origin of replication.

Although the archaeal replication machinery is clearly related to that of eukaryotes, these initial studies in *Pyrococcus* suggested that a 'bacterial-like' mode of replication may be employed by the archaea. However, many archaea possess multiple genes for Orc1/Cdc6 homologues and so it was unclear whether some species may, in fact, harbour more than one replication origin per chromosome. Systematic 2D gel mapping studies in the hyperthermophile, *Sulfolobus solfataricus* (Sso), revealed that the genes for two of the three Orc1/Cdc6 homologues were immediately adjacent to origins of replication.[6] A subsequent report using marker frequency analysis (MFA) confirmed the positioning of these two origins and additionally revealed the presence of a third replication origin about 80 kilobases removed from the gene for the final Orc1/Cdc6 homologue.[7] Furthermore, the MFA analysis produced data supporting the notion that all three replication origins are used in all cells. The presence of the third origin was confirmed by 2D gel mapping approaches, and although not beside an Orc1/Cdc6 homologue, was found to be immediately adjacent to a distant homologue of the eukaryotic replication initiation factor, Cdt1.[8,9] The divergent archaeal Cdt1 homologue has been called WhiP (for winged helix initiator protein).

The multiple origin paradigm has been extended beyond *Sulfolobus* species with the detection of two replication origins in *Aeropyrum pernix* and in the main chromosome of *Haloferax volcanii*.[9,10] The archaeal DNA replication origins in general possess a range of repeated DNA sequence motifs and also usually have an AT-rich region that may serve as a duplex unwinding element (DUE). High-resolution replication initiation point mapping has been performed on a number of the archaeal origins and this has revealed that initiation occurs within, or in the immediate vicinity of, the DUE in all species tested.[6,10,11]

12.2.1 Definition of Archaeal Replication Origins

The presence of archaeal Orc1/Cdc6 homologues suggested that these proteins may play a role in origin definition, and this has been found to be the case. Many archaeal origins of replication contain sequence repeats termed mORB elements that have the consensus sequence (t/c)TNCANNNGAA(a/c), where N is any nucleotide.[6,12] In some species, the consensus is extended by a string of three or more G residues on one side of the mORB;[6] these extended motifs are termed ORB elements. Biochemical studies have revealed that these ORB and mORB elements are bound in a sequence specific manner by homologues of the *Sulfolobus* Orc1-1 protein. However, additional sequence elements in *Sulfolobus* replication origins are bound by Orc1-2 and Orc1-3. Indeed, in *oriC2* of *Sulfolobus*, a mORB element is found immediately adjacent to an Orc1-3 binding site (termed C3) and the two proteins bind this locus with modest positive cooperativity.[6] The complexity of the nucleoprotein architecture is added to by the presence of the WhiP protein which, in *Sulfolobus*, has been found to bind to origins of replication, again showing some cooperativity with Orc1/Cdc6 proteins.

12.2.2 Binding of Orc1/Cdc6 to Origins

Sequence analysis of the archaeal Orc1/Cdc6 proteins predicted the presence of an N-terminal ATPase domain (belonging to the AAA+ superfamily, see Chapter 2) and a C-terminal winged helix (wH) candidate DNA binding domain. The crystal structures of several archaeal Orc1/Cdc6 proteins have been determined, confirming these predictions, and biochemical studies have underscored the importance of the wH domain in mediating DNA binding.[6,13–15]

More recently, considerable insight has been gained into the way in which these proteins function, with the elucidation of the co-crystal structures of a heterodimer of *Sulfolobus* Orc1-1 and Orc1-3 on the *oriC2*-derived hybrid mORB/C3 site and of the *A. pernix* homologue of Sso Orc1-1 on an ORB element.[16,17] Not only do these structures demonstrate the role of the wH domain in DNA recognition, they additionally reveal the unanticipated presence of a second DNA binding site, embedded within the AAA+ domain of the proteins (Figure 12.1). This takes the form of a parallel pair of alpha helices. This motif, termed the initiator specific motif (ISM) is conserved in eukaryotic Orc1, Orc4 and Cdc6. Remarkably, although non-orthologous, the bacterial replication initiator protein, DnaA, also possesses an ISM, albeit a pair of antiparallel alpha helices within its AAA+ domain. The positioning of this DNA binding motif within the AAA+ domain of these initiator proteins suggests that it may reposition upon ATP binding, hydrolysis and release and so may

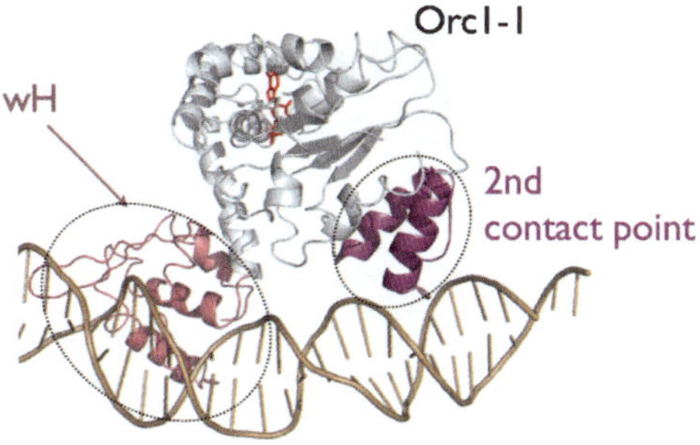

Figure 12.1 **The structure of Orc1-1 from *Sulfolobus solfataricus* bound to DNA**. The winged helix (wH) domain is shown in pink and the initiator specific motif (ISM) in purple; ADP bound by Orc1-1 is in red. The figure was prepared using PyMOL (http://pymol.org/) from PDB coordinates 2QBY.[16]

play a key role in remodelling origin DNA. Notably, binding to the origin induces significant bending of ~30° with respect to the helical axis,[16,17] (see Figure 12.1).

12.3 The Archaeal Replicative Helicase—MCM

The eukaryotic replicative helicase is thought to be the MCM complex, composed of six separate but related polypeptides (MCM2-7)[18,19] (see Chapter 3). Most archaea encode a single MCM homologue, and work in a number of archaeal species has revealed this to be a homohexamer (or double hexamer) with (d)ATP-dependent 3′ to 5′ helicase activity.[20–23] Studies in *Methanothermobacter thermautotrophicum* (Mth) have shown that the Orc1/Cdc6 homologues in that species can physically interact with MCM, leading to a model where the MCM is directly recruited to origins *via* interaction with the initiator factor Orc1/Cdc6.[24] However, to date, origin-dependent recruitment of archaeal MCM has not been reconstituted *in vitro*.

A number of structural studies have been performed on MthMCM. The crystal structure of the N-terminal domain has been solved to 3 Å resolution and revealed a dodecamer in the form of head-to-head stacked double hexameric ring, with each subunit contributing a DNA binding β-hairpin in the centre of the ring.[25] Biochemical analyses of deletion mutants in *Sulfolobus* MCM have revealed that this domain of the protein contributes to the processivity of the helicase on DNA, perhaps by acting as a form of sliding clamp[26] (Figure 12.2). Immediately C-terminal of the processivity domain lies the motor domain of the protein; this is a AAA$^+$ ATPase domain (see Chapter 2) and has been shown to be the minimal functional helicase component *in vitro*.[26,27] This domain contains two motifs key to the helicase function, the first of which is a β-hairpin, a basic residue at the tip of which is required for helicase activity.[28] Additionally, the hairpin is preceded by a β–α–β module that appears to move

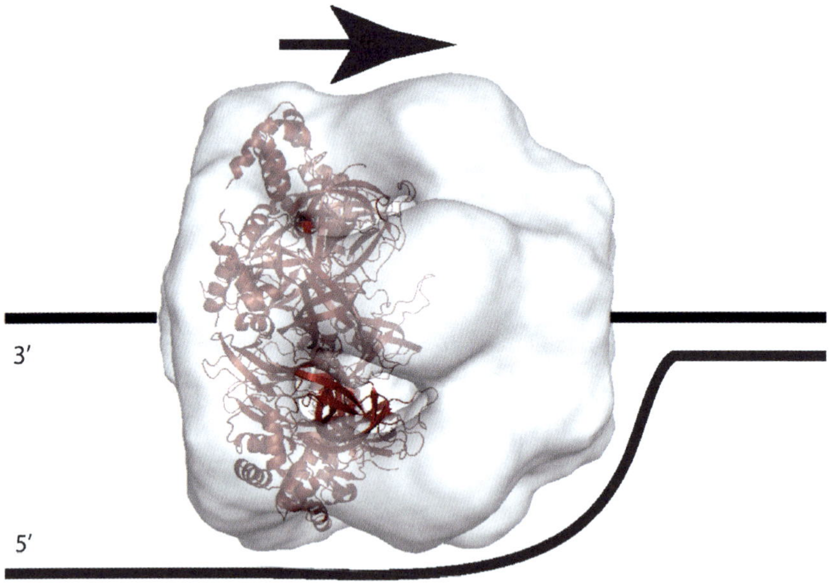

Figure 12.2 **Model of a homohexamer of MCM translocating along DNA.** An electron microscopy (EM) reconstruction of Mth MCM is shown, with the N-terminal domain crystal structure (red) placed in the EM envelope (grey). The coordinates for this structure were kindly provided by Alessandro Costa.[62] The MCM is translocating along the leading strand template in a 3 to 5′ direction, while the displaced strand makes weak contacts with the outer surface of MCM.

in the presence of nucleotide or DNA substrate.[29] Following the AAA$^+$ domain is a wH domain, the role of which is currently unclear, although it has been observed that deletion of this domain elevates the MCM's helicase activity, suggesting it may play a regulatory role.[26,27,29] Both single molecule and ensemble FRET experiments with *Sulfolobus* MCM[28,30] have revealed that MCM loads onto 3′ tailed DNA with the motor domains facing towards the duplex region and with the N-terminal processivity domains trailing. At least with oligonucleotide substrates, the MCM appears to act as a molecular bulldozer, or wire stripper, displacing the 5′ tailed strand as the helicase progresses along DNA (Figure 12.2). Interestingly, the single molecule FRET experiments have revealed that MCM actually reduces the distance between the 5′ and 3′ terminal residues of the two separated strands.[30] This attraction displays an unusual sub-diffusion kinetic behaviour, suggesting that the 5′ displaced strand can occupy multiple equivalent positions on the surface of MCM, perhaps representing a series of channels or grooves that position the displaced strand. This may facilitate the forward motion of MCM and could also direct the displaced lagging strand template to additional factors, such as the GINS-primase assembly discussed in Section 12.4.

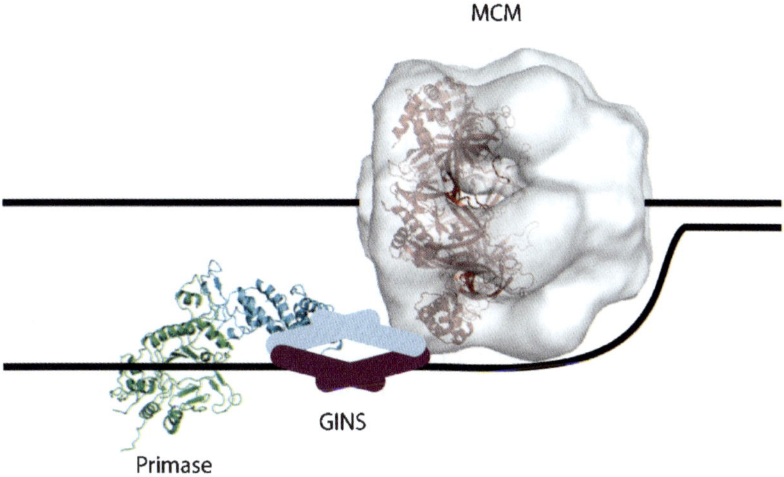

Figure 12.3 **Model of the proposed GINS-mediated connection between MCM and primase.** The Gins23 subunits are shown in blue and Gins15 in purple. The crystal structure of Sso primase is shown (PDB coordinates 1ZT2[37]).

12.4 Replisome Processivity Complex

In eukaryotes, MCMs at the replication fork form a stable complex with Cdc45 and the GINS complex[31] (see also Chapters 1 and 3). Eukaryotic GINS is a heterotetramer composed of one copy of each of Psf1, Psf2, Psf3 and Sld5.[31] Although no archaeal homologue of Cdc45 has yet been detected, archaeal GINS has been identified and shown to be a four subunit complex containing two copies each of Gins15 (related to eukaryotic Sld5 and Psf1) and Gins23 (related to Psf2 and Psf3).[31] This complex, first characterised in the *Sulfolobus* system, was shown to interact with the N-terminal processivity domains of MCM.[32] Although there was no discernable effect of adding *Sulfolobus* GINS back to MCM helicase assays *in vitro*, a recent study has indicated that the very weak helicase activity of *Pyrococcus* MCM is enhanced in the presence of *Pyrococcus* GINS.[33] Interestingly, in *Sulfolobus*, GINS has also been shown to interact with primase.[32]

Although it has not yet been demonstrated biochemically, the interaction data support a model in which archaeal GINS could act as an intermediary between MCM and primase. Perhaps the MCM-displaced strand interactions described above could target the lagging strand template across the outer surface of MCM to the appropriate location of the GINS-primase complex (Figure 12.3). This would provide a mechanism for coupling translocation of MCM along the leading strand template with primase action on the lagging strand.

12.5 The Heterodimeric Archaeal Primase

In archaea, as in eukaryotes, the core primase is a heterodimer with a catalytic and a regulatory subunit.[34,35] In eukaryotes, this core heterodimer associates with two additional factors, the B subunit and DNA polymerase α. These latter two components are not found in archaea. The crystal structure of a truncated derivative of the heterodimeric *Sulfolobus* primase has been solved, as has the structure of the isolated catalytic subunit of *Pyrococus* primase.[36,37] The structure of the *Sulfolobus* enzyme lacks the C-terminal iron-binding module found in the regulatory subunit of other archaeal and eukaryotic primases.[38] Although the iron sulfur module was shown to be essential for the primase activity of the yeast enzyme, it was not absolutely required for activity in the *Sulfolobus* complex.[37] Interestingly, the archaeal primases appear to be considerably more promiscuous than their eukaryotic counterparts. In addition to RNA primase activity, archaeal enzymes from a variety of species have been shown to possess both RNA and DNA primase and polymerase activities, as well as terminal deoxynucleotidyl transferase activity.[34,35,39,40] In light of the extensive range of transactions that the archaeal primase enzymes can perform, it has been proposed that they may play roles in DNA repair processes in addition to functioning purely as primases in DNA replication.

12.6 Archaeal DNA Polymerases

Two principal kingdoms of archaea have been characterised, the Crenarchaea (including *Sulfolobus* and *Aeropyrum*) and the Euryarchaea (including *Haloferax*, *Pyrococcus* and *Methanothermobacter*). Interestingly, the distribution of DNA polymerases shows an intriguing distinction between these two kingdoms. While all archaea characterised to date possess members of the B family of DNA polymerases (*e.g.* the *Pfu* enzyme well known to most molecular biologists for its applications in PCR), the Euryarchaea also possess another novel family of DNA polymerases—family D.[41]

Family D polymerases form a distinct group of DNA polymerases, with no orthologues in eukaryotes. These polymerases are composed of two subunits; the larger of the two, DP2 possesses DNA polymerase activity and the smaller, DP1, has 3′ to 5′ exonuclease activity. Biochemical studies have revealed that in *Pyrococcus*, which possesses both B and D type polymerases, polD is capable of extending DNA from an RNA primer, whereas, in the conditions tested, polB cannot.[42] PolB, however, efficiently extends a DNA primer. This has led to the proposal that polD may be involved in the initial extension of RNA primers and thus could be the main lagging strand polymerase, while polB's main role may be in leading strand replication.[43] This is further supported by the observation that polD can carry out strand displacement synthesis, a role likely to be important for lagging strand maturation, whereas polB cannot. How the hand-off from primase to polD DNA polymerase is effected is currently unclear.

While Crenarchaea do not have D family polymerases, they generally have multiple B family DNA polymerases. For example, *Sulfolobus* has three B family polymerases. It is currently unclear what the individual roles of the multiple B family polymerases are. In light of the possible strand synthesis preferences of polD and polB in the Euryarchaea, it is tempting to speculate that the polB family members may have distinct roles in leading and lagging strand synthesis in the Crenarchaea; there is, however, currently no experimental evidence to support or refute this hypothesis. One intriguing feature that archaeal family B polymerases of both Crenarchaea and Euryarchaea possess is the ability to detect uracil residues in the template ahead of the polymerase active site. This activity is found in both hyperthermophiles and mesophiles, and is due to a small pocket in the leading N-terminal domain of the polymerase that specifically binds to uracil and stalls the polymerase.[44] How this stall is signalled to the DNA repair machinery is not currently known, although archaeal uracil DNA glycosylases have been shown to associate with proliferating cell nuclear antigen (PCNA) and so may be present at the replication fork.[45,46]

12.7 Polymerase Accessory Factors

12.7.1 PCNA

As with eukaryotic replicative DNA polymerases, archaeal DNA polymerases (of both B and D family) have been shown to interact with the sliding clamp, PCNA, and thereby have their processivity enhanced.[47,48] Also as in eukaryotes (see Chapter 3), archaeal PCNAs interact with a range of other client proteins, including the flap endonuclease (FEN1), recombinases and DNA ligase.[48] While all Euryarchaea characterised to date have a single PCNA gene that results in a homotrimeric PCNA ring, a number of Crenarchaea possess multiple PCNA homologues that can form heterotrimeric assemblies. *A. pernix* encodes three PCNA homologues and these can form homotrimers and heterotrimers.[49,50] An even more extreme case is found with *Sulfolobus*, where the three PCNA subunits are monomers individually and only assemble as a 1 : 1 : 1 heterotrimer.[51] Furthermore, the heterotrimer has a defined order of assembly, with PCNA1 first forming a heterodimer with PCNA2. The formation of this heterodimer is a prerequisite for recruitment of PCNA3.[51] Recent structures of the *Sulfolobus* PCNA heterotrimer have revealed the basis of this stereo-specific assembly of the heterotrimer, being manifested by distinct distributions of electrostatic and hydrophobic interactions at the three inter-subunit interfaces.[52–54]

Significantly, the three subunits of *Sulfolobus* PCNA have distinct preferred partner proteins, for example, PCNA1 interacts with FEN1, PCNA2 with DNA polB1 and PCNA3 with DNA lig1. Furthermore, multiple factors can simultaneously occupy a single PCNA heterotrimer.[51] Thus, a simple rotation of PCNA around the double helical axis could result in the coordination of consecutive actions of client proteins. A hypothetical example is shown in Figure 12.4 where DNA polymerase B displaces the primer of a downstream Okazaki fragment, generating a substrate for FEN1. PCNA could rotate, delivering FEN1 to the appropriate site. Following FEN1-dependent cleavage of the flap, PCNA could rotate again to re-position the resultant nicked DNA for ligation by DNA lig 1 (Figure 12.4).

It is likely, however, that in addition to simple rotation of the PCNA ring, conformational changes in the client proteins may also be important for their repositioning into an active configuration. This is perhaps best illustrated with the example of DNA ligase 1. Ellenberger and colleagues solved the crystal structure of human DNA ligase 1 in complex with a nicked DNA substrate.[55] This structure revealed that DNA lig1 encircled DNA and presented an overall structure that could stack on top of a PCNA ring (Figure 12.5B). This would at first glance appear incompatible with PCNA binding any other client proteins other than a single molecule of ligase. However, the structure of DNA-free *Sulfolobus* DNA ligase1 (54% similarity, 33% identical to human ligase) revealed that the ligase adopted an extended configuration.[54] Furthermore, small angle X-ray scattering analyses of the *Sulfolobus* ligase-PCNA complex revealed that the ligase extended out radially from the PCNA ring (Figure 12.5A). Thus, for the ligase to reconfigure around the DNA substrate, large-scale movement between the domains of the ligase will be required. It is tempting to speculate that, as the ligase shuts down on DNA, it may sterically occlude or even actively remove other client proteins bound to PCNA. This could conceivably be important for recycling of the PCNA ring following completion of Okazaki fragment processing.

12.7.2 Clamp Loader

The toroidal nature of PCNA necessitates a mechanism for loading it onto DNA. This task is performed by the ATP-utilising clamp loader complex, RFC. In eukaryotes, RFC has one large subunit, in complex with a hetero-tetramer of small subunits (see Chapter 2). The majority of archaea similarly have an RFC complex composed of one large subunit (RFCL) and a homo-tetramer of a small subunit (RFCS). An interesting exception to this has been described by Cann and colleagues,[56] who revealed that the clamp loader of *Methanosarcina acetivorans* (Mac) in fact has two distinct small subunits (RFCS1 and RFCS2). Reconstituted Mac RFC appears to have a 1 : 3 : 1 stoichiometry of RFCL : $(RFCS1)_3$: RFCS2. It seems, therefore, that this atypical archaeal RFC may represent an evolutionary stepping stone on the route to the eukaryotic hetero-pentameric RFC.[56] The functional organisation of the ATP binding sites in Mac RFC was investigated by mutational analyses, revealing that the RFCS1 ATPase active sites were essential for clamp loading. These data, as well as providing a glimpse into the evolution of multi-subunit systems, also reveal some intriguing mechanistic parallels with the bacterial clamp loader, the γ-complex. This complex has a 1 : 3 : 1 stoichiometry of its δ, γ and δ subunits and the ATPase active sites in the trimeric γ component are, as in Mac RFCS1, essential for clamp loading activity.[47]

A thorough analysis of the roles of ATP binding and hydrolysis has also been performed with the RFC complex of *Archaeoglobus fulgidus*.[57,58] This work has

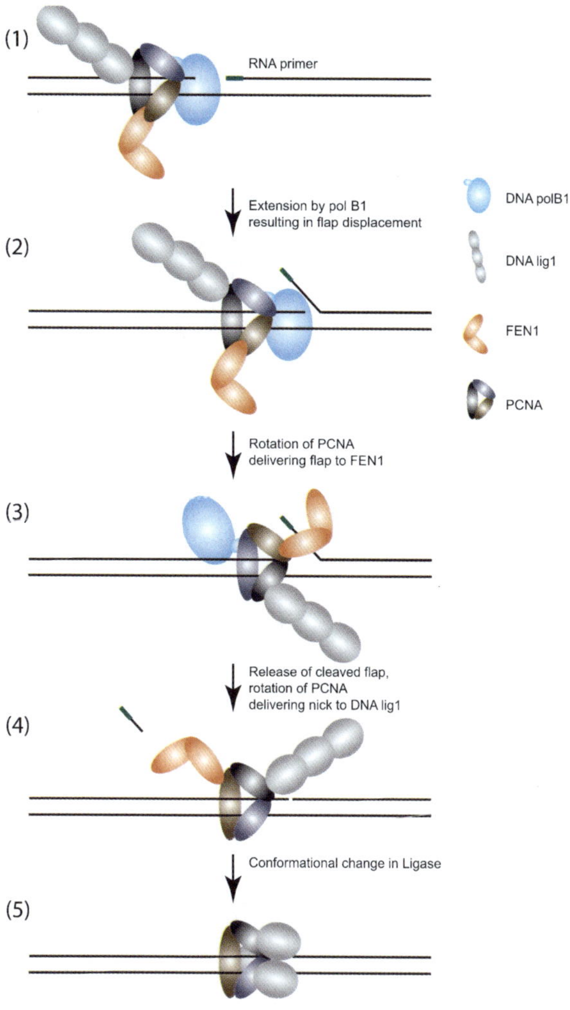

Figure 12.4 **Model for the coordination of client protein activities by PCNA rotation during lagging strand maturation.** DNA polymerase approaches a downstream Okazaki fragment with its 5′ RNA primer shown in green (**1**). The polymerase is shown in pale blue and is bound to a PCNA trimer (the three subunits are in dark grey, purple and brown). PCNA is also bound to FEN1 (orange) and DNA ligase (grey). The polymerase displaces a 5′ flap containing the RNA primer (**2**). PCNA rotates delivering FEN1 to the displaced flap (**3**). FEN1 cleaves the flap and PCNA rotates once again, now delivering ligase to the nick in the DNA (**4**). Finally, ligase undergoes a conformational shift, encircling DNA and displacing FEN1 in the process (**5**).

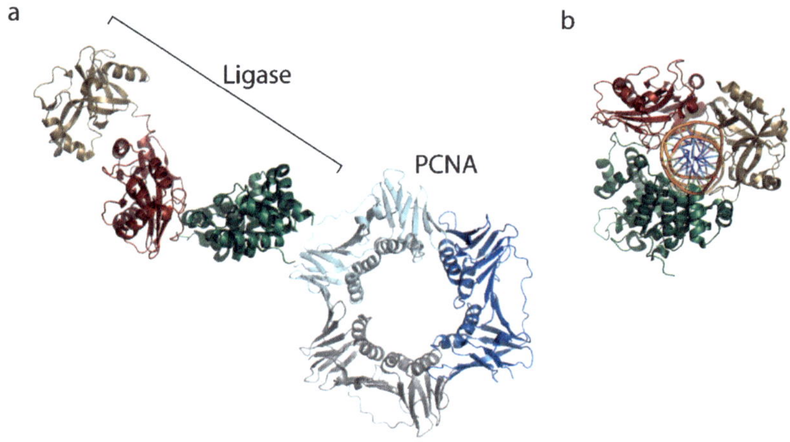

Figure 12.5 **DNA ligase conformation.** Dramatic conformational shifts in DNA ligase 1 occur depending on whether it is bound to PCNA (**a**) or to nicked DNA (**b**). The three principal domains of the ligase are coloured in sand, red and green, and PCNA is shown in blue, dark grey and light grey. The figures were prepared using PyMOL (http://pymol.org/) with PDB files 2HIV (Sso DNA lig1),[54] 2HII (Sso PCNA)[54] and 1X9N (Human DNA lig1).[54]

revealed that ATP binding (but not hydrolysis) by the large subunit and three of the four small subunits was necessary for PCNA loading. However, hydrolysis of ATP bound to the small subunits was required for release of the PCNA from the clamp loader, as also observed in eukaryotes (Chapter 2). Finally, hydrolysis of the ATP bound to the large subunit was required for catalytic loading of the clamp, indicating that this hydrolysis event was required for dissociation of the clamp from the primer-template junction and/or resetting RFC for another round of PCNA loading (Figure 12.6).

Valuable insights into the clamp loading process have also been obtained at the structural level. Crystal structures of the RFC small subunit have been obtained and a 12 Å resolution electron microscopy structure of a complex of *Pyrococcus* RFC and PCNA on DNA has been elucidated.[59] Interestingly, this structure indicated that PCNA was cracked open at one interface. The opening was of only 5 Å, far too small to accommodate double-stranded DNA, so it was proposed that this structure might represent an intermediate in the loading process, formed prior to the hydrolysis of ATP by RFC and the resealing of PCNA around DNA (Figure 12.6). This structure suggests that only a single inter-subunit interface of PCNA needs to be opened to effect clamp loading. This was confirmed in a study that exploited the defined stereochemistry of the *Sulfolobus* heterotrimeric PCNA to covalently shut individual interfaces by fusing subunits with a short linker.[60] This study revealed that only one interface, that between PCNA1 and PCNA3, was required to be opened during the loading process.

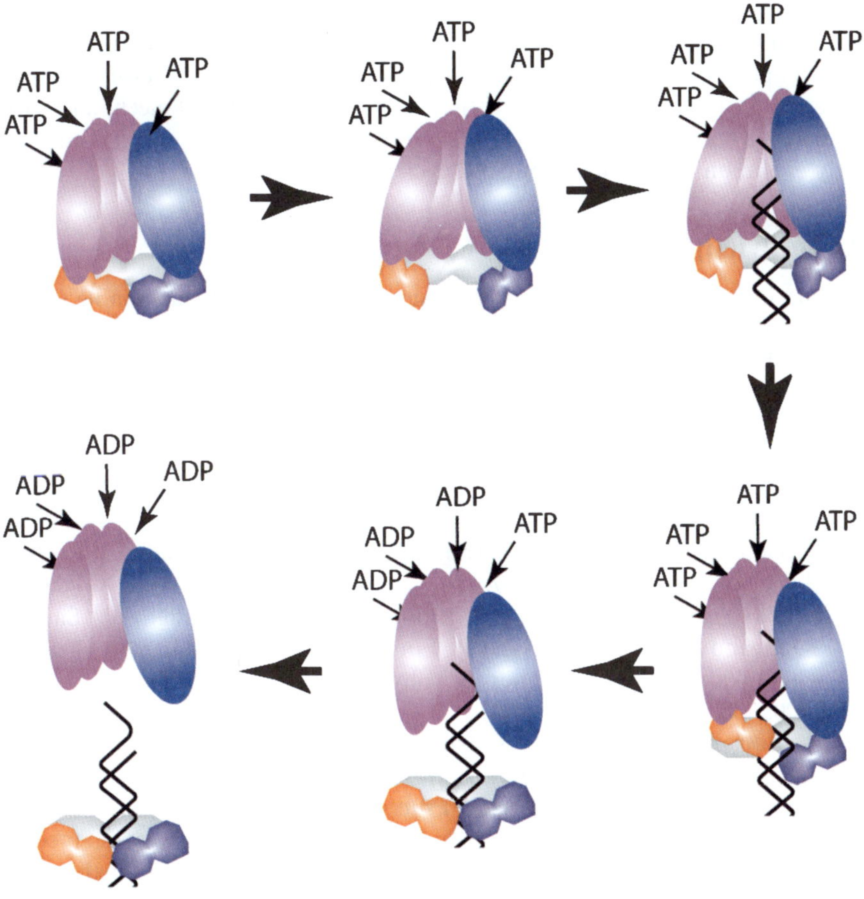

Figure 12.6 **Loading of the PCNA clamp by RFC.** ATP bound RFC (small subunits in purple, large subunit in blue) binds to heterotrimeric PCNA (grey/blue/orange). This opens the PCNA ring, allowing entry of duplex DNA. The clamp partially shuts around DNA prior to hydrolysis of three ATP molecules bound by RFCS. Hydrolysis of ATP by the large subunit of RFC results in dissociation of the clamp loader from the primer template junction.

12.8 Concluding Remarks

Studies of archaeal DNA replication proteins have yielded considerable structural and mechanistic insight into the form and function of these evolutionarily conserved proteins. However, virtually nothing is known about the way in which the activities of these proteins are regulated in the context of the archaeal cell cycle. For example, in the case of *Sulfolobus*, it is known that this organism has a cell cycle oscillating between a very short G1 period (where there is a single copy of its chromosome) and an extensive post-replicative G2 period with two chromosome copies.[61] How the firing of the multiple replication origins in the *Sulfolobus* chromosome is coordinated is not yet known. For example, whether firing is tied to cell division, and whether it may involve post-translational modification of key replication factors, remain undetermined. However, it is likely that studies of the archaeal mechanisms for controlling multiple replication origins may yield significant insight into the early evolution of the eukaryotic cell cycle.

Acknowledgements

I would like to apologise to those colleagues whose work I was unable to cite due to space constraints. Research in my laboratory is supported by the Edward Penley Abraham Trust, the BBSRC and the Wellcome Trust.

References

1. E. R. Barry and S. D. Bell, DNA replication in the Archaea, *Microbiol. Mol. Biol. Rev.*, 2006, **70**, 876–887.
2. Z. Kelman and M. F. White, Archaeal DNA replication and repair, *Curr. Opin. Microbiol.*, 2005, **8**, 669–676.
3. H. Myllykallio, P. Lopez, P. Lopez-Garcia, R. Heilig, W. Saurin, Y. Zivanovic, H. Philippe and P. Forterre, Bacterial mode of replication with eukaryotic-like machinery in a hyperthermophilic archaeon, *Science*, 2000, **288**, 2212–2215.
4. F. Matsunaga, P. Forterre, Y. Ishino and H. Myllykallio, *In vivo* interactions of archaeal Cdc6/Orc1 and minichromosome maintenance proteins with the replication origin, *Proc. Natl. Acad. Sci. U.S.A.*, 2001, **98**, 11152–11157.
5. N. P. Robinson and S. D. Bell, Origins of DNA replication in the three domains of life, *FEBS J.*, 2005, **272**, 3757–3766.
6. N. P. Robinson, I. Dionne, M. Lundgren, V. L. Marsh, R. Bernander and S. D. Bell, Identification of two origins of replication in the single chromosome of the archaeon *Sulfolobus solfataricus*, *Cell*, 2004, **116**, 25–38.
7. M. Lundgren, A. Andersson, L. M. Chen, P. Nilsson and R. Bernander, Three replication origins in *Sulfolobus* species: synchronous initiation of chromosome replication and asynchronous termination, *Proc. Natl. Acad. Sci. U.S.A.*, 2004, **101**, 7046–7051.
8. N. P. Robinson, K. A. Blood, S. A. McCallum, P. A. W. Edwards and S. D. Bell, Sister chromatid junctions in the hyperthermophilic archaeon *Sulfolobus solfataricus*, *EMBO J.*, 2007, **26**, 816–824.
9. N. P. Robinson and S. D. Bell, Extrachromosomal element capture and the evolution of multiple DNA replication origins in archaeal chromosomes, *Proc. Natl. Acad. Sci. U.S.A.*, 2007, **104**, 5806–5811.
10. C. Norais, M. Hawkins, A. L. Hartman, J. A. Eisen, H. Myllykallio and T. Allers, Genetic and physical mapping of DNA replication origins in *Haloferax volcanii*, *PLOS Genet.*, 2007, **3**, e77.

11. I. Grainge, M. Gaudier, B. S. Schuwirth, S. L. Westcott, J. Sandall, N. Atanassova and D. B. Wigley, Biochemical analysis of a DNA replication origin in the archaeon *Aeropyrum pernix*, *J. Mol. Biol.*, 2006, **363**, 355–369.
12. S. A. Capaldi and J. M. Berger, Biochemical characterization of Cdc6/Orc1 binding to the replication origin of the euryarchaeon *Methanothermobacter thermoautotrophicus*, *Nucleic Acids Res.*, 2004, **32**, 4821–4832.
13. J. Y. Liu, C. L. Smith, D. DeRyckere, K. DeAngelis, G. S. Martin and J. M. Berger, Structure and function of Cdc6/Cdc18: implications for origin recognition and checkpoint control, *Mol. Cell*, 2000, **6**, 637–648.
14. M. R. Singleton, R. Morales, I. Grainge, N. Cook, M. N. Isupov and D. B. Wigley, Conformational changes induced by nucleotide binding in Cdc6/ORC from *Aeropyrum pernix*, *J. Mol. Biol.*, 2004, **343**, 547–557.
15. B. Grabowski and Z. Kelman, Autophosphorylation of archaeal Cdc6 homologues is regulated by DNA, *J. Bacteriol.*, 2001, **183**, 5459–5464.
16. E. L. C. Dueber, J. E. Corn, S. D. Bell and J. M. Berger, Replication origin recognition and deformation by a heterodimeric archaeal Orc1 complex, *Science*, 2007, **317**, 1210–1213.
17. M. Gaudier, B. S. Schuwirth, S. L. Westcott and D. B. Wigley, Structural basis of DNA replication origin recognition by an ORC protein, *Science*, 2007, **317**, 1213–1216.
18. K. Labib and J. F. Diffley, Is the MCM2-7 complex the eukaryotic DNA replication fork helicase?, *Curr. Opin. Genet. Dev.*, 2001, **11**, 64–70.
19. M. L. Bochman and A. Schwacha, The Mcm2-7 complex has *in vitro* helicase activity, *Mol. Cell*, 2008, **31**, 287–293.
20. J. P. J. Chong, M. K. Hayashi, M. N. Simon, R. M. Xu and B. Stillman, A double-hexamer archaeal minichromosome maintenance protein is an ATP-dependent DNA helicase, *Proc. Natl. Acad. Sci. U.S.A.*, 2000, **97**, 1530–1535.
21. Z. Kelman, J. K. Lee and J. Hurwitz, The single minichromosome maintenance protein of *Methanobacterium thermoautotrophicum* Delta H contains DNA helicase activity, *Proc. Natl. Acad. Sci. U.S.A.*, 1999, **96**, 14783–14788.
22. F. Carpentieri, M. De Felice, M. De Falco, M. Rossi and F. M. Pisani, Physical and functional interaction between the mini-chromosome maintenance-like DNA helicase and the single-stranded DNA binding protein from the crenarchaeon *Sulfolobus solfataricus*, *J. Biol. Chem.*, 2002, **277**, 12118–12127.
23. I. Grainge, S. Scaife and D. Wigley, Biochemical analysis of components of the pre-replication complex of Archaeoglobus fulgidus, *Nucleic Acids Res.*, 2003, **31**, 4888–4898.
24. J. H. Shin, B. Grabowski, R. Kasiviswanathan, S. D. Bell and Z. Kelman, Regulation of minichromosome maintenance helicase activity by Cdc6, *J. Biol. Chem.*, 2003, **278**, 38059–38067.

25. R. J. Fletcher, B. E. Bishop, R. P. Leon, R. A. Sclafani, C. M. Ogata and X. S. Chen, The structure and function of MCM from archaeal *M. thermoautotrophicum*, *Nat. Struct. Biol.*, 2003, **10**, 160–167.
26. E. R. Barry, A. T. McGeoch, Z. Kelman and S. D. Bell, Archaeal MCM has separable processivity, substrate choice and helicase domains, *Nucleic Acids Res.*, 2007, **35**, 988–998.
27. B. Pucci, M. De Felice, M. Rocco, F. Esposito, M. De Falco, L. Esposito, M. Rossi and F. M. Pisani, Modular organization of the *Sulfolobus solfataricus* mini-chromosome maintenance protein, *J. Biol. Chem.*, 2007, **282**, 12574–12582.
28. A. T. McGeoch, M. A. Trakselis, R. A. Laskey and S. D. Bell, Organization of the archaeal MCM complex on DNA and implications for the helicase mechanism, *Nat. Struct. Mol. Biol.*, 2005, **12**, 756–762.
29. E. R. Jenkinson and J. P. J. Chong, Minichromosome maintenance helicase activity is controlled by N- and C-terminal motifs and requires the ATPase domain helix-2 insert, *Proc. Natl. Acad. Sci. U.S.A.*, 2006, **103**, 7613–7618.
30. E. Rothenberg, M. A. Trakselis, S. D. Bell and T. Ha, MCM forked substrate specificity involves dynamic interaction with the 5′-tail, *J. Biol. Chem.*, 2007, **282**, 34229–34234.
31. A. Gambus and K. Labib, A key role for the GINS complex at DNA replication forks, *Trends Cell Biol.*, 2007, **17**, 271–278.
32. N. Marinsek, E. R. Barry, K. S. Makarova, I. Dionne, E. V. Koonin and S. D. Bell, GINS, a central nexus in the archaeal DNA replication fork, *EMBO Rep.*, 2006, **7**, 539–545.
33. T. Yoshimochi, R. Fujikane, M. Kawanami, F. Matsunaga and Y. Ishino, The GINS complex from *Pyrococcus furiosus* stimulates the MCM helicase activity, *J. Biol. Chem.*, 2007, **283**, 1601–1609.
34. A. A. Bocquier, L. D. Liu, I. K. O. Cann, K. Komori, D. Kohda and Y. Ishino, Archaeal primase: bridging the gap between RNA and DNA polymerases, *Curr. Biol.*, 2001, **11**, 452–456.
35. L. D. Liu, K. Komori, S. Ishino, A. A. Bocquier, I. K. O. Cann, D. Kohda and Y. Ishino, The archaeal DNA primase. Biochemical characterization of the p41-p46 complex from *Pyrococcus furiosus*, *J. Biol. Chem.*, 2001, **276**, 45484–45490.
36. M. A. Augustin, R. Huber and J. T. Kaiser, Crystal structure of a DNA-dependent RNA polymerase (DNA primase), *Nat. Struct. Biol.*, 2001, **8**, 57–61.
37. S. H. Lao-Sirieix, R. K. Nookala, P. Roversi, S. D. Bell and L. Pellegrini, Structure of the heterodimeric core primase, *Nat. Struct. Mol. Biol.*, 2005, **12**, 1137–1144.
38. S. Klinge, J. Hirst, J. D. Maman, T. Krude and L. Pellegrini, An iron-sulfur domain of the eukaryotic primase is essential for RNA primer synthesis, *Nat. Struct. Mol. Biol.*, 2007, **14**, 875–877.
39. S. H. Lao-Sirieix and S. D. Bell, The heterodimeric primase of the hyperthermophilic archaeon *Sulfolobus solfataricus* possesses DNA and

RNA primase, polymerase and 3′-terminal nucleotidyl transferase activities, *J. Mol. Biol.*, 2004, **344**, 1251–1263.
40. M. De Falco, A. Fusco, M. De Felice, M. Rossi and F. M. Pisani, The DNA primase of *Sulfolobus solfataricus* is activated by substrates containing a thymine-rich bubble and has a 3′- terminal nucleotidyl-transferase activity, *Nucleic Acids Res.*, 2004, **32**, 5223–5230.
41. I. K. Cann, K. Komori, H. Toh, S. Kanai and Y. Ishino, A heterodimeric DNA polymerase: evidence that members of Euryarchaeota possess a distinct DNA polymerase, *Proc. Natl. Acad. Sci. U.S.A.*, 1998, **95**, 14250–14255.
42. S. Ishino and Y. Ishino, Comprehensive search for DNA polymerase in the hyperthermophilic archaeon, *Pyrococcus furiosus*, *Nucleosides Nucleotides Nucleic Acids*, 2006, **25**, 681–691.
43. G. Henneke, D. Flament, U. Hubscher, J. Querellou and J. P. Raffin, The hyperthermophilic euryarchaeota *Pyrococcus abyssi* likely requires the two DNA polymerases D and B for DNA replication, *J. Mol. Biol.*, 2005, **350**, 53–64.
44. M. J. Fogg, L. H. Pearl and B. A. Connolly, Structural basis for uracil recognition by archaeal family B DNA polymerases, *Nat. Struct. Biol.*, 2002, **9**, 922–927.
45. H. J. Yang, J. H. Chiang, S. Fitz-Gibbon, M. Lebel, A. A. Sartori, J. Jiricny, M. M. Slupska and J. H. Miller, Direct interaction between uracil-DNA glycosylase and a proliferating cell nuclear antigen homolog in the crenarchaeon *Pyrobaculum aerophilum*, *J. Biol. Chem.*, 2002, **277**, 22271–22278.
46. I. Dionne and S. D. Bell, Characterization of an archaeal family 4 uracil DNA glycosylase and its interaction with PCNA and chromatin proteins, *Biochemical J.*, 2005, **387**, 859–863.
47. C. Indiani and M. O'Donnell, The replication clamp-loading machine at work in the three domains of life, *Nat. Rev. Mol. Cell Biol.*, 2006, **7**, 751–761.
48. J. B. Vivona and Z. Kelman, The diverse spectrum of sliding clamp interacting proteins, *FEBS Lett.*, 2003, **546**, 167–172.
49. K. Daimon, Y. Kawarabayasi, H. Kikuchi, Y. Sako and Y. Ishino, Three proliferating cell nuclear antigen-like proteins found in the hyperthermophilic archaeon *Aeropyrum pernix*: interactions with the two DNA polymerases, *J. Bacteriol.*, 2002, **184**, 687–694.
50. K. Imamura, K. Fukunaga, Y. Kawarabayasi and Y. Ishino, Specific interactions of three proliferating cell nuclear antigens with replication-related proteins in *Aeropyrum pernix*, *Mol. Microbiol.*, 2007, **64**, 308–318.
51. I. Dionne, R. K. Nookala, S. P. Jackson, A. J. Doherty and S. D. Bell, A heterotrimeric PCNA in the hyperthermophilic archaeon *Sulfolobus solfataricus*, *Mol. Cell*, 2003, **11**, 275–282.
52. A. S. Dore, M. L. Kilkenny, S. A. Jones, A. W. Oliver, S. M. Roe, S. D. Bell and L. H. Pearl, Structure of an archaeal PCNA1-PCNA2-FEN1 complex: elucidating PCNA subunit and client enzyme specificity, *Nucleic Acids Res.*, 2006, **34**, 4515–4526.

53. G. J. Williams, K. Johnson, J. Rudolf, S. A. McMahon, L. Carter, M. Oke, H. T. Liu, G. L. Taylor, M. F. White and J. H. Naismith, Structure of the heterotrimeric PCNA from *Sulfolobus solfataricus*, *Acta Crystallogr., Sect. F: Struct. Biol. Cryst. Commun.*, 2006, **62**, 944–948.
54. J. M. Pascal, O. V. Tsodikov, G. L. Hura, W. Song, E. A. Cotner, S. Classen, A. E. Tomkinson, J. A. Tainer and T. Ellenberger, A flexible interface between DNA ligase and PCNA supports conformational switching and efficient ligation of DNA, *Mol. Cell*, 2006, **24**, 279–291.
55. J. M. Pascal, P. J. O'Brien, A. E. Tomkinson and T. Ellenberger, Human DNA ligase I completely encircles and partially unwinds nicked DNA, *Nature*, 2004, **432**, 473–478.
56. Y. H. Chen, S. A. Kocherginskaya, Y. Y. Lin, B. Sriratana, A. M. Lagunas, J. B. Robbins, R. I. Mackie and I. K. O. Cann, Biochemical and mutational analyses of a unique clamp loader complex in the archaeon *Methanosarcina acetivorans*, *J. Biol. Chem.*, 2005, **280**, 41852–41863.
57. A. Seybert, D. J. Scott, S. Scaife, M. R. Singleton and D. B. Wigley, Biochemical characterisation of the clamp/clamp loader proteins from the euryarchaeon *Archaeoglobus fulgidus*, *Nucleic Acids Res.*, 2002, **30**, 4329–4338.
58. A. Seybert and D. B. Wigley, Distinct roles for ATP binding and hydrolysis at individual subunits of an archaeal clamp loader, *EMBO J.*, 2004, **23**, 1360–1371.
59. T. Miyata, H. Suzuki, T. Oyama, K. Mayanagi, Y. Ishino and K. Morikawa, Open clamp structure in the clamp-loading complex visualized by electron microscopic image analysis, *Proc. Natl. Acad. Sci. U.S.A.*, 2005, **102**, 13795–13800.
60. I. Dionne, N. J. Brown, R. Woodgate and S. D. Bell, On the mechanism of loading the PCNA sliding clamp by RFC., *Mol. Microbiol.*, 2008, **68**, 216–222.
61. R. Bernander and A. Poplawski, Cell cycle characteristics of thermophilic archaea, *J. Bacteriol.*, 1997, **179**, 4963–4969.
62. A. Costa, T. Pape, M. van Heel, P. Brick, A. Patwardhan and S. Onesti, Structural basis of the *Methanothermobacter thermautotrophicus* MCM helicase activity, *Nucleic Acids Res.*, 2006, **34**, 5829–5838.

CHAPTER 13

DNA Replication in the Human Malaria Parasite and Potential for Novel Drug Development

JI-LIANG LI

Weatherall Institute of Molecular Medicine, University of Oxford, John Radcliffe Hospital, Headington, Oxford OX3 9DS, UK and School of Biotechnology, Nanfang Medical University, Guangzhou, China

13.1 Introduction

Malaria threatens 40% of the world's population and kills 1–3 million humans, mostly children under five, every year.[1] The lack of effective vaccines and the spreading of drug-resistant parasites has led to a resurgence of malaria in many countries.[2,3] It is therefore imperative that our understanding of the fundamental biology and biochemical processes at different stages of the life cycle of the malaria parasite be improved, to facilitate the identification of new targets for the development of novel drugs.

DNA replication represents one such key biochemical process of the parasite. It takes place in at least five distinct points in the parasite life cycle, two of which occur in the human host (*i.e.* exo-erythrocytic and erythrocytic schizogony in liver and red blood cells, respectively); the remainder take place in the mosquito host (*i.e.* gametogenesis in midgut, pre-meiosis after fertilisation and sporogony in oocysts)[4] (Figure 13.1). Thus the selective disruption of DNA synthesis in the parasite might not only inhibit the disease itself but also block parasite transmission by the mosquito vector. During gametogenesis, three successive rounds of genome replication are completed within 10 minutes, raising the DNA content

Molecular Themes in DNA Replication
Edited by Lynne S. Cox

Published by the Royal Society of Chemistry, www.rsc.org

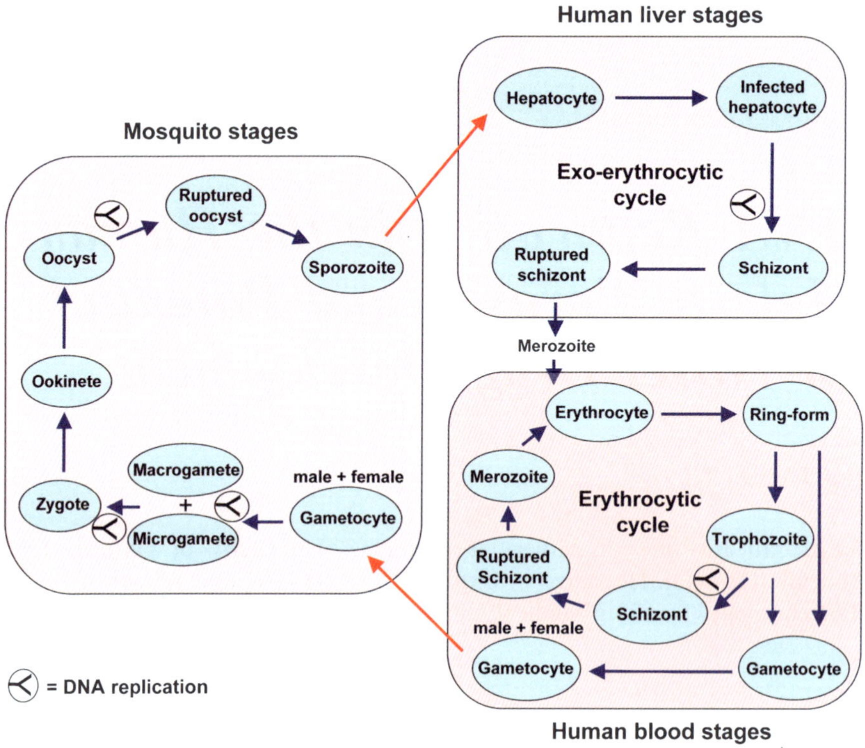

Figure 13.1 ***Plasmodium falciparum* life cycle.** The life cycle of the malarial parasite involves two hosts: mosquito and human. It may be divided into three stages. One is sexual and takes place inside the mosquito, while the other two are asexual and occur inside the human host: the erythrocytic cycle (in red blood cells) and the exo-erythrocytic cycle (in liver cells). During a blood meal, a malaria-infected female *Anopheles* mosquito inoculates sporozoites into the human host. Sporozoites, upon entering the bloodstream, will reach the liver (hepatocytes) where they mature into schizonts, which rupture and release merozoites. After this initial replication in the liver (exo-erythrocytic schizogony), the parasites undergo asexual multiplication in red blood cells (erythrocytic schizogony). Merozoites infect erythrocytes becoming ring-forms. The ring-forms progress to trophozoites and mature into schizonts. The segmented schizonts rupture releasing merozoites. Some parasites in red blood cells differentiate into sexual erythrocytic stages (gametocytes). The gametocytes, including male (microgametocytes) and female (macrogametocytes), are ingested by an *Anopheles* mosquito during a blood meal. The microgametocytes undergo exflagellation in the midgut generating microgametes. The microgametes fertilise the macrogametes producing zygotes. The zygotes in turn become motile and elongated to form ookinetes that invade the midgut wall of the mosquito where they develop into oocysts. The oocysts grow, rupture and release sporozoites, which migrate to the salivary glands. Inoculation of the sporozoites by the mosquito into a new human host continues to transmit the parasites.

to octoploid values just before exflagellation, suggesting that the genome duplication rate of the parasite is extremely high. Assuming that the rate of replication fork movement in the parasite is similar to that in other eukaryotes, at least 1300 replication origins among the 14 chromosomes of *Plasmodium* would be needed to achieve this rate of replication.[5] It is known that at least two replication origins exist within each inverted repeat region of the 35 kb apicoplast genome of the parasite[6–8] but information concerning replication origins on nuclear chromosomes of the parasite remains absent.

This chapter focuses on the important molecules (Tables 13.1, 13.2 and 13.3) involved in the DNA replication initiation and elongation processes, and their potential as targets for development of novel drugs against *Plasmodium falciparum*—the most malignant pathogen of four malarial parasite species that infect humans.

13.2 Replication Initiation Proteins

13.2.1 PfORC

The initiation of DNA replication is tightly regulated such that all chromosomes are replicated precisely at specific points in the parasite life cycle (Figure 13.1). In eukaryotes, ORC is a multi-subunit protein complex composed of six polypeptides (ORC1-6) that binds to replication origins and is essential for the initiation of chromosomal DNA replication (see Chapters 1 and 2). It was originally thought that only two or three PfORC homologues exist in *P. falciparum*.[9–11] However, extensive bioinformatics investigations reveal that there may be four putative homologues (*i.e.* PfORC1, 2, 4 and 5) in the parasite genome.

PfORC1 was the first malarial ORC isolated from *P. falciparum* using vectorette technology.[12] PfORC1 is composed of two distinct domains: a variable N-terminal domain (residue positions 1–783) and a highly conserved C-terminal domain (residue positions 784–1189). The N-terminal domain is the largest extension in the ORC1 family and contains several other unique characteristics:

(i) It is rich in serine/threonine and tyrosine (20%), forming a number of potential phosphorylation sites for a range of known protein kinases such as cyclin-dependent kinases, suggesting that PfORC1 may be regulated by phosphorylation.
(ii) It contains a large proportion (30%) of charged amino acid residues (K, R, E and D) that may be involved in protein–protein interactions.
(iii) It has four copies of a heptamer ISSSLT(S)N repeat.
(iv) It includes two putative nuclear localisation signal (NLS) motifs, suggesting that PfORC1 may be a nuclear protein.
(v) It possesses a leucine zipper motif ($LX_6LX_6LX_6L$) that may be involved in DNA-binding activity.

The C-terminal domain of PfORC1 shares 48–61% similarity and 27–40% identity with members of the ORC1 family from other species, and contains all

Table 13.1 DNA replication initiation proteins in *Plasmodium falciparum*.

Replication initiation protein	*Transcript size (kb)*	*Predicted no. of amino acids*	*Protein molecular weight (kDa)*	*Stage-specific expression*	*Chromosome location*	*Database accession no./Gene ID*	*Reference*
PfORC1	~5.6	1189	138.7	Ase, Gam	12	AF373219	10,12,13
						PFL0150w	
PfORC2		825	97.9		7	MAL7P1.21	
PfORC4		983	117.7		13	PF13_0189	
PfORC5		899	103.9		2	PFB0720c	9,63
PfCDT1		982	114.5		13	PF13_0237	
PfCDC6		979	115		5	PFE0155w	
PfMCM2	~3.8	971	111.4	Ase, Gam	14	AF095948	[a]
						PF14_0177	11
PfMCM3		962	109.7		5	PFE1345c	68
PfMCM4	~4.0	1005	115	Ase, Gam	13	AF083323	11,13,21
						PF13_0095	
PfMCM5	~3.8	758	85.7	Ase, Gam	12	AF139108	[a]
						PFL0580w	62
PfMCM6		929	105.5		13	PF13_0291	11
PfMCM7		821	94.1		7	PF07_0023	11
PfMCM8		1024	119.2		12	PFL0560c	62
PfMCM9		1465	171.2		4	PFD0790c	68

Ase = sexual stages; Gam = gametocyte. [a]Data from unpublished research by J-L. Li.

Table 13.2 Replication elongation factors in *Plasmodium falciparum*.

Elongation factor	*Transcript size (kb)*	*Predicted no. of amino acids*	*Protein molecular weight (kDa)*	*Stage-specific expression*	*Chromosome location*	*Database accession no./Gene ID*	*Reference*
PfPCNA1	~1.6; ~2.2	275	30.6	Ase, Gam	13	P31008	46
						PF13_0328	
PfPCNA2	~1.8; ~2.5	264	30.2	Ase, Gam	12	AF056205	47,48
	~2.0					AF544241	
						PFL1285c	
PfRFC1	~4.0	904	104.2	Ase, Gam	2	AF139827	[a]
						PFB0895c	
PfRFC2	~1.85	330	38.0	Ase, Gam	2	AF071409	[a]
						PFB0840w	63
PfRFC3	~1.8	344	39.2	Ase, Gam	14	AF069296	[a]
						PF14_0601	
PfRFC4	~1.75	336	37.7	Ase, Gam	12	AF126257	[a]
						PFL2005w	
PfRFC5		349	40.3		11	PF11_0117	62
PfFen1	~2.4	672	76.6	Gam	4	AF093702	[a]
	~3.2	650		Ase, Gam		PFD0420c	
						MAL4p2.21	
						AF278764	51
PfRNase H2		288	33.0		6	PFF1150w	68
						MAL6P1.190	
PfRPA1	~6.5	1145	134.1	Tro, Sch	4	AL035475	29
						Ma14p2.32	
						PFD0470c	
PfRPA1′		484	56.1		9	PFI0235w	68
PfRPA2		273	31.6		11	PF11_0332	62
PfRPA2′		191	22.3		11	PF11_0130	62
PfRAP3		135	15.4		7	PF07_0039	

Table 13.2 (*Continued*).

Elongation factor	*Transcript size (kb)*	*Predicted no. of amino acids*	*Protein molecular weight (kDa)*	*Stage-specific expression*	*Chromosome location*	*Database accession no./Gene ID*	*Reference*
PfLigI		912	104		13	MAL13P1.22	52
PfPolα	~7.0	1855	205	Ase, Gam	4	L18785 PFD0590c	37
PfPolaB		539	63.9		14	PF14_0602	
PfPriL		525	62.2		9	PFI0530c	68
PfPriS	~2.1	452	53	Ase	14	X99254	39
PfPolδ	~5.2, ~5.7 ~4.5	1094	126.8	Ase, Gam Tro, Sch	10	PF14_0366 X62423 M64715	41,42
PfPolδS		498	57.6		3	PF10_0165 PFC0340w	69
PfPolε		2907	344.5		6	MAL3P3.4 PFF1470c	68
PfPolεB		624	73.5	Ase	12	MAL6P1.125 PFL1655c	19,70

Ase = sexual stages; Gam = gametocyte; Sch = schizont; Tro = trophozoite. [a]Data from unpublished research by J-L. Li.

Table 13.3 Proteins involved in apicoplast replication in *Plasmodium falciparum*.

Apicoplast DNA replication factor	*Transcript size (kb)*	*Predicted no. of amino acids*	*Protein molecular weight (kDa)*	*Stage-specific expression*	*Chromosome location*	*Database accession no./ Gene ID*	*Reference*
PfPOM1 (PfPREX)	~7.0 (Pc)	2016	235.7	Tro, Sch Gam	14	PF14_0112	61
PfGyrA		1222	143		12	PFL1120c	56,57
PfGyrB		1006	116		12	PFL1915w	56,57

Gam = gametocyte; Sch = schizont; Tro = trophozoite.

conserved sequences of the family including the Walker A (GMPGTGKT, 815–822) and Walker B (DEID, 903–906) motifs of the classical purine nucleotide-binding sites (see Chapter 2), suggestive of ATP binding and hydrolysis activity of PfORC1 (Figure 13.2A). Indeed, ATPase activity of the recombinant C-terminal domain has been recently confirmed *in vitro* and, like other AAA^+ ATPases, is dependent on the K821 residue within the Walker A motif.[10] In addition, PfORC1, like most members of the ORC1 family from other species, contains a putative PCNA binding motif, **QKVLFTLF** (913–920), suggestive of an interaction with PfPCNA (see Chapter 3). In fact, PfORC1 has been shown to interact with PfPCNA1 *in vitro* and to co-localise with the majority of the PfPCNA1 replication foci during trophozoite stages.[9] Expression of PfORC1 has been demonstrated in the parasite nucleus during late trophozoite and early schizont stages and in sexual stages,[9,10,12–15] consistent with its potential role in replication initiation in these stages. Recently, it was reported that PfORC1 can specifically interact with telomeres and with various subtelomeric repeats and, thus, act also as a telomere-associated protein in *P. falciparum*.[15] More than 80% of PfORC1 nuclear foci were shown to co-localise with PfSir2, a telomeric protein that was identified as reversibly associating with the promoter regions of silent but not active subtelomeric *var* genes.[16,17] Therefore, PfORC1 may cooperate with PfSiv2 and contribute to telomeric silencing of virulence-factor genes and antigenic variation in *P. falciparum* by a phenomenon called the telomere position effect.[18] Structural modelling of PfORC1 against archeal ORC/cdc6 suggests very strong conservation such that plasmodium PfORC1 may similarly distort the helical axis of DNA (Figure 13.2B), thought to be important for localised melting of the origin during replication initiation (see also Figure 12.1).

PfORC2, like PfORC1, shares 45% similarity and 26% identity with human ORC2 only at the C-terminal region (residue positions 488–825) (J-L. Li unpublished data). PfORC2 is transcribed during late trophozoite and early schizont stages,[14,19] consistent with a role in DNA replication.

PfORC4 contains the Walker A (GMLGCGKT, 160–167) and Walker B motifs (DEND, 270–273) characteristic of other ORC family members and consistent with a proposed action as an AAA^+ ATPase (see Chapter 2).

Figure 13.2 **PfORC1.** (**A**) The ORC1/Cdc6/Cdc18 domain. Sequences were aligned with the CLUSTAL W (1.85) multiple sequence alignment program,[71,72] using data from GenBank/EMBL/DDJB database accession numbers: *P. falciparum* PfORC1, AF373219; human HsORC1, Q13415; and human HsCdc6, U77949. Six conserved blocks that define PfORC1 as a member of the ORC1/Cdc6/Cdc18 family are highlighted with grey and labelled at the bottom of the sequence. A putative PCNA-binding motif (913–920) is highlighted in black and underlined at the bottom of the sequence. (**B**) Predicted structure of PfORC1. Molecular modelling of PfORC1 (green) was conducted using SWISS-MODEL[73,74] and MacPyMOL (http://delsci.com/macpymol/) with archaeal ORC/Cdc6 (PDB accession number 2QBY) (red in figure) as a template.

A

```
                    10        20        30        40        50        60        70        80
           ----:----|----:----|----:----|----:----|----:----|----:----|----:----|----:----|
Consensus  xExxxxLxxxxVPExLPxREKExxxIxxFLxxxIxxxxxxxxLYISGxPGTGKTAxVxxVIQxLQxxxRxxxxxxFxxxx
HsORC1     EEARLRLHVSAVPESLPCREQEFQDIYNFVESKLLDHTGG-CMYISGVPGTGKTATVHEVIRCLQQAAQANDVPPFQYIE
PfORC1     DKAIRMMQLDVVPKYLPCREKEIKEVHGFLESGIKQSGSNQILYISGMPGTGKTATVYSVIQLLQIKSRKKLLPSFNVFE
HsCdc6     YQQAKLVLNTAVPDRLPAREREMDVIRNFLREHICGKKAG-SLYLSGAPGTGKTACLSRILQDLKKELKG-----FKTIM
                                          Block 1 (Walker A)
                    90       100       110       120       130       140       150       160
           ----:----|----:----|----:----|----:----|----:----|----:----|----:----|----:----|
Consensus  INxMxLxxxxxxYxxxxQQLxxQKxxSxxxxxELLxKxxxxxxxxxxxxxVLIIDELDxLxTxxQxVLYxLFDWPxxxxS
HsORC1     VNGMKLTEPHQVYVQILQKLTGQK-ATANHAAELLAKQFC-TRGSPQETTVLLVDELDLLWTHKQDIMYNLFDWPTHKEA
PfORC1     INGMNVVHPNAAYQVFYKQLFNKKPPNALNSFKIIDRLFNKSQKDNRDVSILIIDEIDYLITKTQKVLFTLFDWPTKINS
HsCdc6     LNCMSLRTAQAVFPAIAQEICQEE-VSRPAGKDMMRKLEKHMTAEKGPMIVLVLDEMDQLDSKGQDVLYTLFEWPWLSNS
                Block 2                                  Block 3 (Walker B)
                   170       180       190       200       210       220       230       240
           ----:----|----:----|----:----|----:----|----:----|----:----|----:----|----:----|
Consensus  xLVLIxIANTMDLxDRILxxRxxSRxxxxxxxLxFxPYxxNQIxxILRxRLxxxxxxxxxDNxAIQxxARKVAxVSGDxR
HsORC1     RLVVLAIANTMDLPERIMMNRVSSRLGLT--RMCFQPYTYSQLQQILRSRLKH--LKAFEDD-AIQLVARKVAALSGDAR
PfORC1     KLILIAISNTMDLPDRLIP-RCRSRLAFG--RLVFSPYKGDEIEKIIKERLEN--CKEIIDHTAIQLCARKVANVSGDIR
HsCdc6     HLVLIGIANTLDLTDRILP-RLQAREKCKPQLLNFPPYTRNQIVTILQDRLNQVSRDQVLDNAAVQFCARKVSAVSGDVR
                Block 4                             Block 5                         Block 6
                   250       260       270       280       290       300       310       320
           ----:----|----:----|----:----|----:----|----:----|----:----|----:----|----:----|
Consensus  KxLxICRRAxxxxxxxxxxxxxIxxxSxxxxxSxxxxPxxxxIxHxxxxxxxxxxxxxxxxxxxxxxxxxxxxxxxxxI
HsORC1     RCLDICRRAT-----------EICEFSQQKPDS----PCLVTIAHSMEAVDEMFSSSYITA-IKNSSVLEQSFLR---AI
PfORC1     KALQICRKAFENKRGHKIVPRDITEATNQLFDS----PLTNAINYLPWAFKIFLTCLIIELRIINEFVIPYKKVVNRYKI
HsCdc6     KALDVCRRAIEIVESDVKSQTILKPLSECKSPSEPLIPKRVGLIHISQVISEVDGNRMTLSQEGAQDSFPLQQKILVCSL

                   330       340       350       360       370       380       390       400
           ----:----|----:----|----:----|----:----|----:----|----:----|----:----|----:----|
Consensus  LxxxxxxxxxxxxxxxLxExxxxKLxxxxxxLxRxxxIxxxxxSExxSxxxxLxxxxxxxxxxxxxxxxxxVxxxxNKxx
HsORC1     LAEFR--------RSGLEEATFQQIYSQHVALCRMEGLPYPTMSETMAVCSHLGSCR-----------LLLVEP--SRND
PfORC1     LIETSGKYIGMCSDNELFKIMLDKLVKMGILLIRPY-IPLENISKNKSKEALLGFNESSKKGNNQKITRAQVSPDIDKES
HsCdc6     MLLIR--------QLKIKEVTLGKLYEAYSKVCRKQQVAAVDQSECLSLSGLLEARG-------------ILGLKRNKET

                   410       420       430       440       450       460       470       480
           ----:----|----:----|----:----|----:----|----:----|----:----|----:----|----:----|
Consensus  xxxxVxxxVxxxxIxxALxxExxxxxxxxxxxx
HsORC1     LLLRVRLNVSQDDVLYALKDE------------
PfORC1     GDMGIELNVETQLIITALMKDPDCSKKLNFY--
HsCdc6     RLTKVFFKIEEKEIEHALKDKALIGNILATGLP
```

B

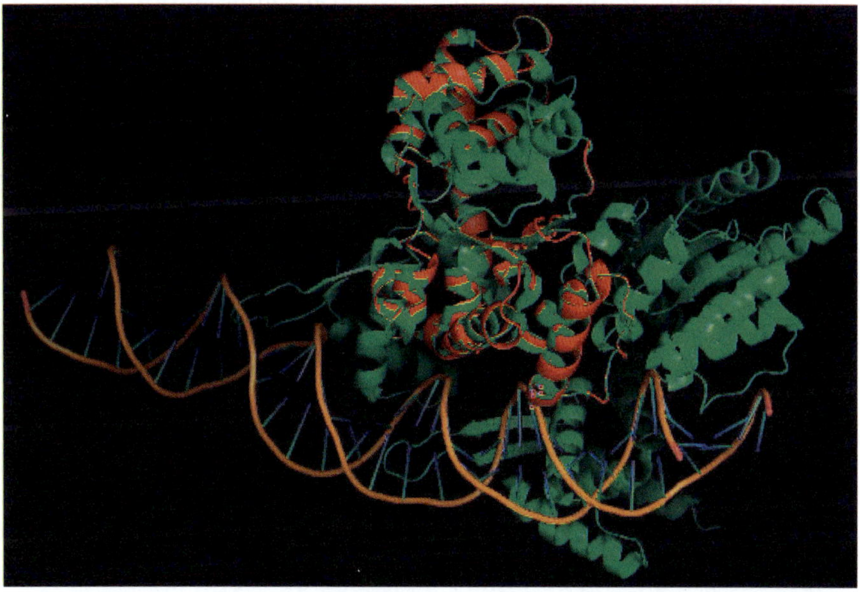

The central region (residues 431–660) of PfORC4 displays up to 31% identity and 50% similarity with other members of ORC4 family. Oligonucleotide array analysis showed that PfORC4 is expressed not only in trophozoite and early schizont stages,[19] but also in sexual and sporozoite stages.[14]

PfORC5, like PfORC1, comprises two distinct domains: a variable N-terminal domain and a highly conserved C-terminal domain. The N-terminal extension contains a putative NLS motif (residue positions 56–72) and a D/N/K-rich repeat (residue positions 81–170). The C-terminal domain of PfORC5 exhibits 25% identity and 46% similarity (between the 572–815 region) with *Drosophila* ORC5. PfORC5, like *Saccharomyces cerevisiae* ORC5 (ScORC5), contains the ATP-binding motif (GLPGMGKT, 303–310) but lacks the Walker B motif designated as a nucleotide hydrolysis domain (see Chapter 2). Between the ATP-binding motif and the C-terminal conserved region, however, five sequence insertions ranging from seven to 56 amino acid residues are found, the largest one having the D/N/K-rich repeat. Surprisingly, PfORC5 has been demonstrated to have ATPase activity, *i.e.* not only binding but also hydrolysis of ATP.[9] Chimeric ORC5 (composed of the N-terminus of ScORC5 and the C-terminus of PfORC5), but not full-length PfORC5, was shown to be able to complement an ScORC5-deficient yeast strain, confirming that PfORC5 is a true homologue of ORC5.[9] PfORC5 is predominantly expressed in trophozoite and early schizont stages, and in sexual stages.[9,14] It was reported that PfORC5 co-immunoprecipitates with PfPCNA1 (see Section 13.3.4) from mixed trophozoite extract, and co-localises with PfPCNA1 replication foci *in vivo* during early-to-mid replicating trophozoite stages, with PfORC5 and PfPCNA1 starting to dissociate from each other during further growth progression and finally separating completely during late schizont stages.[9] PfORC5 was also shown to co-localise with PfORC1 foci from mid-trophozoite to mid-schizont stages until PfORC1 was completely diminished in late schizont stages.[9]

13.2.2 PfCDC6

CDC6 is an ATP-binding protein that plays a crucial role in the assembly of the pre-replication complex (pre-RC) (see Chapters 1 and 2). CDC6 requires ORC to associate with chromatin and is in turn required for MCM2-7 chromatin association. The CDC6 homologue was thought to be absent in *P. falciparum*;[10,11] however, our extensive bioinformatics studies reveal that a gene located on chromosome 5 does encode a PfCDC6-like protein in the parasite. Like PfORC1 (Figure 13.2A), there are six blocks in the central region of PfCDC6 showing strong similarity with members of the ORC1/CDC6/Cdc18 family including the Walker A (GPSGQGKT) and B (DELD) motifs,[12] suggestive of an ATPase activity of PfCDC6. PfCDC6 contains the conserved region corresponding to domains I and II of archaeal Cdc6 and can presumably form a two-lobed, cashew-shaped molecule as does archaeal Cdc6.[20]

Interestingly, between the first two conserved blocks, there is an amino acid insertion of more than 50 residues, which may present a target for development of novel anti-malarial drugs. PfCDC6 seemed to be predominantly expressed in gametocyte but weakly in late trophozoite stages.[14,19]

13.2.3 PfCDT1

CDT1 protein, like CDC6, is required to load the MCM helicase at the replication origin to form the licensed pre-RC (see Chapter 1). Our bioinformatics analysis reveals that there is a putative PfCDT1 protein in the parasite genome. PfCDT1 shares up to 21% identity and 43% similarity with other members of the Cdt1 family between residues 544–700, although two short helices (H1 and H2) and loop L1 in mouse Cdt1 that contact with the N-terminal part of the geminin dimer are absent in PfCDT1. Perhaps this is less surprising in light of the absence of any identifiable geminin-like protein in the parasite genome (as is also the case in yeast; geminin may be a feature of metazoan replication control). Near the C-terminus is a leucine zipper motif ($LX_6LX_6LX_6L$, positions 921–950), suggesting that PfCDT1 may possess DNA-binding activity. It will be interesting to investigate whether PfCDT1 is required in replication initiation as would be predicted from its similarity to Cdt1 from other species. Consistent with a role in DNA replication, PfCDT1 is transcribed in both asexual (trophozoite and schizont) and sexual (gametocyte) stages.[14]

13.2.4 PfMCM

In eukaryotes, the MCM complex is composed of six conserved proteins (MCM2-7) and is recruited to replication origins by Cdc6 and Cdt1 to function as the replicative DNA helicase (see Chapter 3). Although PfMCM4 was the first malarial MCM member isolated from the parasite,[21] all six MCMs (PfMCM2-7) have now been identified in the parasite, the smallest being PfMCM5 (758 amino acid residues, predicted molecular mass 85.7 kDa) and the largest being PfMCM4 (1005 residues with predicted molecular mass 115 kDa) (Table 13.1). Although it has been reported that none of the PfMCM genes contain introns,[11] we have found that four out of six (*i.e.* PfMCM2, 3, 5 and 7) contain between one and three introns (Li *et al.*, unpublished data).

Comparisons of the PfMCM sequences reveal that the central region of approximately 200 amino acids is conserved in all six members, including the Walker A and B motifs and the MCM signature motif (IDEFDKM). In addition, a zinc finger motif ($CX_2CX_{18}CX_{2-4}C$), characteristic of DNA-binding domains, is present in the N-terminal region of four PfMCMs (PfMCM2, 4, 6 and 7), as would be expected for these four members, although there is a slight variation for the motif in PfMCM2 ($CX_2CX_{21}CX_2C$) and in PfMCM4 ($CX_{13}CX_{18}CX_2C$). Moreover, all PfMCMs contain various unique sequence insertions ranging from five to 84 residues compared with members of each

MCM family in other species. For example, PfMCM4 has five unique amino acid insertions with sizes from five to 84 residues located in different places in the sequence[21] (see Figure 13.3).

Co-immunoprecipitation demonstrated that PfMCM2, 6, and 7 are present in a protein complex; among them, PfMCM6 is the only protein that is tightly associated with chromatin, suggesting that PfMCM6 may directly interact with chromatin, while other PfMCMs probably associate *via* protein-protein interactions.[11] PfMCM2-6 all appear to be transcribed during trophozoite, schizont and gametocyte stages[14,19,21] while some (*i.e.* PfMCM3, 4, 5 and 7) are also expressed in sporozoites.[14] At the protein level, PfMCM2 has been demonstrated to peak in late schizont/segmented schizont stages, PfMCM6 in late trophozoite and schizont/segmented schizont stages, and PfMCM7 in schizont/segmented schizont stages.[11] Immunofluorescent staining revealed the PfMCM4 protein in trophozoite, schizont and gametocyte stages.[13] In other eukaryotic systems, the MCM complex is regulated at replication initiation by phosphorylation by cyclin-dependent kinases (*e.g.* Clb5/6-CDK1 in yeast, together with Cdc7-Dbf4). Moreover, association of DNA pol α-primase at the pre-initiation complex is usually mediated by Cdc45.[22–24] However, despite extensive analysis, we have been unable to identify Cdc45 and Dbf4-Cdc7 (DDK) homologues in the parasite genome.

In this context, it is interesting to note that structural changes in yeast MCM5 can remove the need for Dbf4-kinase activation of the MCM complex at initiation.[25] It will therefore be interesting to see whether the PfMCM complex can similarly be activated without the need for Cdc7-Dbf4-dependent phosphorylation.

13.2.5 PfMCM8 and PfMCM9

MCM8 and MCM9 proteins are conserved in a diverse array of eukaryotes but are lacking in most fungi, *Caenorhabditis elegans* and *Giardia lamblia,* and thus were originally thought to be vertebrate-specific proteins[26] (see also Chapter 3). However, we have identified MCM8 and MCM9 homologues in the malarial parasite, and found that they are very similar to even their human orthologues (Li *et al.,* unpublished data): PfMCM8 shares 34% identity and 55% similarity with human hMCM8 in two separate regions and contains all MCM2-7 family

---

Figure 13.3 **Sequence comparisons between *P. falciparum* and human MCM4.** The sequences were aligned with the CLUSTAL W (1.85)[71,72] using data derived from GenBank/EMBL/DDJB database accession numbers: *P. falciparum* PfMCM4, AF083323; human MCM4 (aka CDC21), P33991; and mouse MCM4, P49717. Inserts A–E are labelled and underlined at the bottom of sequence. The zinc-finger motif ($CX_nCX_{18}CX_2C$) is highlighted in grey and the Walker A and B motifs are highlighted and indicated at the bottom of sequence. Note that murine MCM4 is included to demonstrate the very high degree of conservation of MCM4 in mammals.

```
                      10        20        30        40        50        60        70        80
              ----:----|----:----|----:----|----:----|----:----|----:----|----:----|----:----|
Consensus     MxSPxxxxxRRxSxxxxxxxxxSxxxxxSxSxxxxxxxxxxxxSTxxxxxxxTxxxxxLxxxxxxNxxxSxxxxMxxxxx
MCM4_MOUSE    MSSPASTPSRRSSRRGRVTPTQSLRSEESRSSPNRRRRGEDS-STGELLPMPTSPGADLQSPPAQNALFSSPPQMHSLAI
MCM4_HUMAN    MSSPASTPSRRGSRRGRATPAQTPRSEDARSSPSQRRRGEDSTSTGELQPMPTSPGVDLQSPAAQDVLFSSPPQMHSSAI
PfMCM4        MGTPRRRLGQQNNNNNSPFALSSSNIFGSNNEIFGSNFMHTPMSSRRTKNSKSFLNSMLNESRYLNQSNAGSQFIKYGHT

                      90       100       110       120       130       140       150       160
              ----:----|----:----|----:----|----:----|----:----|----:----|----:----|----:----|
Consensus     PLxxxxxxxxxxxxxxxxxExxxxxxxxGxxxxxxxxxxxxxKxLxxxxxxDxxxxxxxxxxDxxxxEQSLxQxLVIxxx
MCM4_MOUSE    PLDFDVSSPLTYGTPSSRVEGTPRSGVRGTPVRQRPDLGSARKGLQVDLQSDGAAAE-----DIVPSEQSLGQKLVIWGT
MCM4_HUMAN    PLDFDVSSPLTYGTPSSRVEGTPRSGVRGTPVRQRPDLGSAQKGLQVDLQSDGAAAE-----DIVASEQSLGQKLVIWGT
PfMCM4        PLAIRRIKCARADIGDVGREAFMEDEESGRLPHFIDSNLEQIKELFNQFFDEFNITNYSDVLDFTDEDRSISEYILLHRD
                                                                      Insert A

                     170       180       190       200       210       220       230       240
              ----:----|----:----|----:----|----:----|----:----|----:----|----:----|----:----|
Consensus     DVxVxxxxxxFQxxLxxFxDxxxKxExxxxNxxxDxxExxxxxxxGxxNIxxxxxxxVNxxHIxxFNKxLYRxLIxYPxE
MCM4_MOUSE    DVNVATCKENFQRFLQCFTDPLAKEE---ENVGIDITQPLYMQQLGEINITGEPFLNVNCEHIKSFSKNLYRQLISYPQE
MCM4_HUMAN    DVNVAACKENFQRFLQRFIDPLAKEE---ENVGIDITEPLYMQRLGEINVIGEPFLNVNCEHIKSFDKNLYRQLISYPQE
PfMCM4        NLKVYLAYYGWK--MIKFIETGRQNECRLNNTNYEDDDENNENSEGIRNLEHIKSFEIDLTHIFFFNKKLYKLIIEYPSD

                     250       260       270       280       290       300       310       320
              ----:----|----:----|----:----|----:----|----:----|----:----|----:----|----:----|
Consensus     xIxxxDxxVNxxFxxxxxxxxxxxxxxxxxDRYPxxSxxExxxQVRxFNxxxxxSxRxLxPxxIDxLIxIxGMVIRxSxL
MCM4_MOUSE    VIPTFDMAVNEIFF----------------DRYP-DSILEHQIQVRPFNALKTKSMRNLNPEDIDQLITISGMVIRTSQL
MCM4_HUMAN    VIPTFDMAVNEIFF----------------DRYP-DSILEHQIQVRPFNALKTKNMRNLNPEDIDQLITISGMVIRTSQL
PfMCM4        CISEIDKIISTKYNSLLALVLEGDTRSSSSDKYPLSSTKQDYCRVRFFNKKHKDTPRKLGPNQIETLVCVKGVIIRCSNI
                             Insert B

                     330       340       350       360       370       380       390       400
              ----:----|----:----|----:----|----:----|----:----|----:----|----:----|----:----|
Consensus     IPEMxxAxFQCxxxxxxxxxxxQxCxxxxxxxIxRGxIxEPxxCxxCHxxHSxxLxHNxxxFSxKQMIKLxExxExMxxG
MCM4_MOUSE    IPEMQEAFFQC-----------QVCAHTTRVEIDRGRIAEPCSCVHCHTTHSMALIHNRSFFSDKQMIKLQESPEDMPAG
MCM4_HUMAN    IPEMQEAFFQC-----------QVCAHTTRVEMDRGRIAEPSVCGRCHTTHSMALIHNRSLFSDKQMIKLQESPEDMPAG
PfMCM4        IPEMTMAAFKCTSKKRIGVNNYEKCNEEVYEHVIQGEVQEPVTCSNCNNKNTFELWHNNCCFSSKQLIKLSEVTEHLKQG
                        Insert C

                     410       420       430       440       450       460       470       480
              ----:----|----:----|----:----|----:----|----:----|----:----|----:----|----:----|
Consensus     QTPxTIxLFAHNDLVDxxQPGDxVxVTGIxRAxPIRVNPRxxxxxSVYKTHIDVIHxRKxDxKRLxxxDExxxxxxxxxx
MCM4_MOUSE    QTPHTIVLFAHNDLVDKVQPGDRVNVTGIYRAVPIRVNPRVSNVKSVYKTHIDVIHYRKTDAKRLHGLDE----------
MCM4_HUMAN    QTPHTVILFAHNDLVDKVQPGDRVNVTGIYRAVPIRVNPRVSNVKSVYKTHIDVIHYRKTDAKRLHGLDE----------
PfMCM4        ETPQSISIYAYDDLIDYTKPGDTVELTGILKASPVRLNPRSRCYNSVHRTYINVIHIKKENKQKMKLTEQNDTANIILKR

                     490       500       510       520       530       540       550       560
              ----:----|----:----|----:----|----:----|----:----|----:----|----:----|----:----|
Consensus     xxxxxxxxxxxxxxExxxxLFSxKxVExLKELSRxPDIYERLxxALAPSIYxxxDIKKGILxQLFGGTRxxxSxxxxxKF
MCM4_MOUSE    --------------EAEQKLFSEKRVKLLKELSRKPDIYERLASALAPSIYEHEDIKKGILLQLFGGTRKDFSHTGRGKF
MCM4_HUMAN    --------------EAEQKLFSEKRVELLKELSRKPDIYERLASALAPSIYEHEDIKKGILLQLFGGTRKDFSHTGRGKF
PfMCM4        NEDGTVEENFEKLNEQGNLLFTTEVIQKMEQLSKDPNIYQRLVDSIAPSIYGRGDIKKGLLCQLFGGSK--ITDKYNNKY
              Insert D

                     570       580       590       600       610       620       630       640
              ----:----|----:----|----:----|----:----|----:----|----:----|----:----|----:----|
Consensus     RAEINILLxGDPxTSKSQLLxYVYxLxPRGxYTSGKGSSAVGLTAYVxKDxETRQxVLQTGALVLSDxGICCIDEFDKMN
MCM4_MOUSE    RAEINILLCGDPGTSKSQLLQYVYNLVPRGQYTSGKGSSAVGLTAYVMKDPETRQLVLQTGALVLSDNGICCIDEFDKMN
MCM4_HUMAN    RAEINILLCGDPGTSKSQLLQYVYNLVPRGQYTSGKGSSAVGLTAYVMKDPETRQLVLQTGALVLSDNGICCIDEFDKMN
PfMCM4        RSEIHILLRGDPSTAKSQLLHYVHKLSPRGIYTSGKGSSSVGLTAFISKDSETKEYILESGAVVLSDKGICCIDEFDKMD
                   Walker A                                                 Walker B

                     650       660       670       680       690       700       710       720
              ----:----|----:----|----:----|----:----|----:----|----:----|----:----|----:----|
Consensus     ESxRSVLHEVMEQQTLSIAKAGIIxxLNARTSVLAAANPIxSQWNxxKxxIENIxLPxTLxSRFDLIFLMLDxxDExxDR
MCM4_MOUSE    ESTRSVLHEVMEQQTLSIAKAGIICQLNARTSVLAAANPIESQWNPKKTTIENIQLPHTLLSRFDLIFLMLDPQDEAYDR
MCM4_HUMAN    ESTRSVLHEVMEQQTLSIAKAGIICQLNARTSVLAAANPIESQWNPKKTTIENIQLPHTLLSRFDLIFLLLDPQDEAYDR
PfMCM4        DSARAILHEVMEQQTVTIAKAGIVATLNARTSILASANPINSRYDKNKAVVENINLPPSLFSRFDLIYLVIDQANEDEDR

                     730       740       750       760       770       780       790       800
              ----:----|----:----|----:----|----:----|----:----|----:----|----:----|----:----|
Consensus     RLAxxLxxxYxxSxEQxEEExxxxxxxxxxxxxxxxxxxxxxxxxxxxxxxxxxxxxxxxxxxxxxxxxxxxxxxxxxxx
MCM4_MOUSE    RLAHHLVSLYYQSEEQVEEE------------------------------------------------------------
MCM4_HUMAN    RLAHHLVALYYQSEEQAEEE------------------------------------------------------------
PfMCM4        KLATVLCKNFSYNPEEEEDEDQEDQEEDEPNYITQQRARKSKGTSRKNERENYYNDGDNDDDDDISNYLNDSNDAQNKRG
                                                                     Insert E

                     810       820       830       840       850       860       870       880
              ----:----|----:----|----:----|----:----|----:----|----:----|----:----|----:----|
Consensus     xxxxxxxxxxxxxxxxxxxxxxxxLDxxxLxxYIAYxxxTxxPxLSxEAxQxLIExYVxMRxxxxSRxxxSAxPRQLEx
MCM4_MOUSE    ------------------------FLDMAVLKDYIAYAHSTIMPRLSEEASQALIEAYVNMRKIGSSRGMVSAYPRQLES
MCM4_HUMAN    ------------------------LLDMAVLKDYIAYAHSTIMPRLSEEASQALIEAYVDMRKIGSSRGMVSAYPRQLES
PfMCM4        SWANVNISYDEYNNSSNKKTSKNYLIDSNTLALYIAYCRITCNPIISLESKKIIIEEYIKMRCKEGTKS-PTASPRQLEG

                     890       900       910       920       930       940       950       960
              ----:----|----:----|----:----|----:----|----:----|----:----|----:----|----:----|
Consensus     LIRLAEAxAKVRxxxxVxxxDxxEAxRLxxxAxxQSxxDPxTGxVDxxxLxxGxSAxxRKRxExLxExLxxxILxKxxTx
MCM4_MOUSE    LIRLAEAHAKVRFSNKVEAIDVEEAKRLHREALKQSATDPRTGIVDISILTTGMSATSRKRKEELAEALRKLILSKGKTP
MCM4_HUMAN    LIRLAEAHAKVRLSNKVEAIDVEEAKRLHREALKQSATDPRTGIVDISILTTGMSATSRKRKEELAEALKKLILSKGKTP
PfMCM4        LVRLSQSLAKMKLKRVVSPEEANEAVRLMNIATFQSLIDPLSGRIDFDQVNLGQTSQHKKKSDLIKDIIMNALVLKNMTK

                     970       980       990      1000      1010      1020      1030      1040
              ----:----|----:----|----:----|----:----|----:----|----:----|----:----|----:----|
Consensus     xxxxxxxxExIxxxxQxxxAIxKxxFEEAxxxLxxxxxLTxxxxxVxxxx
MCM4_MOUSE    ALKYQQLFEDIRG--QSDTAITKDMFEEALRALADDDFLTVTGKTVRLL-
MCM4_HUMAN    ALKYQQLFEDIRG--QSDIAITKDMFEEALRALADDDFLTVTGKTVRLL-
PfMCM4        DELLTHCHETIMNDPQHTTSMDRKSFEEAFYDLEKSQEITRLCSGLYKKK
```

domains including the Walker A (GDPGLGKS) and Walker B (CIDELDKI) motifs as well as the conserved zinc finger motif ($CX_2CX_{18}CX_4C$). However, compared with hMCM8, PfMCM8 has three sequence insertions of six, 26 and 186 amino acid residues, respectively. In addition, near the N-terminus, there are six copies of a hexameric GNKN(Y)G(E)K repeat. PfMCM8 appears to be transcribed in trophozoite and schizont stages, but at slightly higher levels in the sporozoite.[14]

PfMCM9 is the largest of the PfMCM family members, comprising 1465 amino acid residues with a predicted molecular mass of 171.2 kDa. The *PfMCM9* gene is transcribed in late trophozoite and early schizont stages, and in sexual stages.[14,19] In five separate regions between residues 412–1400, PfMCM9 shares 37% identity and 57% similarity with human and mouse MCM9, including the Walker A (GDPGTGKS) and B (CIDEFCLM) motifs as well as the full MCM2-7 family domain. However, there are a number of unique amino acid insertions in PfMCM9. Compared with other members in the MCM9 family, PfMCM9 lacks the C-terminal region but has the large N-terminal extension. In addition, there are two types of amino acid sequence repeats, *i.e.* 13 copies of NN(Y)D(G)DNK at positions 149–226 and 12 copies of INV(G)D(N)D(N) at positions 721–780. How these repeats arose and whether they serve a function in parasite replication remain open questions.

13.3 Replication Elongation Proteins

13.3.1 PfRPA

Replication protein A (RPA) functions as a heterotrimeric complex that stabilises single-stranded DNA (ssDNA), and coordinates the sequential assembly and disassembly of DNA processing proteins on ssDNA during replication, and also during repair and transcription.[27,28] The largest subunit, RPA1 (hRPA70 in humans), contains the primary ssDNA-binding activity, while the two smaller subunits, RPA2 (hRPA32) and RPA3 (hRPA14), stabilise the complex and mediate interactions with replication and repair machinery,[28] (see Chapter 6). In *P. falciparum,* we have identified five putative subunits, *i.e.* two RPA1-type proteins (PfRPA1 and PfRPA1′), two RPA2-like molecules (PfRPA2 and PfRPA2′), and one RPA3 protein (PfRPA3) (Li *et al.*, unpublished data).

PfRPA1 was the first malarial RPA to be experimentally identified, through affinity purification of a 55 kDa factor possessing the major ssDNA-binding activity in asexual stage extracts.[29] Mass spectral analysis of 11 tryptic peptides demonstrated that the 55 kDa protein is the C-terminal fragment of PfRPA1.

PfRPA1 consists of 1145 amino acids with a predicted mass of 134.1 kDa, but it shares 30–39.2% identity with other eukaryotic RPA1 family members only within the C-terminal 466 amino acids. PfRPA1 possesses three ssDNA-binding domains:[30] DBD-A (positions 685–778), DBD-B (817–919) and DBD-C (958–1134); and a consensus zinc finger motif ($CX_2C_{26}CX_2C$, 986-1019) within DBD-C. However, in the N-terminal region of PfRPA1, no significant homology to any known protein was identified; this non-conserved region may provide a suitable target for development of anti-malarial agents. PfRPA1 appears to be expressed in late trophozoite and schizont stages[29] as well as in sexual (gametocyte) and sporozoite stages.[14]

The second malarial RPA1, which we term PfRPA1′, consists of 484 amino acids with a calculated molecular mass of 56.1 kDa, and shares 26% identity and 50% similarity with PfRPA1. Like PfRPA1, PfRPA1′ contains the DNA binding domains DBD-A (10–109), DBD-B (140–247), DBD-C (275–451), and the zinc finger motif ($CX_2C_{28}CX_2C$, 315–350) characteristic of RPA1, but lacks the N-terminal protein interaction domain. PfRPA1′ seems to be expressed in trophozoite and schizont stages, and in sexual (gametocyte) and in sporozoite stages.[14,19]

Two putative malarial RPA2 subunits, PfRPA2 and PfRPA2′, consist of 273 and 191 amino acids with calculated masses of 31.6 and 22.3 kDa, respectively, and share 39% identity and 57% similarity each other in the central region. Importantly, PfRPA2 and PfRPA2′ each contain the consensus DBD-D DNA binding domain, together with several potential interface sites with RPA1 DBD-C and RPA3, and possible phosphorylation sites in their N-terminal regions. PfRPA2 seems to be weakly expressed in sporozoite and in early schizont and gametocyte stages, while PfRPA2′ is expressed in gametocyte and early trophozoite stages.[14]

The smallest malarial subunit, PfRPA3, consists of 135 amino acids with a predicted molecular mass of 15.4 kDa, and contains several putative interface sites with RPA2 DBD-D and RPA1 DBD-C. PfRPA3 appears to be expressed very weakly in gametocytes.[14]

13.3.2 DNA Polymerases

DNA polymerases α, δ and ε are the three established polymerases responsible for the bulk of DNA replication (see Chapter 4) that are sensitive to aphidicolin but insensitive to dideoxynucleotide analogues. There have been several reports on stage-dependent, aphidicolin-sensitive and aphidicolin-resistant DNA polymerase activities in the asexual stage extracts of *P. falciparum*.[31–33] DNA polymerase activity was undetectable in ring-form extracts, but increased in trophozoites and peaked in schizonts. Seven DNA polymerase activities have been identified and partially purified from *P. berghei*, five of which are aphidicolin-sensitive, while two are resistant.[34]

13.3.2.1 PfPolα

The pol α-primase complex is responsible for making a chimeric RNA–DNA primer of ~40 nucleotides for initiating DNA synthesis, and is composed of a large catalytic subunit (Polα/p180), an intermediate B subunit (p70) and two small subunits (PriL/p55 and PriS/p48).[35] The largest subunit possesses the polymerase catalytic activity, while the two small subunits, PriS and PriL, together function as the core primase for creating an RNA primer of 6–15 nucleotides.[36] The malarial Polα catalytic subunit (PfPolα), B subunit (PfPolαB) and two small primase subunits (PfPriS and PfPriL) have been found in the parasite genome.

PfPolα shares 14–17% overall identity only with other members of the Polα family but contains all seven motifs I–VII conserved in DNA polymerases, together with four of five Polα-specific motifs (A–E). However, PfPolα has a number of unique features. Firstly, motif A, present in the human counterpart, is absent in PfPolα. Secondly, in motif D, a highly conserved glycine (residue G493 in ScPolα), implicated in the interaction with DNA primase in yeast, is replaced by Leu in *P. falciparum*. Thirdly, there are at least six sequence insertions with sizes ranged from 10 to 91 amino acids interspersed between the polymerase-specific motifs, two of which contain repeated sequences. Insertion 2 has seven degenerate copies of repeat QQSVVS, while insertion 4 has four copies of the degenerate repeat KNIHSD. Oddly, the QQSVVS motif is shared with an RNA-directed RNA polymerase from bovine viral diarrhoea virus (Swissprot accession P19711). Fourthly, there are four Asn-rich tracts, three in the N-terminal region and one in the C-terminal region. Finally, PfPolα has a novel C-terminal extension of 98 residues compared with the human counterpart. PfPolα is transcribed in asexual and sexual stages.[37]

PfPolαB shares 24% identity and 48% similarity with its human counterpart in the region of 148–514. However, PfPolαB contains two unique sequence insertions with nine and 12 residues, respectively, compared with the human orthologue. PfPolαB seems to be expressed in asexual and sexual (gametocyte) stages, and the sporozoite stage.[14,19]

Primase in the parasite is encoded by two genes, *PfPriL* and *PfPriS*. PfPriL shares 33% identity and 53% similarity with its human orthologue, including a putative iron–sulfur cluster coordinated by four conserved cysteine residues (C362, 438, 456 and 496) that contributes to enzymatic activity of the core primase.[38]

PfPriS, encoded by a gene containing 15 introns, displays 36% identity and 53% similarity with its human counterpart. It contains all five domains (I–V) conserved in eukaryotic primase small subunits, including three catalytic aspartates (D130, 132 and 347) and the consensus sequence SGXRG (residues 181–185) involved in nucleotide binding. Additional conserved regions, termed Ia, VI and VII, are found in mammalian and yeast homologues.[39] However, PfPriS has three unique sequence insertions, two flanking domain Ia and one before domain V. Primase activity has been confirmed for recombinant PfPriS expressed in insect cells.[39] Both PfPriL and PfPriS appear to be expressed in trophozoites, schizonts, gametocytes and sporozoites.[14]

13.3.2.2 *PfPolδ*

DNA polymerase δ functions as a heterotetramer in fission yeast and mammals (see Chapter 4), and consists of a tightly associated dimer of a large catalytic subunit (p125) and a small subunit (p50), which is associated with two other subunits (p68 and p12). However, in budding yeast, only three subunits have been identified.[40] In *P. falciparum*, two subunits, the large catalytic subunit (PfPolδ) and the small regulatory subunit (PfPolδS) have been identified.

PfPolδ shares 46% identity and 67% similarity with *Schizosaccharomyces pombe* Polδ between amino acids 97–1013.[41,42] PfPolδ contains all seven major motifs (I–VII) used to identify DNA polymerases in a correct spatial order, five additional regions (δ1-5) conserved in members of the Polδ family, and two putative zinc finger motifs ($CX_2CX_9CX_2C$, 1003-1019; $CX_2CX_9CX_4C$, 1049-1067). In addition, PfPolδ has the 3′–5′ exonuclease domain (residues 126–471) responsible for conferring proofreading activity, and retains several conserved residues (D308, I309, E310, Y390, D396 and D509) that presumably form the exonuclease catalytic site. PfPolδ is transcribed in asexual (trophozoite and schizont) and sexual (gametocyte) stages[41,42] as well as in sporozoites.[14] Western blotting analysis revealed that PfPolδ protein is expressed in trophozoite and schizont stages, but not in the ring stage.[43]

PfPolδS is slightly less conserved, showing up to 33% identity and 56% similarity with other members of the Polδ2 family. PfPolδS seems to be transcribed in asexual (trophozoite, schizont), sexual (gametocyte) and sporozoite stages.[14]

13.3.2.3 PfPolε

Pol ε comprises four different subunits in vertebrates (Polε A–D) and budding yeast (Pol2/256kDa, Dpb2/79kDa, Dpb3/23kDa, and Dpb4/22kDa).[40] The DNA polymerase and 3′–5′ exonuclease of polymerase ε reside in the largest subunit. In *P. falciparum,* genes encoding both the catalytic subunit (PfPolε) and the second largest subunit (PfPolεB) have been found in the parasite genome. PfPolε displays significant homology to its human counterpart in four different regions:

- 30% identity and 54% similarity in the region of 27–175;
- 37% identity and 57% similarity in the region of 259–1406;
- 28% identity and 49% similarity in the region of 1729–1841;
- 25% identity and 46% similarity in the region of 2714–2899.

PfPolε contains all seven ordered motifs (I–VII) conserved in DNA polymerases in the N-terminal region, the exonuclease domain in the region of 337–635 including several conserved residues (D344, I345, E346, Y433, D439 and D533) that form the exonuclease catalytic site, plus two putative zinc finger motifs ($CX_2CX_{26}CX_2C$, 2782-2815; $CX_2CX_{11}CXC$, 2845-2863) conserved in the Polε family at the C-terminal side. In addition, a putative PCNA-binding motif, **QKKITSFF** (1312-1319), is found in the middle part of PfPolε, suggestive of an interaction with PfPCNA. It is noted that PfPolε also possesses numerous sequence insertions compared with its human counterpart; such additional sequences may prove useful in directed drug design.

PfPolεB shares 23% identity and 44% similarity with human PolεB in the region of 168–575. PfPolεB appears to be expressed in trophozoite and schizont stages.[14,19]

13.3.3 PfRFC

RFC consists of five non-identical subunits (RFC1–5) and is a DNA-dependent ATPase that functions as the clamp loader for PCNA (see Chapters 2 and 3) to confer processivity on DNA Pol δ and ε. We have identified all five subunits in the parasite and our Northern blot hybridisation shows that PfRFC1-5 are all expressed in asexual and sexual stages (J-L. Li and A. Goldsmith, unpublished data).

Each PfRFC subunit contains three domains (domains I–III) and the primary sequence of these domains is homologous among the five subunits; only PfRFC1, the large subunit, has additional N- and C-terminal regions. There is an overall similarity between the parasite PfRFC and corresponding orthologues in most other eukaryotes. All RFC boxes (I–VIII) occur in PfRFC, ranged from the N-terminus to the C-terminus.

- Box I is present only in PfRFC1 and consists of about 95 amino acids from positions 185–280. This region shows homology to the DNA ligase homology domain and BRCT, the BRCA1 C-terminus domain.
- Box II, at the N-terminus of PfRFC2-5, shows the WV(I,L)E-KYR(S)PXXL(I) consensus sequence.
- Box III, contains the phosphate-binding motif (P-loop) with the consensus sequence GXXGXGKT(S). It is notable that a classic GKT(S) sequence in the consensus P-loop is replaced by GKK in both human hRFC38 and yeast ScRFC5; however, PfRFC5 retains the motif (GKS) in the P-loop.
- Box IVcontains the L(I)EL(F)NASD sequence although a motif variation (LELQCFE) occurs in PfRFC5.
- Box V has the DEA(V)D motif, which for PfRFC5 is replaced by KDAE, suggesting that PfRFC5 might not be able to hydrolyse ATP.
- Box VI is different in the small and the large PfRFC subunits, as would be expected.
- Box VIa, only present in PfRFC1, has the consensus sequence of GMSsGDKGGstaI while box VIb, present in the PfRFC2-5 subunits, shows xM(L)T(S)xxAQxxL(M)RRI(t)M(I/L)E.
- Box VII, the arginine finger motif (SRC), is conserved in all small subunits (PfRFC2-5) but only the cysteine is present in PfRFC1, as would be expected for members of the RFC1 family.
- Box VIII has the consensus sequence GDL(M/I)RxxA(M/I)L(V)xxLQ, which for PfRFC5 is replaced by THGRNLRKVI.

These sequence characteristics suggest that all PfRFC subunits may bind ATP (*via* Walker A boxes), but that only PfRFC2-4 may effectively hydrolyse the bound ATP by virtue of their Walker B boxes.

13.3.4 PfPCNA

Proliferating cell nuclear antigen (PCNA) functions as a DNA sliding clamp that tethers the replicative DNA polymerases to the template, and also interacts with and regulates the activities of numerous proteins involved in DNA and chromatin processing during DNA synthesis (see Chapter 3). In yeast and mammals, only one form of PCNA has been found; this associates into a homotrimer with pseudo-sixfold symmetry.[44,45] By contrast, two different PCNAs have been described in the malarial parasite[46–48] (note that three different forms with subtly different interaction profiles have been characterised in Crenarchaea—see Chapter 12).

PfPCNA1 shares 29% identity and 53% similarity with PfPCNA2. PfPCNA1 and 2 are not only expressed in asexual trophozoite and schizont stages,[43,47,48] but also in the sexual gametocyte stage,[47] consistent with a role in rapidly proliferating cells. Both PfPCNA1 and PfPCNA2 contain all conserved motifs of members of the PCNA family, including the potential DNA-binding and protein interaction domains. PfPCNA2 may form a homotrimeric structure and PfPCNA1 may multimerise, but whether it forms dimers or trimers is unclear.[47,48]

Compared with human hPCNA, both PfPCNAs have a unique sequence insertion (PfPCNA1 with nine amino acid residues and PfPCNA2 with seven residues) just before the I2 motif. In addition, PfPCNA2 contains a second insertion with four amino acid residues located just before the βH1 region. We have conducted structural modelling of PfPCNA2 based on the published crystal structure for human PCNA,[44] and as shown in Figure 13.4, the additional regions present in PfPCNA2 but absent in hPCNA (white arrows in Figure 13.4) may be solvent-exposed and thus accessible to drugs.

13.3.5 PfFen1

Fen1, a structure-specific 5′ endo/exonuclease, cleaves a 5′-unannealed flap and degrades nucleotides from a nick or a gap on dsDNA. It plays a key role in the removal of RNA primers during Okazaki fragment maturation in lagging strand DNA synthesis (see Chapter 5), long-patch base excision repair and maintaining genome stability.

The malarial Fen1, PfFen1, shares up to 54% identity and 72% similarity with other members of the Fen1 family and consists of five regions. The N and I regions, like other members in the Fen1 family, constitute the nuclease core domain (residues 1–345) which contains the helical clamp region (residues 89–136), the helix-three turn-helix motif (residues 234–266), and seven critical acidic residues that cluster to form two active sites. By analogy to human FEN1, four of the acidic amino acids in PfFen1 (D34, D88, E160 and E162) are predicted to form the first metal ion-binding site, and three others (D181, D183 and D246) are likely to be involved in forming the second metal ion-binding site. The C region, located internally, contains the conserved PCNA-binding motif (**Q**RR**L**DN**FF**, 350–357) that in fact forms a part of the βA–αA–βB

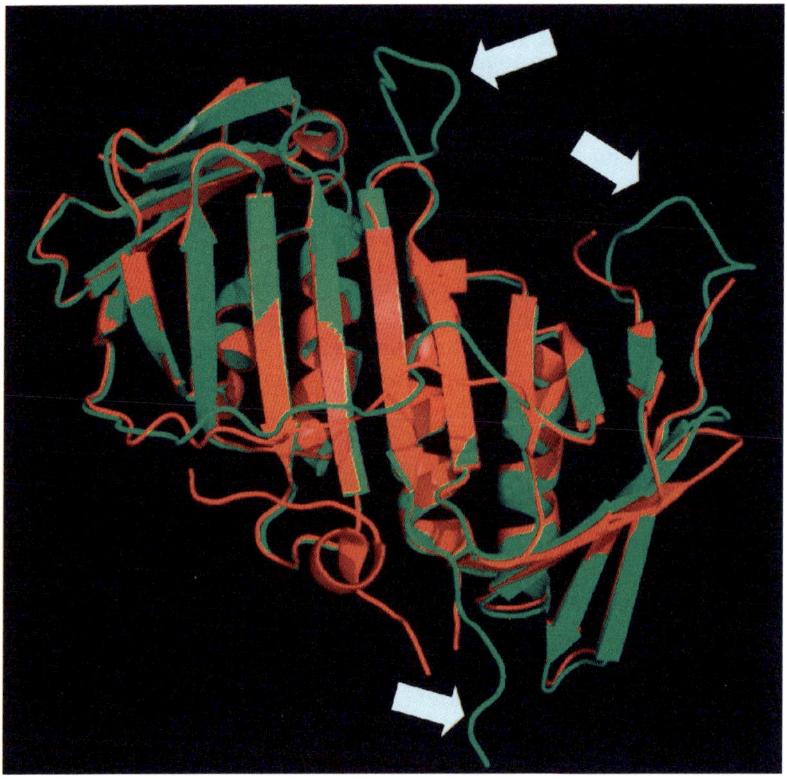

Figure 13.4 **Predicted structural model of PfPCNA2 monomer.** Homology modelling of PfPCNA2 (green) was conducted using SWISS-MODEL[73,74] using the coordinates obtained for yeast PCNA[44] (PDB accession number 1PLR) as a template, and superimposed on human PCNA (red in figure) using MacPyMOL (http://delsci.com/macpymol/). The additional regions present in PfPCNA2 but absent in hPCNA are highlighted by white arrows.

structure, the main interacting interface with PCNA.[49,50] PfFen1 can generate a nicked DNA substrate that can be ligated by PfLigI.[51,52] PfFen1 seems to be transcribed in both asexual and sexual stages (J-L. Li, unpublished data). Western blot analysis of parasite lysates from the erythrocytic stages revealed that PfFen1 protein is present in all asexual stages.[51]

The R region, unique to PfFen1, has the DdeKXX hexamer repeated 12 times (residues 421–492). The E region (residues 493–672), the unique large-extended C-terminal segment, is present in a natively disordered state with no discernible secondary structure but possesses a putative NLS motif (residues 645–667). Although such unique extensions may appear to be prime targets for designing agents to inhibit malarial DNA replication without impacting on the function of essential host enzymes, functional analysis is critical in determining target validity. For example, the R and E regions of PfFen1 negatively influence the enzyme's activities, since a PfFen1 C-terminal truncation consisting of about 400 amino acids and lacking the R and E regions possesses endonuclease and exonuclease activities 300-fold and 30-fold higher than full-length PfFen1, respectively.[51] Thus, these domains serve a regulatory role and their disruption may promote rather than block parasite DNA replication.

13.3.6 PfRNase H

RNase H specifically degrades the RNA moiety in RNA–DNA hybrids and plays an important role in DNA replication by removal of the RNA primer of Okazaki fragments (see Chapter 5). There are two major types of RNase H: RNase H1 is the smaller enzyme with major activity in prokaryotes, while RNase H2 is the RNase H large subunit which is most abundant in eukaryotes and archaea.[53]

Bioinformatics investigations failed to detect any RNase H1 candidate in the parasite genome (Li *et al.*, unpublished data), suggesting that RNase H1 might not be conserved in *P. falciparum*. However, a malarial orthologue of RNase H2, PfRNase H2, has been found in the parasite genome. PfRNase H2 is highly similar to the human RNase H large subunit, sharing 51% identity and 69% similarity; it retains all the conserved domains, including the active site triad residues (D22, D130 and D158) and the nearby serine (S168) and glutamate (E23) discovered in the archaeal orthologue.[53] An effort to define the functional capacity of PfRNase H2 is urgently needed.

13.3.7 PfLigase I

DNA ligase I constitutes the primary ligase utilised in DNA replication and plays an essential role in the joining of Okazaki fragments and nick sealing. The malarial DNA ligase I, PfLigI, shares 30% overall identity with human DNA ligase I, with even greater homology (60% identity) in the C-terminal region. PfLigI retains all of the conserved domains common to the ATP-dependent DNA ligase family, including important residues from motifs I–V that constitute the nucleotide binding pocket and motif VI which, together with motifs I–V, comprises the catalytic core of DNA ligase molecules.

In addition, PfLigI contains a bipartite nuclear localisation motif (residues 100–117), suggesting that PfLigI may be a nuclear protein. However, a unique apicoplast signal sequence, consisting of a signal peptide and a transit peptide, occurs in the N-terminus of PfLigI instead of the PCNA binding domain that normally exists in mammalian counterparts, suggesting that PfLigI might also be transported into apicoplasts (see Section 13.4.1). In addition, in the conserved region there is a sequence insertion consisting of about 100 amino acids compared with mammalian counterparts. PfLigI protein appears to be expressed in the asexual schizont stage.[52] Recombinant PfLigI expressed in HEK293 cells catalysed phosphodiester bond formation on a singly nicked

DNA substrate in an ATP-dependent manner, and joined RNA-DNA substrates only when the RNA sequence was upstream of the nick.[52]

13.4 Potential Targets for Novel Drug Development

P. falciparum contains most of the essential DNA replication components that are conserved in eukaryotes, suggestive of a similar core replication machinery of the parasite to that of the human host. However, the parasite seems to lack a number of core replication proteins including ORC3, ORC6, Cdc45, Dbf4-Cdc7, geminin and DNA2 helicase, suggestive of some differences in replication and regulatory pathways between the parasite and the human host, thereby presenting excellent targets for development of novel anti-malarial drugs.

13.4.1 Targeting Unique Replication Pathways: Apicoplast DNA Replication

The apicoplast, a chloroplast-like plastid present in *P. falciparum* and absent in humans, is no longer photosynthetic, but like chloroplasts, it contains its own genome and this is essential for parasite survival.[54] Apicoplast DNA replication is dependent on nuclear-encoded apicoplast-targeted proteins (Table 13.3). DNA gyrase, a typical type II topisomerase that can introduce negative supercoils in DNA, is required for parasite apicoplast DNA replication.[54,55] The malarial gyrase A and gyrase B, PfGyrA and PfGyrB, are encoded in the nucleus and then transported into the apicoplast. PfGyrB contains strong intrinsic ATPase activity and, together with the N-terminal domain of PfGyrA, can efficiently cleave supercoiled DNA.[56,57] The fluoroquinolone antibiotic, ciprofloxacin, that targets bacterial DNA gyrase, has been demonstrated to inhibit apicoplast DNA replication and growth of the parasite *in vitro*.[8,58,59] Novobiocin, a specific inhibitor of bacterial GyrB, has been shown to inhibit the ATPase activity of PfGyrB and caused parasite death in culture.[57,60] Thus, PfGyrA and PfGyrB have become attractive targets for use of existing antimicrobials and for development of novel anti-malarial therapeutics.

PfPOM1, known as *P. falciparum* plastidic DNA replication/repair enzyme complex (PfPREX), contains multiple domains with DNA primase, DNA helicase, DNA polymerase and 3′–5′ exonuclease activities. The N-terminal sequence has been demonstrated to direct PfPOM1 to the apicoplast, suggesting that PfPOM1 may have an important role in apicoplast DNA replication.[61] As the polymerase and primase subunits of PfPOM1 have no direct orthologues in human host cells, PfPOM1 presents a perfect target for new drug development.

13.4.2 Targeting Unique Protein Sequences

The sequences of malarial replication proteins exhibit high homology to their human counterparts, but also contain a number of unique features such as large

N- or C-terminal extensions, numerous sequence insertions within highly conserved regions, and various amino acid sequence repeats. These unique properties of the replication proteins offer potentially ideal targets for new drug therapy. In particular, the extra regions in PfPCNA, and the 12 repeats in the R region of PfFen1 pose very exciting potential targets, since high resolution crystallographic data exist on human Fen1–PCNA complexes[49] (see also Figure 3.8), which can be used to model similarities and differences in the protein folds and interactions. Moreover, the genes encoding these malarial proteins have been cloned and recombinant protein expressed and assayed *in vitro*.[46–48,51] Relatively simple *in vitro* assays for Fen1 nuclease activity[51] and PCNA binding already exist,[48,50] both of which should be adaptable to high throughout methodologies, to allow rapid screening of compounds that specifically and selectively inhibit malarial but not human Fen1. Given more investment in the field of malarial research, the identification and characterisation of *Plasmodium* replication components should provide a veritable gold mine of 'druggable' targets.

13.5 Concluding Remarks

With the completion of the *P. falciparum* genome sequencing project,[62,63] our knowledge of the DNA replication machinery in the human malarial parasite has developed rapidly in the past decade. The parasite contains most DNA replication components conserved in eukaryotes. Nearly every replication protein described in *P. falciparum* has a close orthologue in other malarial species such as *P. vivax, P. knowlesi, P. yoelii, P. berghei,* and *P. chabaudi.*[64–66] However, some differences between *P. falciparum* and other eukaryotes (particularly humans) have been exploited, particularly at the replication initiation phase. Based on protein sequence comparisons, many proteins that are absolutely essential for eukaryotic DNA replication cannot be found in the parasite genome. In addition, a large proportion of replication proteins of the parasite display unique and interesting features such as terminal sequence extensions, amino acid sequence insertions and sequence repeats. These attractive characteristics present ideal potential targets for development of novel anti-malarial drugs.

Unfortunately, the majority of DNA replication components of the parasite are only at this stage 'predicted proteins', being revealed from the parasite genome using data-mining bioinformatics approaches. It has been shown that approximately 24% of the parasite genes in current databases were predicted incorrectly, including one or more additional introns found in some so-called 'intron-free' genes, different sizes and locations of introns in numerous predicted 'intron-containing' genes, and alternative splicings for many genes available in current databases.[67] Therefore it is essential to verify amino acid sequences and real identities of the replication proteins experimentally in the near future.

The challenges now are to express replication proteins in heterologous systems and to purify sufficient amounts of the recombinant proteins for more detailed biochemical and three-dimensional structural studies. In addition, to reconstitute a functional malarial replication system *in vitro* will greatly facilitate our understanding of the replication mechanisms in detail, and permit large-scale screening of natural product and chemical compound libraries for development of new anti-malarial drugs.

Acknowledgements

I thank Dr. Lynne S. Cox very much for her critical reading and comments on the chapter and her tremendous help in preparing Figures 13.2B and 13.4.

References

1. R. W. Snow, C. A. Guerra, A. M. Noor, H. Y. Myint and S. I. Hay, The global distribution of clinical episodes of *Plasmodium falciparum* malaria, *Nature*, 2005, **434**, 214–217.
2. J. K. Baird, Effectiveness of antimalarial drugs, *N. Engl. J. Med.*, 2005, **352**, 1565–1577.
3. S. M. Todryk and A. V. Hill, Malaria vaccines: the stage we are at, *Nat. Rev. Microbiol.*, 2007, **5**, 487–489.
4. J. H. White and B. J. Kilbey, DNA replication in the malaria parasite, *Parasitol. Today*, 1996, **12**, 151–155.
5. C. J. Janse, P. F. van der Klooster, H. J. van der Kaay, M. van der Ploeg and J. P. Overdulve, DNA synthesis in *Plasmodium berghei* during asexual and sexual development, *Mol. Biochem. Parasitol.*, 1986, **20**, 173–182.
6. D. Singh, S. Chaubey and S. Habib, Replication of the *Plasmodium falciparum* apicoplast DNA initiates within the inverted repeat region, *Mol. Biochem. Parasitol.*, 2003, **126**, 9–14.
7. D. Singh, A. Kumar, E. V. Raghu Ram and S. Habib, Multiple replication origins within the inverted repeat region of the *Plasmodium falciparum* apicoplast genome are differentially activated, *Mol. Biochem. Parasitol.*, 2005, **139**, 99–106.
8. D. H. Williamson, P. R. Preiser, P. W. Moore, S. McCready, M. Strath and R. J. Wilson, The plastid DNA of the malaria parasite *Plasmodium falciparum* is replicated by two mechanisms, *Mol. Microbiol.*, 2002, **45**, 533–542.
9. A. Gupta, P. Mehra and S. K. Dhar, Plasmodium falciparum origin recognition complex subunit 5: functional characterization and role in DNA replication foci formation, *Mol. Microbiol.*, 2008, **69**, 646–665.
10. P. Mehra, A. K. Biswas, A. Gupta, S. Gourinath, C. E. Chitnis and S. K. Dhar, Expression and characterization of human malaria parasite *Plasmodium falciparum* origin recognition complex subunit 1, *Biochem. Biophys. Res. Commun.*, 2005, **337**, 955–966.

11. S. Patterson, C. Robert, C. Whittle, R. Chakrabarti, C. Doerig and D. Chakrabarti, Pre-replication complex organization in the atypical DNA replication cycle of *Plasmodium falciparum*: characterization of the mini-chromosome maintenance (MCM) complex formation, *Mol. Biochem. Parasitol.*, 2006, **145**, 50–59.
12. J. L. Li and L. S. Cox, Characterisation of a sexual stage-specific gene encoding ORC1 homologue in the human malaria parasite *Plasmodium falciparum*, *Parasitol. Int.*, 2003, **52**, 41–52.
13. A. Gupta, P. Mehra, R. Nitharwal, A. Sharma, A. K. Biswas and S. K. Dhar, Analogous expression pattern of *Plasmodium falciparum* replication initiation proteins PfMCM4 and PfORC1 during the asexual and sexual stages of intraerythrocytic developmental cycle, *FEMS Microbiol. Lett.*, 2006, **261**, 12–18.
14. K. G. Le Roch, Y. Zhou, P. L. Blair, M. Grainger, J. K. Moch, J. D. Haynes, P. De La Vega, A. A. Holder, S. Batalov, D. J. Carucci and E. A. Winzeler, Discovery of gene function by expression profiling of the malaria parasite life cycle, *Science*, 2003, **301**, 1503–1508.
15. L. Mancio-Silva, A. P. Rojas-Meza, M. Vargas, A. Scherf and R. Hernandez-Rivas, Differential association of Orc1 and Sir2 proteins to telomeric domains in *Plasmodium falciparum*, *J. Cell Sci.*, 2008, **121**, 2046–2053.
16. M. T. Duraisingh, T. S. Voss, A. J. Marty, M. F. Duffy, R. T. Good, J. K. Thompson, L. H. Freitas-Junior, A. Scherf, B. S. Crabb and A. F. Cowman, Heterochromatin silencing and locus repositioning linked to regulation of virulence genes in *Plasmodium falciparum*, *Cell*, 2005, **121**, 13–24.
17. L. H. Freitas-Junior, R. Hernandez-Rivas, S. A. Ralph, D. Montiel-Condado, O. K. Ruvalcaba-Salazar, A. P. Rojas-Meza, L. Mancio-Silva, R. J. Leal-Silvestre, A. M. Gontijo, S. Shorte and A. Scherf, Telomeric heterochromatin propagation and histone acetylation control mutually exclusive expression of antigenic variation genes in malaria parasites, *Cell*, 2005, **121**, 25–36.
18. J. A. Baur, Y. Zou, J. W. Shay and W. E. Wright, Telomere position effect in human cells, *Science*, 2001, **292**, 2075–2077.
19. Z. Bozdech, M. Llinas, B. L. Pulliam, E. D. Wong, J. Zhu and J. L. DeRisi, The transcriptome of the intraerythrocytic developmental cycle of *Plasmodium falciparum*, *PLoS Biol.*, 2003, **1**, E5.
20. J. Liu, C. L. Smith, D. DeRyckere, K. DeAngelis, G. S. Martin and J. M. Berger, Structure and function of Cdc6/Cdc18: implications for origin recognition and checkpoint control, *Mol. Cell*, 2000, **6**, 637–648.
21. J. L. Li and L. S. Cox, Identification of an MCM4 homologue expressed specifically in the sexual stage of *Plasmodium falciparum*, *Int. J. Parasitol.*, 2001, **31**, 1246–1252.
22. A. Gambus, R. C. Jones, A. Sanchez-Diaz, M. Kanemaki, F. van Deursen, R. D. Edmondson and K. Labib, GINS maintains association of Cdc45

with MCM in replisome progression complexes at eukaryotic DNA replication forks, *Nat. Cell Biol.*, 2006, **8**, 358–366.
23. S. E. Moyer, P. W. Lewis and M. R. Botchan, Isolation of the Cdc45/Mcm2-7/GINS (CMG) complex, a candidate for the eukaryotic DNA replication fork helicase, *Proc Natl Acad Sci U S A*, 2006, **103**, 10236–10241.
24. M. Pacek, A. V. Tutter, Y. Kubota, H. Takisawa and J. C. Walter, Localization of MCM2-7, Cdc45, and GINS to the site of DNA unwinding during eukaryotic DNA replication, *Mol. Cell*, 2006, **21**, 581–587.
25. M. L. Hoang, R. P. Leon, L. Pessoa-Brandao, S. Hunt, M. K. Raghuraman, W. L. Fangman, B. J. Brewer and R. A. Sclafani, Structural changes in Mcm5 protein bypass Cdc7-Dbf4 function and reduce replication origin efficiency in *Saccharomyces cerevisiae*, *Mol. Cell. Biol.*, 2007, **27**, 7594–7602.
26. M. Lutzmann, D. Maiorano and M. Mechali, Identification of full genes and proteins of MCM9, a novel, vertebrate-specific member of the MCM2-8 protein family, *Gene*, 2005, **362**, 51–56.
27. E. Fanning, V. Klimovich and A. R. Nager, A dynamic model for replication protein A (RPA) function in DNA processing pathways, *Nucleic Acids Res.*, 2006, **34**, 4126–4137.
28. Y. Zou, Y. Liu, X. Wu and S. M. Shell, Functions of human replication protein A (RPA): from DNA replication to DNA damage and stress responses, *J. Cell. Physiol.*, 2006, **208**, 267–273.
29. T. S. Voss, T. Mini, P. Jenoe and H. P. Beck, *Plasmodium falciparum* possesses a cell cycle-regulated short type replication protein A large subunit encoded by an unusual transcript, *J. Biol. Chem.*, 2002, **277**, 17493–17501.
30. E. Bochkareva, S. Korolev, S. P. Lees-Miller and A. Bochkarev, Structure of the RPA trimerization core and its role in the multistep DNA-binding mechanism of RPA, *EMBO J.*, 2002, **21**, 1855–1863.
31. L. Abu-Elheiga, D. T. Spira and U. Bachrach, *Plasmodium falciparum*: properties of an alpha-like DNA polymerase, the key enzyme in DNA synthesis, *Exp. Parasitol.*, 1990, **71**, 21–26.
32. P. Chavalitshewinkoon, E. de Vries, J. G. Stam, F. F. Franssen, P. C. van der Vliet and J. P. Overdulve, Purification and characterization of DNA polymerases from *Plasmodium falciparum*, *Mol. Biochem. Parasitol.*, 1993, **61**, 243–253.
33. I. Choi and R. B. Mikkelsen, Cell cycle-dependent biosynthesis of *Plasmodium falciparum* DNA polymerase-alpha, *Exp. Parasitol.*, 1991, **73**, 93–100.
34. E. de Vries, J. G. Stam, F. F. Franssen, P. C. van der Vliet and J. P. Overdulve, Purification and characterization of DNA polymerases from Plasmodium berghei, *Mol. Biochem. Parasitol.*, 1991, **45**, 223–232.
35. D. N. Frick and C. C. Richardson, DNA primases, *Annu. Rev. Biochem.*, 2001, **70**, 39–80.
36. S. H. Lao-Sirieix, L. Pellegrini and S. D. Bell, The promiscuous primase, *Trends Genet.*, 2005, **21**, 568–572.

37. J. H. White, B. J. Kilbey, E. de Vries, M. Goman, P. Alano, S. Cheesman, S. McAleese and R. G. Ridley, The gene encoding DNA polymerase alpha from *Plasmodium falciparum*, *Nucleic Acids Res.*, 1993, **21**, 3643–3646.
38. B. E. Weiner, H. Huang, B. M. Dattilo, M. J. Nilges, E. Fanning and W. J. Chazin, An iron-sulfur cluster in the C-terminal domain of the p58 subunit of human DNA primase, *J. Biol. Chem.*, 2007, **282**, 33444–33451.
39. S. Prasartkaew, N. M. Zijlstra, P. Wilairat, J. P. Overdulve and E. de Vries, Molecular cloning of a *Plasmodium falciparum* gene interrupted by 15 introns encoding a functional primase 53 kDa subunit as demonstrated by expression in a baculovirus system, *Nucleic Acids Res.*, 1996, **24**, 3934–3941.
40. A. Johnson and M. O'Donnell, Cellular DNA replicases: components and dynamics at the replication fork, *Annu. Rev. Biochem.*, 2005, **74**, 283–315.
41. B. A. Fox and D. J. Bzik, The primary structure of *Plasmodium falciparum* DNA polymerase delta is similar to drug sensitive delta-like viral DNA polymerases, *Mol. Biochem. Parasitol.*, 1991, **49**, 289–296.
42. R. G. Ridley, J. H. White, S. M. McAleese, M. Goman, P. Alano, E. de Vries and B. J. Kilbey, DNA polymerase delta: gene sequences from *Plasmodium falciparum* indicate that this enzyme is more highly conserved than DNA polymerase alpha, *Nucleic Acids Res.*, 1991, **19**, 6731–6736.
43. P. Horrocks, M. Jackson, S. Cheesman, J. H. White and B. J. Kilbey, Stage specific expression of proliferating cell nuclear antigen and DNA polymerase delta from *Plasmodium falciparum*, *Mol. Biochem. Parasitol.*, 1996, **79**, 177–182.
44. T. S. Krishna, X. P. Kong, S. Gary, P. M. Burgers and J. Kuriyan, Crystal structure of the eukaryotic DNA polymerase processivity factor PCNA, *Cell*, 1994, **79**, 1233–1243.
45. J. M. Gulbis, Z. Kelman, J. Hurwitz, M. O'Donnell and J. Kuriyan, Structure of the C-terminal region of p21(WAF1/CIP1) complexed with human PCNA, *Cell*, 1996, **87**, 297–306.
46. B. J. Kilbey, I. Fraser, S. McAleese, M. Goman and R. G. Ridley, Molecular characterisation and stage-specific expression of proliferating cell nuclear antigen (PCNA) from the malarial parasite, *Plasmodium falciparum*, *Nucleic Acids Res.*, 1993, **21**, 239–243.
47. J. L. Li, A. V. Warren and L. S. Cox, Identification of a second proliferating cell nuclear antigen in the human malarial pathogen *Plasmodium falciparum*, *Int. J. Parasitol.*, 2002, **32**, 1683–1692.
48. S. Patterson, C. Whittle, C. Robert and D. Chakrabarti, Molecular characterization and expression of an alternate proliferating cell nuclear antigen homologue, PfPCNA2, in *Plasmodium falciparum*, *Biochem. Biophys. Res. Commun.*, 2002, **298**, 371–376.
49. S. Sakurai, K. Kitano, H. Yamaguchi, K. Hamada, K. Okada, K. Fukuda, M. Uchida, E. Ohtsuka, H. Morioka and T. Hakoshima, Structural basis for recruitment of human flap endonuclease 1 to PCNA, *EMBO J.*, 2005, **24**, 683–693.
50. E. Warbrick, D. P. Lane, D. M. Glover and L. S. Cox, Homologous regions of Fen1 and p21Cip1 compete for binding to the same site on

PCNA: a potential mechanism to co-ordinate DNA replication and repair, *Oncogene*, 1997, **14**, 2313–2321.
51. L. J. Casta, J. S. Buguliskis, Y. Matsumoto and T. F. Taraschi, Expression and biochemical characterization of the *Plasmodium falciparum* DNA repair enzyme, flap endonuclease-1 (PfFEN-1), *Mol. Biochem. Parasitol.*, 2008, **157**, 1–12.
52. J. S. Buguliskis, L. J. Casta, C. E. Butz, Y. Matsumoto and T. F. Taraschi, Expression and biochemical characterization of *Plasmodium falciparum* DNA ligase I, *Mol. Biochem. Parasitol.*, 2007, **155**, 128–137.
53. L. Lai, H. Yokota, L. W. Hung, R. Kim and S. H. Kim, Crystal structure of archaeal RNase HII: a homologue of human major RNase H, *Structure*, 2000, **8**, 897–904.
54. M. E. Fichera and D. S. Roos, A plastid organelle as a drug target in apicomplexan parasites, *Nature*, 1997, **390**, 407–409.
55. V. Weissig, T. S. Vetro-Widenhouse and T. C. Rowe, Topoisomerase II inhibitors induce cleavage of nuclear and 35-kb plastid DNAs in the malarial parasite *Plasmodium falciparum*, *DNA Cell Biol.*, 1997, **16**, 1483–1492.
56. M. A. Dar, A. Sharma, N. Mondal and S. K. Dhar, Molecular cloning of apicoplast-targeted *Plasmodium falciparum* DNA gyrase genes: unique intrinsic ATPase activity and ATP-independent dimerization of PfGyrB subunit, *Eukaryot. Cell*, 2007, **6**, 398–412.
57. E. V. Raghu Ram, A. Kumar, S. Biswas, A. Kumar, S. Chaubey, M. I. Siddiqi and S. Habib, Nuclear gyrB encodes a functional subunit of the *Plasmodium falciparum* gyrase that is involved in apicoplast DNA replication, *Mol. Biochem. Parasitol.*, 2007, **154**, 30–39.
58. E. L. Dahl and P. J. Rosenthal, Multiple antibiotics exert delayed effects against the *Plasmodium falciparum* apicoplast, *Antimicrob. Agents Chemother.*, 2007, **51**, 3485–3490.
59. C. D. Goodman, V. Su and G. I. McFadden, The effects of anti-bacterials on the malaria parasite *Plasmodium falciparum*, *Mol. Biochem. Parasitol.*, 2007, **152**, 181–191.
60. A. A. Divo, A. C. Sartorelli, C. L. Patton and F. J. Bia, Activity of fluoroquinolone antibiotics against *Plasmodium falciparum in vitro*, *Antimicrob. Agents Chemother.*, 1988, **32**, 1182–1186.
61. F. Seow, S. Sato, C. S. Janssen, M. O. Riehle, A. Mukhopadhyay, R. S. Phillips, R. J. Wilson and M. P. Barrett, The plastidic DNA replication enzyme complex of *Plasmodium falciparum*, *Mol. Biochem. Parasitol.*, 2005, **141**, 145–153.
62. M. J. Gardner, N. Hall, E. Fung, O. White, M. Berriman, R. W. Hyman, J. M. Carlton, A. Pain, K. E. Nelson, S. Bowman, I. T. Paulsen, K. James, J. A. Eisen, K. Rutherford, S. L. Salzberg, A. Craig, S. Kyes, M. S. Chan, V. Nene, S. J. Shallom, B. Suh, J. Peterson, S. Angiuoli, M. Pertea, J. Allen, J. Selengut, D. Haft, M. W. Mather, A. B. Vaidya, D. M. Martin, A. H. Fairlamb, M. J. Fraunholz, D. S. Roos, S. A. Ralph, G. I. McFadden, L. M. Cummings, G. M. Subramanian, C. Mungall, J. C.

Venter, D. J. Carucci, S. L. Hoffman, C. Newbold, R. W. Davis, C. M. Fraser and B. Barrell, Genome sequence of the human malaria parasite *Plasmodium falciparum*, *Nature*, 2002, **419**, 498–511.

63. M. J. Gardner, H. Tettelin, D. J. Carucci, L. M. Cummings, L. Aravind, E. V. Koonin, S. Shallom, T. Mason, K. Yu, C. Fujii, J. Pederson, K. Shen, J. Jing, C. Aston, Z. Lai, D. C. Schwartz, M. Pertea, S. Salzberg, L. Zhou, G. G. Sutton, R. Clayton, O. White, H. O. Smith, C. M. Fraser, M. D. Adams, J. C. Venter and S. L. Hoffman, Chromosome 2 sequence of the human malaria parasite *Plasmodium falciparum*, *Science*, 1998, **282**, 1126–1132.
64. J. M. Carlton, J. H. Adams, J. C. Silva, S. L. Bidwell, H. Lorenzi, E. Caler, J. Crabtree, S. V. Angiuoli, E. F. Merino, P. Amedeo, Q. Cheng, R. M. Coulson, B. S. Crabb, H. A. Del Portillo, K. Essien, T. V. Feldblyum, C. Fernandez-Becerra, P. R. Gilson, A. H. Gueye, X. Guo, S. Kang'a, T. W. Kooij, M. Korsinczky, E. V. Meyer, V. Nenc, I. Paulsen, O. White, S. A. Ralph, Q. Ren, T. J. Sargeant, S. L. Salzberg, C. J. Stoeckert, S. A. Sullivan, M. M. Yamamoto, S. L. Hoffman, J. R. Wortman, M. J. Gardner, M. R. Galinski, J. W. Barnwell and C. M. Fraser-Liggett, Comparative genomics of the neglected human malaria parasite *Plasmodium vivax*, *Nature*, 2008, **455**, 757–763.
65. J. M. Carlton, S. V. Angiuoli, B. B. Suh, T. W. Kooij, M. Pertea, J. C. Silva, M. D. Ermolaeva, J. E. Allen, J. D. Selengut, H. L. Koo, J. D. Peterson, M. Pop, D. S. Kosack, M. F. Shumway, S. L. Bidwell, S. J. Shallom, S. E. van Aken, S. B. Riedmuller, T. V. Feldblyum, J. K. Cho, J. Quackenbush, M. Sedegah, A. Shoaibi, L. M. Cummings, L. Florens, J. R. Yates, J. D. Raine, R. E. Sinden, M. A. Harris, D. A. Cunningham, P. R. Preiser, L. W. Bergman, A. B. Vaidya, L. H. van Lin, C. J. Janse, A. P. Waters, H. O. Smith, O. R. White, S. L. Salzberg, J. C. Venter, C. M. Fraser, S. L. Hoffman, M. J. Gardner and D. J. Carucci, Genome sequence and comparative analysis of the model rodent malaria parasite *Plasmodium yoelii* yoelii, *Nature*, 2002, **419**, 512–519.
66. A. Pain, U. Bohme, A. E. Berry, K. Mungall, R. D. Finn, A. P. Jackson, T. Mourier, J. Mistry, E. M. Pasini, M. A. Aslett, S. Balasubrammaniam, K. Borgwardt, K. Brooks, C. Carret, T. J. Carver, I. Cherevach, T. Chillingworth, T. G. Clark, M. R. Galinski, N. Hall, D. Harper, D. Harris, H. Hauser, A. Ivens, C. S. Janssen, T. Keane, N. Larke, S. Lapp, M. Marti, S. Moule, I. M. Meyer, D. Ormond, N. Peters, M. Sanders, S. Sanders, T. J. Sargeant, M. Simmonds, F. Smith, R. Squares, S. Thurston, A. R. Tivey, D. Walker, B. White, E. Zuiderwijk, C. Churcher, M. A. Quail, A. F. Cowman, C. M. Turner, M. A. Rajandream, C. H. Kocken, A. W. Thomas, C. I. Newbold, B. G. Barrell and M. Berriman, The genome of the simian and human malaria parasite *Plasmodium knowlesi*, *Nature*, 2008, **455**, 799–803.
67. F. Lu, H. Jiang, J. Ding, J. Mu, J. G. Valenzuela, J. M. Ribeiro and X. Z. Su, cDNA sequences reveal considerable gene prediction inaccuracy in the *Plasmodium falciparum* genome, *BMC Genomics*, 2007, **8**, 255.

68. N. Hall, A. Pain, M. Berriman, C. Churcher, B. Harris, D. Harris, K. Mungall, S. Bowman, R. Atkin, S. Baker, A. Barron, K. Brooks, C. O. Buckee, C. Burrows, I. Cherevach, C. Chillingworth, T. Chillingworth, Z. Christodoulou, L. Clark, R. Clark, C. Corton, A. Cronin, R. Davies, P. Davis, P. Dear, F. Dearden, J. Doggett, T. Feltwell, A. Goble, I. Goodhead, R. Gwilliam, N. Hamlin, Z. Hance, D. Harper, H. Hauser, T. Hornsby, S. Holroyd, P. Horrocks, S. Humphray, K. Jagels, K. D. James, D. Johnson, A. Kerhornou, A. Knights, B. Konfortov, S. Kyes, N. Larke, D. Lawson, N. Lennard, A. Line, M. Maddison, J. McLean, P. Mooney, S. Moule, L. Murphy, K. Oliver, D. Ormond, C. Price, M. A. Quail, E. Rabbinowitsch, M. A. Rajandream, S. Rutter, K. M. Rutherford, M. Sanders, M. Simmonds, K. Seeger, S. Sharp, R. Smith, R. Squares, S. Squares, K. Stevens, K. Taylor, A. Tivey, L. Unwin, S. Whitehead, J. Woodward, J. E. Sulston, A. Craig, C. Newbold and B. G. Barrell, Sequence of *Plasmodium falciparum* chromosomes 1, 3–9 and 13, *Nature*, 2002, **419**, 527–531.
69. S. Bowman, D. Lawson, D. Basham, D. Brown, T. Chillingworth, C. M. Churcher, A. Craig, R. M. Davies, K. Devlin, T. Feltwell, S. Gentles, R. Gwilliam, N. Hamlin, D. Harris, S. Holroyd, T. Hornsby, P. Horrocks, K. Jagels, B. Jassal, S. Kyes, J. McLean, S. Moule, K. Mungall, L. Murphy, K. Oliver, M. A. Quail, M. A. Rajandream, S. Rutter, J. Skelton, R. Squares, S. Squares, J. E. Sulston, S. Whitehead, J. R. Woodward, C. Newbold and B. G. Barrell, The complete nucleotide sequence of chromosome 3 of *Plasmodium falciparum*, *Nature*, 1999, **400**, 532–538.
70. Z. Bozdech, J. Zhu, M. P. Joachimiak, F. E. Cohen, B. Pulliam and J. L. DeRisi, Expression profiling of the schizont and trophozoite stages of *Plasmodium falciparum* with a long-oligonucleotide microarray, *Genome Biol.*, 2003, **4**, R9.
71. R. Chenna, H. Sugawara, T. Koike, R. Lopez, T. Gibson, D. Higgins and J. Thompson, Multiple sequence alignment with the Clustal series of programs, *Nucleic Acids Res.*, 2003, **31**, 3497–34500.
72. M. A. Larkin, G. Blackshields, N. P. Brown, R. Chenna, P. A. McGettigan, H. McWilliam, F. Valentin, I. M. Wallace, A. Wilm, R. Lopez, J. D. Thompson, T. J. Gibson and D. G. Higgins, Clustal W and Clustal X version 2.0, *Bioinformatics*, 2007, **23**, 2947–2948.
73. K. Arnold, L. Bordoli, J. Kopp and T. Schwede, The SWISS-MODEL workspace: a web-based environment for protein structure homology modelling, *Bioinformatics*, 2006, **22**, 195–201.
74. F. Kiefer, K. Arnold, M. Kunzli, L. Bordoli and T. Schwede, The SWISS-MODEL repository and associated resources, *Nucleic Acids Res.*, 2009, **37**, D387–D392.

CHAPTER 14

Drug Targets in DNA Replication

ALISON D. WALTERS AND JAMES P.J. CHONG

Department of Biology (Area 5), University of York, York, YO10 5YW, UK

14.1 Introduction

The process of DNA replication is the result of the convergence of multiple signalling pathways that control cell proliferation. Our increasing understanding of the structures and functions of the multiple proteins involved in DNA replication has revealed many possible drug targets for the treatment of hyperproliferative diseases such as cancer. Targeting the key proteins involved in DNA replication in pathogenic viruses and eukaryotic pathogens such as fungi and parasites (for example, see Chapter 13) is a promising strategy for disease treatment. The differences between the DNA replication mechanisms in these organisms and in humans can be exploited to directly target the DNA replication in pathogens without affecting the equivalent processes in human cells. For example, foscarnet is a small molecule that binds to and inhibits retroviral DNA polymerases at concentrations that do not affect the equivalent human nuclear enzymes.[1]

14.1.1 Therapy

Cancer is a disease caused by uncontrolled cell proliferation. The inhibition of DNA replication, which is central to proliferation, has been shown to be an effective anti-cancer treatment. Currently, most drug therapies used to treat

Molecular Themes in DNA Replication
Edited by Lynne S. Cox

Published by the Royal Society of Chemistry, www.rsc.org

cancer patients inhibit dNTP synthesis or nucleotide chain elongation (*e.g.* by binding irreversibly to the DNA template or by causing DNA damage). The only DNA replication proteins currently targeted directly by cancer therapy drugs are DNA topoisomerases I and II, which are the targets of the drugs toptecan, irinotecan (topoisomerase I) and etoposide and amsacrine (topoisomerase II).[2–4] Despite the fact that DNA replication proteins are often overexpressed in cancer cells and are usually downregulated in normal differentiated cells, side effects ranging from nausea to loss of bone marrow caused by the inhibition of DNA replication in normal cells are still a significant problem with current drugs. Therefore, the challenge of identifying possible drug targets in DNA replication for the treatment of cancer lies not only in identifying proteins that could be inhibited effectively by small molecules, but also in finding a way of targeting the inhibitory effects specifically to cancer cells.

14.1.2 Prognosis and Diagnosis

A number of DNA replication proteins also show promise as disease markers for use in diagnosis and prognosis of various human diseases. For example, anti-RPA and anti-topoisomerase antibodies have been identified as markers of some autoimmune diseases.[5,6] Both proliferating cell nuclear antigen (PCNA)[7] and Ki67[8,9] are proteins used routinely in diagnosis of cancer cells by immunohistochemistry. However, both markers have significant disadvantages.

PCNA is highly expressed in cancer cells, which has led to its widespread use as a marker for cancer diagnosis.[7] It is also an auxiliary factor for DNA polymerases δ and ε, and is important in DNA repair as well as DNA replication.[10,11] The role of PCNA in DNA repair means that it often results in false positives when used in cancer diagnosis because it is not specific to proliferating cells. In fact, PCNA has been found in quiescent cells long after DNA replication has ceased.[12] In addition, PCNA expression is high in S phase but low in G1, G2 and M phases,[13] so it is cells that are actively replicating their DNA, not cells with the potential to proliferate, that are detected.

The function of Ki67, which is commonly used as a proliferation marker, remains unclear, although it has been implicated in ribosomal biosynthesis.[14] The lack of an identified function for Ki67, and the fact that it is only an effective marker for a subgroup of cancers,[15] limits its usefulness in cancer diagnosis.

The disadvantages of current cancer cell markers has led to a search for more effective alternatives, many of which are proteins essential for the initiation and elongation phases of DNA replication. Several key DNA replication proteins, such as MCM2-7 have recently been identified as more effective cancer cell markers and are discussed below (for review, see ref. 15).

14.2 Replication Initiation Factors—Markers or Targets?

14.2.1 MCMs

Recent studies have suggested that several subunits of the MCM2-7 complex may be useful diagnostic and prognostic biomarkers in a wide range of human cancers. MCMs are good potential biomarkers because, unlike PCNA and Ki67, they are not only seen in actively proliferating cells but in any cells that are licensed for DNA replication. This means they can be used to identify pre-cancerous cells before they become malignant. MCMs are absent in quiescent, differentiated and senescent cells,[16,17] but are highly abundant, stable and immunogenic in cycling cells.

MCMs 2, 3, 5 and 7 have all been shown to be good biomarkers in a wide range of cancers.[18–20] In particular, antibodies to MCM3 have been shown to be more specific to a wide range of different cancer cells than PCNA, and the levels of MCM3 detected are a good indicator of time to recurrence in some cancers.[21,22] MCM2 and MCM5 have also been shown to be more sensitive markers for cancer diagnosis than PCNA and Ki67,[21] with staining of these proteins correlating directly with the severity of the cancer, thus potentially facilitating prognosis.

The fact that MCMs are overexpressed in a cell cycle specific manner in a wide range of cancers means that they are not only potentially useful markers for cancer diagnosis, but that they may also be potential anti-cancer drug targets. Liang and colleagues[23] showed that MCM2 could be effectively silenced in human cells by RNAi or antisense oligonucleotides (ODNs), and that this led to the inhibition of DNA replication. Importantly, silencing of MCM2 *in vitro* resulted in cancer cells entering an abortive S phase, leading to apoptosis, whereas the same treatment in normal cells resulted in a viable arrest prior to initiation of DNA replication.[23] In this case, the cancer-cell-specific apoptosis is independent of the tumour suppressor protein p53, which is important as p53 is inactivated in a large proportion of cancers.[24] The observation that normal cells do not apoptose under these conditions should limit potential side effects caused by the death of healthy cells, making MCM2 inhibition an attractive prospect for cancer treatment. However, tests have so far only been conducted *in vitro*, and although several RNAi-based therapies targeting other proteins are currently in clinical trials, delivering the antisense ODN/RNAi molecules to cancerous cells *in vivo* remains a significant problem (for review, see ref. 25).

Inhibition of MCMs using small molecule drugs, which can be delivered relatively easily, is an appealing strategy for cancer treatment. The roles of individual MCM subunits in humans remain unclear and no crystal structure of the active MCM2-7 complex is currently available. However, the structure of the N-terminus of the archaeal hexameric MCM complex from *Methanothermobacter thermautotrophicus* (Mth) has been solved[26] (see Chapters 3 and 12). The strict conservation of functional domains between archaea and

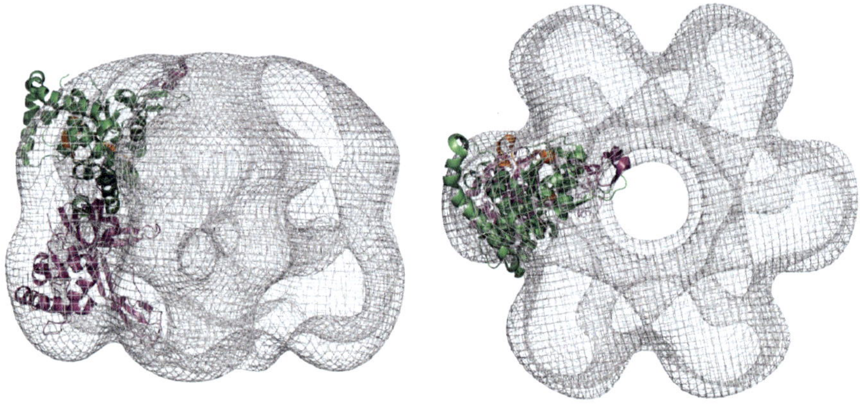

Figure 14.1 Side and top views of *Mth*MCM showing the position of the helix-2 insert relative to a hexameric ring. The N-terminal structure of *Mth*MCM (PDB accession 1LTL[26]) and an edited version of the structure of the ATPase domain of magnesium chelatase (PDB accession 1G8P[91]) were modelled into an electron density map of full-length *Mth*MCM[92] using MacPyMOL (http://delsci.com/macpymol/). The helix-2 insert is shown in magenta. Images taken from ref. 27 with permission.

eukaryotes means that the archaeal structure provides a good source of potential sites where small molecules may be used to inhibit function of the complex in human cancer cells. Site-directed mutagenesis studies of the *Mth*MCM complex suggest that the helix-2 insert is essential for helicase activity[27] and this may represent a possible accessible site for an inhibitory drug (Figure 14.1). Sites essential for DNA binding or ATPase activity associated with the MCM2-7 complex are also potential drug targets. However, without a full structure of a eukaryotic MCM complex, rational design of small molecule inhibitors will remain challenging.

Modification of the MCM complex by MCM-interacting proteins may provide clues as to possible mechanisms of inhibition. Cyclin A/Cdk2 phosphorylates MCM4 and inactivates the MCM4,6,7 complex helicase activity *in vitro*.[28,29] MCMs 2, 3 and 7 are phosphorylated by the ATM/ATR checkpoint kinases in response to DNA damage[30,31] and overexpression of an MCM3 acetylating protein (MCM3AP) inhibits the initiation of DNA replication.[32,33] If the mechanism(s) by which MCM3 phosphorylation or acetylation inhibit MCM2-7 activity could be ascertained, it may be possible to mimic this inhibition using a small molecule drug.

14.2.2 ORC and Cdc6

ORC, unlike MCM, shows little potential as either a marker or drug target in cancer. It is expressed during quiescence,[34] and may have essential roles in cellular processes other than DNA replication such as transcriptional silencing through chromatin remodelling[35] and chromatid cohesion. Thus, ORC inhibition is likely to have dramatic effects on normal as well as malignant cells. However, the ORC-interacting protein Cdc6 shows some potential as both a cancer marker and therapeutic target. Cdc6, like many DNA replication proteins, has been shown to be overexpressed in malignant cells.[36] Anti-Cdc6 antibodies have been shown to detect more abnormal cells in cervical cancer and brain tumours that the classical markers PCNA and Ki67.[37,38] However, Cdc6 is far less abundant than MCM and is also distributed between the cytoplasm and nucleus,[39] so is less easy to use as a cancer cell marker. Mutational analysis of the yeast Cdc6 homologue (Cdc18 in *Saccharomyces pombe*), in combination with a high resolution structure of Cdc6 from the archaeon *Pyrococcus aerophilum* has been used to identify regions of the protein essential for its function.[40] One particularly interesting residue (Asn-320 in *S. pombe*), which leads to loss of function when mutated, lies on the surface of the protein close to the nucleotide-binding pocket (Figure 14.2). This residue lies in a well-conserved patch on the surface of Cdc6 and may represent a possible accessible site for small molecule inhibition.

14.3 Targeting the Pre-initiation Complex

14.3.1 GINS

GINS is a heterotetrameric complex first discovered in yeast.[41] The four subunits of the tetramer, Sld5, Psf1, Psf2 and Psf3, are all closely related and appear to be important in the activation of MCM2-7-mediated DNA unwinding in DNA replication[42] (see Chapters 1 and 3). A large complex containing Cdc45 and MCM2-7 and GINS (CMG) was recently isolated from *Drosophila* and found to have helicase activity *in vitro*[43] (see Chapter 3). It is thought that GINS mediates the interaction between MCM2-7 and Cdc45, which is essential in recruiting DNA polymerase to the origin of replication, and has also been shown to stimulate DNA polymerase α-primase and DNA polymerase ε directly.[44,45]

The apparent central role of GINS in mediating interactions between proteins in the replisome makes it an attractive drug target. The crystal structure of the GINS complex has recently been solved and provides some possible sites for small molecule inhibition[46,47] (Figure 14.3). Most notable in this structure is the huge predominance of alpha helical regions (Figure 14.3A), together with the very even distribution of positively and negatively charged amino acids on the surface (blue and red, respectively, in Figure 14.3B and C). Several disordered regions occur on the surface of the GINS tetramer, which probably represent sites of interaction with other replication proteins. Interfaces between the GINS subunits are highly conserved in eukaryotes, but these sites are not attractive drug targets because contacts between subunits are extensive and small molecule inhibition is unlikely to perturb them. However, several highly

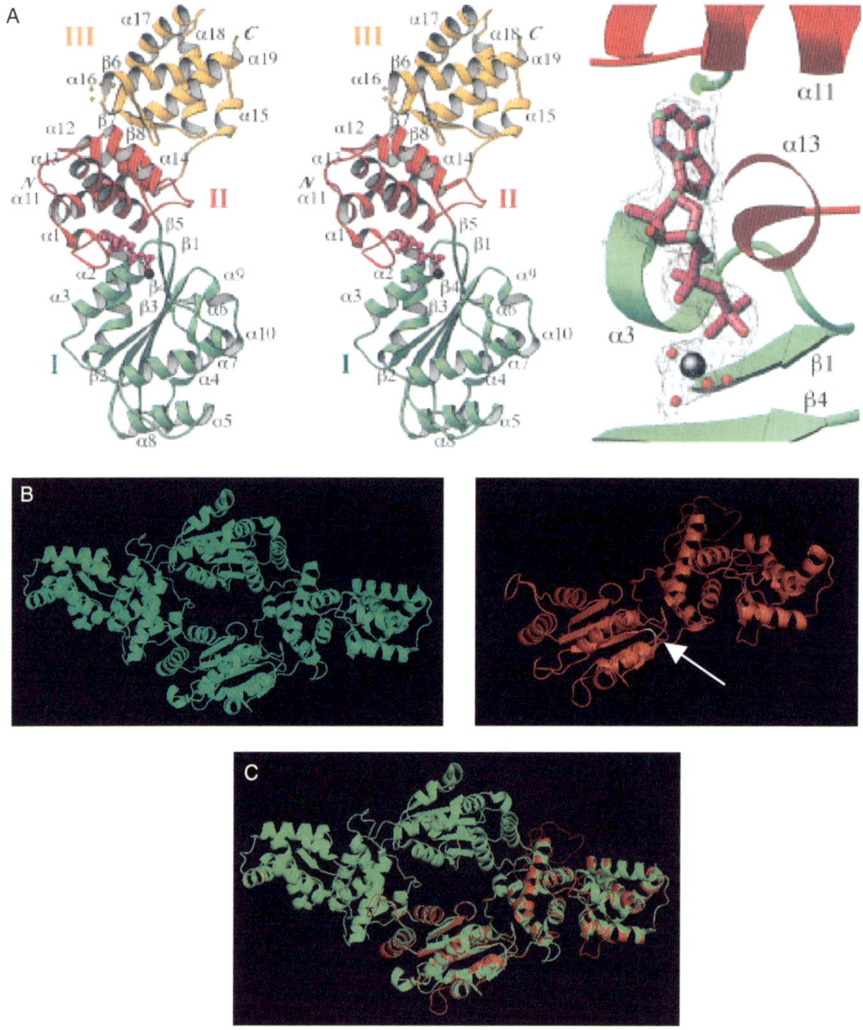

Figure 14.2 **Modelling the structure of Cdc6/Cdc18.** (**A**) Cartoon structure of Cdc6 from *Pyrobaculum aerophilum* cdc6 (1FNN).[40] Left panel shows all three domains (I, II, III) while the right panel shows ATP (magenta) and Mg^{2+} (blue sphere) with water (small red spheres). Reprinted from: J. Liu, C. L. Smith, D. DeRyckere, K. DeAngelis, G. S. Martin and J. M. Berger, Structure and function of Cdc6/Cdc18: implications for origin recognition and checkpoint control, *Molecular Cell*, **6**, 637–648, copyright (2000), with permission from Elsevier.[40] (**B**) First pass molecular modelling of *S. pombe* Cdc18 (orthologne of Cdc6) was carried out using SWISS-MODEL against 1FNN. The white arrow highlights Asn-320 in the putative nucleotide binding loop. (**C**) The known structure of *Pyrobaculum* Cdc6 and the modelled structure of *S. pombe* Cdc18 were merged using MacPyMOL (http://delsci.com/macpymol/).

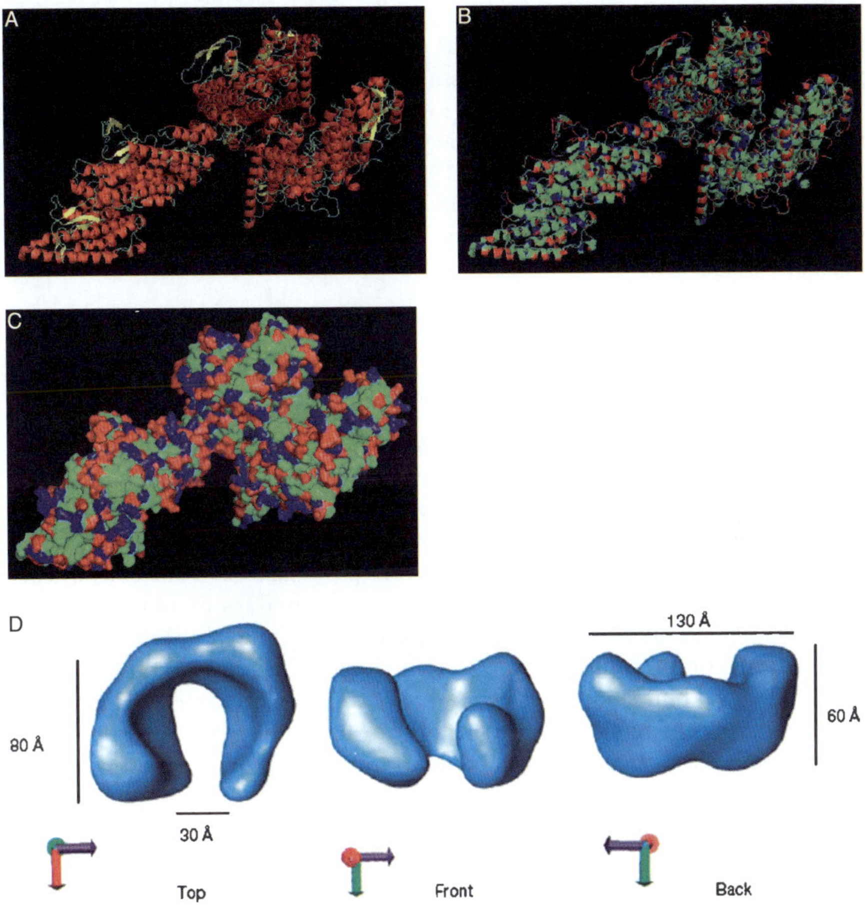

Figure 14.3 **GINS structure.** (**A**) Cartoon of GINS structure (using coordinates from PDB accession number 2EHO[47]) showing the high percentage of alpha helical secondary structure (alpha helices are shown in red, beta sheets in yellow and loops in green). (**B**) Cartoon and (**C**) surface of the GINS complex showing the distribution of acidic amino acids (in red: Asp, Asn, Glu and Gln), and positively charged amino acids (in blue: Lys, Arg and His). (**D**) Horseshoe structure of the GINS complex. Reprinted by permission from Macmillan Publishers Ltd: J. Boskovic, J. Coloma, T. Aparicio, M. Zhou, C. V. Robinson, J. Mendez and G. Montoya, Molecular architecture of the human GINS complex, *EMBO Report*, **8**, 678–684, copyright (2007).[48]

conserved patches have been identified on the surface of the Sld5 and Psf2 subunits of the human complex, which affected cell viability when mutated in yeast.[47] GINS mutants that showed decreased viability were still able to form tetramers, and therefore it is likely that the loss in viability is because mutations interfered with key interactions between GINS and Cdc45, MCM2-7 and possibly other replication/checkpoint factors. Drugs targeted to these interaction sites may be effective in inhibiting DNA replication, but further dissection of which proteins interact at these sites and the exact nature of the protein–protein interactions are necessary before inhibitor design can be considered. In addition to its important interaction with other DNA replication proteins, it has been demonstrated that the GINS complex forms a horseshoe shape that can bind directly to DNA[48] (Figure 14.3D). It is possible that the DNA interaction is essential for the function of the Cdc45-MCM-GINS (CMG) complex and that the DNA interaction site may offer another small molecule target.

14.3.2 Cdc45

Cdc45 is essential in both the initiation and elongation stages of DNA replication[49,50] (see also Chapter 1), and interacts with a large number of other DNA replication proteins. Cdc45 has been shown to be essential for recruiting DNA polymerases to the site of DNA replication and may play a crucial role in mediating interactions between several proteins within the replisome, including MCM2-7, GINS and DNA polymerases during S phase.[51–53] The convergence of signals from the Cdk and DDK pathways is required for chromatin binding of Cdc45, which may be a key regulatory factor in the switch from pre-RC to pre-initiation complex (pre-IC).[52,54] Recent studies have shown that Cdc45 has a relatively low abundance in replicating cells and may be a rate-limiting factor for the initiation of replication.[12] There is no structural information currently available for Cdc45, but its role as mediator of several key interactions and its central regulatory role in origin firing make it a good potential target for drugs inhibiting DNA replication.

14.3.3 Geminin and Cdt1

Geminin is a negative regulator of DNA replication (see Chapter 1) that shows promise as a useful biomarker for cell cycle progression and as a basis for the design of small molecule inhibitors of replication. Geminin inhibits origin licensing in S, G2 and M phases through tight binding to Cdt1,[55] preventing the interaction between Cdt1 and MCM which is required for pre-RC formation. The inhibition of licensing ensures that origins fire only once per cell cycle, preventing re-replication of DNA which can lead to genomic instability.

Geminin is expressed in the cell cycle in S, G2 and M phases but not in G1 or G0. Therefore, geminin staining shows the number of cells that have progressed past G1 but not exited mitosis, giving an indication of cell cycle 'rate', which can be useful in making a prognosis from cancer cell samples.[56] High levels of geminin staining in malignant cells correlate with a poor survival rate.

Expression of a non-degradable form of geminin causes apoptosis in human cancer cell lines, but not in normal primary fibroblasts.[57] This cancer-cell specific 'dominant negative' action indicates that a geminin mimic would make an effective anti-cancer drug with limited side effects. The crystal structures of

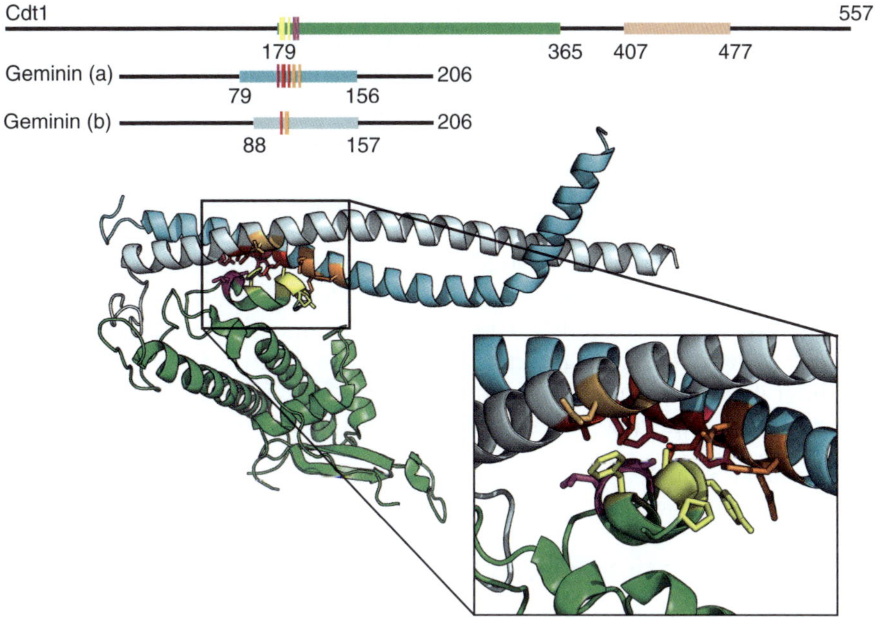

Figure 14.4 **Structure showing interactions between Cdt1 and a geminin dimer**. The structure of Cdt1-geminin was solved using truncated proteins. Regions of the proteins represented in the structure are shown in the upper panel diagrams in green and blue. The region of Cdt1 reported to interact with MCMs is shown in light brown, but no structure is available for this part of the protein. Key interacting residues are highlighted in different colours, and side chains for these residues are shown in the expanded region. Association between these two truncated proteins can be substantially reduced by mutating a handful of residues, which might make a good small molecule target but for the fact that a more C-terminal region of Cdt1 has also been shown to interact with geminin in experiments using full-length proteins. The geminin dimer shows an asymmetric interaction with Cdt1. Adapted with permission from Macmillan Publishers Ltd: C. Lee, B. Hong, J. M. Choi, Y. Kim, S. Watanabe, Y. Ishimi, T. Enomoto, S. Tada, Y. Kim and Y. Cho, Structural basis for inhibition of the replication licensing factor Cdt1 by geminin, *Nature*, 430, 913–917, copyright (2004).[58]

human geminin and of the mouse geminin-Cdt1 complex have been solved and may be used as a basis for designing drugs[58,59] (Figure 14.4). Previous observations that MCM and geminin bind to different regions of Cdt1[60] raised the question of how geminin inhibited the interaction between MCM and Cdt1. The crystal structure reveals that geminin probably inhibits the Cdt1-MCM interaction through its extended coiled-coil structure, which allows the C-terminal region of geminin to block MCM binding to Cdt1 by steric hindrance.

There are at least two regions of the geminin homodimer that interact independently with Cdt1. The main site of interaction appears to be through negatively charged residues on the surface of the dimerised coiled-coil domain of geminin. In addition, residues 70–92 of geminin interact with the N-terminus of Cdt1, which is essential for replication inhibition.[59]

Designing a drug mimicking geminin may prove difficult because of the multiple interactions sites between Cdt1 and geminin, as well as the relatively bulky region of geminin required to inhibit the Cdt1–MCM interaction. However, there are alternative approaches to inhibiting Cdt1 such as the delivery of a non-degradable form of geminin itself directly into cancer cells. Okuyama and colleagues have recently shown that attaching a small molecule mimic of the alpha-helical peptide protein transduction domain to geminin allows it to be taken up by cells *in vitro*.[61] When geminin was delivered to human cancer cells by this method, it had a significant anti-proliferation effect.[61] Alternatively, if geminin mimics are ineffective, a small molecule inhibitor could be designed to bind directly to the C-terminal region of Cdt1, which has been shown to be essential for its interaction with the MCM complex in *Xenopus*.[62]

14.4 Replication Elongation Factors

14.4.1 PCNA

PCNA is a replication protein routinely used as a biomarker in cancer diagnosis,[7] but has more recently been explored as a possible target for anti-cancer drugs. PCNA is a DNA sliding clamp that is essential for replication (see Chapters 3 and 7). PCNA tethers the relevant DNA polymerase to the replication template and also binds to a number of replication factors involved in sequential processing of the nascent DNA (for reviews, see refs. 63–65; see also Chapters 3, 6 and 7). Binding of synthetic peptides derived from the C-terminus of p21 to PCNA inhibits DNA replication *in vitro*,[66] while antisense oligonucleotide ablation of PCNA kills cancer cells in culture.[67]

The structure of native human PCNA has been solved and shows it forms a trimer, with a central hole through which double-stranded DNA passes[68] (see Figure 3.8). The structure of the PCNA/p21 complex has also been solved and reveals that there is a 'linker strand' between the N- and C-termini of each PCNA monomer that is important in p21 recognition.[69] This structure and information regarding the interaction between PCNA and p21 have been used to design a peptide that binds to PCNA close to the 'linker strand' with a similar affinity to p21. The synthetic peptide inhibits DNA replication *in vitro* and the site at which it binds is adjacent to the cyclin binding site.[69] However, since PCNA has an essential role in the stimulation of DNA polymerase δ in DNA repair, as well as in DNA replication, drugs targeted to PCNA are likely to affect healthy cells as well as malignant cells. The possible disruption caused to DNA repair in normal cells by targeting PCNA would at first sight suggest that it is a less attractive target than other replication proteins, such as the MCM complex. DNA repair *in vitro* was initially shown to be insensitive to p21,[70,71] though other studies have shown that the C-terminus of p21 can inhibit nucleotide excision repair (NER) both *in vitro* and *in vivo*.[72] Thus, the utility of p21-based anti-PCNA agents in the cancer clinic awaits definition of their activity and therapeutic index in animal models and clinical trials.

14.4.2 Ciz1

Ciz1 was first identified as a unique p21-interacting protein found in higher eukaryotes.[73] Ciz1 contains three zinc-finger motifs, polyglutamine tracts and a matrin-3 homologous domain 3 (MH3 domain), indicating that it may interact with DNA and nuclear matrix proteins.[73] Ciz1 binds DNA at a specific consensus sequence[74] and promotes DNA replication in cell-free extracts and intact human cells.[75] Stimulation of replication occurs even in the absence of the p21 inhibitor of DNA replication, indicating that Ciz1 has a direct role in promoting DNA replication and does not simply reduce the inhibition of replication by binding and sequestering p21. In fact, a reduction in Ciz1 protein levels inhibits progression through S phase.[75] Ciz1 has been shown to localise to 'replication factories' during S phase.[76] Alternatively-spliced forms of Ciz1 are found in some cancers and it has been shown that these alternative forms are able to stimulate DNA replication but do not localise to replication foci.[77] It may be possible to specifically target these alternatively-spliced forms of Ciz1 using small molecules. However, a great deal more needs to be known about the role of Ciz1 in DNA replication before it could be considered a good target for anti-proliferative drugs.

14.4.3 DNA Polymerases

DNA polymerases are one of the few classes of replication proteins to which anti-viral drugs are already targeted (reviewed in ref. 78). Foscarnet is a commonly used anti-viral drug, which binds to DNA polymerase at the pyrophosphate-binding site, preventing the cleavage of pyrophosphates from nucleoside triphosphates in phosphodiester bond formation (see Figure 1.1). Foscarnet can be used as an effective anti-viral drug in humans because it affects viral polymerase at concentrations that do not affect human nuclear DNA polymerases,[1] though it may adversely affect mitochondrial DNA polymerase γ (see Chapter 11). Other commonly used anti-viral drugs are nucleoside analogues, including acyclovir, which is used to treat Herpes simplex, among other viruses.[79] Acyclovir causes DNA chain termination and has a high affinity for viral DNA polymerase but binds human DNA polymerase very weakly.[80] However, there are still significant side effects from acyclovir and its derivatives, including nausea and vomiting due to inhibition of host DNA polymerases (reviewed in ref. 78). Nucleoside analogues such as azidothymidine (AZT) that bind and inhibit reverse transcriptase are commonly used to treat HIV. Whilst nucleoside analogues bind at the active site of reverse transcriptase, another class of anti-HIV drugs, the non-nucleoside RT inhibitors (*e.g.* efavirenz) bind to an allosteric site 10Å from the active site and inhibit enzymatic activity by causing distortions in several sub-domains of the protein.[81]

DNA polymerase α-primase, ε and δ are all involved in DNA replication, although DNA polymerase α-primase is the only one that can start replication *de novo* (for review, see ref. 82). Several compounds have been found that directly bind and inhibit human DNA polymerases. Aphidicolin is a tetracyclic diterpene that inhibits viral and eukaryotic DNA polymerases α, δ and ε. It binds to polymerase-DNA complexes and inhibits dCTP incorporation, leading to chain termination.[83] However, clinical trials using aphidocolin as an anti-cancer therapy have shown it has a very low toxicity to malignant cells and is ineffective in cancer cell killing. Another group of compounds more recently discovered as DNA polymerase inhibitors are the sulfolipids. Sulfoquinovosylmonoacylglycerol (SQMG) and sulfoquinovosyldiacylglycerol (SQDG) have both been shown to be potent inhibitors of DNA polymerases α, β (a repair enzyme) and ε *in vitro*.[84,85] Both of these compounds contain a 6-sulfate moiety, a monosaccharide, a glyceride and either one or two long chain fatty acids (Figure 14.5A). SQMG, but not SQDG, inhibited growth of stomach cancer cell lines *in vitro*.[84] It was found that increased fatty acid chain length led to increased polymerase inhibition (Figure 14.5B). NMR analysis and cross-linking studies indicate that the inhibition by SQMG/SQDG may depend on the binding of the fatty acid ester to the polymerase.[86] Elenic acid (EA) is another compound containing a long chain fatty acid that has been identified as a potent inhibitor of DNA polymerases α and β as well as the DNA topoismerases.[87] The long fatty acid side chain of EA binds to the DNA binding site of DNA polymerases α and β with stronger affinity than DNA, preventing primer-template binding. Both the sulfolipids and elenic acid show some promise as the basis for designing anti-cancer drugs. However, the compounds in their current form would inhibit DNA repair (through DNA polymerase β) as well as replication, which could lead to significant side effects, including genomic instability, which in turn may lead to the formation of further tumours. Specifically targeting inhibitory molecules to DNA polymerase α may prove difficult because there is a lack of structural information regarding the eukaryotic replicative polymerases (though see Chapter 4), and therefore identifying small differences between α and β that can be exploited in drug design is not yet possible.

14.4.4 RNase HI

RNase HI is a protein found in eukaryotes that recognises RNA-DNA duplexes and cleaves the RNA strand. It is required to remove the RNA primer of Okazaki fragments from the lagging strand during elongation[88] (see Chapter 5) and is essential in DNA replication. RNase HI is of medical significance because reverse transcriptase, an enzyme found in retroviruses such as HIV, contains an RNase HI domain. Since the RNase HI domain is essential for HIV to convert its RNA genome to double-stranded DNA, it is a potential

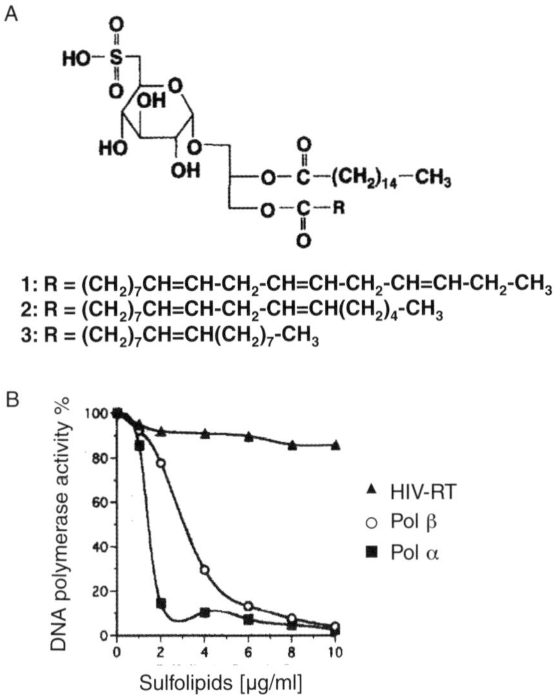

Figure 14.5 **Sulfolipid inhibitors of DNA replication associated with DNA polymerase α.** (**A**) The structural formulae of the three sulfolipids tested by Mizushina *et al.* 1998[93] for inhibition of DNA polymerase α. (**B**) Degree of inhibition of polymerase activity by the sulfolipids. Adapted from: Y. Mizushina, I. Watanabe, K. Ohta, M. Takemura, H.Sahara, N. Takahashi, S. Gasa, F. Sugawara, A. Matsukage, S. Yoshida and K. Sakaguchi, Studies on inhibitors of mammalian DNA polymerase α and β: sulfolipids from a pteridophyte, *Athyrium niponicum*, *Biochemical Pharmacology*, **55**, 537–541, copyright (1998), with permission from Elsevier.[93]

target for anti-HIV drugs. However, in order to avoid unwanted side-effects, the anti-HIV drug must be specifically targeted to the viral RNase HI domain and not affect the rather similar RNase HI protein in human cells. The recently solved structure of human RNase HI complexed with a DNA-RNA hybrid indicates there are some key differences between the catalytic domains of the RNase HI from HIV and RNase HI from humans,[89] which could be exploited in designing virus-specific small molecule inhibitors. The major differences between the human RNase HI and HIV RNase HI lie in the substrate binding pockets of the enzymes. In particular, the phosphate-binding pocket in HIV RNase HI is extended and contains a hydrophobic cleft not found in the human equivalent, to which a small molecule inhibitor could possibly be designed.[89]

14.4.5 Topoisomerases

There are two major classes of topoisomerases in humans: one that can cleave a single strand of DNA (type I), and another that cleaves double-stranded DNA (type II). Both classes of topoisomerases are required to relax DNA by removing supercoiling to allow DNA replication to occur. Topoisomerases I and II are currently targets of several anti-cancer drugs that have been in use for many years. Some of the drugs targeted to topoisomerase, such as etoposide, 'trap' topoisomerase-DNA intermediates by inhibiting the religation of DNA by topoisomerase after it has cleaved the DNA strand(s) (for review, see ref. 2). This 'trapping' leads to the accumulation of single- or double-stranded breaks in DNA (single-stranded breaks can lead to formation of DSBs following passage of a replication fork), which eventually causes cell death. The mechanism by which the religation is inhibited is not fully understood. Other anti-cancer drugs that target topoisomerase such as the bisdioxopiperazines, which inhibit topoisomerase II, do so by preventing the initial catalytic activity of the enzyme.[90] Despite the fact that topoisomerase inhibitors have been used to treat many types of cancer effectively, resistance occurs in some malignancies due to mutations in topoisomerase in cancer cells. Drugs that cause accumulation of breaks in DNA can sometimes lead to secondary tumours because they cause genomic instability.

14.5 Conclusions

A number of drugs exist that target the enzymatic components of the DNA replication fork and these have been exploited particularly in the treatment of viral infections. However, in most of these cases, these drugs are not suitable for treatment of human hyperproliferative diseases such as cancer as they do not specifically target cancerous cells. Recent attention has been directed to proteins involved in DNA replication initiation. The initiation process is subject to multiple layers of strict cell cycle control, and some of these proteins may also trigger or react to checkpoint controls when replication forks encounter DNA damage. Therefore, replication initiation proteins may be good targets for small molecule therapies—particularly if used in combination with more traditional DNA-damage inducing drugs—which might induce cancerous cells to continue DNA replication under conditions that result in genomic toxicity and inviable cells; the ideal outcome would be apoptotic death specifically of the cancer cells with no attendant damage to normal cells. Such a goal, while not yet achieved, is potentially possible given recent advances in understanding the process of DNA replication.

References

1. P. S. Lietman, Clinical pharmacology: foscarnet, *Am. J. Med.*, 1992, **92**, 8S–11S.
2. K. R. Hande, Etoposide: four decades of development of a topoisomerase II inhibitor, *Eur. J. Cancer*, 1998, **34**, 1514–1521.

3. K. R. Hande, Clinical applications of anticancer drugs targeted to topoisomerase II, *Biochim. Biophys. Acta*, 1998, **1400**, 173–184.
4. H. Ulukan and P. W. Swaan, Camptothecins: a review of their chemotherapeutic potential, *Drugs*, 2002, **62**, 2039–2057.
5. R. Garcia-Lozano, F. Gonzalez-Escribano, J. Sanchez-Roman, I. Wichmann and A. Nunez-Roldan, Presence of antibodies to different subunits of replication protein A in autoimmune sera, *Proc. Natl. Acad. Sci. U.S.A.*, 1995, **92**, 5116–5120.
6. H. L. Weiner, Z. J. Zhang, S. J. Khoury, A. Miller, A. al-Sabbagh, S. A. Brod, O. Lider, P. Higgins, R. Sobel and R. B. Nussenblatt, *et.al.*, Antigen-driven peripheral immune tolerance. Suppression of organ-specific autoimmune diseases by oral administration of autoantigens, *Ann. N. Y. Acad. Sci.*, 1991, **636**, 227–232.
7. O. W. Kamel, D. P. LeBrun, R. E. Davis, G. J. Berry and R. A. Warnke, Growth fraction estimation of malignant lymphomas in formalin-fixed paraffin-embedded tissue using anti-PCNA/Cyclin 19A2. Correlation with Ki-67 labeling, *Am. J. Pathol.*, 1991, **138**, 1471–1477.
8. J. Gerdes, F. Dallenbach, K. Lennert, H. Lemke and H. Stein, Growth fractions in malignant non-Hodgkin's lymphomas (NHL) as determined *in situ* with the monoclonal antibody Ki-67, *Hematol. Oncol.*, 1984, **2**, 365–371.
9. R. D. Mashal, S. Lester, C. Corless, J. P. Richie, R. Chandra, K. J. Propert and A. Dutta, Expression of cell cycle-regulated proteins in prostate cancer, *Cancer Res.*, 1996, **56**, 4159–4163.
10. G. Prelich, C. K. Tan, M. Kostura, M. B. Mathews, A. G. So, K. M. Downey and B. Stillman, Functional identity of proliferating cell nuclear antigen and a DNA polymerase-delta auxiliary protein, *Nature*, 1987, **326**, 517–520.
11. L. Toschi and R. Bravo, Changes in cyclin/proliferating cell nuclear antigen distribution during DNA repair synthesis, *J. Cell Biol.*, 1988, **107**, 1623–1628.
12. S. Pollok, C. Bauerschmidt, J. Sanger, H. P. Nasheuer and F. Grosse, Human Cdc45 is a proliferation-associated antigen, *FEBS J.*, 2007, **274**, 3669–3684.
13. J. E. Celis and A. Celis, Cell cycle-dependent variations in the distribution of the nuclear protein cyclin proliferating cell nuclear antigen in cultured cells: subdivision of S phase, *Proc. Natl. Acad. Sci. U.S.A.*, 1985, **82**, 3262–3266.
14. D. E. MacCallum and P. A. Hall, The location of pKi67 in the outer dense fibrillary compartment of the nucleolus points to a role in ribosome biogenesis during the cell division cycle, *J. Pathol.*, 2000, **190**, 537–544.
15. J. W. Semple and B. P. Duncker, ORC-associated replication factors as biomarkers for cancer, *Biotechnol Adv*, 2004, **22**, 621–631.
16. C. Musahl, H. P. Holthoff, R. Lesch and R. Knippers, Stability of the replicative Mcm3 protein in proliferating and differentiating human cells, *Exp. Cell Res.*, 1998, **241**, 260–264.

17. K. Stoeber, T. D. Tlsty, L. Happerfield, G. A. Thomas, S. Romanov, L. Bobrow, E. D. Williams and G. H. Williams, DNA replication licensing and human cell proliferation, *J. Cell Sci.*, 2001, **114**, 2027–2041.
18. A. Freeman, L. S. Morris, A. D. Mills, K. Stoeber, R. A. Laskey, G. H. Williams and N. Coleman, Minichromosome maintenance proteins as biological markers of dysplasia and malignancy, *Clin. Cancer Res.*, 1999, **5**, 2121–2132.
19. G. Mukherjee, B. Muralidhar, U. D. Bafna, R. A. Laskey and N. Coleman, MCM immunocytochemistry as a first line cervical screening test in developing countries: a prospective cohort study in a regional cancer centre in India, *Br. J. Cancer*, 2007, **96**, 1107–1111.
20. C. Scarpini, V. White, B. Muralidhar, A. Patterson, N. Hickey, N. Singh, J. Mullerat, M. Winslet, R. J. Davies, M. L. Phillips, P. Stacey, R. A. Laskey, R. Miller, M. Nathan and N. Coleman, Improved screening for anal neoplasia by immunocytochemical detection of minichromosome maintenance proteins, *Cancer Epidemiol. Biomarkers Prev.*, 2008, **17**, 2855–2864.
21. S. A. Ha, S. M. Shin, H. Namkoong, H. Lee, G. W. Cho, S. Y. Hur, T. E. Kim and J. W. Kim, Cancer-associated expression of minichromosome maintenance 3 gene in several human cancers and its involvement in tumorigenesis, *Clin. Cancer Res.*, 2004, **10**, 8386–8395.
22. A. Soling, M. Sackewitz, M. Volkmar, D. Schaarschmidt, R. Jacob, H. J. Holzhausen and N. G. Rainov, Minichromosome maintenance protein 3 elicits a cancer-restricted immune response in patients with brain malignancies and is a strong independent predictor of survival in patients with anaplastic astrocytoma, *Clin. Cancer Res.*, 2005, **11**, 249–258.
23. D. Feng, Z. Tu, W. Wu and C. Liang, Inhibiting the expression of DNA replication-initiation proteins induces apoptosis in human cancer cells, *Cancer Res.*, 2003, **63**, 7356–7364.
24. T. Soussi, p53 alterations in human cancer: more questions than answers, *Oncogene*, 2007, **26**, 2145–2156.
25. A. R. de Fougerolles, Delivery vehicles for small interfering RNA *in vivo*, *Hum. Gene Ther.*, 2008, **19**, 125–132.
26. R. J. Fletcher, B. E. Bishop, R. P. Leon, R. A. Sclafani, C. M. Ogata and X. S. Chen, The structure and function of MCM from archaeal *M. thermoautotrophicum*, *Nat. Struct. Biol.*, 2003, **10**, 160–167.
27. E. R. Jenkinson and J. P. Chong, Minichromosome maintenance helicase activity is controlled by N- and C-terminal motifs and requires the ATPase domain helix-2 insert, *Proc. Natl. Acad. Sci. U.S.A.*, 2006, **103**, 7613–7618.
28. Y. Ishimi, Y. Komamura-Kohno, H. J. Kwon, K. Yamada and M. Nakanishi, Identification of MCM4 as a target of the DNA replication block checkpoint system, *J. Biol. Chem.*, 2003, **278**, 24644–24650.
29. Y. Ishimi, Y. Komamura-Kohno, Z. You, A. Omori and M. Kitagawa, Inhibition of Mcm4,6,7 helicase activity by phosphorylation with cyclin A/Cdk2, *J. Biol. Chem.*, 2000, **275**, 16235–16241.

30. D. Cortez, G. Glick and S. J. Elledge, Minichromosome maintenance proteins are direct targets of the ATM and ATR checkpoint kinases, *Proc. Natl. Acad. Sci. U.S.A.*, 2004, **101**, 10078–10083.
31. Y. Shi, G. E. Dodson, P. S. Mukhopadhyay, N. P. Shanware, A. T. Trinh and R. S. Tibbetts, Identification of carboxyl-terminal MCM3 phosphorylation sites using polyreactive phosphospecific antibodies, *J. Biol. Chem.*, 2007, **282**, 9236–9243.
32. Y. Takei, M. Assenberg, G. Tsujimoto and R. Laskey, The MCM3 acetylase MCM3AP inhibits initiation, but not elongation, of DNA replication *via* interaction with MCM3, *J. Biol. Chem.*, 2002, **277**, 43121–43125.
33. Y. Takei, M. Swietlik, A. Tanoue, G. Tsujimoto, T. Kouzarides and R. Laskey, MCM3AP, a novel acetyltransferase that acetylates replication protein MCM3, *EMBO Rep.*, 2001, **2**, 119–123.
34. K. C. Thome, S. K. Dhar, D. G. Quintana, L. Delmolino, A. Shahsafaei and A. Dutta, Subsets of human origin recognition complex (ORC) subunits are expressed in non-proliferating cells and associate with non-ORC proteins, *J. Biol. Chem.*, 2000, **275**, 35233–35241.
35. S. P. Bell, J. Mitchell, J. Leber, R. Kobayashi and B. Stillman, The multidomain structure of Orc1p reveals similarity to regulators of DNA replication and transcriptional silencing, *Cell*, 1995, **83**, 563–568.
36. M. Liontos, M. Koutsami, M. Sideridou, K. Evangelou, D. Kletsas, B. Levy, A. Kotsinas, O. Nahum, V. Zoumpourlis, M. Kouloukoussa, Z. Lygerou, S. Taraviras, C. Kittas, J. Bartkova, A. G. Papavassiliou, J. Bartek, T. D. Halazonetis and V. G. Gorgoulis, Deregulated overexpression of hCdt1 and hCdc6 promotes malignant behavior, *Cancer Res.*, 2007, **67**, 10899–10909.
37. S. Ohta, M. Koide, T. Tokuyama, N. Yokota, S. Nishizawa and H. Namba, Cdc6 expression as a marker of proliferative activity in brain tumors, *Oncol. Rep.*, 2001, **8**, 1063–1066.
38. G. H. Williams, P. Romanowski, L. Morris, M. Madine, A. D. Mills, K. Stoeber, J. Marr, R. A. Laskey and N. Coleman, Improved cervical smear assessment using antibodies against proteins that regulate DNA replication, *Proc. Natl. Acad. Sci. U.S.A.*, 1998, **95**, 14932–14937.
39. B. O. Petersen, J. Lukas, C. S. Sorensen, J. Bartek and K. Helin, Phosphorylation of mammalian CDC6 by cyclin A/CDK2 regulates its subcellular localization, *EMBO J.*, 1999, **18**, 396–410.
40. J. Liu, C. L. Smith, D. DeRyckere, K. DeAngelis, G. S. Martin and J. M. Berger, Structure and function of Cdc6/Cdc18: implications for origin recognition and checkpoint control, *Mol. Cell*, 2000, **6**, 637–648.
41. Y. Takayama, Y. Kamimura, M. Okawa, S. Muramatsu, A. Sugino and H. Araki, GINS, a novel multiprotein complex required for chromosomal DNA replication in budding yeast, *Genes Dev.*, 2003, **17**, 1153–1165.
42. K. Labib and A. Gambus, A key role for the GINS complex at DNA replication forks, *Trends Cell Biol.*, 2007, **17**, 271–278.
43. S. E. Moyer, P. W. Lewis and M. R. Botchan, Isolation of the Cdc45/Mcm2-7/GINS (CMG) complex a candidate for the eukaryotic DNA

replication fork helicase, *Proc. Natl. Acad. Sci. U.S.A.*, 2006, **103**, 10236–10241.

44. M. De Falco, E. Ferrari, M. De Felice, M. Rossi, U. Hubscher and F. M. Pisani, The human GINS complex binds to and specifically stimulates human DNA polymerase alpha-primase, *EMBO Rep.*, 2007, **8**, 99–103.
45. T. Seki, M. Akita, Y. Kamimura, S. Muramatsu, H. Araki and A. Sugino, GINS is a DNA polymerase epsilon accessory factor during chromosomal DNA replication in budding yeast, *J. Biol. Chem.*, 2006, **281**, 21422–21432.
46. Y. P. Chang, G. Wang, V. Bermudez, J. Hurwitz and X. S. Chen, Crystal structure of the GINS complex and functional insights into its role in DNA replication, *Proc. Natl. Acad. Sci. U.S.A.*, 2007, **104**, 12685–12690.
47. J. M. Choi, H. S. Lim, J. J. Kim, O. K. Song and Y. Cho, Crystal structure of the human GINS complex, *Genes Dev.*, 2007, **21**, 1316–1321.
48. J. Boskovic, J. Coloma, T. Aparicio, M. Zhou, C. V. Robinson, J. Mendez and G. Montoya, Molecular architecture of the human GINS complex, *EMBO Rep.*, 2007, **8**, 678–684.
49. O. M. Aparicio, A. M. Stout and S. P. Bell, Differential assembly of Cdc45p and DNA polymerases at early and late origins of DNA replication, *Proc. Natl. Acad. Sci. U.S.A.*, 1999, **96**, 9130–9135.
50. L. Zou, J. Mitchell and B. Stillman, CDC45, a novel yeast gene that functions with the origin recognition complex and Mcm proteins in initiation of DNA replication, *Mol. Cell. Biol.*, 1997, **17**, 553–563.
51. P. Saha, K. C. Thome, R. Yamaguchi, Z. Hou, S. Weremowicz and A. Dutta, The human homolog of Saccharomyces cerevisiae CDC45, *J. Biol. Chem.*, 1998, **273**, 18205–18209.
52. L. Zou and B. Stillman, Assembly of a complex containing Cdc45p, replication protein A, and Mcm2p at replication origins controlled by S-phase cyclin-dependent kinases and Cdc7p-Dbf4p kinase, *Mol. Cell. Biol.*, 2000, **20**, 3086–3096.
53. C. Bauerschmidt, S. Pollok, E. Kremmer, H. P. Nasheuer and F. Grosse, Interactions of human Cdc45 with the Mcm2-7 complex, the GINS complex, and DNA polymerases delta and epsilon during S phase, *Genes Cells*, 2007, **12**, 745–758.
54. S. Mimura and H. Takisawa, Xenopus Cdc45-dependent loading of DNA polymerase alpha onto chromatin under the control of S-phase Cdk, *EMBO J.*, 1998, **17**, 5699–5707.
55. J. A. Wohlschlegel, B. T. Dwyer, S. K. Dhar, C. Cvetic, J. C. Walter and A. Dutta, Inhibition of eukaryotic DNA replication by geminin binding to Cdt1, *Science*, 2000, **290**, 2309–2312.
56. M. A. Gonzalez, K. E. Tachibana, S. F. Chin, G. Callagy, M. A. Madine, S. L. Vowler, S. E. Pinder, R. A. Laskey and N. Coleman, Geminin predicts adverse clinical outcome in breast cancer by reflecting cell-cycle progression, *J. Pathol.*, 2004, **204**, 121–130.
57. S. Shreeram, A. Sparks, D. P. Lane and J. J. Blow, Cell type-specific responses of human cells to inhibition of replication licensing, *Oncogene*, 2002, **21**, 6624–6632.

58. C. Lee, B. Hong, J. M. Choi, Y. Kim, S. Watanabe, Y. Ishimi, T. Enomoto, S. Tada, Y. Kim and Y. Cho, Structural basis for inhibition of the replication licensing factor Cdt1 by geminin, *Nature*, 2004, **430**, 913–917.
59. S. Saxena, P. Yuan, S. K. Dhar, T. Senga, D. Takeda, H. Robinson, S. Kornbluth, K. Swaminathan and A. Dutta, A dimerized coiled-coil domain and an adjoining part of geminin interact with two sites on Cdt1 for replication inhibition, *Mol. Cell*, 2004, **15**, 245–258.
60. K. Yanagi, T. Mizuno, Z. You and F. Hanaoka, Mouse geminin inhibits not only Cdt1-MCM6 interactions but also a novel intrinsic Cdt1 DNA binding activity, *J. Biol. Chem.*, 2002, **277**, 40871–40880.
61. M. Okuyama, H. Laman, S. R. Kingsbury, C. Visintin, E. Leo, K. L. Eward, K. Stoeber, C. Boshoff, G. H. Williams and D. L. Selwood, Small-molecule mimics of an alpha-helix for efficient transport of proteins into cells, *Nat. Methods*, 2007, **4**, 153–159.
62. A. Ferenbach, A. Li, M. Brito-Martins and J. J. Blow, Functional domains of the *Xenopus* replication licensing factor Cdt1, *Nucleic Acids Res.*, 2005, **33**, 316–324.
63. Z. Kelman, PCNA: structure, functions and interactions, *Oncogene*, 1997, **14**, 629–640.
64. L. S. Cox, Who binds wins: competition for PCNA rings out cell-cycle changes, *Trends Cell Biol.*, 1997, **7**, 493–498.
65. E. Warbrick, The puzzle of PCNA's many partners, *Bioessays*, 2000, **22**, 997–1006.
66. E. Warbrick, D. P. Lane, D. M. Glover and L. S. Cox, A small peptide inhibitor of DNA replication defines the site of interaction between the cyclin-dependent kinase inhibitor p21WAF1 and proliferating cell nuclear antigen, *Curr. Biol.*, 1995, **5**, 275–282.
67. C. Sakakura, A. Hagiwara, H. Tsujimoto, K. Ozaki, T. Sakakibara, T. Oyama, M. Ogaki and T. Takahashi, Inhibition of gastric cancer cell proliferation by antisense oligonucleotides targeting the messenger RNA encoding proliferating cell nuclear antigen, *Br. J. Cancer*, 1994, **70**, 1060–1066.
68. J. M. Gulbis, Z. Kelman, J. Hurwitz, M. O'Donnell and J. Kuriyan, Structure of the C-terminal region of p21(WAF1/CIP1) complexed with human PCNA, *Cell*, 1996, **87**, 297–306.
69. G. Kontopidis, S. Y. Wu, D. I. Zheleva, P. Taylor, C. McInnes, D. P. Lane, P. M. Fischer and M. D. Walkinshaw, Structural and biochemical studies of human proliferating cell nuclear antigen complexes provide a rationale for cyclin association and inhibitor design, *Proc. Natl. Acad. Sci. U.S.A.*, 2005, **102**, 1871–1876.
70. M. K. Shivji, E. Ferrari, K. Ball, U. Hubscher and R. D. Wood, Resistance of human nucleotide excision repair synthesis *in vitro* to p21Cdn1, *Oncogene*, 1998, **17**, 2827–2838.
71. M. K. K. Shivji, S. J. Grey, U. P. Strausfeld, R. D. Wood and J. J. Blow, Cip1 inhibits DNA replication but not PCNA-dependent nucleotide excision repair, *Curr. Biol.*, 1994, **4**, 1062–1068.

72. M. P. Cooper, A. S. Balajee and V. A. Bohr, The C-terminal domain of p21 inhibits nucleotide excision repair *in vitro* and *in vivo*, *Mol. Biol. Cell*, 1999, **10**, 2119–2129.
73. K. Mitsui, A. Matsumoto, S. Ohtsuka, M. Ohtsubo and A. Yoshimura, Cloning and characterization of a novel p21(Cip1/Waf1)-interacting zinc finger protein, ciz1, *Biochem. Biophys. Res. Commun.*, 1999, **264**, 457–464.
74. D. E. Warder and M. J. Keherly, Ciz1, Cip1 interacting zinc finger protein 1 binds the consensus DNA sequence ARYSR(0-2)YYAC, *J. Biomed. Sci.*, 2003, **10**, 406–417.
75. D. Coverley, J. Marr and J. Ainscough, Ciz1 promotes mammalian DNA replication, *J. Cell Sci.*, 2005, **118**, 101–112.
76. J. F. Ainscough, F. A. Rahman, H. Sercombe, A. Sedo, B. Gerlach and D. Coverley, C-terminal domains deliver the DNA replication factor Ciz1 to the nuclear matrix, *J. Cell Sci.*, 2007, **120**, 115–124.
77. F. A. Rahman, J. F. Ainscough, N. Copeland and D. Coverley, Cancer-associated missplicing of exon 4 influences the subnuclear distribution of the DNA replication factor CIZ1, *Hum. Mutat.*, 2007, **28**, 993–1004.
78. A. J. Berdis, DNA polymerases as therapeutic targets, *Biochemistry*, 2008, **47**, 8253–8260.
79. L. Corey, A. J. Nahmias, M. E. Guinan, J. K. Benedetti, C. W. Critchlow and K. K. Holmes, A trial of topical acyclovir in genital herpes simplex virus infections, *N. Engl. J. Med.*, 1982, **306**, 1313–1319.
80. P. A. Furman, M. H. St Clair and T. Spector, Acyclovir triphosphate is a suicide inactivator of the herpes simplex virus DNA polymerase, *J. Biol. Chem.*, 1984, **259**, 9575–9579.
81. L. A. Kohlstaedt, J. Wang, J. M. Friedman, P. A. Rice and T. A. Steitz, Crystal structure at 3.5 A resolution of HIV-1 reverse transcriptase complexed with an inhibitor, *Science*, 1992, **256**, 1783–1790.
82. E. C. Friedberg, W. J. Feaver and V. L. Gerlach, The many faces of DNA polymerases: strategies for mutagenesis and for mutational avoidance, *Proc. Natl. Acad. Sci. U.S.A.*, 2000, **97**, 5681–5683.
83. R. Sheaff, D. Ilsley and R. Kuchta, Mechanism of DNA polymerase alpha inhibition by aphidicolin, *Biochemistry*, 1991, **30**, 8590–8597.
84. S. Hanashima, Y. Mizushina, K. Ohta, T. Yamazaki, F. Sugawara and K. Sakaguchi, Structure-activity relationship of a novel group of mammalian DNA polymerase inhibitors, synthetic sulfoquinovosylacylglycerols, *Jpn. J. Cancer Res.*, 2000, **91**, 1073–1083.
85. Y. Mizushina, X. Xu, H. Asahara, R. Takeuchi, M. Oshige, N. Shimazaki, M. Takemura, T. Yamaguchi, K. Kuroda, S. Linn, H. Yoshida, O. Koiwai, M. Saneyoshi, F. Sugawara and K. Sakaguchi, A sulphoquinovosyl diacylglycerol is a DNA polymerase epsilon inhibitor, *Biochem. J*, 2003, **370**, 299–305.
86. Y. Mizushina, T. Ohkubo, T. Date, T. Yamaguchi, M. Saneyoshi, F. Sugawara and K. Sakaguchi, Mode analysis of a fatty acid molecule binding to the N-terminal 8-kDa domain of DNA polymerase beta. A 1:1 complex and binding surface, *J. Biol. Chem.*, 1999, **274**, 25599–25607.

87. Y. Mizushina, C. Murakami, K. Ohta, H. Takikawa, K. Mori, H. Yoshida, F. Sugawara and K. Sakaguchi, Selective inhibition of the activities of both eukaryotic DNA polymerases and DNA topoisomerases by elenic acid, *Biochem. Pharmacol.*, 2002, **63**, 399–407.
88. J. J. Turchi, L. Huang, R. S. Murante, Y. Kim and R. A. Bambara, Enzymatic completion of mammalian lagging-strand DNA replication, *Proc. Natl. Acad. Sci. U.S.A.*, 1994, **91**, 9803–9807.
89. M. Nowotny, S. A. Gaidamakov, R. Ghirlando, S. M. Cerritelli, R. J. Crouch and W. Yang, Structure of human RNase H1 complexed with an RNA/DNA hybrid: insight into HIV reverse transcription, *Mol. Cell*, 2007, **28**, 264–276.
90. J. Roca, R. Ishida, J. M. Berger, T. Andoh and J. C. Wang, Antitumor bisdioxopiperazines inhibit yeast DNA topoisomerase II by trapping the enzyme in the form of a closed protein clamp, *Proc. Natl. Acad. Sci. U.S.A.*, 1994, **91**, 1781–1785.
91. M. N. Fodje, A. Hansson, M. Hansson, J. G. Olsen, S. Gough, R. D. Willows and S. Al-Karadaghi, Interplay between an AAA module and an integrin I domain may regulate the function of magnesium chelatase, *J. Mol. Biol.*, 2001, **311**, 111–122.
92. T. Pape, H. Meka, S. Chen, G. Vicentini, M. van Heel and S. Onesti, Hexameric ring structure of the full-length archaeal MCM protein complex, *EMBO Rep.*, 2003, **4**, 1079–1083.
93. Y. Mizushina, I. Watanabe, K. Ohta, M. Takemura, H. Sahara, N. Takahashi, S. Gasa, F. Sugawara, A. Matsukage, S. Yoshida and K. Sakaguchi, Studies on inhibitors of mammalian DNA polymerase α and β: sulfolipids from a pteridophyte, *Athyrium niponicum, Biochem. Pharmacol.*, 1998, **55**, 537–541.

Subject Index

Italic page numbers refer to figures and tables. Abbreviated terms are used, as in the text. Species names are given in full.